Tranquillitas Ordinis

Tranquillitas Ordinis

The Present Failure and Future Promise of American Catholic Thought on War and Peace

GEORGE WEIGEL

Oxford New York
OXFORD UNIVERSITY PRESS
1987

Oxford University Press

Oxford New York Toronto
Delhi Bombay Calcutta Madras Karachi
Petaling Jaya Singapore Hong Kong Tokyo
Nairobi Dar es Salaam Cape Town
Melbourne Auckland

and associated companies in
Beirut Berlin Ibadan Nicosia

Published by Oxford University Press, Inc.,
200 Madison Avenue, New York, New York 10016

Oxford is a registered trademark of Oxford University Press

Library of Congress Cataloging-in-Publication Data
Weigel, George.
Tranquillitas ordinis.
Includes index.
1. Peace—Religious aspects—Catholic Church.
2. War—Religious aspects—Catholic Church.
3. Catholic Church—Doctrines. I. Title.
BX1795.W37W45 1987 261.8'73'08822 86-12742
ISBN 0-19-504193-3

9 8 7 6 5 4 3 2 1

Printed in the United States of America

In memory of
John Courtney Murray, S.J.
and
in gratitude to
Robert Pickus

Acknowledgments

Grateful acknowledgment is made for permission to quote from the following sources:

Excerpts from *The Documents of Vatican II*, copyright 1966, all rights reserved, by America Press, Inc., 106 West 56th Street, New York, NY 10019, and reprinted by their permission.

Excerpts from "Totalitarianism" in *The Origins of Totalitarianism*, copyright 1951 by Hannah Arendt; renewed 1979 by Mary McCarthy West. Reprinted by permission of Harcourt Brace Jovanovich, Inc. and by André Deutsch Ltd. for British markets.

Excerpts from Roland Bainton, *Christian Attitudes Toward War and Peace*, copyright 1960 by Abingdon Press, and reprinted by their permission.

Excerpts from *The Trial of the Catonsville Nine*, copyright 1970 by Rev. Daniel Berrigan, S.J., and reprinted by his permission.

Excerpts from Albert Camus, *Neither Victims Nor Executioners*, reprinted by permission of The Continuum Publishing Company.

Excerpts from Herbert A. Deane, *The Political and Social Ideas of St. Augustine*, copyright 1963 by Columbia University Press and reprinted by their permission.

Excerpts from Paul Fussell, *The Great War and Modern Memory*, copyright 1975 by Oxford University Press and reprinted by their permission.

Excerpts from *Divine Disobedience*, copyright 1969, 1970 by Francine du Plessix Gray. Reprinted by permission of Alfred A. Knopf, Inc.

Excerpts from *Modern Times*, copyright 1983 by Paul Johnson. Reprinted by permission of Harper & Row.

Excerpts from John Langan, S.J., "The Elements of St. Augustine's Just War Theory," reprinted, with permission, from *The Journal of Religious Ethics*, vol. 12, no. 1 (Spring 1984), pp. 19–38.

Excerpts from Thomas Merton, "Peace: Christian Duties and Responsibilities," "The Christian in World Crisis: Reflections on the Moral Climate of the 1960s," "Note on Civil Disobedience and Nonviolent Revolution," "Christian Ethics and Nuclear War," "Christianity and Defense in the Nuclear Age," and "Note for *Ave Maria*," reprinted by permission of New Directions Publishing Corp. as agents for the Thomas Merton estate.

Excerpts from William D. Miller, *A Harsh and Dreadful Love, Dorothy Day and the Catholic Worker Movement*, by permission of Liveright Publishing Corporation. Copyright © 1973 by William D. Miller.

Excerpts from John Courtney Murray, S.J., *We Hold These Truths*, used by permission of Sheed & Ward, 115 E. Armour Blvd., Kansas City, MO 64141.

Excerpts from H. Richard Niebuhr, *Christ and Culture*, copyright 1951 by Harper & Row, and reprinted by their permission.

Excerpts from Paul Ramsey, "The Vatican Council of Modern War," reprinted, with permission, from *Theological Studies* 27, no. 2 (June 1966).

Excerpts from Paul Ramsey, "Pacem in terris," reprinted by permission of the author.

Preface

Those who write long books are under a moral obligation to write short prefaces. As Thomas More said to the headsman, I shall be brief.

This book takes up what I might call, to paraphrase Robert Frost, a "lover's quarrel with my church."

I am convinced that Catholic social theory, particularly as it developed in the United States through the Second Vatican Council and the work of John Courtney Murray, S.J., has been, and could once again be, of great importance to a world that sought both peace and freedom.

I am equally convinced that the Church in the United States has not only failed to develop its heritage of thought over the past generation; the Church's most influential teaching centers have, in the main, largely abandoned their heritage. Both the Church and American political culture have suffered as a consequence of this failure. The task for today, then, is to reclaim *and develop* the heritage abandoned by many of American Catholicism's intellectual and religious leaders in the years since Vatican II.

Some of the abandoners argue, of course, that there was nothing to abandon. I hope that Part One of this study will put that canard to rest. Reclamation and development require an understanding of the themes of the abandonment that has taken place, and of how those themes eventually influenced the Church's official commentary on issues of U.S. foreign policy; thus, Part Two. And as it would be irresponsible to criticize others without taking my own shot at reclamation and development, Part Three is a sketch toward a new theology and politics of peace and freedom: one that I hope is faithful to, even as it stretches, the classic Catholic theory that has been unhappily forgotten. I also hope that this new theology and politics of peace and freedom has broad ecumenical appeal.

One qualification is important at the outset. The temptation to commit the logical fallacy of *post concilium ergo propter concilium*—to think that everything that happened *after* Vatican II happened *because* of Vatican II—is rarely resisted in American Catholicism today. There is a complex, substantive connection between the work of the Council and the transformation of American Catholic thought on war and peace since 1965, a connection to be explored here in some detail. In most cases, however, I have used the phrase "in the years after Vatican II" as a temporal reference point, not as an explanatory principle. I shall be grateful if my readers, of all theological and political persuasions, keep that distinction in mind.

This has been a collaborative work in more ways that I can briefly acknowledge. Richard John Neuhaus, Michael Novak, and Robert Pickus have influenced my thought decisively, as well as being close friends and colleagues in a variety of enterprises seeking to advance peace and freedom. Other friends and associates may also find ideas suspiciously similar to their own woven into what follows: William A.

Douglas, Nick Eberstadt, James Finn, Max M. Kampelman, Penn Kemble, Roy Prosterman, and Robert Woito.

Dean C. Curry; Thomas M. Gannon, S.J.; Marguerite Green, R.S.C.J.; Francis Kane; William J. Lee, S.S.; Michael Novak; Robert Pickus; and Philip Siegelman read my first draft and generously took the time to make detailed comments.

This is a better (and shorter) volume because of the good counsel of my editor, Cynthia Read, whose faith in this project has been an important span in the bridge between a lumpish manuscript and a book.

A year at the Woodrow Wilson International Center for Scholars in Washington, D.C., gave me the opportunity to conduct my research and prepare a first draft in the flag tower of the old Smithsonian Castle, with the Mall at my feet—an office the likes of which I never expect to be favored with again. I am very grateful to the Wilson Center Board of Trustees, to the Center's Director, James H. Billington, and to Michael J. Lacey, the Secretary of the Center's Program on American Society and Politics, for taking the chance they did on me. Rodger Potocki, who played the Library of Congress the way Brooks Robinson played third base, was my Wilson Center research assistant from September 1984 through August 1985; Noah Pickus joined the team from June through August 1985. Both were indispensable. Fellow fellows Bohdan Bociurkiw, James Childress, Joseph Komonchak, and Menachem Milson were particularly helpful colleagues.

I must also thank the Earhart Foundation for the fellowship research grant that supported the completion of the book; the Homeland Foundation, whose grant made possible my introduction to computers; and the board of directors of the James Madison Foundation, who have seen this book as an important part of our work together.

My wife Joan, and my daughters Gwyneth and Monica, each helped, in their distinctive ways, to maintain a sufficiency of *tranquillitas ordinis* on the home front so that the composition and editing I did there were a pleasure.

This book is dedicated to the memory of American Catholicism's finest public theologian and in honor of the work of a Jewish peace activist who defies all the stereotypes associated with that vocation. The former was a brilliant exponent of just-war theory; the latter is a principled pacifist, certainly the most thoughtful I have ever known. These two considerable figures never met. They should have. It would have been a fine, and needed, argument. I am happy to be able to bring them together at last, this side of the New Jerusalem, if only on a page.

July 4, 1986
Washington, D.C.

G. W.

Contents

Tranquillitas Ordinis

PROLOGUE

Between the Fire and the Pit: Moral Imagination in the Modern World

Bent double, like old beggars under sacks,
Knock-kneed, coughing like hags, we cursed through sludge,
Till on the haunting flares we turned our backs
And towards our distant rest began to trudge.
Men marched asleep. Many had lost their boots
But limped on, blood-shod. All went lame; all blind;
Drunk with fatigue; deaf even to the hoots
Of gas shells dropping softly behind.

Gas! GAS! Quick, boys!—An ecstasy of fumbling,
Fitting the clumsy helmets just in time;
But someone still was yelling out and stumbling,
And flound'ring like a man in fire or lime . . .
Dim, through the misty panes and thick green light,
As under a green sea, I saw him drowning.

In all my dreams, before my helpless sight,
He plunges at me, guttering, choking, drowning.

If in some smothering dreams you too could pace
Behind the wagon we flung him in,
And watch the white eyes writhing in his face,
His hanging face, like a devil's sick of sin;
If you could hear, at every jolt, the blood
Come gargling from the froth-corrupted lungs,
Obscene as cancer, bitter as the cud
Of vile, incurable sores on innocent tongues—
My friend, you would not tell with such high zest
To children ardent for some desperate glory,
The old Lie: *Dulce et decorum est,*
Pro patria mori.

Wilfred Owen, during World War I

I plead guilty to having followed sentimental impulses, and in so doing have been led into contradiction with historical necessity. I have lent my ear to the laments of sacrifice, and thus become deaf to the arguments which proved the necessity of sacrificing them. I plead guilty to having rated the question of guilt and innocence higher than that of utility and harmfulness. Finally, I plead guilty to having placed the idea of man above the idea of mankind.

Rubashov's confession, in *Darkness at Noon*

When did the modern world begin? If we think of modernity as the history of nation-states, we might say that the modern world began in 1648, when the Peace of Westphalia ended the European wars of religion. If we think of modernity as man's philosophical self-awareness as subject, we might conclude that the modern world began when Descartes suggested that human consciousness was the measure of things knowable. If we think of science as a defining characteristic of modernity, we might answer that the modern world began on May 29, 1919, when photographs of a solar eclipse confirmed Einstein's special theory of relativity, which taught that the previously assumed absolutes of "time" and "space" were in fact relative to the circumstances of the beholder.[1] Those who believe the modern world is most starkly distinguished by the threat of thermonuclear holocaust have an equally precise date from which all things have been made new: August 6, 1945, the day a primitive atomic bomb was dropped on the headquarters of the Japanese army in Hiroshima.

The question is unanswerable in any final way; there is truth in all these identifications of the beginning of modernity. But for the purposes of this inquiry, the modern world began on August 27, 1914, and November 7, 1917. On the first date, according to Sir Basil Liddell Hart,[2] the German advance toward Paris under the famous Schlieffen Plan ground to a halt. All hopes of a quick, "civilized" war on the nineteenth-century model soon evaporated, and the European world hunkered down in trenches to conduct a slaughter of unprecedented proportions.[3] On the second date, Lenin's Bolsheviks expropriated the Russian revolution and began to erect the first totalitarian state in the long, sad history of human tyranny. The first date began to reveal the catastrophe that the technology of modern war could wreak on flesh and earth; the second date began to reveal the catastrophe that tyrannical forces using modern means of social control could wreak on the human spirit.

Much of the world's history since then has been an expression of the dynamics let loose by these twin dangers—modern war and totalitarianism. Born together in the carnage of World War I, they have been inseparable ever since, hard as the well-intentioned and the malicious have tried to divide them. Moreover, the threat posed to the very fabric of civilization by modern means of destruction and by totalitarianism has put previously unimagined strains on the traditional Catholic moral theory of war and peace. How did this happen? What new terrors entered the moral imagination of the human spirit in the events at the Marne, the Somme, and Passchendaele, and in the streets of Petrograd and Moscow?

One of the characteristic conceits of our age is to fancy that we are dealing with unprecedented problems; so much of the recent past has been described as a time of supreme emergency that we often lose our grip on what is genuinely new, and what is only terribly frightening. But in this case an intuition of the unprecedented is true: something did shift in the fields of Flanders and France during World War I, and on the streets and steppes of Russia during the Bolshevik consolidation of totalitarian power. The "normality" of war as a natural, if regrettable, part of the human condition was fundamentally challenged; but so, too, were the norms that had hitherto guided civilized life in society. "War" and "tyranny" acquired a new bitterness. Ever since, Catholic moral imagination has struggled with the pressures of a human story lived out between the fire of modern war and the pit of modern totalitarianism. How can there be both peace and freedom? The only way to begin fashioning an answer to this defining question of our age is to understand the trapgate through which human history hurtled in 1914 and 1917.

THE GREAT WAR

There have always been wars and rumors of wars, Scripture tells us. Why, then, should we have been so shaken by the Great War, World War I? Perhaps it was because we had thought that we knew better. Coming flush on the heels of the Edwardian era of good feeling, the Great War was a bloody reminder that original sin had not been expunged from human affairs, and that progress was a work of our hands, not a given in the structure of the universe. Sheer carnage, on an almost unimaginable scale and cutting across traditional lines of class, nation, race, and creed, had its effect. Then there was the aftermath of it all: a peace settlement unsatisfactory to virtually all concerned, and which set in motion forces that would lead, within a generation, to another struggle of even more sanguinary proportions.

However one unravels the mystery that the Great War posed for the moral imagination of modernity, it seems clear that it *was* a mystery. Churchill, ever-prescient, sensed that something had changed, and in a drastic way. During his time at the War Office (1919–21), he made the following note about the slaughter just past:

> All the horrors of all the ages were brought together, and not only armies but whole populations were thrust into the midst of them. The mighty educated States involved conceived—not without reason—that their very existence was at stake. Neither peoples nor rulers drew the line at any deed which they thought could help them to win. Germany, having let Hell loose, kept well in the van of terror; but she was followed step by step by the desperate and ultimately avenging nations she had assailed. Every outrage against humanity or international law was repaid with reprisals—often of a greater scale and of longer duration. No truce or parley mitigated the strife of the armies. The wounded died between the lines: the dead mouldered in the soil. Merchant ships and neutral ships and hospital ships were sunk on the seas and all on board left to their fate, or killed as they swam. Every effort was made to starve whole nations into submission without regard to age or sex. Cities and monuments were smashed by artillery. Bombs from the air were cast down indiscriminately. Poison gas in many forms stifled or seared the soldiers. Liquid fire was projected upon their bodies. Men fell from the air in flames, or were smothered often slowly in the dark recesses of the sea. The fighting strength of armies was limited only by the manhood of their countries. Europe and large parts of Asia and Africa became one vast battlefield on which after years of struggle not armies but nations broke and ran. When all was over, Torture and Cannibalism were the only two expedients that the civilized, scientific, Christian states had been able to deny themselves: and they were of doubtful utility.[4]

Writing over sixty years later, Aleksandr Solzhenitsyn saw the war through a similar prism. The moral imagination of the civilized world had failed, and the results were still with us:

> The failings of human consciousness, deprived of its divine dimension, have been a determining factor in all the major crimes of this century. The first of these was

> World War I, and much of our present predicament can be traced back to it. That war (the memory of which seems to be fading) took place when Europe, bursting with health and abundance, fell into a rage of self-mutilation that could not but sap its strength for a century or more, and perhaps forever. The only possible explanation for this war is a mental eclipse among the leaders of Europe due to their lost awareness of a Supreme Power above them.[5]

But the results, Solzhenitsyn continued, were to be found not only in Flanders fields: "Only the loss of that higher intuition which comes from God could have allowed the West to accept calmly, after World War I, the protracted agony of Russia as she was being torn apart by a band of cannibals. . . . The West did not perceive that this was in fact the beginning of a lengthy process that spells disaster for the whole world."[6] The Great War, and the Great Terror, were demonic twins from the same womb. We live under the shadow of them still.

The results of the Great War cannot be measured solely in terms of the subsequent history of nations and states. As Churchill intuited and Solzhenitsyn makes plain, the trapgate through which humanity was thrust by the Great War was even more a matter of human self-understanding, or of what we might call "moral imagination." The questions raised were of first principles: What can be expected of the human beast after this revelation of unmitigated cruelty? Could there be progress, an evolution of the human spirit toward higher forms of community, when the moral coordinates to guide such a development seemed so tattered and frayed?

If these were the kinds of questions raised by the Great War, we should not look first to geopolitical analyses for an understanding of how this war reshaped human consciousness. Rather, we ought to look to more textured modes of thought, toward literature. Such a literary exploration has been mounted by Paul Fussell in his contemporary classic, *The Great War and Modern Memory*.[7] As Fussell demonstrates with compassion and elegance, it is in the poetry and prose of the Great War that we begin to glimpse the shadow essence of its meaning.

A primary victim of the Great War was moral certitude: an intuition that, whatever the vicissitudes of personal and social history, meanings and values endured. That certitude was as much a casualty of the Great War as the Hapsburg Empire:

> The Great War was perhaps the last to be conceived as taking place within a seamless, purposeful "history" involving a coherent stream of time running from past through present to future. The shrewd recruiting poster depicting a worried father of the future being asked by his children, "Daddy, what did *you* do in the Great War?" assumes a future whose moral and social pressures are identical with those of the past. Today, when each day's experience seems notably *ad hoc*, no such appeal would shame the most stupid to the recruiting office. But the Great War took place in what was, compared to ours, a static world, where the values appeared stable and where the meanings of abstractions seemed permanent and reliable. Everyone knew what Glory was, and what Honor meant. It was not until eleven years after the war that Hemingway could declare in *A Farewell to Arms* that "abstract words such as glory, honor, courage, or hallow were obscene beside the concrete names of villages, the numbers of roads, the names of rivers, the numbers of regiments and the dates." In the summer of 1914 no one would have understood what on earth he was talking about.[8]

Alongside moral certitude fled a higher vocabulary that had lent meaning to the experience of combat. The adjectives, nouns, and verbs used by Tennyson seemed, not quaint, but vaguely offensive, after the experiences of the Great War. Were words like "valiant," "gallant," "plucky," "the fallen," "ardent," "conquer," "the legion," even "warrior," ever to have the same resonances again? Chivalry and romance fell beside moral certitude, and even a war widely held to be just would be fought with the grim determination characteristic of the GI in World War II; no banners of glory, only a miserable, necessary piece of dirty work that had to be done.[9]

This alteration of consciousness, in moral imagination, could be understood only in archetypical biblical terms; humanity had entered upon "the knowledge of good and evil."[10] And the results of eating that particular fruit paralleled the experience of being cast from Eden: irony, not moral certitude, became the characteristic mode of human understanding. For the soldiers of the western front, the supreme symbol of this ironic imagination was the Golden Madonna of Albert. The basilica of Albert was crowned by a golden statue of the Virgin, clutching her Child in her arms. In an earlier age, it stood as a symbol of the old verities; "But now the whole statue was bent down below the horizontal, giving the effect of a mother about to throw her child—in disgust? in sacrifice?—into the debris-littered street below."[11] The Blessed Virgin transformed into a sacrificient to Moloch: Could it be that human evolution was retrogressing, that "progress" was a chimera, that we were no better than the child-sacrificers of Carthage?

This fracturing of moral certitude shows up in other literary phenomena of the period. The majesty of death was defiled by its ubiquity in the trenches; as a later poet, Randall Jarrell, would write of World War II, "It was not dying; everybody died."[12] Classic poetic imagery was inverted: "Dawn has never recovered from what the Great War did to it. Writing four years after the Armistice, Eliot accumulates the new, modern associations of dawn: cold, the death of multitudes, insensate marching in files, battle, and corpses too shallowly interred."[13] Nor were human relationships spared: "gross dichotomizing is a persisting imaginative habit of modern times, traceable, it would seem, to the actualities of the Great War. 'We' are all here on this side; 'the enemy' is over there. 'We' are individuals with names and personal identities; 'he' is a mere collective entity. We are visible; he is invisible. We are normal; he is grotesque. Our appurtenances are natural; his, bizarrre. He is not as good as we are."[14]

Such dichotomizing, however necessary to maintain the public morale of a nation-at-arms in a struggle to the death, eroded moral imagination; it suggests not only a breakdown of human community, but the impossibility of its re-creation. Synthesis is impossible; thesis and antithesis, in constant combat, are the only realities. The sense of the ambiguous that characterizes compassionate and civilized moral discourse gives way to "simple distinction, simplification, and opposition. If truth is the main casualty of war, ambiguity is another."[15] Experience cannot be made into a coherent whole. Wilfred Owen, writing to his mother before combat at the Somme in January 1917, told her that "there is a fine, heroic feeling about being in France, and I am in perfect spirits." But, "sixteen days later, everything has changed: 'I can see no excuse for deceiving you about these four days. I have suffered seventh hell. I have not been at the front. I have been in front of it.' There is no dialectic capable of synthesizing those two moments in Owen's experience."[16] Orwell's later image for this fracture in human moral imagination was typically apt; the Spanish civil war had made men rush back to "the mental slum of 1915." Degradation of the body had, under these unprecedented circumstances, yielded degradation, not exaltation, of the spirit.

The result, Fussell suggests, is that the horror of the front becomes relativized, neutralized; "horror, truthfully described, weakens to the merely clinical."[17] This led to a further distortion of truth through the profligate use of euphemism. "The crucial cable sent home by His Britannic Majesty's Consul in Sarajevo on June 28, 1914, was couched in these terms: 'Heir apparent and his consort assassinated this morning by means of an explosive nature.'" Servia became Serbia; "German shepherds" became "Alsatians." French mutinies were acts of "collective indiscipline." Even personal identity had to be conformed to the new realities of euphemism. Most famous were the royal rechristenings: the Saxe-Coburg-Gotha dynasty reborn as the "Windsors," the Battenbergs as the "Mountbattens." But lesser mortals were not spared: "When the young Evelyn Waugh returned to school in September, 1914, he found several boys who 'turned up with names suddenly anglicized—one, unfortunately named Kaiser, appeared as "Kingsley".'"[18] And yet, in a bizarre paradox, this masking of the realities of war by the language of euphemism (with all that that implied for moral imagination), was paralleled by a militarization of the language of civilian life: raincoats became "trenchcoats," and the members of trades unions the "rank and file."

If the experience of the Great War had collapsed the myth of Great Men in charge of human affairs (how could anyone believe the myth of the great when their leadership had brought such unspeakable disaster?); if the Great War had fractured an orderly sense of history, so broken the continuity of past and present that the future seemed unimaginable; if irony had become the characteristic mode of understanding; if transcendent symbols (such as the Virgin of Albert) had become grotesques, revealing the obverse of that for which they were intended; if death had itself been lost, communities of discourse broken, and the spectre of endless struggle revived—how, then, would moral imagination cope? Could one make any sense of it all, or was the absurd the final measure of the human experience? The literature of the Great War reflects several strategies for the recovery of moral imagination.

One, particularly favored by combatants from the British literary classes, was an Arcadianism in which the past was romantically glorified as a means of coping with present horrors. "Home" had already developed Arcadian overtones for Britons out in the empire; now, under the pressures of the Great War, "home" became an image for recollecting and creating meaning in the trenches along the western front. Predictably, this often took an ironic turn. One anonymous trench-dwelling wit wrote a parody letter to the editor of the trench paper, the *Wipers* (as in "Ypres") *Times*: "Sir: Whilst on my nocturnal rambles along the Menin Road last night, I am prepared to swear that I heard the cuckoo. Surely I am the first to hear it this season. Can any of your readers claim the same distinction?" The letter was signed "A Lover of Nature."[19]

But the return to pastoral imagery served a deeper purpose: "recourse to the pastoral is an English mode of both fully gauging the calamities of the Great War and imaginatively protecting oneself against them. Pastoral reference, whether to literature or to actual rural localities and objects, is a way of invoking a code to hint by antithesis at the indescribable; at the same time, it is a comfort in itself, like rum, a deep dugout, or a wooly vest."[20] Isaac Rosenberg's "Break of Day in the Trenches" brilliantly employs that most poignant of Great War pastoral images—the poppy—but with a harsher twist than could be imagined in the pastoral elegiac tradition of, say, a Wordsworth:

> The darkness crumbles away—
> It is the same old druid Time as ever.
> Only a live thing leaps my hand—
> A queer sardonic rat—
> As I pull the parapet's poppy
> To stick behind my ear.
> Droll rat, they would shoot you if they knew
> Your cosmopolitan sympathies.
> Now you have touched this English hand
> You will do the same to a German—
> Soon, no doubt, if it be your pleasure
> To cross the sleeping green between.
> It seems you inwardly grin as you pass
> Strong eyes, fine limbs, haughty athletes
> Less chanced than you for life,
> Bonds to the whims of murder,
> Sprawled in the bowels of the earth,
> The torn fields of France.
> What do you see in our eyes
> At the shrieking iron and flame
> Hurled through still heavens?
> What quaver—what heart aghast?
> Poppies whose roots are in man's veins
> Drop, and are ever dropping;
> But mine in my ear is safe,
> Just a little white with the dust.[21]

Rosenberg's bittersweet poem illustrates a common use of elegiac symbols in the literature of the Great War: to reconstruct a memory that gives the present a semblance of meaning. But pastoral imagery could also be turned to purposes of propaganda, as in the most popular poem of the war:

> In Flanders fields the poppies blow
> Between the crosses, row on row,
> That mark our place; and in the sky
> The larks, still bravely singing, fly
> Scarce heard amid the guns below.
>
> We are the Dead. Short days ago
> We lived, felt dawn, saw sunset glow,
> Loved and were loved, and now we lie
> in Flanders fields.
>
> Take up the quarrel with the foe:
> To you from failing hands we throw
> The torch; be yours to hold it high.

If ye break faith with us who die
We shall not sleep, though poppies grow
in Flanders fields.[22]

John McCrae's first two stanzas offer what Fussell calls "familiar triggers of emotion": red flowers, crosses, larks in the sky counterposed to the roar of artillery, dawn and sunset. Here, we think, is horror transformed through pastoral imagery into blessed, if poignant, sacrifice. But in the jarring third stanza, all is turned to furthering the carnage, with the ghostly reminder that those who will not grasp the torch condemn the dead to no rest—that is, to hell.

Juxtaposed, the Rosenberg and McCrae poems, written only seven months apart, suggest both the power of traditional pastoral imagery and its ultimate futility as a means for making final moral sense out of the Great War. When pastoral leads either to propaganda (as in McCrae) or back into irony (as in Rosenberg), appropriate moral coordinates that would redeem the poets' experiences have yet to be found. The return to an Arcadian past could not, in the final analysis, give meaning to the horror of the present; it only highlighted the grossness of the Great War. The pastoral beauty of "home" did not provide imagery for a new moral imagination capable of enlightening the trenches; it only reminded us of how bad things really were.

A parallel attempt to take the full measure of the horror of the Great War and yet avoid the despair of the absurd appears in the religious imagery used by writers of the trenches. The proliferation of references to *Pilgrim's Progress* in the English literature of World War I is striking: "It would be impossible to count the number of times 'the Slough of Despond' is invoked as the only adequate designation for churned-up mud morasses pummeled by icy rain and heavy shells. It becomes one of the inevitable clichés of memory. So does 'the Valley of the Shadow of Death,' where, in Bunyan, 'lay blood, bones, ashes, and mangled bodies of men, even of Pilgrims that had gone this way formerly.'"[23] What could better capture the terror of the narrow, murderous paths between trenches than the "straight and narrow" path so familiar from *Pilgrim's Progress*? If, as Fussell suggests, "scenes of hazardous journeying constitute the essence of *Pilgrim's Progress*,"[24] what could be more natural than to see in the horrors of the trenches the contemporary living out of Christian's journey in Bunyan?

And yet one suspects that something more was going on here than a recourse to familiar imagery. No doubt the resort to Bunyon's familiar images was a mechanism for trying to put meaning onto the essentially meaningless. But if, as the literature of the war repeatedly suggests, the Great War pushed human beings to a new kind of limit experience—on the far boundaries of the possibly meaningful, where the line between beatitude and terror is thin indeed—then it can be argued that the experiences of World War I reopened, ironically or providentially (and perhaps both), the question of the moral imagination and the problem of war and peace in ways that would have been impossible in more "civilized" circumstances. The perception of the war as "anxiety without end";[25] the commonality of shared victimhood, not only among combatants but among all caught up in the nation-at-arms[26]—these searing experiences suggest that Churchill and Solzhenitsyn were right. The Great War caused a fundamental shift in moral imagination, or, to be more precise, created circumstances in which a new moral imagination about the problem of war was required. After Verdun and Passchendaele, it was no longer possible to think of war as a "natural" phenomenon—unless we were to think of ourselves, in the starkest Hobbesian terms, as beasts. And if Fussell is right in claiming that the Great War is not just a matter of the past, but is in fact "the essential condition of consciousness in the twentieth century,"[27] because of its pervasive impacts on our understanding of the human condition, then in truth something new entered human affairs in those four terrible years.

But the Beast let loose by the Great War was not only a matter of the awesomely destructive capabilities of modern weapons; the horror of the trenches was not the only result of World War I. The twin horror, totalitarianism, born, as Solzhenitsyn suggests, from the same demonic womb, is as much a part of the problem of war's relationship to modern moral imagination as Wilfred Owen's gassed comrades.

THE GREAT TERROR

Over the past decade, there has been a largely frustrating argument over the meaningfulness of distinguishing between "authoritarian" and "totalitarian" forms of despotism. Those who argue against the distinction follow one of two lines of reasoning. First, they claim, correctly, that it makes little difference to the victim of torture whether the hand that wields the truncheon is labeled "authoritarian" or "totalitarian;" what counts is the pain, and what ought to be done is to prevent it. Secondly, questions are raised about the very possibility of a "totalitarian" regime. Is it feasible to construct a system of social control so draconian as to be total and complete in its capacity to hammer human beings into the desired mold?

These are serious questions, but they are also answerable. To the first, one ought to admit that the victim's perspective is both true and limited. If the purposes of humane men and women include minimizing the number of torture victims, it is reasonable to distinguish between regimes that torture you if you overtly challenge their political authority, and regimes that also torture you if you compose the "wrong" kind of music, write the "wrong" books, or hold religious opinions. There is a difference, in other words, between bad and worse, and failing to make that distinction can lead to policies that compound misery rather than relieve it. To the second question, whether totalitarianism exists in a pure form, one can answer no (perhaps excepting Albania); but that does not mean that the totalitarian model is not alive and well, and that totalitarian tyrants have not tried their mightiest to achieve their despotic ends. The twentieth century is filled with the corpses of too many of the victims of Lenin, Stalin, Hitler, Mao, Pol Pot, Ho Chi Minh, and Enver Hoxha to validate the claim that totalitarianism is only an "ideal type."

So there is, then, a new thing in our midst; it differs from traditional forms of tyranny. How?

Part of the immense difficulty twentieth-century minds have had in coming to grips with this new thing is that totalitarianism began with a moral imperative: it arose from a passion to close the painful gap between things as they are and things as they ought to be. That gap was described by Emile Durkheim in his classic study *Suicide* as the problem of anomie: a distinctively modern experience of feeling cut off from the bonds of human community that had sustained individuals in the pre-modern world. The transition from a society composed of organic, communal relationships to a society of contractual, legal relationships—the defining characteristic of "modernity" for Durkheim—created conditions of unparalleled social stress. Anomie was the resultant social disease.[28] Hannah Arendt made the connection between these pains of modernity and the rise of totalitarianism: "What prepares men for totalitarian domination in the non-totalitarian world is the fact that loneliness, once a borderline experience usually suffered in certain marginal social conditions like old age, has become an everyday experience of the evergrowing masses of our century. The merciless process into which totalitarianism drives and organizes the masses looks like a suicidal escape from this reality."[29]

Loneliness leads to anomie, which prepares men and women for the totalitarian temptation; but is there not more going on here? Arendt also argues that the conceptual ground for totalitarianism was laid in the nineteenth century:

> The tremendous intellectual change which took place in the middle of the last century consisted in the refusal to view or accept anything "as it is" and in the consistent interpretation of everything as being only a stage of some further development. Whether the driving force of this development was called nature or history is relatively secondary. In these ideologies, the term "law" itself changed its meaning: from expressing the framework of stability within which human actions and motions can take place, it became the expression of the motion itself.[30]

This drive for change took the totalitarian turn when it closed the windows of the human experience to any transcendent dimension, and sought warrants for change in the immutable laws of history or nature (race, in the Nazi case), before which all must bend or be broken.

Lenin, the father of the first totalitarian state, exemplified the problem of totalitarianism as a deformed moral impulse. Paul Johnson writes:

> We have to assume that what drove Lenin on to do what he did was a burning humanitarianism, akin to the love of the saints for God, for he had none of the customary blemishes of the politically ambitious: no vanity, no self-consciousness, no obvious relish for the exercise of authority. But his humanitarianism was a very abstract passion. It embraced humanity in general but he seems to have had little love for, or even interest in, humanity in particular. He saw the people with whom he dealt, his comrades, not as individuals but as receptacles for his ideas. On that basis, and on no other, they were judged. So he had no hierarchy of friendships; no friendships, in fact, merely ideological alliances. He judged men not by their moral qualities but by their views, or rather the degree to which they accepted his. . . . Lenin was the first of a new species: the professional organizer of totalitarian politics. It never seems to have occurred to him, from adolescence onwards, that any other kind of human activity was worth doing. Like an anchorite, he turned his back on the ordinary world.[31]

That totalitarianism is much more, and much worse, than a grandiose form of traditional human ambition—that it in fact originated in a defective moral impulse—is crucial to understanding both the full measure of its perverse cruelty, and the otherwise inexplicable hold a society organized by men like Lenin and Stalin could have had on the moral and political imaginations of Western intellectuals. That attraction was experienced, and then rejected, by a quartet of novelists before it was analyzed by a theoretician of the gifts of Hannah Arendt—and the results include some of the most penetrating literary works of the century.

In André Malraux's *Man's Fate*, for example, we see a soul struggling in the midst of revolutionary upheaval with the moral imperative to change the impossible present, and yet do so in ways that do not lead to an even more impossible future. Ignazio Silone, in *Bread and Wine*, also defends the necessity of individual social concern and political activism, even as he wrestles with the totalitarian temptation. Arthur Koestler's *Darkness at Noon* confronts the terrible moral costs involved in combining utopianism with amoral politics. George Orwell's *1984* captured, for a popular audience, the degradation of individuality that is the essence of totalitarian control.[32] These works have an important commonality: they were all written by men of the Left, passionately committed to social, political, and economic reform, who had come to understand the horrible possibility of a transition from bad to worse during the Spanish civil war. They rejected the totalitarian temptation, as expressed in Stalin's U.S.S.R., not out of fondness for the *ancien régime*, but because of a moral intuition that no good could come from a system which utterly destroyed the moral integrity of the human person.

And that, as Arendt and other political theorists came to understand, was the essence of the totalitarian project.

The terror that marks all totalitarian regimes is neither random nor purposeless; its objective is the elimination of the moral imagination. In Arendt's analysis:

> Terror in totalitarian government has ceased to be a mere means for the suppression of opposition. . . . Its chief aim is to make it possible for the force of nature or of history to race freely through mankind, unhindered by any spontaneous human action. . . . Terror as the execution of a law of movement whose ultimate goal is not the welfare of men or the interest of one man but the fabrication of mankind, eliminates individuals for the sake of the species, sacrifices the "parts" for the sake of the "whole". . . . [Terror] substitutes for the boundaries and channels of communication between individual men a band of iron which holds them so tightly together that it is as though their plurality had disappeared into One Man of gigantic dimensions. . . . By pressing men against each other, total terror destroys the space between them; compared to the condition within its iron band, even the desert of tyranny, insofar as it is still some kind of space, appears like a guarantee of freedom. Totalitarian government does not just curtail liberties or abolish essential freedoms; nor does it, at least to our limited knowledge, succeed in eradicating the love for freedom from the hearts of man. It destroys the one essential prerequisite of all freedom which is simply the capacity of motion which cannot exist without space. . . . In the iron band of terror . . . a device has been found not only to liberate the historical and natural forces, but to accelerate them to a speed they would never reach if left to themselves. Practically speaking, this means that terror executes on the spot the death sentences which Nature is supposed to have pronounced on races of individuals who are "unfit to live," or History on "dying classes," without waiting for the slower and less efficient processes of nature or history themselves.[33]

The results of this impulse to accelerate the human journey to the promised future have been staggering. By 1920, Lenin's Cheka had carried out over fifty thousand political executions. But even this bloodbath was a warm-up for what would follow.

The famine caused by Leninist agricultural policies is estimated to have caused 3 million deaths during the winter of 1921–22. The purge of the wealthy peasantry during the collectivization campaign of 1930–31 killed 22 million more. Stalin's party purges of the 1930s led to a further 19 million deaths. Hitler's race-madness laid waste at least 6 million Jews, and uncountable millions of Gypsies, Poles, Russians, and other "inferior" beings. Mao's slaughter in China, during the revolution and the later Cultural Revolution, added tens of millions of corpses to the pyre. And then there was Pol Pot's genocide of the Cambodians, for sheer efficiency the most horrid totalitarian terror campaign of an already blood-drenched century. On statistical grounds alone, totalitarianism has proven every bit as deadly as modern war.[34]

And yet the deeper one digs into the literature of totalitarianism, the more the dangers multiply. As Paul Johnson writes:

> The most disturbing and, from the historical point of view, important characteristic of the Lenin terror was not the quantity of the victims but the principle on which they were selected. Within a few months of seizing power, Lenin had abandoned the notion of individual guilt, and with it the whole Judeo-Christian ethic of personal responsibility. He was ceasing to be interested in *what* a man did or had done—let alone *why* he had done it—and was first encouraging, then commanding, his repressive apparatus to hunt down people, and destroy them, not on the basis of crimes, real or imaginery, but on the basis of generalizations, hearsay, rumors. First came condemned categories. . . . Following quickly, however, came entire occupational groups. The watershed was Lenin's decree of January 1918 calling on the agencies of the state to "purge the Russian land of all kinds of harmful insects." This was not a judicial act: it was an invitation to mass murder. . . . Once Lenin had abolished the idea of personal guilt, and had started to "exterminate" (a word he frequently employed) whole classes, merely on account of occupation or parentage, there was no limit to which this deadly principle might be carried. Might not entire categories of people be classed as "enemies" and condemned to imprisonment or slaughter merely on account of the color of their skin, or their racial origins, or indeed, their nationality? There is no essential moral difference between class-warfare and race-warfare, between destroying a class and destroying a race. Thus the modern practice of genocide was born.[35]

The abolition of personal moral responsibility amid totalitarian terror had functions beyond the elimination of "enemies of the state," however; it was an essential component of the totalitarian project of reconstructing the human person. The justification for such an enterprise was ideological; but ideological here meant more than philosophical. Arendt makes the key distinction when she writes that totalitarian ideologies "are never interested in the miracle of being." The function of ideology in a totalitarian system, be it Leninist or Hitlerite (and here the essential similarity of the two can be seen), is to smother human experience, to create a world without windows or doors, a world devoid of transcendent purpose, meaning, or values; as Arendt puts it, "to fit each of [its subjects] equally well for the role of executioner and the role of victim."[36] The *content* of ideology soon evaporates; "in essentials," Johnson writes, "[Lenin] was not a Marxist at all."[37] Dialectical materialism is useful insofar as it relates the terror, and the abandonment of individuality, to a larger scheme of things. But the real purpose of totalitarian ideology in a Leninist state is that of an instrument of social control, a perverted antitranscendence to replace the transcendent horizon against which former life was lived (the hammer and sickle as a crippled cross).

Another essential component of the totalitarian project is its denial of the distinction between "public" and "private" life, and its purposeful destruction of both: public life—politics as it has been understood since the Greeks—is destroyed by the cancer of the terror; private life, by the collapse of individual moral responsibility. In the process of this joint destruction, human sociability is lost; as Solzhenitsyn puts it, in the Gulag Archipelago "men were encouraged to survive at the cost of the lives of others."[38]

And so, by a perverse logic, totalitarianism, with roots in the quest for community against the frightening prospect of anomie, becomes, through its deliberate abandonment of individual moral responsibility and its corrosion of human sociability, the engine of an even greater, more terrifying loneliness. As Arendt puts it, "What makes loneliness so unbearable is the loss of one's own self which can be realized in solitude, but confirmed in its identity only by the trusting and trustworthy company of my equals. In this situation [i.e., totalitarianism], man loses trust in himself as the partner of his thoughts and that elementary confidence in the world which is necessary to make experiences at all. Self and world, capacity for thought and experience are lost at the same time."[39]

What Arendt describes in the cool prose of the political theorist, Orwell captured in fictional form in *1984.* At the climax of the novel, Winston Smith, terrorized to the roots of his soul by O'Brien's carnivorous rats, screams: "Do it to Julia! Do it to Julia! Not me! Julia! I don't care what you do to her. Tear her face off, strip her to the bones. Not me! Julia! Not me!" With this, resistance ends. Self, others, world are all gone. Moral responsibility abandoned leads to the degradation of self that is the meaning of hell. The totalitarian will to power, when brought to bear with full force on an individual life, can be resisted only by a power of the spirit that draws its strength from a Source that transcends it. Drawing from such a Source, the Austrian peasant Franz Jaegerstaetter could face down Hitlerite totalitarianism alone, as Solzhenitsyn won out over contemporary Soviet terror and Armando Valladares over the Stygian tortures of Fidel Castro's political prisons. Lacking it, Winston Smith is reduced to a gelatinous shell.[40]

This explains one facet of the totalitarian aversion to religion. Because the Judeo-Christian tradition insists that a core of privileged privacy, a soul, rests at the center of every person, it poses the sharpest challenge to the absolutist pretensions of totalitarianism. Lenin, whom Johnson describes as a "clerical" revolutionary, typified this strain in totalitarianism:

> Religion was important to him in the sense that he hated it. Unlike Marx, who despised it and treated it as marginal, Lenin saw it as a powerful and ubiquitous enemy. . . . "There can be nothing more abominable," he wrote, "than religion." From the start, the state he created set up and maintains to this day an enormous academic propaganda machine against religion. . . . Lenin had no real feelings about corrupt priests, because they were easily beaten. The men he really feared and hated, and later persecuted, were the saints. The purer the religion, the more dangerous. . . . The clergy most in need of suppression were not those committed to the defense of exploitation but those who expressed their solidarity with the proletariat

> and the peasants. It was as though he recognized in the true man of God the same zeal and spirit which animated himself, and wished to expropriate it and enlist it in his own cause. No man personifies better the replacement of the religious impulse by the will to power.[41]

Although Lenin's heirs learned to co-opt religious leaders for their political purposes, the essential antipathy between transcendent religion and modern totalitarianism remains to this day. The opposition is final and, if you will, total: at every level—the nature of the human person, of society, and of history itself—the contest is joined. Modern totalitarianism, a defective transcendence, expropriates religious imagery and symbolism, as in the Lenin mausoleum in Moscow. But the real meaning of the contest has never been denied, at least by totalitarians.

The paradox is that in its fundamental challenge to the religious imagination, totalitarianism opens up anew the question of religious experience. Human material has not proven as malleable as totalitarians expected. Solzhenitsyn's novel, *One Day in the Life of Ivan Denisovich*, is an exploration of how, even in the Gulag, the human impulse toward community and a measure of decency cannot finally be crushed, though many will fall by the wayside in the process. At the end of a day that by civilized standards would have been thought unbearable, Ivan Denisovich Shukhov rests:

> Shukhov went to sleep, and he was very happy. He'd had a lot of luck today. They hadn't put him in the cooler. The gang hadn't been chased out to work in the Socialist Community Development. He'd finagled an extra bowl of mush at noon. The boss had gotten them good rates for their work. He'd felt good making that wall. They hadn't found that piece of steel in the frisk. Caesar had paid him off in the evening. He'd bought some tobacco. And he'd gotten over that sickness. Nothing had spoiled the day, and it had been almost happy. There were three thousand six hundred and fifty-three days like this in his sentence, from reveille to lights out. The three extra ones were because of the leap years.[42]

Simple things, surely, but indicative of an iron in the human soul that cannot be melted. In Peter Berger's happy image, the experiences of Ivan Denisovich were "rumors of angels,"[43] hints and traces of a transcendence to human experience that points, the faithful believe, to God. It does not matter that Ivan Denisovich is not a Christian (and in fact rejects the evangelizing of his fellow prisoner, Alyoshka); what matters is that the totalitarian project cannot nail all of the windows shut.

Which does not mean that totalitarians have not tried, and are not still trying, to do so. Totalitarianism, like modern war, is a new kind of limit experience. Like the trenches of World War I, totalitarianism reveals the face of human cruelty and depravity in its starkest form. But, again like war, totalitarianism reopens a set of moral questions: about the human person, about power, about violence (for Leninist and fascist totalitarians, the final arbiter of the great struggle),[44] about personal conscience and social responsibility, about the possibilities of political community in the modern world. Just as the fire of modern war and the pit of totalitarianism evolved from the same historical circumstances, so they raise similar questions, particularly pertinent for those who wish to think about the morality of war and peace today. The questions, like the phenomena that gave rise to them, are not separable.

A NEW MORAL IMAGINATION: NEITHER VICTIMS NOR EXECUTIONERS

Between the fire of war and the pit of totalitarianism, moral imagination in the modern world is in schism. Our choices seem reduced to either/or propositions: either resist totalitarian aggression, even by war, or run the risk of a world in Gulag; either end the threat of war, even by appeasing totalitarians, or run the risk of global holocaust.

In 1946, a young veteran of the French Resistance, Albert Camus, faced the either/or quality of the modern moral imagination, and tried to transcend it. Writing in the Resistance journal *Combat*, Camus argued that the task of moral men and women in the modern world was to deny neither dimension of our dilemma, but to think clearly about how we could be "neither victims nor executioners."[45]

Camus began with the stark fact that "our twentieth century is the century of fear." Pervasive fear had had a terrible impact on human consciousness: "the most striking fact of the world we live in is that most of its inhabitants—with the exception of pietists of various sorts—are cut off from the future."[46] The future from which we are cut off is not an abstraction called "tomorrow" but the very future of the human prospect. "The years we have gone through," Camus wrote, "have killed something in us. And that something is simply the old confidence man had in himself, which led him to believe that he could always elicit human reactions from another man if he spoke to him in the language of a common humanity."[47] What had died at the Somme, the Marne, and Passchendaele, in Petrograd and Moscow, in Vorkuta and the *katorga* camps and on all the other islands of the Gulag Archipelago, at Auschwitz and Dachau and Theresianstadt, over London, Coventry, Dresden, Hamburg, Berlin, Tokyo, Hiroshima, and Nagasaki, was the very idea of Man, as that had come to be understood in the tradition of the West. What was left was either the collective man of totalitarian ideology, or the atomic man of Durkheim's anomie. What had died between the fire and the pit was the very possibility of history understood as a human conversation: "Mankind's long dialogue has just come to an end."[48] And the result of that paralysis of conversation, a paralysis of moral imagination, was The Fear.

Fear cannot be avoided; it must be confronted by intelligence and imagination, which means understanding the precise nature of fear. "To come to terms, one must understand what fear means: what it implies and what it rejects. It implies and rejects the same fact: a world where murder is legitimate, and where human life is considered trifling."[49] Moreover, the issue will not be resolved in the realm of personal psychology, as an answer to the question, How shall I cope with my personal fear? No, "this is the great *political* question of our times, and before dealing with other issues, one must take a position on it. Before anything can be done, two questions must be put: Do you or do you not, directly or indirectly, want to be killed or assaulted? Do you or do you not, directly or indirectly, want to kill or assault? All who say No to both these questions are automatically committed to a series of consequences which must modify their way of posing the problem."[50] Those who are morally repelled by the either/or character of the fear that has put moral imagination into schism must take responsibility for thinking through the implications of their claim that a both/and answer is possible. Synthesis, given the risks involved on both sides of the either/or propositions, must be an act of moral imagination and responsibility, not a flight from either.

What should those committed to a new moral imagination, one that has faced the dangers of *both* war and totalitarianism *and* seeks a morally sound path to peace *and* freedom because it understands their indivisibility—what ought they seek in this world? Camus's answer is suggestive: "People like myself want not a world in which murder no longer exists (we are not so crazy as that!) but rather one in which murder is not legitimate."[51] But is this not utopian, and has not this bloody century taught us the terrible dangers of utopian schemes to reform the human condition?

> Our refusal to legitimate murder forces us to reconsider our whole idea of Utopia. This much seems clear: Utopia is whatever is in contradiction with reality. From this standpoint, it would be completely Utopian to wish that men should no longer kill each other. That would be absolute Utopia. But a much sounder Utopia is that which insists that murder be no longer legitimized. . . . We may therefore conclude, practically, that in the next few years the struggle will be not between the forces of Utopia and the forces of reality, but between different Utopias which are attempting to be born into reality. It will be simply a matter of choosing the least costly among them. I am convinced that we can no longer reasonably hope to save everything, but that we can at least propose to save our skins, so that *a* future, if not *the* future, remains a possibility.[52]

Camus's objective of "saving our skins" must be distinguished from contemporary survivalism. Camus knew, as his life in the Resistance demonstrated, that there were things worth dying for, and that freedom from totalitarian tyranny was one of them. But Camus had become convinced, in his bitter struggle with the French Left over the post-occupation future of his country, that violence provided no real way out; it only kept the modern beast at bay, temporarily. No doubt temporary expedients are necessary on occasion. But for those who want to live like human beings (and not "like dogs," in Camus's graphic phrase), the way to a humane future lies not with war as an answer to the totalitarian temptation, but with an approach to politics that squarely addresses both war *and* totalitarianism. This is Camus's "sounder Utopia."

The path to that "sounder utopia" will not be charted by those who have let fear come to dominate, and ultimately fracture, moral imagination. What is needed is "a political position that is modest, i.e., free of messianism and disencumbered of nostalgia for an earthly paradise."[53] The job is not to bring the Kingdom of God to earth; the job is to create a human future fit, morally and politically, for human beings and not for dogs. The beginning of that enterprise is clear thinking. "Sincerity," Camus saw acutely, "is not in itself a virtue: some kinds are so confused that they are worse than lies."[54] And in thinking clearly about a world caught between the fire and the pit, the first thing that must be confronted is that there is no escape from the rigors of history:

> We know today that there are no more islands, that frontiers are just lines on a map. We know that in a steadily accelerating world, where the Atlantic is crossed in less than a day and Moscow speaks to Washington in a few minutes, we are forced into fraternity—or complicity. The forties have taught us that an injury done a student in Prague strikes down simultaneoulsy a worker in Clichy, that blood shed on the banks of a Central European river brings a Texas farmer to spill his own blood in the Ardennes, which he sees for the first time. There is no suffering, no torture anywhere

> in the world which does not affect our everyday lives. . . . Today, tragedy is collective.[55]

The dean of St. Paul's was right, then, in warning us not to ask for whom the bell was tolling; but unlike Donne's warning in Elizabethan days, Camus's call to face the collective (i.e., "both/and") nature of the contemporary dilemma involves choices with the gravest world-historical consequences. Nor should Camus's call for "fraternity" be seen as a vague, pious appeal to fellow-feeling. For what Camus had in mind was not a falsification of the great divide that separates the free from the tyrants (that would be "complicity") but a facing of it that concurrently asks, in a tough-minded way, about the possibilities of political community, a distinctive form of "fraternity," as a means to work at both the problem of war and the problem of totalitarianism.

Camus's vision was of an "international democracy," in which "law has authority over those governed, law being the expression of the common will."[56] The answer to the twinned dilemma of war and totalitarianism was *democratic* political community, genuinely international in scope. The United Nations, Camus saw as early as 1946, was not what was needed. But rather than go on to prescribe specific institutional arrangements, Camus proposed the conditions for the possibility of an evolving international political community faithful to democratic values. Those conditions were essentially moral and cultural. What was needed, to deal with the schismatic condition of the world and the moral imagination, was the re-creation of a *civilisation du dialogue*, which Dwight Macdonald translated as a "sociable culture."[57] The needed dialogue would begin where all political dialogue worthy of the name begins: with the moral question of Man.

The angle of vision of the dialogue would be that of a humane skeptic, with emphasis on both adjective and noun: "I have always held," Camus noted, "that, if he who bases his hopes on human nature is a fool, he who gives up in the face of circumstances is a coward."[58] The dialogue would be conducted with no illusions about possible outcomes, with no false optimisms. "Between the forces of terror and the forces of dialogue," Camus told the monks at the Dominican monastery of Latour-Maubourg, "a great, unequal battle has begun. I have nothing but reasonable illusions as to the outcome of that battle. But I believe it must be fought."[59] The chances of re-creating a *civilisation du dialogue*, as the basis of a world where we need be neither victims nor executioners, may be slim. But the battle must be fought.

Ever since Camus's death, commentators have argued over what Karl Rahner might have called Camus's "anonymous Christianity." However that particular critical controversy is settled, Camus was thinking down a remarkably similar track to that laid out over centuries of Catholic moral reflection on the problem of war and peace. Like Camus, Catholic moral thought is both resolutely realistic about the limits of the human condition, and yet convinced that the battle for peace and freedom must be fought: not on Stoic grounds, for the sake of good form, but because those who make wars are the same human beings who build political communities and resolve political conflicts without mass violence, when the means are available. Even under the abiding effects of original sin, human beings are human beings, not dogs. Like Camus, Catholic moral thought claims that history is not the plaything of impersonal forces, but an arena of personal and social responsibility. Even amid its grotesqueries and cruelties, human history is still the material of a history of salvation and grace. Camus sensed this; he wanted Christians (and, in his address to the Dominicans, Catholics specifically) to stand with him in creating a moral imagination ready to deal with the demands of both peace and freedom. Why? Because Catholic Christianity, Camus intuited, understood the moral imperative implied in his claim that the battle must be fought. "Perhaps we cannot prevent this world from being a world in which children are tortured," he told the Dominicans. "But we can reduce the number of tortured children. And if you don't help us, who else in the world can help us do this?"[60]

The trapgate through which human history rushed in 1914 and 1917 put the question of moral imagination at the forefront of political thinking with an urgency that demands an answer. Caught between the fire of modern war and the pit of totalitarianism, one possibility is to simply accept, as a stringent given of the age, the either/or character of the choices that confront us; managing the chaos wisely, with as much, but no more, decency as is safe seems not just politically expedient, but morally sound. Larger visions, pursued without regard for human consequences, have cost so much in pain, violence, and torture. A second possibility is to simply deny the totalitarian side of the dilemma, minimizing its threat or acquiescing to its purposes. Against these two possible strategies, Camus poses a third option: to face squarely the inextricable linkage between the moral problem of mass violence and the moral problem of mass terror, and to think clearly about the ways in which we might be, not angels, but human beings who are neither victims nor executioners.

It is this third option, whatever its difficulties of accomplishment, that seems most faithful to the victims of the trenches and the *katorga* camps, and to the hope that springs from a recognition of fallen human beings as still, for all our weaknesses, the images of God in history.

THE CHALLENGE OF PEACE AND FREEDOM

The balance of this study is a portrait of how American Catholicism has tried, and largely failed, to shape a moral imagination that would meet the standards set by Albert Camus in *Neither Victims Nor Executioners*; and it is a prescription for how we ought to try again.

Camus was right in his intuition that the Catholic moral tradition has rich and deep resources to bring to this task. Its heritage includes the biblical witness of ancient Hebrews and early Christians; but, over almost two millennia of reflection, these precious gifts have been woven together with insights from classical, medieval, and modern political philosophy to form a complex tapestry. That tapestry has been bound together by an "incarnational" or "analogical" view of the relationship between human intention and action, and divine providence; the result was a style of moral thinking eminently suitable for moral action between the fire and the pit, one that held principle and practice in a creative tension, mediated by the virtue of prudence. Roman Catholicism, itself a highly structured, multivalent community, has understood, at least since Augustine, the potential of political community for providing a this-worldly alternative to mass violence in resolving human conflict, even as it has, particularly in recent years, come to emphasize the centrality of freedom in any morally worthy City.

Within the wider family of Catholic Christianity, American Catholicism might have been expected to have played a central role in shaping a moral imagination adequate to the task of charting a way to peace and freedom. American Catholicism had enjoyed the privileges of a *civilisation du dialogue* in a democratic society. Moreover, American Catholicism seemed, by the time of the Second Vatican Council, to have been the part of the wider Catholic world that had thought best about the problems of pluralism, power, and political community; its own experience of democratic pluralism had opened the door to the Council's development of doctrine on religious liberty. Seemingly freed from the prejudices often suffered by an immigrant church, American Catholicism, by the end of the Council, in 1965, also seemed to be in a distinctive position to help the wider American polity come to grips with its international responsibilities for peace and freedom in a way that would have been impossible even a generation before. By the end of Vatican II, great expectations were in order.

What follows is a portrait of how these hopes were, at best, frustrated in the very generation in which they might have been expected to flower: the generation following the election of John F. Kennedy and the work of the Second Vatican Council.

Three theses shape this study. The first is that American Catholicism's elites—its bishops, priests, religious, intellectuals, and publicists—were the bearers of a heritage well equipped to define, with wisdom, a moral imagination capable of facing the threats of modern war and totalitarianism. Exploring this thesis requires an examination of the Catholic heritage on the moral question of war and peace up through Vatican II.

The second thesis is that this heritage was largely abandoned by the most influential sectors of the American Catholic elite in a short decade after Vatican II, and that, under the influence of this shift, the official organs of the Church in the United States became, not the shapers of a new and wiser moral argument, but antagonists in old ones. Exploring this thesis will involve a detailed examination of the war/peace argument conducted within American Catholic intellectual and activist circles during the Vietnam and post-Vietnam periods, and an analysis of the impacts of the dominant themes in the argument on the official positions taken by Church leaders.

And the third thesis is a wager: that there yet remain within the Catholic tradition, especially as it has evolved through the American Catholic experience, rich resources for a reclamation and expansion of the Catholic heritage in ways more adequate to the central political task of the age—the task of securing both peace and freedom.

PART ONE

The Heritage

Pax est tranquillitas ordinis.

St. Augustine, *The City of God*, XIX, 13

We believe that our country's heroes were the instruments of the God of Nations in establishing this home of freedom.

The Catholic bishops of the United States,
at the Third Plenary Council of Baltimore, 1884

Neither as a doctrine nor as a project is the American Proposition a finished thing. . . . the Proposition itself requires development on pain of decadence. . . . In a moment of national crisis Lincoln asserted the imperilled part of the theorem and gave impetus to the impeded part of the project in the noble utterance, at once declaratory and imperative, "All men are created equal." Today, when civil war has become the basic fact of world society, there is no element of the theorem that is not menaced by active negation, and no thrust of the project that does not meet powerful opposition. Today, therefore, thoughtful men among us are saying that America must be more clearly conscious of what it proposes, more articulate in proposing, more purposeful in the realization of the project proposed.

John Courtney Murray, S.J., *We Hold These Truths:*
Catholic Reflections on the American Proposition, 1960

CHAPTER 1

The Catholic Tradition of Moderate Realism

John Courtney Murray's story of his difficult entrance into the argument over morality and foreign policy in the early 1950s is one of the most memorable vignettes in his magisterial work, *We Hold These Truths: Catholic Reflections on the American Proposition*:[1]

> My introduction to the state of the problem took place . . . in a conversation with a distinguished journalist who is now dead. In public affairs he was immensely knowledgeable; he was also greatly puzzled over the new issue that was being raised. What, he asked, has the Sermon on the Mount got to do with foreign policy? I was not a little taken aback by this statement of the issue. What, I asked, makes you think that morality is identical with the Sermon on the Mount? Innocently and earnestly he replied: "Isn't it?" And that in effect was the end of the conversation. We floundered a while in the shallows and miseries of mutual misunderstanding, and then changed the subject to the tactics of the war going on in Korea.[2]

The dangers of moralism that Murray decried in *We Hold These Truths* are just as threatening in the American Catholicism of the 1980s as they were in the Catholicism (and cultural Protestantism) of the 1950s. It is an ever-present danger in the Church, over the centuries and down to the present: the temptation to conduct the debate over the morality of war and peace as if the only arguable points were on the "war and peace" (i.e., public policy) side of the equation. In fact, as Murray knew full well, the real source of the "miseries of mutual misunderstanding" on these questions lay in disagreement over the nature of moral reasoning. For even if we might agree that the operant virtue in these matters is prudence, we are still left with the question of the moral content of prudence under the conditions of our historical moment, conditions that have changed with alarming rapidity in a twentieth century whose hallmarks include Passchendaele, nuclear weaponry, and the Gulag Archipelago.

A thoughtful judgment on the present condition of the American Catholic debate on war and peace requires more than a survey of recent analyses and positions; it requires a sense of the *status quo ante*. What was the tradition borne by American Catholics as they entered full force into the national debate over peace, security, and freedom? Part One of this study will sketch the broad outlines of the Catholic tradition of moral reasoning on war and peace: a rich, not-to-be-oversimplified heritage that seemed to John Courtney Murray in 1960, on the eve of Vatican II and the election of the first Catholic president of the United States, singularly well equipped to help shape a moral imagination capable of reasoning between the fire of modern war and the pit of totalitarianism.

THE AUGUSTINIAN HERITAGE: PEACE AS PUBLIC ORDER

St. Augustine of Hippo was not the first Christian to reflect on the moral problem of war and peace. If one defines "the moral problem of war and peace" with some breadth—including, for instance, not only questions about the morally responsible use of armed force, but also questions of the relationship between personal conscience and public responsibility—Christian reflection on these matters can be traced straight back into the earliest strata of the New Testament. Jesus' warning that "those who live by the sword will die by the sword" (Matt. 26:52), set in the context of the personal moral code of the Sermon on the Mount, establishes one dimension of the testamental discussion; St. Paul's injunction to obey public authority in Romans 13, and Jesus' teaching that Caesar is to be given what is owed him (Matt. 22:21), suggest another, perhaps more complicating, dimension to moral choice.[3]

But it would be unfair, and in fact mistaken, to look to the New Testament for a systematic explication of Christian moral teaching on the ethics of war and peace. During much of the period in which the New Testament was composed, the Christian community lived in expectation of the imminent return of its Lord. This hope for a decisive, world-ending act of God in history colored much of the preaching of Jesus as it has been preserved for us in the Gospels. In these circumstances, and given the sociological context of primitive Christianity—a small sect, primarily located on the social fringes of the Roman Empire—it would be anachronistic to expect the New Testament to include systematic thought on moral problems that emerged only when the eschatological hope of primitive Christianity had been transformed by the realization that the coming of the Lord in glory would be a matter for the unforeseeable future. This is not to say that the New Testament has nothing to offer to moral reflection on war and peace; the recovery of its biblical heritage has had a profound (and complex) impact on the war/peace debate in American Catholicism since the Second Vatican Council. Still, expeditions into the New Testament in search of a systematic ethics of war and peace are fundamentally misconceived.

More deliberate Christian thought on these issues, particularly on the question of personal participation in military life, emerged only in the patristic period, when the Church had become a larger influence in the life of the empire.[4] It is in this historical context that St. Augustine's just-war thinking should be located.

Augustine's thought on these matters, as on virtually every other topic he addressed, was forged in the crucible of theological controversy. Augustine's theology grew, not from serene reflection amid the peace of a Christendom assumed to be the natural ordering of society, but in the fire of argument and schism, charges and countercharges of heresy. The preliminary sketch of just-war theory in *The City of God* and in Augustine's letters is of a piece with the controversial origins of much of the rest of his work. In this case, the controversy was conducted, not only within the Church, but between the Church and the wider society of his time.

The fifth century of the common era was not a tranquil time in the western regions of the Roman empire. The internal collapse of the old order was compounded by the peril of invasion from without. Given its post-Constantinian position within the empire, Christianity had to face questions of social ethics and political theory that would have been inconceivable before the Edict of Milan. Augustine was specifically confronted with the charge (native to his time and given classic form by Edward Gibbon) that the collapse of the western empire before the invading barbarian hordes had something to do with the Christian ethos. In its narrower form, the charge was that traditional Christian pacifism had helped create untenable military circumstances for the western empire; the broader charge was that Christian other-worldliness, the Church's concern to prepare its members for the next life, had led to an ahistorical and socially irresponsible approach to the inescapable problems of individuals and societies in history. Rebutting these indictments required the systematic exposition of a Christian social ethic and a more nuanced answer to the moral problem of war and peace than had been given before.

Augustine's answer to these questions reflected the distinctively autobiographical character of his theology. As the *Confessions* reveal, Augustine was an accomplished student of the weaknesses of human nature, beginning with himself. But it would be wrong to depict his social and political thought as utterly soured by his resolute realism: "for all of his pessimism about the history of Rome, Augustine was concerned to illuminate principles of leadership in a practical way. He spoke of the limitations of human beings, especially Romans. [But] he also spoke of human potential and of men who had accomplished some of the ideals of human association that the Romans had always claimed to seek."[5] The strategy of Augustine's social and political argument was thus two-pronged: first, to turn the charge of social demoralization back against those who had made it, in a critique of the lack of a philosophy of public virtue in late-classical antiquity; and secondly, to re-create a theory of public virtue in society and in history by weaving together earlier classical themes with his own specifically Christian reflections. Augustine's project was not "an escape from the world, but a start at living more freely within it."[6]

To see Augustine in this "conversionist" light, as argued elegantly by H. Richard Niebuhr in *Christ and Culture*,[7] is not to minimize the dourness that is a hallmark of Augustine's social and political reflections. As Roland Bainton observes, one distinctive characteristic of Augustine's thought, especially on issues of war and peace, is its "mournful mood."[8] This is one reason why Augustine seems a familiar figure; like the poets of the Great War, Augustine understood that public life, particularly in the limit case of war, was full of ambiguity and irony. Still, the tasks of constructing a moral framework for social and political life and providing ethical coordinates for personal moral decisions had to be taken up; and thus it can be argued that Augustine's fundamental concern was to establish the realistic conditions for the possibility of a politics of virtue, even in a fallen world.[9]

Four basic concepts shaped Augustine's social and political thought; sorting them out is essential to understanding his political philosophy, and his teaching on the morality of war and peace. These four concepts were the Church, the state, the heavenly City, and the earthly City. Charles McCoy summarizes their nature and relationship:

> The Church, a divinely established society, is the guardian of the Sacred Scriptures, of the Law of the Old and New Testaments by which men are directed to eternal beatitude in the vision of God, Whom St. Augustine calls the Common Good of the universe. The State is concerned with "just dealing and all the things belonging to good manners": in the measure that men organize for a life of the political virtues as well as the virtues of the mind . . . they compose the political community, the State. The Church and the State are thus recognizable, "visible" societies for the good.
>
> But since indeed God alone sounds the heart and plumbs the depths of the mind, there are two invisible "cities": the heavenly City of the predestined and the Earthly City of the damned. Those who, availing themselves as far as possible of the grace given through the Church, direct their just dealings and all the things belonging to good manners to final beatitude, are, the world over, members of the Heavenly City. Finally, those who seek no good beyond the present life, or on the other hand, seeking this good condemn the other, that is "the honesty of virtue; the love of country; the faith of friendship; just dealing and all the things belonging to good manners" are, the world over, members of the Earthly City and destined to eternal death.[10]

Unlike later Catholic theorists, notably St. Thomas Aquinas, Augustine did not conceive the state as a "natural" institution, an expression of innate human sociability. Rather, the state exists because of man's weakened nature after the Fall. The state is a necessary institution because without it the reign of evil in the world would be unfettered. Yet the state is itself problematic, given the fallen nature of its citizens; the sins of those who inhabit the earthly City (i.e., those who seek no good beyond this present life) injure the state by turning virtue into vice through excess. Unless virtue is ordered to its true end, the God who is the Common Good of the universe, it necessarily becomes vice; "the inordinateness of seeking virtue on its own account is of the essence of . . . impiety because [it] leads to an insane search for a substitute infinity—as in emperor worship—and destroys the proper forms by which human life is well lived on this earth."[11]

In this fourfold analysis of the structure of social and political life, Augustine tried to turn the tables on those who blamed the fall of Rome on Christianity: Augustine argued that late-antiquity's positing of "substitute infinities," not the catechesis of the Church, had destroyed the basis on which social life could be well lived. But the responsibility of the citizens of the heavenly City is not simply to point out the errors of past and present; Augustine was not a sectarian in his view of the relationship between the believing community and the wider society. Rather, it is "the members of the Heavenly City who ought to perfect the State, healing its wounded nature and restoring the free character of its rule."[12] It is precisely those whose lives are ordered to God as the Common Good of the universe who are most capable of restoring to the state the virtues that are its precondition for existence. "The saints," in Augustine's view, cannot escape from responsibility in history; they should be the citizens most impelled to work for temporal peace.

Augustine's understanding of the state as a barrier against the evil rampant in a fallen world led naturally to his definition of peace as *tranquillitas ordinis*, the "tranquillity of order."[13] Anarchy was the worst possible this-worldly condition; order was the condition for the possibility of virtue in public life. The peace of this world was but the "peace of Babylon"; but such a peace was not to be deprecated: it allowed fallen human beings to "live and work together and attain the objects that are necessary for their earthly existence."[14] Moreover, earthly peace makes an important contribution to the well-being and progress of those called to heavenly peace.[15] The role of the state was to nurture *tranquillitas ordinis*, and coercion could be employed to that end.[16] Within this "realistic political theory"[17] Augustine shaped his theory of the just war.

Augustine was no militarist; there is no glorification of war in *The City of God*, and Augustine held in contempt those who saw conquest as a noble aspiration (Alaric's sack of Rome in 410 surely had an effect here).[18] But Augustine broke decisively with the pacifism and antimilitarism of earlier Church fathers such as Tertullian, Origen, and Lactantius.[19] In this fallen world, war was inevitable. Like the authority of the state, war exists in God's providential plan as an instrument for punishment, so that a minimum of justice—that is, order—may be maintained. Just as the damned are part of God's plan for the heavenly City, so war could be a part of peace.[20] On this view, pacifism was a concession to evil and injustice. The task of moral reasoning was to determine which wars were "just," and therefore morally permissible.[21]

In Augustine's theory, three kinds of war were morally defensible: a defensive war against aggression, a war to gain just reparations for a previous wrong, and a war to recover stolen property.[22] But historical circumstances alone did not create the conditions for a "just" war. A properly constituted authority must decide that the resort to war is necessary. Once that decision has been made, it is the soldier's duty to obey. Further, war must be waged with a "right intention": not for revenge, or lust for glory and spoils, but with a sober commitment to conduct the war as an instrument of punishment necessary to protect the innocent and to prevent malefactors from doing worse.[23] Right intentions would be expressed, not only in the decision to go to war, but also in the just conduct of a war. Here, "the rules were taken from classical antiquity. Faith must be kept with the enemy. There should be no wanton violence, profanation of temples, looting, massacre, or conflagration. Vengeance, atrocities, and reprisals were excluded, though ambush was allowed."[24] Thus, Augustine, who counseled against self-defense in the case of individual threat,[25] outlined the skeleton of a just-war theory rooted in a moral passion for peace as the tranquillity of order.

John Langan, S.J., has noted the key elements in this Augustinian formulation of just-war theory. Augustine's conception of war was *punitive*: wars were to be fought as punishments for evil. War itself was not a good; but Augustine judged the evil of war in terms of the interior attitudes of the participants in it rather than in terms of evil deeds in war or evil consequences of fighting. Augustine interpreted the evangelical norms of the Gospel (e.g., love of neighbor) in terms of *interior personal dispositions* rather than overt actions. *Properly constituted authority* was also a key concept in the Augustinian formulation of the just war theory. Rooted in Augustine's passion for *tranquillitas ordinis*, this led to a certain social passivity or quiessence in his discussion of the relationship of personal conscience to public duty. Once properly constituted authority had declared the necessity of war, the Christian's duty was to obey (always, of course, with the right interior attitude and motivation). In sum Augustine was "really interested in the preservation of a moral order which is fundamentally a right internal order of dispositions and desires, and in which the question of whether action is violent or not is not fundamental. The restoration of that order constitutes a sufficient justification for resort to violence."[26]

The exigencies of the twentieth century may make us look askance at Augustine's starkly realistic political theory, and his application of that theory to the moral problem of war and peace. Modern means of destruction are so devastating that we may wonder at Augustine's reluctance to articulate a fully detailed theory of the *ius in bello* (the moral rules to be observed during warfare), with guidelines to judge (and set limits on) particular military strategies and tactics. Moreover, Augustine's failure to develop moral criteria for distinguishing among governments—for determining what is a *properly constituted* authority—can seem negligent in times such as our own, when the despotic power of governments can be as much a threat to human freedom as are anarchy and chaos.[27] Langan also argues persuasively that Augustine's concept of punitive war is not very helpful under contemporary conditions;[28] that modern weapons of mass destruction pose a challenge to Augustine's resolute insistence on the priority of interior attitudes;[29] that Augustine's attempt to "export responsibility for violence onto the accounts of higher powers (human or devine)" is not only incongruent with contemporary understandings about human moral agency but contradicts the opinion of a wide range of authorities on the morality of war and peace;[30] and that Augustine's exegesis of the Gospel's evangelical counsels "threatens to sever the connection between virtuous attitude and right action."[31] These are not insignificant charges.

Augustine was without doubt taking a formidable risk in crossing the theological Rubicon of just-war theory. Could Christianity be lifted from its accustomed sectarian/pacifist context and, without doing irreparable damage to its essential religious and moral insights, engage fully with the ambiguities of a world where the resort to violence was often the lesser of evils? As Herbert Deane puts it:

> Augustine's conception of war . . . his attempt to distinguish between just and unjust wars, and his flat rejection of pacifism and merely passive resistance to evil may be more realistic and more adequate than the ethic of love and non-resistance. They represent, however, a pronounced change from the beliefs of the early Christian Church, and they mark a significant milestone in the process of relativizing and accommodating Christ's teachings to the imperatives of earthly existence and to the views of right and wrong that were generally accepted in the world into which Jesus came.[32]

Are "relativizing and accommodating" the right terms, though? The process of "accommodating" to an ecclesiastical situation in which expectations of the imminent return of the Lord have been frustrated did not begin with Augustine, but appears within the New Testament itself (in, for example, the arguments in Acts on whether gentile Christians should observe Mosaic laws on circumcision and diet). These were not matters of "church and state"; but they reflect an understanding, evident in the first post-Easter generation of Christians, that a detailed set of rules for public Christian life in the interim between the Resurrection and the Second Coming are not to be found in the sayings of Jesus. Viewed from this angle, Augustine's project was not so much a "relativization" and an "accommodation" as it was a necessary development in Christian self-understanding, once the decision had been made to leave the religious ghetto and undertake a transforming mission in the world.

Augustine's formulation of just-war theory need not be accepted in its entirety; but Augustinian political theory contained important resources for the subsequent reflections of a Church that would perennially have to cope with problems of moral understanding similar to those addressed by Augustine. Foremost among these resources is the notion of peace as *tranquillitas ordinis*, the tranquillity of order. For Augustine, *tranquillitas ordinis* was a negative conception—order keeps things from getting worse than they would be under conditions of chaos and anarchy. Yet Augustine's recognition of the place of public order in creating the conditions for the possibility of virtue is of cardinal importance to virtually all subsequent Catholic reflection on war and peace. There *is* a this-worldly alternative to a Hobbesian state of nature, where all are at war with all; that alternative is the order of a governed community. Later Christian reflection would have to develop this conception considerably. Medieval thought, more comfortable with the notion of the state as a "natural" institution, took a good measure of the curse of coercion off the Augustinian conception of "order," and came to see political community as a good in its own right. But the first step in this direction was taken by Augustine, with his resolute (if dour) introduction of Christianity into the travail of history and society. The popes of the twentieth century have continued to argue that political community, constructed around norms of freedom and justice, is an available alternative to mass violence in the defense of rights and the adjudication of conflict; Pope John XXIII went so far as to envision international political community as the answer to the problem of war this side of the Kingdom of Heaven. The seeds of this contemporary papal teaching were, clearly, planted by Augustine.*

Even the dimension of Augustine's thought that most worries those accustomed to liberal categories of political philosophy—his insistence on the coercive prerogatives of the state—can be seen as an essential contribution to a Christian theology of peace. For implicit in Augustine's emphasis on coercion is a recognition that power is the central reality of every community, and particularly of every political community. Perhaps Augustine too readily equated power with the capacity to inflict pain and punishment. But he also seemed to understand, even if he did not emphasize, that power in its classical political sense meant the ability to achieve common purpose. Augustine could not reasonably have been expected to develop this theme, given the demands of his polemic against late-classical antiquity. But in a less chaotic period, the Middle Ages, Christian thinkers would recover this component of the classical political heritage, and thus add an important dimension to Christian reflection on the problem of war and peace. For, if peace as *tranquillitas ordinis*, peace as order within political community, is the only this-worldly alternative to war, as Augustine suggests, then moral reasoning must address the question of an ethics of power within and among political communities. A theological ethics of peace must include, as a core and not peripheral issue, a theological ethics of power. Augustine did not do this in any significant detail, and it would be anachronistic to expect him to have done so. But it was more than sufficient that (however paradoxically through his reflections on coercion) he raised the issue.

*Today, when the genitive case is not widely recognized, it must be emphasized that *tranquillitas ordinis* does *not* translate as "tranquil order." Given the complex meaning of *tranquillitas ordinis* in post-Augustinian Catholic theory, I will translate the phrase variously in the rest of this study, but primarily as "the peace of public order in dynamic political community." Despite grammatical, theological, and political confusions to the contrary, there is nothing "static" about the concept of *tranquillitas ordinis* as it evolved after Augustine.

Herbert Deane concludes his survey of St. Augustine's social and political thought with this observation:

> In our century, when, once more, men have been compelled to recognize the almost incredible brutalities of which human beings are capable, especially when they struggle for political power and military domination, it is no accident that Augustinian pessimism and realism have enjoyed a considerable revival among both theologians and secular thinkers. . . . We may have learned our lessons by reading Freud and by observing the new barbarism of our century rather than by listening to Christian realists. Nevertheless, the optimistic beliefs of many nineteenth-century liberals and Marxists . . . strike us as hopelessly irrelevant as guides to present and future action and shamelessly hypocritical if offered as descriptions of present reality . . . in our age of war, terror, and sharp anxiety about man's future, when, again, a major epoch in human history may be drawing to a close, we cannot afford to ignore Augustine's sharply etched, dark portrait of the human condition. . . . The intellectual equipment that we employ as we face our dilemmas will be needlessly restricted if it has no place for Augustine's powerful and somber vision.[33]

This is one useful way to appropriate the Augustinian heritage: to pose Augustinian realism as a prophylaxis against excessive optimism about the human prospect. But it is too restricted a view. Whatever its dourness, Augustine's thought on the morality of war and peace contains more than helpful warnings against theological and political enthusiasms.[34] For the Augustinian conception of peace as *tranquillitas ordinis*—if located in a more incarnational, less dialectical, theological context and completed by a political ethics that gives considerably more weight to human freedom than Augustine was prepared to entertain—can, in fact, become a positive understanding of the possibilities of peace in this world, even under the conditions of the Fall.[35] Elements of such an appropriation of the Augustinian heritage would come to the fore in medieval Catholic thought.

THE MEDIEVAL HERITAGE: PEACE AS HUMAN POSSIBILITY

St. Augustine's key teaching—that war and peace are in the hands of God, who may permit war as part of the design of a human history that is also the material of salvation history—continued to orient Catholic thought as the patristic period gave way to the Middle Ages, the era of Christendom.[36] But new historical circumstances created the possibility of a development of Catholic doctrine on war and peace, through the reappropriation of elements in the classical heritage that, to Augustine, seemed dross in light of the chaotic collapse of the Roman Empire during the barbarian invasions of Europe and North Africa.

Medieval Catholic thought on war and peace was part and parcel of a wider reflection on the relationship between the Church and the world (as, indeed, was Augustine's theory of the just war). But, unlike Augustine who, as H. Richard Niebuhr argues, began with an "incarnational" or "conversionist" perspective on this larger set of issues, only to abandon it under the theological pressures of his resolute insistence on the pervasive effects of original sin, medieval Catholic thought took the "incarnational imagination" with full seriousness. For medieval Christendom, the world was not just a City of godlessness, for God the creator is also the sustainer and governor of creation.[37] Medieval Catholic thought thus represented a model of the Church/world problem which saw that the essence of that complex relationship was not "either-or," but "both-and."[38]

What Niebuhr terms the "synthetic" model of medieval Christendom had antecedents in patristic thought. Clement of Alexandria, for instance, exemplified a theological perspective that saw human achievement as an expression of the sustaining creativity of God who not only begins all that is, but holds it in being.[39] It was in medieval Christendom, though, that the "synthetic" model hinted at by Clement came to full flower, most especially in the thought of St. Thomas Aquinas.

Aquinas on Political Community

Aquinas, whom Niebuhr regards as the greatest of all the synthesists in Christian history, represented "a Christianity that has achieved or accepted full social responsibility for all the great institutions."[40] St. Thomas knew that the "both-and" approach of the synthesists ought not reduce Christian revelation to mere human wisdom. Thomas was, after all, a monk "faithful to the vows of poverty, celibacy, and obedience. With the radical Christians, he has rejected the secular world. But he is a monk in the church which has become the guardian of culture, the fosterer of learning, the judge of the nations, the protector of the family, the governor of social religion."[41] New historical circumstances had created the conditions for the possibility of a development of doctrine on the relationship between Church and world, which would have important implications for Catholic moral reasoning on the problem of war and peace.

St. Thomas's theological project was universal in scope:

> Thomas Aquinas . . . like Plato and Aristotle before him . . . came at the end of the social development whose inner rationale he set forth. . . . In his system of thought he combined without confusing philosophy and theology, state and church, civic and Christian virtues, natural and divine laws, Christ and culture. Out of these various elements he built a great structure of theoretical and practical wisdom, which like a cathedral was solidly planted among the streets and marketplaces, the houses, palaces, and universities that represent human culture, but which, when one had passed through its doors, presented a strange new world of quiet spaciousness, of sounds and colors, actions and figures, symbolic of a life beyond all secular concerns.[42]

The key to this all-encompassing Thomistic project was the doctrine of the Trinity, the conviction that "the Creator of nature and Jesus Christ and the immanent spirit are of one essence."[43] This trinitarian conviction put Aquinas in a theological position to acknowledge human fallenness and yet insist that human beings remained open to a Godly word in history. Even under the conditions of original sin, human nature was not totally depraved. And "when we regard this nature of ours with the reason that is both God's gift and human activity, then we discern, Thomas is certain, that the purpose implicit in our existence—since we are made as intelligent, willing beings—is to realize our potentialities completely, as intellects in the presence of universal truth and wills in the presence of universal good."[44]

St. Thomas insisted that "God alone can fill the heart of man."[45] But human reason, understanding, and will can know, even under the conditions of original sin, that God is our proper end. In this sense, Thomas took Augustine's famous cry in the *Confessions*—"Thou hast made us for Thee, and our hearts are restless until they rest in Thee"—and made the realization of that rest in God a matter of human possibility, under grace. In truth, grace is always required; but it is a grace that fulfills the God-given yearnings of our human nature.

Thomas was thus considerably more willing than his great North African predecessor to look into human nature and human history for hints and traces of a Godliness that would lead human beings to live well—live gracefully—in this life. The essential, God-touched structure of the world, established by the Creator, had not been negated by original sin; and in human nature and human experience, Aquinas argued, we can find, through reason, "broad principles [of] a natural law which all reasonable men living human lives under the given conditions of common human existence can discern, and which is based ultimately on the eternal law in the mind of God, the creator and ruler of all."[46] The application of these natural law principles in the common life of the human community can, because of the ever-present grace of God, result in social institutions conformed to the best instincts of human nature: the instincts for truth and *caritas* (that love which transcends the requirements of justice).

Governance, for Aquinas, was not the strictly remedial or punitive reality it was for Augustine; governance could be an expression of that human nature which has God for its origin. Earthly governance, and the rules and laws that shape it, cannot raise us to the level of supernatural life with God, which is our true end; for this, the divine law revealed by the prophets and by Christ was necessary.[47] St. Thomas did not collapse Christ into culture. But his trinitarian vision of reality, and his conviction that human reason can know with conviction the natural law that orders us (even after Adam) toward our proper end of communion with God, allowed Aquinas to develop a theory of social and political life focused on human possibility rather than depravity; on human creativity as an expression of our God-given natural gifts, rather than on obedience to a social order whose primary function is to keep evil at bay.

St. Thomas's fundamental theological convictions, influenced and shaped (although not necessarily determined) by his appropriation of Aristotelian philosophy, led him to a view of society as "natural" rather than "remedial"—the Augustinian conviction. But Aquinas also taught that organized political or civil community, not just society in general, was "natural."[48] If society is the expression of innate human sociability, if we need society in order to be fully human, then, as Frederick Copleston writes:

> Organized government is a natural institution. Even if men had never sinned, there would still be need for some control over their activities directed to the common good. In short, a civil society and civil government are natural institutions in the sense that they are necessary for the fulfillment of men's natural needs and for the leading of a full human life: there is no essential connexion between the State and sin, and man would still have required the State even if he had never fallen. Further, if the State is a natural institution, it is willed by God, the Author of nature.[49]

This Thomistic/Aristotelian concept of the state as a natural rather than remedial institution was the foundation on which other key themes were developed. The state was not, for Thomas, simply a set of mechanical arrangements for the ordering of public life; the state was oriented, by its nature, to an end, the common good of the citizens. And "by 'common good' he did not mean simply their temporal welfare in the material sense, but, more fully, the leading of the good life, which is defined, in Aristotelian fashion, as a life according to virtue."[50] Governance had a positive function: the achievement of the common good, understood as the virtuous life.

As Thomas himself puts it, the final end of life in society is "not merely to live in virtue, but rather through virtuous life to attain to the enjoyment of God."[51] For the achievement of this ultimate end, divine grace was necessary. But the governed community was not simply the structure of civil order through which individuals could fulfill their responsibilities to God. The life of the political community was itself an important dimension of the life of virtue, which led, under grace, to the beatific vision of the blessed. St. Thomas did not believe that "man has, as it were, two final ends, a temporal end which is catered for by the State and a supernatural, eternal end which is catered for by the Church: he says that man has one final end, a supernatural end, and that the business of the monarch, in his direction of earthly affairs, is to facilitate the attainment of that end."[52]

This sense of the governor's responsibilities for the right ordering of public life for the common good shaped Aquinas's "constitutionalism," his claim that sovereignty does not belong to the monarch by divine right through birth, but through the consent of the governed, who delegate the sovereignty that is given to them in community by God. This is not to confuse Aquinas with Jefferson; but it is to understand, as Copleston puts it, that, for Aquinas, "the ruler possesses legislative power only in so far as he stands in place (*gerit personam*) of the people."[53] The monarch who tries to rule without such consent, Aquinas held, is a tyrant.[54] The roots of a Catholic theory of democratic governance can be found here; and the key to the whole construct was Thomas's insistence on the "natural" character of society and government as implied by the creaturely sociability of man.

Complementing Thomas's "high" view of the responsibilities of the natural state was his firm insistence that the citizen, the individual human being, is a *person* with a unique dignity and value.[55] Thomas's Christianity modified his Aristotelian understanding of the state:

> Sometimes he speaks of the individual as being ordered to the community as a part to a whole and of the private good of the individual as being subordinated to the common good; but his Christian conception of the individual, as a spiritual person whose final end is supernatural, rendered any complete subordination of the individual to the State entirely unacceptable. Anything savouring of political "totalitarianism" was quite out of the question for Aquinas, as for any other Christian medieval thinker. . . . Government is willed by God, but the government has a trust to fulfill.[56]

The measure of a government's goodness is thus the degree to which it creates the conditions for the possibility of the fulfillment of this trust for each of its citizens. To adopt a more contemporary usage, "human rights" inhere in persons; human rights are not benefices distributed by the state at its pleasure.

Finally, Aquinas taught that the civil peace that is constitutive of the common good is not simply a matter of contractual justice, of minimally fulfilling social obligations for the sake of a remedial order. "It is not enough," he wrote, "for peace and concord to be preserved among men by the precepts of justice, unless there be a further consolidation of mutual love. Justice provides for men to the extent that one shall not get in the way of another, but not to the extent of helping another in his need. . . . Thus there came to be need for an additional precept of mutual love among men so that one should aid another even beyond the obligations of justice"[57]—not justice, narrowly construed as minimal contractual obligation, but *caritas*, was the instrument by which civil peace could flower into virtue in the governed community. As one commentator puts it in a summary of Aquinas's view of Christianity and politics, "the civil law may not force the highest virtues on men or force men to perform virtuously the acts which it prescribes. The perfection of liberty must come through a law that, by reaching the interior movements of the soul, forbids and prescribes, rewards and punishes without compelling. And this perfection of freedom is the end at which every lawgiver aims."[58]

Aquinas on the Just War

St. Thomas addressed the morality of war in his discussion of *caritas* in the *Summa Theologiae*.[59] The key article begins with the question, "Is it always a sin to wage war?" In the stately pattern of medieval theological disputation, Aquinas suggests the reasons why we might think that war is always sinful. First, there are the words of the Lord himself, in Matthew 26:52: "All who live by the sword will die by the sword." Then there is the witness of St. Paul, who admonished his fellow Christians in Romans 12:19, "Beloved, never avenge yourselves, but leave it to the wrath of God." There are also theological reasons for concern: if peace is a virtue, then war, the opposite of peace, would seem to be always sinful. Finally, there is the customary law of the Church: if the Church forbids practicing for war in tournaments, war itself must be wrong. Thomas then presents the opposing case, and cites Augustine's sermon on the healing of the centurion's son in which the bishop of Hippo wrote, "If Christian teaching forbade war altogether, those looking for the salutary advice of the Gospel would have been told to get rid of their arms and give up soldiering. But instead they were told, Rob no one by violence or by false accusation, and be content with your wages (Luke 3:14). If the Gospel ordered them to be satisfied with their pay, then it did not forbid a military career."

Having set out the case for and against the proposition that war is always sinful, St. Thomas then offered his own judgment:

> Three things are required for any war to be just. The first is the authority of the sovereign on whose command war is waged. . . . Secondly, a just cause is required, namely that those who are attacked are attacked because they deserve it on account of some wrong they have done. So Augustine wrote, "We usually describe a just war as one that avenges wrongs, that is, when a nation or state has to be punished either for refusing to make amends for outrages done by its subjects, or to restore what has been siezed injuriously." . . . Thirdly, the right intention of those waging war is required, that is, they must intend to promote the good and to avoid evil. . . . Now it can happen that even given a legitimate authority and a just cause for declaring war, it may yet be wrong because of a perverse intention. So again Augustine says, "The craving to hurt people, the cruel thirst for revenge, the unappeased and unrelenting spirit, the savageness of fighting on, the lust to dominate, and such like—all these are rightly condemned in wars."

Aquinas then concludes his judgment with this reminder: "Even those who wage a just war intend peace. They are not then hostile to peace, except that evil peace which Our Lord 'did not come to send on the earth' (Matthew 31:34). So again Augustine says, 'We do not seek peace in order to wage war, but we go to war to gain peace. Therefore be peaceful even while you are at war, that you may overcome your enemy and bring him to the prosperity of peace.'"

Aquinas did not regard war as a positive moral good. The presumption is always for peace, and the burden of moral reasoning lies with those who argue for the justness of a particular resort to war. Yet, and this is of equal importance, Aquinas did not regard war and peace as disjunct realities; thus the importance of the citation from Augustine, that "we go to war to gain peace." Although the burden of proof may lie on those who would resort to war, it was clear to Aquinas that that burden could be met. It is met not only in terms of just cause and proper authority, but in terms of the final goal to be sought: the reestablishment of a bond of community that will allow both attacker and attacked to resume their responsibilities for the pursuit of the common good of their peoples. The just war, then, is not only a remedy for grievances suffered; like all morally sound public life, the just war is ordered to the common good, both of one's own community and of the enemy.

Thomistic just-war theory is "minimalist": its purpose is not to glorify combat, but to determine the minimal conditions under which the resort to armed force may be morally justifiable. There is a "presumption for peace" at work here; but there is also a parallel concern for discerning how the proportionate and discriminate use of armed force can be ordered to peace, and thus contribute to the common good that is the end of all governance. Such an approach to the morality of war is fully consistent with Thomas's "synthetic" project. Rather than see war as outside the boundaries of the moral order, Aquinas insisted that even war—surely a limit case for moral theory—could be understood and conducted in ways that contribute to the working out of salvation in history. The morality of war is part and parcel of the larger question of the right ordering of society, the natural habitat for our pilgrimage to God.

It would be an exaggeration to suggest that St. Thomas's development of just war theory is a major theme in his theology. Later medieval and scholastic theologians discussed the just resort to and conduct of war in considerably more detail, whereas Aquinas seemed satisfied with Augustine's strictures against lustful intentions and wartime reprisals. The importance of Thomas's contribution lies not so much in his discussion of the morality of war as in his more developed theory of peace as a human possibility within a rightly ordered political community. In St. Thomas's constitutionalism, we see the outlines of a Catholic theory of peace that takes the notion of *tranquillitas ordinis* beyond the punitive and remedial, and locates the question of peace within the larger context of political community as our natural home in this world. It would not be an exaggeration, then, to argue that Aquinas transformed *tranquillitas ordinis* from a negative to a positive concept. The peace of rightly ordered political community is an expression of the goal of society—the common good that is the life of virtue for all citizens. The distinguishing mark of such a community of virtue will be not only its just conduct of affairs, understood in contractual terms of minimal obligation, but its *caritas*, its capacity to create those conditions under which human beings can act on their best instincts for truth and goodness.

Reformers, Crusaders, and Neo-Scholastic Commentators

Other themes in medieval political theory helped shape the complex Catholic heritage of thought on war and peace.

Medieval Christendom was not so static as it is often portrayed. Reform movements were common, and some of them took up questions of the morality of war and peace. Bainton notes:

> The first half of the eleventh century was marked by a great campaign—mainly in France, but also in Germany—to promote the Peace of God and the Truce of God. The first category limited those involved in war by increasing enormously the category of the exempt. The Council of Narbonne in 1054, for example, decreed that there should be no attack on clerics [a medieval canonical category considerably broader than our contemporary notion of "clergy" as sacramentally ordained deacons, priests, and bishops], monks, nuns, women, pilgrims, merchants, peasants, visitors to councils, churches and their surrounding grounds to thirty feet (provided that they did not house arms), cemeteries and cloisters to sixty feet, the lands of the clergy, shepherds and their flocks, agricultural animals, wagons in the fields, and olive trees.
>
> The Truce of God limited the times for military operations. There should be no fighting from Advent through Epiphany, nor from Septuagesima until the eighth day after Pentecost, nor on Sundays, Fridays, and every one of the holy days throughout the year.[60]

This severely circumscribed the time for legitimate military operations.

Arbitration of disputes was another medieval method for limiting the resort to war; here the Church's role was often more admonitory than judicial.[61] But the most radical reform movements in the medieval world—certain of the Franciscans, the Waldensians, the Chelciky branch of the followers of Huss (the Bohemian Brethren)—tried to take the argument far beyond the restrictions of the Peace and Truce of God and to resurrect the tradition of Christian pacifism. A tendency to biblical literalism, a countercultural ethos, and a sectarian view of the Church/world question marked each of these movements (in different ways, and to different degrees). Much of their concern over the morality of war and peace came from a passion for the repristination of the Church, rather than for the reform of civil society. And, for the radical reformers of the Middle Ages, one great expression of the Church's corruption was its enthusiasm for the Crusades.[62]

Although most medieval theologians addressed the Crusades within the categories of just war theory, Bainton seems more faithful to the historical record in arguing that the Crusades represented a new voice in the Christian debate over war and peace. The primitive Church and some later sectarians had argued "no war," or pacifism; the mainstream of the tradition, following Augustine, had argued "just war." The Crusades represented a new answer: "holy war." As Bainton puts it:

> Here was a war inaugurated by the Church. Service was volunteered rather than exacted by a ruler from his retainers. The code of the just war, which was being elaborated and refined by the secular ideas of chivalry and the Church's ideal of the

> Truce and the Peace of God, was largely in abeyance in fighting the infidel. Crucifixion, ripping open those who had swallowed coins, mutilation—Bohemond of Antioch sent to the Greek Emperor a whole cargo of noses and thumbs sliced from the Saracens—such exploits the chronicles of the crusades recount without qualm. A favorite text was a verse in Jeremiah, "Cursed be he that keepeth back his hand from blood." There was no residue here of the Augustinian mournfulness in combat. The mood was strangely compounded of barbarian lust for combat and Christian zeal for faith.[63]

Nor was the crusading passion limited to the laity; "monastic pacifism collapsed and there came to be monastic military orders, the Templars, the Hospitalers, and the Knights of St. John."[64] St. Francis of Assisi, often taken as the paradigm of Christian gentleness and the model of Christian pacifism, accompanied and did not condemn the fifth Crusade; other Franciscans, notably St. John Capistrano, in whose name the swallows still return to California, preached the Crusade.[65]

The Crusade is now regarded throughout the Christian world as a grotesquerie and an aberration. But inasmuch as elements of the theory of the Crusade—that war can be a positive moral good, that violence can be redemptive—have reemerged in contemporary debate through some themes in the theologies of liberation, it is well to pause and ask precisely what allowed the medievals to embrace crusading with such religious enthusiasm, and with what seems to have been so little moral scruple.

Bainton traces the root of the problem to the attempt of post-Augustinian theology to link war and peace within a common moral framework. The key text was Augustine's claim, quoted by Aquinas in the *Summa*, that "we go to war to gain peace." This theory had led to the medieval attempt to enforce peace through "peace militia . . . in which the clergy participated in their church banners. [And] implicit in these attempts to enforce the peace was the idea of a crusade, that is to say of a war conducted under the auspices of the Church for a holy cause—the cause of peace."[66] Because many of these efforts had failed in Europe, the Church tried to displace them into a foreign arena. If it could not be very successful in mitigating violence in medieval Europe, perhaps the Church could channel violent human energies into war against the infidel. Evidence for such an analysis is ready to hand in Pope Urban II's speech at the Council of Clermont in 1095, which began the Crusades:

> Oh race of the Franks, we learn that in some of your provinces no one can venture on the road by day or by night without injury or attack by highwaymen, and no one is secure even at home. Let us then re-enact the law of our ancestors known as the Truce of God. And now that you have promised to maintain the peace among yourselves you are obligated to succour your brethren in the East, menaced by an accursed race, utterly alienated from God. The Holy Sepulchre of our Lord is polluted by the filthiness of an unclean nation. Recall the greatness of Charlemagne. O most valiant soldiers, descendants of invincible ancestors, be not degenerate. Let all hatred depart from you, all quarrels end, all wars cease. Start upon the road to the Holy Sepulchre to wrest that land from the wicked race and subject it to yourselves.[67]

As Bainton notes, "the assembly cried *Deus vult*. Peace should thus be achieved at home by diverting bellicosity to a foreign adventure."[68]

Sociological and demographic factors also played their part in the rise of the Crusades. As Paul Johnson puts it:

> The idea that Europe was a Christian entity, which had acquired certain inherent rights over the rest of the world by virtue of its faith, and its duty to spread it, married perfectly with the need to find some outlet both for its addiction to violence and its surplus population. . . . The crusades were thus to some extent a weird half-way house between the tribal movements of the fourth and fifth centuries and the mass trans-Atlantic migration of the poor in the nineteenth.[69]

But even these factors could not provide much of a theological or even political rationale for crusading. "The West was never very happy in its justification of the Islamic Crusade," according to R. W. Southern; "the violence of its advocates covered some very weak points in their arguments."[70]

The Crusades may also reflect the institutional Church's perennial difficulty in coping with religious enthusiasm. "The central problem of the institutional Church," Johnson writes, "was always how to control the manifestations of religious enthusiasm, and divert them into orthodox and constructive channels."[71] The fragmentation of medieval Europe did not allow for channeling those constructive energies into new political communities more adequately expressing St. Thomas's constitutionalism. Larger-scale polities, better able to realize in practice the values contained in the Thomistic principles of personalism, the common good, pluralism, and subsidiarity, would be needed. The Crusades, then, embarrassing as they may be, do not invalidate the Thomistic development of Catholic political theory and its relevance to moral reasoning about war and peace.

This development continued for centuries, most prominently in the work of the neo-scholastic commentators Francisco de Vitoria (ca. 1492–1546) and Francisco Suárez (1548–1617);[72] their work added several important themes to a Thomistic framework that, in its essentials, remained intact. The neo-scholastics broadened the community of moral discourse and responsibility to include the "community of nations," not merely one's own political community; this would remain a constant theme in official Catholic teaching on the morality of war and peace down to our own time. The neo-scholastics also continued the Thomistic retreat from Augustine's negative view of *tranquillitas ordinis*, but they added to St. Thomas's positive reinterpretation of the peace-of-order particular diplomatic mechanisms (e.g., arbitration). Vitoria and Suárez placed a considerable burden of proof on rulers for determining whether the resort to violence served the good of a community now considered to include the adversary. Neo-scholastic theologians gave greater weight to the conscience of individual soldiers than did Thomas (and certainly more than Augustine).

Finally, the neo-scholastics insisted that the criterion of proportionality—Does the good to be achieved by the resort to violence outweigh the undoubted damage to be done, both to individuals and to the "community of nations"?—applied to the decision to go to war, just as it did to conduct in war. The proportionate use of military force had to be justified in a larger context; more was involved than the claims of the two contending parties. For the neo-scholastics, potential combatants were not only bound into a community of mutual moral responsibility; they were, together, part of an even more complex fabric and tapestry of social life in the "community of nations." The claims of that wider community had to bear on the decision to go to war.[73]

Summary

The classic Catholic approach to the moral problem of war and peace can be properly termed "Augustinian/Thomistic." This is not to deny that there were significant differences in tone and content to the Augustinian and medieval components of the Catholic heritage on these questions. But it is to suggest, following H. Richard Niebuhr's analysis of Augustine's theological intention, that the core theological understandings of the bishop of Hippo and the *doctor communis* were, on the matter of war and peace, congruent, and form two moments in an evolving, unified tradition. John Eppstein provides a useful schematic summary of this Augustinian/Thomistic tradition:

A. There is a greater merit in preventing war by peaceful negotiation and conciliation than in vindicating rights by bloodshed.
B. Peace attained by conciliation is better than peace attained by victory.
C. There is a natural society of mankind which gives rise to certain rights and duties relevant to the morality of war.
D. The absence of a superior tribunal before which a prince can seek redress can alone justify him in making war, except when resisting actual attack.
E. Saving the direct intervention of God, the following conditions must in addition be fulfilled before a war can be just:
   1. It must have a *just cause*: this can only be a grave injury received (e.g., actual invasion; unlawful annexation of territory; grave harm to citizens or their property; denial of peaceful trade and travel) or a great injustice perpetrated upon others whom it is a duty to help (e.g., the same injuries as above, violation of religious rights).
   2. It must be *necessary*: i.e., the only available means of restoring justice or preventing the continued violation of justice.
   3. It must be the consequence of a *formal warning* to the offending state and must be formally *declared.*
   4. It must be declared and waged only by the *sovereign authority* in the state (i.e., one who has no political superior) and, if the defense of religious rights are involved, with the *consent of the Church.*
   5. The good to be attained by war must be reasonably supposed to be *greater than the certain evils*, material and spiritual, which war entails.
   6. A *right intention* must actuate . . . the declaration, conduct, and conclusion of war. That intention can only be the restoration or attainment of true peace.
   7. *Only so much violence* may be used *as is necessary*: in the case of defense, only so much as is necessary to repel the violence of the aggressor.
F. The *moral responsibility* for war lies upon the *sovereign authority*, not upon the *individual soldier* or citizen: his duty is to obey, except in a war which he is certainly convinced is wrong.
G. Priests may not fight even in just war.
H. The duty of *repelling injury inflicted upon another* is the common obligation of all rulers and peoples.[74]

By the time of the neo-scholastics, then, Catholicism had developed a rich body of thought on the larger question of the right ordering of society, and on the specific limit case of the morality of war and peace. *These two dimensions of the problem, its wider and narrower aspects, cannot be separated in Catholic theory*, which is as interested in the nature of a righteous peace as it is in the specific moral issues of the *ius ad bellum* (the norms guiding the decision to go to war) and the *ius in bello* (the norms of just conduct to be observed in warfare). Moreover, there is a theological trajectory evident in the tradition—in its personalism, its constitutionalism, its (medieval) concept of "two authorities" (i.e., the civil and religious powers, yet operating within one society), and its positive understanding of peace as *tranquillitas ordinis*—that points toward the development of *democratic* forms of political community as this-worldly alternatives to war in the resolution of conflict, the vindications of rights, and the protection of the innocent. Neither St. Thomas nor the neo-scholastics would have phrased the issue this way. But the seeds of a distinctively American Catholic contribution to the continuing evolution of Roman Catholic theory on the morality of war and peace are clearly evident in the riches of the medieval heritage and its positive construction and elaboration of Augustine's *tranquillitas ordinis.*

MODERATE REALISM AND POLITICAL COMMUNITY: THE DISTINCTIVENESS OF CATHOLIC THEORY

The great Renaissance humanist Erasmus of Rotterdam (ca. 1466–1536) should be mentioned, if only briefly, in closing this initial survey of the formative currents in Catholic moral reasoning on war and peace. Erasmus's intellectual interests included the political theory of the Greek and Roman classicists; thus we find in his passionate personal rejection of war "a fine blending of . . . classical and . . . Christian themes."[75] Erasmus also lived at a time when Renaissance optimism about the possibilities of the human adventure intersected with the political intemperateness of a Europe struggling to make the often bloody transition from city-states to nation-states. The result was a volatile brew: although remaining theoretically committed to just-war theory, Erasmus inveighed against the institution of war, the stupidity and bloodthirstiness of the princes whose rashness and greed led to war, and the failures of the Church to bridle the human passion for violence and conquest.

Unlike his friend Thomas More, whose *Utopia* had room within it for the proportionate use of armed force in defense of the innocent and the injured, Erasmus's polemics were essentially antiwar. The themes he used for his rhetorical ends—"the recital of the horror, cost, and folly of war . . . the stress on the fatherhood of God and the brotherhood of believers, and the passionate concern that swords be beaten into plowshares"[76]—would reappear time and again in subsequent Christian, and especially Catholic, pacifism, particularly after the Second Vatican Council (of whose optimism Erasmus may be considered a distant ancestor).

Renaissance humanism, particularly under the chaotic political conditions of the fifteenth and sixteenth centuries, led the Catholic Erasmus into a position that was pacifist in rhetoric if not in theory. Similar interests led the Protestant Dutchman Hugo Grotius (1583–1645) to a further development of the Augustinian/Thomistic approach to the problem of war. Grotius, like other humanists, appealed to classical sources for his arguments (particularly the Stoics), but there is a clear resonance between Grotius's notion that "war may be waged justly, but only to achieve or to re-establish the natural end of man, which is peace or the condition of tranquil social life,"[77] and the Thomistic interpretation of Augustine's *tranquillitas ordinis.*

Grotius also continued the neo-scholastic tradition of moral and political reflection on the "community of nations," and applied the concept of law to international life in a way that would have been agreeable, one suspects, to Aquinas. In *De iure belli ac pacis*, for example, Grotius wrote that "if no association of men can be maintained without law . . . surely also that association which binds together the human race, or binds many nations together, has need of law; this was perceived by him who said that shameful deeds ought not to be committed even for the sake of one's country."[78]

Grotius also reflected the Thomistic "minimalist" tradition of just-war theory when he argued that "war ought not be undertaken except for the enforcement of rights; [and] when once undertaken, it should be carried on only within the bounds of law and good faith."[79] There is also a Thomistic echo in Grotius's humanistic assertion that "those who are enemies do not in fact cease to be men."[80] With the medievals and the neo-scholastics, Grotius claimed that war was justifiable, not as an instrument of imperial ambition or territorial greed, but only as a means for defending rights in the absence of a superior tribunal or authority to which grievances and conflicts could be referred. On the question of *ius in bello*, Grotius argued, again following the neo-scholastics, that the "law of nations," not merely utilitarian considerations, determined the moral legitimacy of particular acts in war.[81]

It is a curiosity of history that the Dutch Protestant Grotius extended the trajectory of reflection first established by the Augustinian/Thomistic tradition, rather than the Catholic reformer Erasmus. But the very fact of Grotius's work suggests the power of the Catholic heritage of moderate realism, particularly as the nation-state system began to form in Europe. That heritage was realistic: in its view of human nature and the inevitable tensions of social and political life, and in its conviction that the Kingdom of God would not be established by the works of our hands. But it was also a moderate tradition: convinced that human nature had not fallen completely under the shadow of depravity; that society and politics were not merely remedial, but were part of the warp and woof of a human life ordered to beatitude; that a this-worldly peace was possible (if not assured) if political communities could be rightly ordered and justly governed. This Catholic tradition of moderate realism was built around three core convictions.

1. *Politics* is an arena of rationality and moral responsibility. The source of this conviction is the rejection of any "total-depravity" interpretation of the fall of Adam and Eve. Even under original sin, and granting the ever-present need for God's grace, the Catholic tradition of moderate realism contended that human beings could build and sustain political communities worthy of their dignity as images of God in history. Coupled with, and related to, the rejection of a total-depravity view of human nature was the conviction that society is a "natural" phenomenon, and that governance has a positive, and not merely punitive, character and function. In the Catholic tradition of moderate realism, political community is a good in its own right, an institutional expression of the sociability that is part of the God-given texture of human life.

The Augustinian dimension of the Catholic tradition is a continuing reminder that sin marks all the days of our lives, and that politics is always subject to corruption. But the basic stance of the tradition, faced with the question of human nature and political life, is sober optimism, with equal emphasis on adjective and noun. Human reason can grasp an "ought" under the conditions of original sin and amid the complexities of political life. The grasping of that "ought," no matter how dimly it may at times be perceived, means that politics is not an amoral or irredeemably immoral exercise, but an exercise in virtue, in moral responsibility.

2. *Power* should be understood in classical terms: as the ability to achieve a corporate purpose, for the common good. Power, in the Catholic tradition of moderate realism, is not to be reduced (or better, traduced) to violence; violence is a limit case defining the margins of a rationally ethical politics. Thus, power, like governance, has a positive dimension: it is an expression of the human creativity that is analogically related to the abiding creativity of God the Father, Jesus the Logos, and the Spirit brooding over the waters of the created order.

Moreover, power and governance are related: power is the central reality of political community, and without the exercise of power—the achievement of purpose for the common good—there is anarchy, chaos, and, inevitably, violence. But to say that power and governance are related is to acknowledge that the exercise of power is under moral judgment. The issue for Catholic moderate realism, then, is not *whether* power shall be exercised, but *how*: To what ends, under whose authority, by what means? If peace is one goal of governance, and power is central to the reality of governance for the common good, then a major dimension of any theology or political theory of peace will be its understanding of power.

3. Peace should be understood as *tranquillitas ordinis*, a dynamic, not static, concept. Peace provides the stability that allows social life to grow and develop. But peace, properly understood, is itself a dimension of that growth and development, which always aims toward *caritas* as its most worthy end. Given human divisions, and considering the mainstream understanding of revelation on the question, a peace fully comprehending the biblical vision of *shalom* will never be achieved in this world. But there need not be agreement on every potentially controverted point of justice before a peace worthy of human rationality and morality can be constructed. The Catholic tradition of moderate realism, holding in tension both human possibility and human limitation, affirmed that morally worthy political community could be built around the principles of personalism, the common good, and pluralism, and that such a political community could achieve a morally sound peace.

Justice is one component of *tranquillitas ordinis*. But justice is not the precondition to peace, if by "justice" we mean the final and complete achievement of that *shalom* vision of human concord where all conflict and strife have ceased. The dimensions of justice that are most relevant to the creation of a morally worthy peace of *tranquillitas ordinis* are those that derive from Aquinas's constitutionalism: concepts of the limited authority of the state, and of the importance of consent for the proper exercise of authority.

The Catholic tradition of moderate realism knew that there could be a morally unworthy peace—under a particularly efficient tyrant, for example. But between the peace of tyranny and the eschatological peace of *shalom* is the peace of dynamic, rightly ordered political community. And it is to this third option that the Catholic tradition patiently but resolutely points us, urging that we never lose heart, but that we not lose our sense, either. If a peace of political community can be rationally imagined, then it can be created; and if it can be created in ways that satisfy basic standards for the right ordering of society, then we are under a moral obligation to make the effort.

To say that the Catholic tradition of moral reflection on the problem of war and peace is a tradition of "moderate realism" is to say, in sum, that it is neither Hobbesian nor utopian. Recognizing the beast in the human heart, the tradition yet affirmed that we are not, at bottom, beasts, even under the pressures of the quest for political power. Understanding that conflict is a constant of the human condition—the political meaning of the doctrine of original sin—the tradition still claimed that political community, rightly ordered, provides a morally worthy means for resolving conflict on this side of the Kingdom of God. Political community and the Kingdom must never be confused; all the works of our hands, and particularly our political works, which are so fraught with ambiguity, stand under judgment. But the sinner who takes up the burden of creating and sustaining political community ordered to the common good is, simultaneously, a spark struck from the creativity of the Godhead. We are, in the end, the image of God in history, and the task of history is one we cannot lay down. The issue is to take it up wisely, gracefully, well.

CHAPTER 2

American Catholicism and the Tradition Received: From John Carroll to the Second Vatican Council

A classic historiographic theme in the study of American Catholicism holds that Church leadership in the United States was unswervingly devoted to the cause of American arms from the time of George Washington to the latter days of Lyndon Johnson. Whatever dissent from national security policy individual bishops may have felt was registered privately; the public face of official American Catholicism was always one of vigorous and firm support for the foreign and military policy of the United States.[1] In recent years, a codicil has been added to this analysis: that at certain key moments, American Church leaders went far beyond support and, abandoning their critical faculties, became jingoists of the worst sort.[2] The current foreign policy activism of the American bishops is thus perceived, by some, as an act of expiation for the uncritical stance of the bishops' predecessors.

The classic historiography offers three answers to the question of why the American bishops crafted "this almost unvaried policy of support . . . to the public authorities in time of war, a policy that continued practically unbroken to the 1960s."[3] John Tracy Ellis writes:

> First, it was virtually universal Catholic teaching that Catholics should uphold the hands of legitimately constituted governments. Secondly, the general antipathy of the American people for Catholicism which was a colonial inheritance and which again lasted down to the 1960s, was another powerful influence. The point was neatly summarized in a remark made to the present writer by the late Arthur M. Schlesinger, Sr., when he said, "I regard the prejudice against your Church as the deepest bias in the history of the American people."
>
> A third factor that should be kept in mind was the arrival in the United States between 1790 and 1920 of approximately 9,395,000 Catholic immigrants who by their sheer numbers quickly overshadowed the native Catholic element, prompted Americans to regard the Church as a "foreign" institution, and, in turn, caused many of the Catholic immigrants and their children to seek refuge in urban ghettoes where they took on a siege mentality. Their new cast of mind was never better illustrated than by the special pains they took to show their patriotism, and that in a way that left a deep imprint on the psyche of the general Catholic community.[4]

Three reasons, then, purport to explain the pre-Vietnam adherence of the American Church to the national security policies of the United States: the Pauline/Augustinian tradition of obedience to properly constituted authority; the need to combat nativist anti-Catholicism; and the exigencies of assimilating an immigrant population into mainstream American life. The second and third points in this analysis are related. The classic analysis also explains, within its own framework, the occasional quirks in this pattern of support for governmental policy: Catholic converts with unimpeachably American backgrounds like Orestes Brownson and James McMaster, or old-native Catholics like Bishop John Lancaster Spalding, whose forebears settled in Maryland around 1658, could "express views that other Catholics would hardly have dared to utter."[5]

The impression left by this historiography is that the leadership of the Catholic Church in the United States thought very little about the moral problem of war and peace prior to our own times; this changed only after American Catholics were politically "emancipated" in the presidential election of 1960. The conclusion often drawn by both scholars and popular translators of historical scholarship is that the only tradition of American Catholic thought relevant to the dilemmas of war and peace today is the tradition formulated after 1960, and particularly after the Second Vatican Council. (An exception is usually made for the Catholic Worker movement, which, by definition, took a position on the morality of war and peace considerably different from that of the American hierarchy.)[6] One gets the impression that many contemporary Catholic commentators in the moral debate over war and peace, security and freedom, are rather embarrassed by their American Catholic ancestors.[7]

Such an attitude is both unwarranted and unwise. It is unwise because it reflects a tendency to limit "thought about the moral problem of war and peace" to "critical analysis of particular weapons," or "criticism of particular wars" (or both), and especially to criticism leveled at one's own government. Criticism of the Hitler-Stalin pact of 1939, for example, would not count under the rubric of "thought about the moral problem of war and peace," nor would support for the League of Nations in the aftermath of World War I. The notion that thinking about and acting on the problem of war is basically a question of weapons is widespread in American political culture today.[8] But why should a Church whose tradition includes a rich analysis of the meaning of peace, and of the possibilities of political community as a means toward peace, contribute to such an unwarranted constriction of the argument, particularly in considering its own history?

The classic historiography is correct in its assertion that there was a consistent pattern of official American Catholic support for the foreign and military policies of the United States from John Carroll to Francis Spellman; the documentary evidence overwhelmingly supports that claim. But it does not necessarily follow that little thought about the problem of war and peace had gone on in American Catholicism prior to the Second Vatican Council. American Catholicism (and in particular its official leadership) had important things to say about the moral problem of war and peace prior to Vatican II. This heritage of thought has been largely lost, or ignored, because of a post-Vietnam tendency to identify "the moral problem of war" with the problem of contemporary weapons systems (especially nuclear weapons), to the detriment of thought about larger *ad bellum* questions (e.g., What is the nature of the threat to which nuclear weapons are a deterrent response?). Classic Catholic questions about the possibilities of political community as an alternative to war or its threat in the resolution of conflict have also been ignored.

The classic historiography is not so much wrong as it is incomplete. Recollecting the American Catholic tradition of thought on the moral problem of war and peace—in all the complexity of that problem—is a necessary first step toward understanding the present state of the war/peace debate within American Catholicism.

FROM JOHN CARROLL TO JAMES GIBBONS: AMERICAN CATHOLICISM AND THE MORAL PROBLEM OF WAR FROM THE REVOLUTION TO THE FIRST WORLD WAR

Nativist suspicion of all matters Catholic was a serious problem at the time of the American Founding. A letter that future President John Adams wrote his wife Abigail during the 1774 session of the Continental Congress illustrates the bias that was at work. For amusement, Adams spent a day visiting various churches in Philadelphia. He wrote Abigail about his impressions of Mass in a local Catholic parish:

> This afternoon, led by curiousity and good company, I strolled away to mother church, or rather grandmother church. I mean the Romish chapel . . . [the] entertainment was to me most awful and affecting: the poor wretches fingering their beads, chanting Latin, not a word of which they understood; their pater nosters and ave Marias; their holy water; their crossing themselves perpetually; their bowing to the name of Jesus, whenever they hear it; their bowings, kneelings and genuflections before the altar. The dress of the priest was rich white lace. His pulpit was velvet and gold. The altar piece was very rich, little images and crucifixes about; wax candles lighted up. But how shall I describe the picture of our Saviour in a frame of marble over the altar, at full length, upon the cross in the agonies, and the blood dropping and streaming from his wounds! The music, consisting of an organ and a choir of singers, went all the afternoon except sermon time, and the assembly chanted most sweetly and exquisitely.
>
> Here is everything which can lay hold of the eye, ear, and imagination—everything which can charm and bewitch the simple and ignorant. I wonder how Luther ever broke the spell.[9]

Adams's sentiments were hardly unique in his time; the concern of John Carroll, founder of the American hierarchy, to establish his fellow Catholics' commitment to the cause of the new nation was the concern of a realistic and prudent man. About the time of his appointment as first bishop of Baltimore, Carroll (along with his cousin Charles Carroll of Carrollton, signer of the Declaration of Independence, and two other distinguished laymen) wrote George Washington a letter of congratulations on his first inauguration. After thanking Washington for his role in launching the new republic, Carroll noted:

> This prospect of national prosperity is peculiarly pleasing to us on another account; because whilst our country preserves her freedom and independence, we shall have a well-founded titled to claim from her justice equal rights of citizenship, as the price of our blood spilt under your eyes, and of our common exertions for her defense, under your auspicious conduct, rights rendered more dear to us by the remembrance of former hardships.[10]

Washington, mindful of John Carroll's heroic role in the abortive attempt to draw Canada into the revolution and Charles Carroll's leadership in the nascent Federalist Party, and aware of the deep residue of anti-Catholic bias in the new United States, was quick to respond in kind:

> As mankind become more liberal, they will be more apt to allow, that all those who conduct themselves as worthy members of the community are equally entitled to the protection of civil government. I hope ever to see America among the foremost nations in examples of justice and liberality. And I presume that your fellow-citizens will not forget the patriotic part which you took in the accomplishment of their Revolution, and the establishment of their government; or the important assistance which they received from a nation in which the Roman Catholic religion is professed. . . . May the members of your Society in America, animated alone by the pure spirit of christianity, and still conducting themselves as the faithful subjects of our free government, enjoy every temporal and spiritual felicity.[11]

Carroll's letter to Washington was of a piece with his attempts to get the Vatican to understand the peculiar condition of Catholics in a nation whose religious circumstances did not fit the pattern of those European confessional states with which the Vatican was considerably more familiar. Ellis tells of another initiative undertaken by Carroll prior to his appointment as first bishop of Baltimore:

> Carroll informed the Holy See of the advantages for the Church to be derived from the new nation's separation of Church and State with its accompanying religious freedom for all. It was the Catholics' duty, said Carroll, to use "the utmost prudence" to preserve these advantages by "demeaning ourselves on all occasions as subjects zealously attached to our government and avoiding to give any jealousies on account of any dependence on foreign jurisdiction more than that which is essential to our religion, an acknowledgement of the Pope's spiritual supremacy over the whole Christian world."[12]

The difficulties that Carroll, and subsequent generations of American bishops, had with Vatican misunderstandings of the American denominational society were not finally resolved to the Americans' satisfaction until the Second Vatican Council's *Declaration on Religious Liberty* vindicated the Americans' approach to religious freedom and church/state theory.[13] But the degree to which Carroll followed his own advice was evident during the War of 1812. By then archbishop of Baltimore, Carroll privately hoped that an accommodation could be reached between the American and British governments on the impressing of seamen and the other issues separating the two countries at the time of Napoleon. But Carroll feared that "our American cabinet, and a majority of Congress, seem to be infatuated with a blind predilection for France and an unconquerable hostility to England,"[14] as he wrote to an old English friend. Still, once war had been declared, in June 1812, Carroll believed his public duty lay in giving clear support to the American government:

> Thus in a sermon preached on August 20 of that year he exhorted his people to contribute to the nation's preservation since, he maintained, it was a just war. Turning to the role that had been played by President Madison, he added: "We have witnessed the unremitting endeavors of our chief magistrate to continue to us the blessings of peace; that he has allowed no sentiments of ambition or revenge, no ardor for retaliation . . . to withhold him from bearing in his hands the olive branch of peace."[15]

Whether Carroll would have said the same about the Congressional war hawks is unclear. But on the first occasion when the Catholic leadership of the United States was called upon to judge the claim that the nation was engaged in a just war, Carroll "set the pattern for his successors in the hierarchy."[16]

Those successors were not primarily exercised about the righteousness or evil of their country's actions in war between the Revolution and the First World War. Although some bishops had their qualms about President Polk's war in Mexico, the bishops as a whole and Catholic lay magazines like the *Freeman's Journal* publicly supported Polk's policy; a similar pattern of private nervousness and public approbation marked Catholic commentary during the Spanish-American War.[17] Bishops in both the North and the South supported their respective governments during the Civil War (without, however, suffering the breach in communion that marked virtually every other Christian denomination in the country). Catholic bishops undertook diplomatic missions on behalf of both the Union and the Confederacy.[18]

But the principal civic concern of the American hierarchy up to the time of the First World War was not U.S. foreign policy. The nativist question continued to fester throughout this 150-year period, and the bishops corporately addressed it with some regularity.

Here, as on the question of war, Archbishop John Carroll's pattern was adopted by his successors. Succeeding generations of bishops affirmed the American democratic experiment, argued for its congruence with Catholic social theory (to the occasionally harsh discomfiture of the Vatican), and claimed for American Catholics a full place as citizens in American society. This pattern continued, quite remarkably, throughout the nineteenth century even as the tides of nativist bigotry ebbed and flowed with depressing ubiquity—even, on occasion, to the point of anti-Catholic rioting and convent burnings. The constant theme of the American bishops' pastoral letters during this time of trial was that nativism, not the American experiment, was the aberration. The harshest nativist attacks did not result in an American Catholic rejection of the essential moral worthiness of what Murray would later call the "American Proposition."

The "Pastoral Letter to the Laity" of the First Provincial Council of Baltimore in 1829 set this pattern.[19] The bishops gave thanks that "owing to our admirable civil and political institutions," the life of the Church in the United States continued to prosper, even though "amongst the various misfortunes to which we have been exposed, one of the greatest is misrepresentation of the tenets, the principles, and the practices of our church."[20] In the face of nativism, Catholics should not break faith with the American experiment but ought to recall that it was the Catholic settlers of the Maryland colony who had "ventured to introduce a milder, a better, a more Christian-like principle: that of genuine religious liberty. . . . If our brethren of other denominations have, since that period, adopted the principle, and now cherish it, they will not be displeased with our gratification that it emanated from the body to which we belong, and at our inculcating upon you, to preserve the same spirit that those good men manifested not only in our civil and political, but also in your social relations with your separated brethren."[21]

Similar themes issued from the Second, Third, and Fourth Provincial Councils of Baltimore in 1833, 1837, and 1840, and from the First Plenary Council of Baltimore in 1852. The answer to nativism was neither to match calumny for calumny, nor to question the principle of religious liberty on which the American experiment rested.[22] Catholicism did not threaten democracy; nativism menaced an otherwise worthy experiment in self-government. Catholics ought to call the nation back to its best instincts, and to remind one and all that "our religious rights are secured to us by those same instruments which secure to our fellow-citizens and to ourselves, all those other valuable possessions which have been acquired, for them and for us, by the lives, by the fortunes, and by the sacred honour of that devoted assembly who though widely differing in religion, yet were in love of country, a band of brothers."[23] Nativism would be met by holding the American experiment accountable to the convictions of its most sacred founding text, the Declaration of Independence.[24] The bishops affirmed the truth of Jefferson's claims. American Catholics, rather than fading into a placid sectarianism, were to be civically active in maintaining "the respectability of our land, the stability of our constitution, the perpetuation of our liberties, and the preservation of pure and undefiled religion."[25]

The Civil War put a temporary end to nativist attacks on Catholicism. But nativism proved harder to subdue than the Confederacy, and the Third Plenary Council of Baltimore took up the question of Catholicism and the American experiment again in 1884, in a statement that remains the high-water mark of the bishops' affirmation of American democracy.

Meeting in Baltimore's Cathedral of the Assumption and at the first American seminary, St. Mary's, Baltimore III brought together "fourteen archbishops, sixty American bishops and five visiting bishops from Canada and Japan, and ninety theologians."[26] The Council is best remembered today for mandating the parochial school system and founding the Catholic University of America. But the Council also addressed itself, in refutation of nativism, to the question of Catholicism and American democracy. After affirming the teachings of the First Vatican Council (which, in 1870, had defined the infallibility of the pope when teaching *ex cathedra* on matters of faith and morals, to the combined consternation and glee of American nativists), the Council fathers made the strongest possible statement about the compatibility of the American democratic experiment and the social teachings of their Church:[27]

> We think we can claim to be acquainted both with the laws, institutions and spirit of the Catholic Church, and with the laws, institutions, and spirit of our country; and we emphatically declare that there is no antagonism between them. A Catholic finds himself at home in the United States; for the influence of his Church has constantly been exercised in behalf of individual rights and popular liberties. And the right-minded American nowhere finds himself more at home than in the Catholic Church, for nowhere else can he breathe more freely that atmosphere of Divine truth, which alone can make us free.
>
> We repudiate with equal earnestness the assertion that we need to lay aside any of our devotedness to our Church, to be true Americans; the insinuation that we need to abate any of our love for our country's principles and institutions, to be faithful Catholics. To argue that the Catholic Church is hostile to our great Republic, because she teaches that "there is no power but from God" (Romans 13:1); because, therefore back of the events which led to the formation of the Republic, she sees the Providence of God leading to that issue, and back of our country's laws the authority of God as their sanction,—this is evidently so illogical and contradictory an accusation, that we are astonished to hear it advanced by persons of ordinary intelligence. *We believe that our country's heroes were the instruments of the God of Nations in*

establishing this home of freedom; to both the Almighty and to His instruments in the work, we look with grateful reverence; and to maintain the inheritance of freedom which they have left us, should it ever—which God forbid—be imperilled, our Catholic citizens will be found to stand forward as one man ready to pledge anew "their lives, their fortunes, and their sacred honor."[28]

This episcopal pledge would soon be tested in the Spanish-American War and World War I. But before considering the bishops' reactions to the emergence of America as a global power, it is well to pause a moment over this remarkable affirmation of Baltimore III: that the American experiment was not only congruent with Catholic social theory, but was in fact rooted in a providential design for human history.

Is this jingoism, as some contemporary American Catholic commentators charge? No doubt our more chastened ears blanch at the extreme patriotic rhetoric that marked John Ireland's and Denis O'Connell's reactions to the Spanish-American War;[29] but something more substantive was at work in the claims of Baltimore III. Despite the fact that the bishops' affirmation was historically demanded by a renascent nativism, and thus might be interpreted as defensive in character, it can also be argued that the American bishops were developing a positive intuition: that Roman Catholicism's traditionally incarnational (or "conversionist," in Niebuhr's terminology) imagination about human history, and the social ethic derived from it, were not merely congruent with an experiment in democratic pluralism, *but in fact implied it.*

On this positive construction of the bishops' intention at Baltimore III, the American hierarchy stood squarely within the tradition of Thomistic constitutionalism even as they developed its meaning. Their approach to church/state theory was not universally shared in the Church of the late nineteenth century; Pope Leo XIII's 1895 encyclical to the American hierarchy, *Longinqua Oceani*, warned against absolutizing the theory of democratic pluralism and religious liberty in a confessionally neutral state.[30] But the vindication of the American approach at the Second Vatican Council suggests that the bishops of 1884, rather than being marginally orthodox on the issue of democracy and religious liberty, were the true inheritors of the Thomistic constitutionalist tradition.

The implications of this for Catholic moral reflection on war and peace are profound. If the classic Catholic "answer" to the problem of war is *tranquillitas ordinis* in a rightly ordered political community; if the democratic experiment underway in America was not merely a tolerable exception to the preferred confessional state; if in fact it was the most appropriate modern expression of the Augustinian/Thomistic tradition on what constituted a rightly ordered society because it maximized the opportunities to create a public community of virtue—then the affirmations of Baltimore III can be seen, not simply as a patriotic salute to a particular national experience, but as a genuine moment in the development of doctrine on the moral question of war and peace. Peace and political community had been linked in Catholic theory since Augustine. In the Third Plenary Council, hints and traces of a more developed theory emerged: that peace and *democratic* political community were necessarily connected in the history of this world.

This distinctive American development of the Church's moral reflection on the problem of war and peace was brought to bear on the post–World War I conditions of international society in the bishops' joint pastoral letter of 1919. American entry into the war coincided with the annual meeting of the hierarchy in 1917. The bishops immediately wrote President Wilson, pledging that "Our people, now as ever, will rise as one man to serve the nation," thus fulfilling the pledge of the Third Plenary Council.[31] The vigor with which the American hierarchy threw itself into the national war effort need not detain us here.[32] What is of interest is the interpretation that the American bishops put on the national experience of the war, and the lessons for moral reflection on war and peace that they drew from that experience. This was the task of the joint pastoral letter of 1919.

The bishops addressed themselves to American Catholics who were "citizens of the Republic on whose preservation the future of humanity so largely depends."[33] After surveying the spiritual, disciplinary, and pastoral needs of the Church in the United States, the bishops turned their attention to the great issues that had just sent millions to their deaths. They began by noting that the world had undergone radical changes since their last joint pastoral letter, in 1884.[34] Among the most encouraging changes was "the idea of a human weal for whose promotion all should strive and by whose attainment all should profit."[35] There were serious obstacles to progress but the idea was loose in the world that human beings, "however they differed in race, tradition, and language . . . were humanly one in the demand for freedom with equal right and opportunity."[36] The bishops connected this general aspiration of humanity and the American experiment: "As this consciousness developed in mankind at large, the example of our own country grew in meaning and influence. For a century and more, it had taught the world that men could live and prosper under free institutions."[37]

It was in this context that the bishops explored the American entry into World War I. It was, the bishops conceded, a war whose "ultimate meaning" would remain unclear for generations.[38]

> But even now we can recognize the import of this conspicuous fact [about U.S. involvement in the war]: a great nation conscious of power yet wholly given to peace and unskilled in the making of war, gathered its might and put forth its strength in behalf of freedom and right as the inalienable endowment of mankind. When its aims were accomplished, it laid down its arms, without gain or acqusition, save in the clearer understanding of its own ideals and the fuller appreciation of the blessings which freedom alone can bestow."[39]

The moral rationale for the American role in World War I was clear to the bishops of the United States: American had fought a just war, for freedom:

> We entered the War with the highest of objects, proclaiming at every step that we battled for the right and pointing to our country as a model for the world's imitation. We accepted therewith the responsibility of leadership in accomplishing the task that lies before mankind. The world awaits our fulfillment. Pope Benedict himself has declared that our people, "retaining a most firm hold on the principles of reasonable liberty and of Christian civilization, are destined to have the chief role in the restoration of peace and order on the basis of these same principles, when the violence of these tempestuous days shall have passed."[40]

The American role had not ended on Armistice Day; the responsibilities of a democratic power in history were not completed. But how should the exercise of those responsibilities be guided?

The bishops suggested that only a comprehensive view of human nature, in the Catholic tradition of moderate realism, would suffice to bring order out of the disorder that four years of world war had wrought:

> Instructed by His example, the Church deals with men as they really are, recognizing the capacities for good and the inclinations to evil that are in every human being. Exaggeration in either direction is an error. That the world has progessed in many respects, is obviously true; but it is equally plain that the nature of man is what it was twenty centuries ago. Those who overlooked this fact, were amazed at the outbreak of war among nations that were foremost in progress.[41]

As Solzhenitsyn argued sixty-four years later, the root cause of the war's evil was that God had been forgotten: "God, from whom all things are and on whom all things depend . . . has, practically at least, disappeared from the whole conception of life so far as this is dominated by a certain type of modern thought."[42] Original sin, the denial of creatureliness, remained a pertinent fact of social and political life, even in an age dedicated to progress. Denying this, the bishops noted, led to the totalitarian temptation:

> It lies in the very nature of man that something must be supreme, something must take the place of the divine when this has been excluded; and this substitute for God, according to a predominant philosophy, is the State. Possessed of unlimited power to establish rights and impose obligations, the State becomes the sovereign ruler in human affairs; its will is the last word in justice, its welfare the determinant of moral values, its service the final aim of man's existence and action. When such an estimate of life and its purpose is accepted, it is idle to speak of the supreme value of righteousness, the sacredness of justice, or the sanctity of conscience. Nevertheless, these are things that must be retained, in name and in reality: the only alternative is that supremacy of force against which humanity protests.[43]

The first step in reconstructing a world worthy of humanity entailed a confession of faith:

> We are not the authors of our own being. . . . God has established, by the very constitution of our nature, the end for which He created us, giving us life as a sacred trust to be administered according to His design. Thereby He has also established the norm of our individual worth, and the basis of our real independence. . . . In light of this central truth, we can understand and appreciate the principle on which our American liberties are founded—"that all men are endowed by their Creator with certain inalienable rights." These are conferred by God with equal bounty upon every human being, and therefore, in respect of life, liberty, and the pursuit of happiness, the same rights belong to all men and for the same reason. Not by mutual concession or covenant, not by warrant or grant from the State, are these rights established; they are the gift and bestowal of God.[44]

Over half a century before it became the public policy of the United States, the American bishops taught that human rights, understood as belonging to the human person by the very fact of his or her existence and not by fiat of the state, were an essential foundation of the kind of international order that would be worthy of human beings and capable of resolving conflict without mass violence.[45]

In addressing the new problems of international relations brought on by the war, the bishops argued that the growth of democracy was essential to the preservation of peace,[46] and, without specifically mentioning the League of Nations, joined with Pope Benedict XV in calling for an international organization "whereby the peace of the world would be secured."[47]

The bishops' approach to the moral problem of war and peace, in the immediate aftermath of the carnage of World War I, was an expression of themes that have appeared time and again in this survey of the classic Catholic heritage on these central issues of political ethics. The insistence that human nature be understood within a "moderately realistic" framework; the sober faith in political community as an alternative to violence in adjudicating conflict; the attempt to relate proportionate and discriminate use of military force to the quest for peace, so that even war, a limit case, falls within the bounds of moral reflection—all these themes appear in the 1919 bishops' pastoral letter. But two new teachings, one specifically drawn from the bishops' American experience, were added to the framework. Following Pope Benedict XV, the bishops now argued that *tranquillitas ordinis* had an international dimension; it could not be considered solely in national terms. The bishops also suggested, as did their predecessors in 1884, that peace in the modern world involved the advance and defense of democratic values and institutions.

Far from being absent bystanders in the moral debate over war and peace, the American bishops had evolved, by the First World War, a body of thought that combined central elements in the Augustinian/Thomistic heritage with their own experience of democracy. The result, seen in embryonic form at the Third Plenary Council of Baltimore, and in a more developed manner in the 1919 pastoral letter, was a framework for moral analysis that recognized both the danger of war and the threat posed by an absolutist ideology of the state—the twin demons of the modern age. Given the failures of the Versailles Treaty and the League of Nations, and the rise of Hitler, Stalin, and Tojo, the bishops would soon have further need to test their moral theory against the cruel realities of their times.

THE AMERICAN BISHOPS AND WORLD WAR II

The twilight period between the First and Second World Wars—a breathing spell between what Churchill called a new Thirty Years' War[48]—saw fissures open in the common front behind which the American bishops customarily addressed U.S. foreign policy issues. Like the rest of the population, American Catholics were divided on the Spanish civil war. "Popular Catholic feeling was not as one-sided as some thought," James Hennesey has written, "but support for Franco's Nationalists was virtually unanimous in nearly two hundred fifty Catholic publications in the United States and among bishops and opinion makers." The journalistic exceptions were *The Catholic Worker* and *Commonweal*, both of which seemed more perturbed by the prospects of a fascist Spain than by the possibility of a Soviet-leaning regime in Iberia.[49] Hennesey cites the "notable" difference between Catholic and Protestant public opinion on the war; according to a December 1938 Gallup poll, 83 percent of American Protestants were pro-Loyalist, as compared with 42 percent of American Catholics. But it seems equally noteworthy that Catholics themselves were badly divided, in a 3:2 ratio of support for Franco versus support for the Spanish regime.[50] If the Gallup figures are trustworthy, the split between the general Catholic population

(42 percent pro-Loyalist) and the overwhelming support for Franco among Catholic elites is itself notable; four out of ten American Catholics differentiated themselves from the apparent willingness of their leadership to countenance a fascist counterpoise to the threat of communism.

The interwar period also saw the rise of an isolationist caucus within the American hierarchy, including such major figures as Archbishops John McNicholas of Cincinnati and Michael Curley of Baltimore.[51] McNicholas's isolationism was related to his general nervousness about Franklin D. Roosevelt's domestic policy; Curley, born in Ireland, had a not-untypical Irish-American aversion to aiding John Bull.[52] Bishop Gerald Shaughnessy of Seattle was another isolationist voice, who carried his prewar dissent into the war itself in the only episcopal critique of strategic bombing.[53]

But these were not the only voices in the hierarchy as debate over the Neutrality Acts, the draft, and Lend-Lease sharpened. Archbishop Robert Lucey of San Antonio (later a strong supporter of Lyndon Johnson during the Vietnam War) and Msgr. John A. Ryan of Catholic University and the National Catholic Welfare Conference (NCWC) were charter members of the interventionist Committee to Defend America by Aiding the Allies. Archbishop Joseph Schrembs of Cleveland, and Bishops Edwin V. O'Hara of Kansas City, Joseph P. Hurley of St. Augustine, and James H. Ryan of Omaha all supported Roosevelt's policies.[54] Lucey and Ryan had been vocal supporters of the president's domestic program and prolabor record, and reinforce the impression that the isolationist/interventionist debate among the American bishops was not so much an argument over first principles as a referendum on Roosevelt himself (or, in the case of men like Curley, the expression of ancient ethnic feuds). Roosevelt's principal champion in the American hierarchy was the archbishop of Chicago, Cardinal George Mundelein, who remained on terms of some intimacy with the White House until his death in 1939.[55]

If such contention within the American hierarchy seems odd in light of the bishops' common purpose during World War I, several factors should be kept in mind. First, the hierarchy no longer had a figure of Cardinal Gibbons's overwhelming stature to moderate disputes, work out compromises, and act as a public voice for the unified hierarchy. Gibbons had died in 1921, and his successor in the primatial see of Baltimore, Archbishop Curley, was more a voice for exuberant contention than for moderation and accommodation.[56] Secondly, the conciliar tradition within American Catholicism had dried up, largely due to the Vatican's concern to maintain more direct control over the American hierarchy. The bishops still met annually, but their business in the 1920s and 1930s tended more toward ecclesiastical housekeeping than during the contest with nineteenth-century nativism.

Finally, the Vatican itself had a complex relationship to the rise of fascism. Pius XI originally welcomed Mussolini and concluded a concordat with Hitler shortly after his appointment to the German chancellorship. But the pope had turned bitterly against the fascists by the end of his life, issuing an anti-Mussolini encyclical (*Non abbiamo bisogno*) in 1931, and a biting attack on Hitler's violations of the concordat in the encyclical *Mit brennender Sorge* of 1937. The latter was followed five days later by Pius XI's equally strong critique of communism in the encyclical *Divini Redemptoris*; the moral war against totalitarianism was being declared.[57] Still, the Vatican position during the 1930s was sufficiently muddled to afford ample ground for disagreement among the bishops of the United States.

The bishops began to close ranks as war broke out in Europe with the German/Soviet invasion of Poland in September 1939. At their annual meeting in November of that year, the hierarchy endorsed the government's intention to promote peace while maintaining U.S. security short of warfare. America's primary duty, the bishops wrote, was to attend to its own strength, stability, and security, although this should not be done "in a spirit of selfish isolation, but rather in a spirit of justice and charity to those people whose welfare is our first and chief responsibility."[58] The bishops also voiced their concern over "the fact that [in modern warfare] tragic sufferings are visited not only upon combatants, but as well upon women and children, the weak and infirm, young and old. No longer do armies alone march in battle, but whole nations are mobilized for total and unrestricted warfare."[59] The bishops also expressed fraternal solidarity with the Polish hierarchy.[60]

Five months later, the administrative board of the NCWC (then responsible for the bishops' commentary on public issues between the annual meetings of the hierarchy) issued a statement supporting the defense program of the Roosevelt administration, and pledged to work with the USO and other private agencies concerned with the spiritual and moral well-being of American servicemen. The administrative board also endorsed the five-point peace program of Pope Pius XII: protection of the national independence of all states; progressive disarmament; international legal institutions to resolve conflicts; satisfaction of the "fair demands" of national and racial minorities; and respect for "the precepts of justice and charity among men and nations."[61]

At their annual meeting in November 1940, the bishops renewed "their most sacred and sincere loyalty to our government and to the basic ideals of the American Republic," hearkening back to the affirmations of the Third Plenary Council of Baltimore. "Standing ever securely upon their unswerving allegiance to our nation since its foundation," American Catholics were "again resolved to give themselves unstintingly to [the nation's] defense and its lasting endurance and welfare."[62]

On the eve of America's formal entry into World War II, the National Catholic Welfare Conference, acting on instructions of the annual bishops' meeting, issued a statement, "The Crisis of Christianity," dated November 14, 1941. The bishops argued that the "two greatest evils" of the day were Nazism and communism. Noting Pius XI's condemnation of both, they also recalled the late pope's distinction between the evils of the Nazi and Stalinist systems and the peoples of Germany and the Soviet Union. The bishops repeated their endorsement of Pius XII's five-point peace plan and declared that "we cannot too strongly condemn the inhuman treatment to which the Jewish people have been subjected in many countries."[63]

The bishops' immediate reaction to the American declarations of war the next month followed the pattern set by Cardinal Gibbons in 1917. Archbishop Edward Mooney of Detroit, chairman of the NCWC administrative board, wrote President Roosevelt on December 22, 1941: "We, the Catholic bishops of the United States, spiritual leaders of more than twenty million Americans, wish to assure you, Mr. President, that we are keenly conscious of our responsibilities in the hour of our nation's testing. With a patriotism that is guided and sustained by the Christian virtues of faith, hope, and charity, we will marshall the spiritual forces at our command to render secure our God-given blessings of freedom."[64] The bishops wished not only to redeem the pledge of the Third Plenary Council of Baltimore, but to assert that the democratic experiment to be defended was "God-given" in its origins.

Archbishop Mooney then summarized what, in the bishops' judgment, America's war aims should be:

> We will do our full part in the national effort to transmute the impressive material and spiritual resources of our country into effective strength, *not for vengeance but for the common good, not for national aggrandizement but for common security in a world in which individual human rights shall be safeguarded, and the will to live on the part of all nations great or small shall be respected*—a world in which the eternal principles of justice and charity shall prevail.[65]

Mooney concluded his letter with a direct citation of the Third Plenary Council's affirmation of the providential nature of the American experiment, and a prayer that God may "strengthen us all to win a victory that will be a blessing not for our nation alone, but for the whole world."[66]

That American Catholics redeemed the bishops' pledge is neither disputable nor surprising. "Catholic patriotism [during the war] was unalloyed," according to Hennesey.[67] Catholics comprised between 25 and 35 percent of the American armed forces; 3,036 priests served as chaplains.[68] Of the 11,887 conscientious objectors of World War II, some 135 were Catholics (although their subsequent influence on the war/peace debate in American Catholicism would be substantial).[69] Archbishop Spellman took an active role in the American war effort as the head of the military vicariate (the "diocese" of American servicemen), and a less public role as a diplomat for both the Roosevelt administration and the Vatican.[70] The war allowed President Roosevelt to break a long-standing set of obstacles in Congress and appoint Myron C. Taylor as his "personal representative" to the Holy See. Taylor, a Protestant, held the personal title of ambassador but his appointment was not submitted to the Senate for confirmation. Diplomatic traffic between the White House and the Vatican was intense throughout the war.[71]

These patterns of activity seem to confirm the classic historiography in its assertion that American Catholicism, and especially its leadership, has traditionally followed a quiescent line toward U.S. foreign and military policy (at least until Vietnam). In fact, however, the American bishops' running commentary on World War II raises important questions about that historiographical model. In five remarkable statements issued during and immediately after the war, statements almost completely ignored in the contemporary American Catholic debate on war and peace, the bishops both challenged the Roosevelt and Truman administrations to make real the promises implicit in the origins of U.S. involvement in the war (particularly the program laid out in the Atlantic Charter), and contributed to the further development of the Catholic theory of peace as dynamic political community.

On November 11, 1943, the administrative board of the NCWC, on behalf of the entire hierarchy, issued a statement entitled "Essentials of a Good Peace."[72] Peace, the bishops insisted, would not automatically follow the triumph of arms. In fact, "unless we have the vision of a good peace and the will to demand it, victory can be an empty, even a tragic thing."[73] Encouraged that the recently completed Moscow conference had opened the prospect of international cooperation for peace as in war, the bishops were still gravely concerned that "compromises on the ideals of the Atlantic Charter are in prospect."[74] The bishops did not intend to enter the realm of statesmanship; but they did think it their obligation to "go deeper into the Catholic heritage of truth in faith and reason and to indicate the application of primary religious truth to the problems of peace and the planning of a right social order."[75]

The first truth that had to be faced was the truth about human nature that, forgotten or ignored, had helped bring on the current war: "Without doubt, the root of the maladies which afflict modern society and have brought on the catastrophe of world war is the social forgetfulness and even the rejection of the sovereignty of God and of the moral law."[76] The irrationality of totalitarianism, the crude glorification of race in Nazi ideology, and the passions that had been let loose in the world came from the rejection of a fundamental postulate of Catholic (indeed, classic) moral theory:

> There is written in human reason the law of good and evil. . . . When this moral law is cast aside in social life, every principle, every right, every virtue rests on the shifting sands of mere human conventions. Human dignity, human solidarity become, then, not endowments from the Creator, but mere fictions of man-made systems. In the quest for some principle of social stability, the authority of the State is exaggerated and its function of protecting and defending the rights of the citizens in the pursuit of the common good cedes to a tyrannical violation and invasion of these rights. In the name of realism the rights of the weak and the helpless are sacrified.[77]

But recognizing the God-given dignity of the individual human person did not carry implications for domestic social order only: "the recognition of the sovereignty of God and of the moral law . . . is basic to the right ordering of international relations. In creation God gave to the human race its essential unity and bound all men together in a brotherhood as comprehensive as humanity itself. In the plan of Divine Providence the human family was divided into nations and races, but this division in no way impaired the essential unity of mankind."[78] This "natural" unity was more than a piety of faith, however; the ideal of human fellowship in community ought to shape the actual conduct of affairs among nations.[79]

The fact of human unity was not to be confused with a surrender of national sovereignty; to point to this unity "does not mean that national rights and national sovereignties, rightly interpreted, must be surrendered to a world government."[80] The key phrase here is "rightly interpreted." The bishops immediately went on to argue that the providentially given unity of the human race meant that "every nation and every people must recognize and satisfy its obligations in the family of nations. . . . The discharge of these duties is entirely compatible with national differences which give a happy variety of cultural treasures in human unity. Indeed these differences must be respected and defended, and every effort made to assist peoples now in tutelage to a full juridic status among the nations of the world."[81]

The means for fulfilling these obligations were clear to the bishops: "In the circumstances of our times it is imperative that the nations . . . unite in setting up international institutions for the preservation of world peace and mutual assistance."[82] But such international institutions could not be mere mechanisms for the conduct of business-as-usual. The bishops recognized that "circumstances of history, differences of culture, and economic inequalities" had created a multiplicity of national forms of political community.[83] Yet the bishops' natural law ethic did not allow them to invoke a vague principle of pluralism to paper over the differences between types of states. "All nations," the bishops insisted, "must embody in their political structures the guaranty of the free exercise of native human rights, encouragement in the practice of virtue, an honest concern for the common good, and a recognition of the inviolability of the human person."[84] Sovereignty had limits, defined by right reason: "No nation has under God authority to invade family freedom, abrogate private ownership, or impede, to the detriment of the common good, economic enterprise, cooperative undertakings for mutual welfare, and organized works of charity sponsored by groups of citizens."[85] Moreover, human rights violations within national borders were a legitimate international issue; "it is only when nations adhere to right principles in their domestic administration that they will cooperate for the common good of the family of nations."[86]

The bishops concluded their 1943 statement by expressing concern over the rising domestic crime rate and the impact of the war on family stability; they also urged full civil rights for American blacks and Hispanics. But the real import of the 1943 statement was to outline themes that continued to draw the bishops' attention throughout the war. The bishops did not doubt that U.S. involvement in World War II satisfied the criteria for a just war. Any *in bello* reservations individual bishops may have had about tactics such as strategic bombing were not expressed, and the full weight of the statement was put behind the notion that the American cause was just *ad bellum*. But the bishops also insisted that the war would be morally worthy in the final analysis only if it served morally worthy ends: if it contributed to the development of institutions of international political community that concretized the vision of human rights that gave birth to the Atlantic Charter. Such a development, the bishops argued, was a practical necessity in the modern world and a requirement of morality and right reason.

The bishops developed these themes further in their November 1944 "Statement on International Order," which followed the Dumbarton Oaks conference, called to begin sketching the outlines of the postwar international system. As American armed forces continued their triumphant, if bloody, progress through Europe and the Pacific, the bishops began with an assertion and a question: "We have met the challenge of war. Shall we meet the challenge of peace?"[87] Claiming that only a "sane realism" could help forge an answer to their question, the bishops immediately got down to cases: "We have no confidence in a peace which does not carry into effect, without reservations or qualifications, the principles of the Atlantic Charter."[88] A peace worthy of the sacrifices being made would provide reconstruction assistance for those countries whose economic, social, and political fabric was being shredded by the war.[89] But more than reconstruction was required. Peace requires the recognition that "there *is* an international community of nations. God Himself has made the nations interdependent for their full life and growth. *It is not, therefore, a question of creating an international community but of organizing it.* To do this we must repudiate absolutely the tragic fallacies of 'power politics' with its balance of power, spheres of influence in a system of puppet governments, and the resort to war as a means of settling international difficulties."[90]

The bishops were aware of a previous generation's failure to complete this task: "After the last world war an attempt was made to organize the international community. It failed not because its objective was mistaken but because of inherent defects in its charter and more especially perhaps because the nations were not disposed to recognize their duty to work together for the common good of the world." What was needed, and what the bishops feared was missing, was a common recognition of the moral norms that should guide the affairs of nations in the international community. "International law must govern international relations. An international institution, *based on the recognition of an objective moral obligation and not on the binding force of covenant alone*, is needed for the preservation of a just peace and the promotion of international cooperation for the common good of the international community."[91]

What would be the form and function of such an institution?

> The international institution must be universal. It must seek to include, with due regard to basic equality of rights, all the nations, large and small, strong and weak. Its constitution must be democratic. While it is reasonable to set up a Security Council with limited membership, this Council must not be an instrument for imperialistic domination by a few powerful nations. Before it every nation must stand on its rights and not on its power. It must not allow any nation to sit in judgment in its own case. Frankly it must recognize that for nations as well as individuals life is not static. It must therefore provide in its charter for the revision of treaties in the interest of justice and the common good of international community, as well as for the recognition of a people's coming of age in the family of nations.
>
> The function of the international organization must be the maintenance of international peace and security, the promotion of international cooperation and the adoption of common policies for the solution of common economic, social, and other humanitarian problems. In the maintenance of peace it is reasonable that the organization have at its disposal resources for coercing outlaw nations even by military measures.[92]

Such an international organization, the bishops continued, should have a "world court to which justiciable disputes among nations must be submitted. Its authority should not be merely advisory but strictly judicial." But to make such a court work, international law had to be developed and codified. This required that "competent international authority . . . enact into positive law the principles of the moral law in their international references." Nations that refuse to submit international disputes involving a threat to peace or to the common good of the international community should be regarded as "outlaw nations." Compulsory arbitration of disputes "would mark a signal advance in international relations."[93]

The bishops then tackled the root question of national sovereignty. "The international organization must never violate the rightful sovereignty of nations [because] sovereignty is a right which comes from the juridical personality of a nation and which the international organization must safeguard and defend." But national sovereignty was limited; it did not absolve a nation from its obligations to the international community and it was constrained within states by the prior, innate rights of persons and families. The bishops did not hesitate to press this case to the limit: "The ideology of a nation in its internal life is a concern of the international community." To reject this was the same as arguing that human rights violations had no relationship to world peace. How could anyone hold this, when "just at this moment, in the interest of world peace, our nation is exerting itself to root out some ideologies which violate human rights in the countries we are liberating."

The bishops summarized their case by making human rights performance the criterion by which a lasting peace could be built: "We hold that if there is to be a genuine and lasting world peace, the international organization should demand as a condition of membership that every nation guarantee in law and respect in fact the innate rights of men, families, and minority groups in their civil and religious life."[94] The world had seen too much over the past decades to do otherwise: "Surely our generation should know that tyranny in any nation menaces world peace."[95]

The bishops ended with an affirmation of human possibility: "We have it in our power to introduce a new era, the era for which peoples have been longing through the centuries, the era in which nations will live together in justice and charity . . . a world of free men and free nations with their freedom secured by law."[96] The creation of such a world would give meaning to the sufferings and sacrifices of the war.[97]

The bishops' November 1944 statement, even granting its nervousness over the accommodations of Dumbarton Oaks, was the high-water mark of American Catholic internationalism. The bishops soon found their hopes for international order dashed on the hard rock of reality, particularly the postwar posture of the Soviet Union. But before turning to that melancholy tale, it is well to pause for a moment and see just what the American bishops had wrought—perhaps unintentionally, perhaps not—by the winter of 1944.

The bishops' statements of November 1943 and November 1944 pose a decisive challenge to the historiographic assertion that the American hierarchy had essentially "gone along" with the foreign and military policy of the United States until Vietnam. Although it is true that the bishops did not publicly protest Allied military tactics that seemed to fall under traditional *ius in bello* proscription,[98] the evidence of the 1943 and 1944 statements strongly suggests that the American hierarchy was increasingly critical of the postwar planning of the Roosevelt administration—and not solely for fear of the increasing power of the Soviet Union (although that surely concerned the bishops) but for a more positive reason. The bishops had thought their way through the classic Catholic heritage and the American experience of democratic pluralism to a vision of peace that emphasized the inextricable relationship of Atlantic Charter human rights to problems of international political community. There is a direct, if lengthy, line that runs from the Augustinian/Thomistic tradition of peace as rightly ordered political community, through the neo-scholastic commentators, Baltimore III's sense of the providential nature of democratic institutions, and the bishops' post–World War I reflections, to the bishops' commentary on the Second World War.

In their 1944 statement, the American bishops claimed the mantle of Thomistic optimism about peace as a human possibility, but expanded it to take account of the global dimensions of the problem of war and peace, while implying, if not flatly asserting, that the American experience of democratic pluralism was a precious heritage of immense importance in organizing the world community that was already a fact, if rife with dangers. The bishops sensed, correctly as it turned out, that the Roosevelt administration's view of the emerging United Nations system was fundamentally flawed: because of the nature of the great power condominium suggested in the Security Council veto and the policy of permanent membership, but also because of the administration's unwillingness to hold fast to the Atlantic Charter principles of human rights so as not to create further difficulties with Stalin.

In 1944 the bishops did not offer specific diplomatic prescriptions; but they insisted, as religious and moral teachers, that the United States was under moral and rational obligation to hold fast to several goals at once, even if historical circumstance dictated that those goals would be in tension: peace as *tranquillitas ordinis* through the mediation of international organization; the defense and advance of human rights;

and the growth of democracy, both within nations and in the conduct of international affairs. Those goals were to be sought with "sane realism," the bishops taught; but unless there was progress on this comprehensive international agenda, there would be no peace worthy of the sacrifices that were being made on the battlefields of the world. Forty years after Yalta, one may argue that the bishops expected too much, too soon; but one cannot argue that the American hierarchy was in lockstep with government policy of the day, or that the bishops had not made a valiant attempt to bring the tradition of Thomistic constitutionalism, expanded and deepened by the American democratic experience, to bear on the central foreign policy questions of their day.

If the bishops in 1943 and 1944 expressed a Thomistic optimism about peace as a human possibility, they quickly, under the pressure of events, reclaimed a complementary sense of Augustinian realism.

This shift was apparent in the April 1945 statement of the NCWC administrative board, "World Peace." The bishops began on a note of affirmation: "A sound international organization is not a utopian dream," and achieving it will "test the fullness of our victory." To "yield to the fear that this thing cannot be done is defeatism. In nations, as well as in individuals, we must indeed face the fact of human weakness, but we must face it in order to conquer it; we must not accept it in a spirit of paralyzing fatalism. An opportunity is here, as in every world crisis, to begin a new era of genuine progress in the community of nations."[99] Failing to seize the moment, the bishops argued, would only result in a rebirth of isolationism and international fragmentation.

This sobered optimism, this call to gather ourselves to make a peace worthy of the war's sacrifices, was immediately, and rather sharply, tempered: the bishops were not encouraged by the results of the San Francisco conference charged with drafting the U.N. Charter. The problems they previously addressed, on the composition of and veto power within the Security Council, had not been satisfactorily resolved. Nor had the U.N. Charter been linked to an international bill of rights, agreement to which ought to be the fundamental requirement for admission into the new international organization.[100]

But the bishops' worries extended beyond the nascent U.N. "The solution of the Polish question agreed upon by the representatives of the Three Great Victorious Powers in the Crimean Conference was a disappointment to all who had built their hopes on the Atlantic Charter."[101] Peace demanded a "free, independent, democratic Poland," and no foreign power could be allowed to usurp the Polish people's choice of their new government.[102] The bishops were also "struck by the ominous silence of the Three Great Powers on Lithuania, Estonia, and Latvia. . . . We hope that when the final peace treaty is framed and approved, it will not be recorded that our country condoned the enslavement of these freedom-loving people."[103] The bishops hoped that "our government will discharge its full responsibility in reestablishing all the liberated nations of Europe under genuine democratic regimes" and urged that, in dealing with the vanquished, "the common good of the whole world must be kept in mind."[104] Those who were once foes "must be freed from tyranny and oppression, and must be given the opportunity to reconstruct their institutions on the foundations of genuine democracy."[105]

Nor, even on the eve of Allied victory in Europe, were the bishops reluctant to identify the source of the chasm they feared was opening again in the world:

> Every day makes more evident the fact that two strong essentially incompatible ways of life will divide the loyalties of men and nations in the political world of tomorrow. They are genuine democracy and Marxist totalitarianism. Democracy is built on respect for the dignity of the human person with its God-given inalienable rights. It achieves unity and strength in the intelligent cooperation of all citizens for the common good under governments chosen and supported by the people.[106]

Marxist totalitarianism, on the other hand, "herds the masses under dictatorial leadership, insults their intelligence with its propaganda and controlled press, and tyrannically violates innate human rights."[107]

The bishops closed on a cautionary note: "We entered this war to defend our democracy. It is our solemn responsibility, in the reconstruction, to use our full influence in safeguarding the freedoms of all peoples. This, we are convinced, is the only way to an enduring peace."[108]

The sobriety of the bishops' April 1945 statement gave way, seven months later, to downright depression. "The war is over but there is no peace in the world,"[109] began the American hierarchy's first postwar statement:

> In the Atlantic Charter we were given the broad outline of the peace for which we fought and bled and, at an incalculable price, won a great martial victory. It was that ideal of peace which sustained us through the war, which inspired the heroic defense of liberty by millions driven underground in enslaved countries. It made small, oppressed countries confide in us as the trustee of their freedoms. It was the broad outline of a good peace. Are we going to give up this ideal of peace? If, under the pretext of a false realism, we do so, then we shall stand face to face with the awful catastrophe of atomic war.[110]

The "false realism" the bishops condemned was not one that argued against the possibilities of democratic international organization under contemporary conditions; the bishops knew the difficulties involved, as their previous statements had shown. The real "false realism" was that which failed to face the challenge that totalitarian power posed to the peace that must be created, because it was morally and rationally required. In 1945 the American bishops were not premature McCarthyites, as some exponents of the classic historiography seem to claim; nor should their position be understood as but the logical extension of some bishops' resistance to diplomatic recognition of and Lend-Lease aid to the Soviet Union.[111] The bishops' opposition to Soviet totalitarianism should be understood as a function of their commitment to democracy, and to an international peace of *tranquillitas ordinis* through political and legal institutions capable of resolving international conflict without war.

The last episcopal commentary on World War II was issued in November 1946, "Man and the Peace." The bishops began with an analysis that, thirty-three years later, would reflect the concerns of Pope John Paul II:

> At the bottom of all problems of the world today is the problem of man. Unless those who bear the responsibility of world leadership are in basic agreement on what man is, there is no way out of the confusion and conflict which block the road to peace. Clashes on the question of boundaries, national security, minority safeguards, free movement of trade, easy access to raw materials, progressive disarmament, and the control of the atomic bomb, important as these are, take a second place to the need of unity in protecting man in the enjoyment of his God-given native rights.

Such a fundamental moral problem did not admit of realpolitik solutions: "The struggle of the small nations for their indisputable rights and the stalemate among the strong nations in a contest of power would admit of bearable, even though hard, compromise if the fate of man, as man, did not hang in the balance."[112]

The basic issue was human rights. The fundamental nature of the division in the world and the basic moral question involved in that division meant that human rights could not be disentangled from the problem of international organization for peace. The bishops reviewed the great power negotiations on the fate of defeated nations and the disposition of borders and of persons displaced by the ravages of war. "In so difficult a task," the bishops admitted, "it is understandable that there should be differences and a clash of interests."[113] Moreover, given the requirements of peace, security, and freedom, "some sort of sacrifice of particular national advantages for the common good of the international community, and therefore for the ultimate good of all nations, must be made."[114] Such sacrifices could be made in good conscience if there were agreement on the question of human rights. "But the tragic fact is that the cleavage [in world politics] touches on issues on which there can be no compromise."[115]

The American bishops' continuous commentary on the moral meaning of World War II ended on a note of sorrow, bordering on anger. "Throughout the war our battle cry was the defense of native freedoms against Nazi and Fascist totalitarianism. The aftermath of the war has revealed victorious Soviet totalitarianism no less aggressive against these freedoms in the countries it has occupied."[116] The answer was not to abandon the task of creating a peace of *tranquillitas ordinis* through international organizations founded on democratic principles; it was to insist that such organizations, and their national members, abide by those principles:

> Before we can hope for a good peace there must come an agreement among the peacemakers on the basic question of man, as man. If this agreement is reached, then secondary, though important, defects in the peace may be tolerable in the hope of their eventual correction. . . . In the charter of the United Nations the signatories have contracted to cooperate "in promoting and encouraging respect for human rights and for fundamental freedoms for all without distinction as to race, language, or religion." Let the nations in the making of the peace do even more and in solemn covenants actually secure men everywhere in the enjoyment of their native rights. Then there will be the beginning of peace, and the fear of war will be banished from men's minds.[117]

Perhaps the American bishops should have more forthrightly judged, and criticized, certain *in bello* practices of the Allies during World War II. Was strategic bombing congruent with the just-war principles of proportionality and discrimination? Was "unconditional surrender" a wise *ad bellum* policy? The bishops did criticize certain domestic consequences of the war effort; they also resisted governmental proposals for universal military training that were bruited about toward the end of the war.[118] Still, the bishops maintained a discreet and unworthy silence on the great domestic travesty of World War II, the forced internment of American citizens of Japanese descent. On all these questions, the bishops could have been more vocal; they surely could have done so within the boundaries of their claim that the Allied cause was a just one.

On the other hand, the five statements just reviewed demonstrate that the bishops were neither passively nor silently acquiescent to governmental policy. Their critique of government planning for the postwar world suggests that, by 1944, the bishops of the United States had evolved a sophisticated and measured approach to the dual dilemma of war and totalitarianism with several notable features.

First, the bishops recognized that these two evils of the age were inextricably connected. Work for peace must, the bishops argued, be work for human freedom. In this light, the bishops' resistance to Soviet totalitarianism, the second key feature of their approach to war and peace, had evolved far beyond the visceral anticommunism of the 1930s, and was located within a positive statement of human rights and democratic values as the only possible foundation on which a lasting peace could be built.

The third distinctive feature of the bishops' approach was its careful insistence on the need for a democratically responsible international organization as the guarantor of peace. This concern evolved from a conjunction of the bishops' appropriation of the classic Catholic tradition of Thomistic constitutionalism (rooted in the Augustinian definition of peace as *tranquillitas ordinis*) with the bishops' development of the American hierarchy's traditional sense of the moral significance of the American experiment in democratic pluralism. The bishops' experience of democratic pluralism as an achievable, this-worldly form of *tranquillitas ordinis* led them to speculate, with care and precision, on the possibilities of such forms of political community in the international sphere.

In all of this, the bishops were walking a path parallel to that laid out by Pope Pius XII;[119] but their formulation of the issues had an American flavor and nuance. The universal Catholic heritage of which the bishops were the trustees had not only been received; it had been developed and applied to contemporary circumstances in an impressively thoughtful and distinctively American way.

No claim is made here that the American bishops' commentary on World War II seized and shaped the political and moral imagination of the great mass of American Catholics; it is questionable just how much it seized the moral and political imagination of the broad body of the episcopate. The Catholic internationalism of the American bishops' address to World War II may have had little impact on subsequent Catholic education, or even subsequent episcopal statements; it was argued about in Catholic intellectual journals and quality popular magazines such as *Commonweal* and *America*, and it profoundly influenced the work of the Catholic Association for International Peace. All of these qualifications about the impact of the bishops' World War II statements are appropriate.

And yet the statements were made. They were drafted by senior members of the American hierarchy and issued in the name of the entire episcopate of the United States.[120] It therefore seems reasonable to suggest that they expressed an important consensus within the official leadership of the American Church on the proper formulation of the moral question of war and peace in its broadest dimensions. These five remarkable statements, now virtually forgotten, are a notable part of the patrimony of American Catholicism, not only because they refute the charge that the American bishops were quietly acquiescent to American foreign and military policy prior to Vietnam, but even more importantly because they stand as a rich and important resource for contemporary American Catholic reflection on the moral problem of war and peace. They are a crucial part of the American Catholic development of the classic Catholic concept of peace as *tranquillitas ordinis*.

PRELUDE TO THE COUNCIL: THE CATHOLIC 1950s

The 1950s, often remembered as a somnambulant period in American Catholicism,[121] were in fact a time of heated controversy over the issue of Catholicism and democracy. A renascent secular nativism, symbolized by Paul Blanshard,[122] raised all the old shibboleths about the alleged incompatibility of Catholic social theory and the American experiment. Domestic polemics were further complicated by the fact that the Vatican was expressing its own concerns over the church/state theory being developed (partially in response to Blanshard) by John Courtney Murray. Curial conservatives were not without their American Catholic allies in these debates,[123] which will be considered later in examining the Murray Project.[124]

The 1950s, the time of Pope Pius XII, were also a period of profound Catholic concern about communism: a concern on which the Vatican wished to elicit the strongest possible support from the American hierarchy.[125] Given the sophisticated framework the bishops had developed during World War II, they might have been expected to provide a continuing commentary on world affairs, applying the principles sketched in the episcopal statements of 1943 to 1946 and further developing the theory of American Catholic internationalism. In fact, while the bishops were not silent, their commentary during the 1950s did not meet the high standard set during the war.

For five years, the hierarchy and the NCWC administrative board issued a series of statements and resolutions on the persecution of the Church in Eastern Europe and the Soviet Union. In November 1950 the bishops noted "with dismay . . . the apathy of Christian nations" in the face of a "Godless persecution."[126] In November 1951 the bishops again deplored the "indifference" that had met persecution in Eastern Europe and China, in a "Resolution of Sympathy for the Victims of Iron Curtain Persecutions."[127] Five months later, the NCWC administrative board condemned the persecution of the Church in Czechoslovakia, Rumania, Lithuania, Albania, Poland, and the Peoples Republic of China.[128] In November 1952 the bishops "bow[ed] their heads in reverent homage to the multitude of contemporary martyrs and confessors" in communist countries.[129] In September 1953 Archbishop Patrick O'Boyle of Washington, D.C., acting chairman of the NCWC administrative board, issued a statement in support of Cardinal Stefan Wyszynski, who had recently been placed under arrest; the struggle in which the Polish primate was caught up was "between good and evil, between freedom and slavery."[130]

The bishops issued a statement, "Man's Dignity," at the end of their November 1953 annual meeting, that was reminiscent of the more developed declarations of the 1940s:

> The practical social theory of the last century enthroned the individual but not the person. An individual can be a thing: as for instance an individual tree; but in virtue of his rational soul, a person is more than a thing. Yet the depersonalized view of man gained ascendancy, and generated a society which was a criss-cross of individual

> egotisms, and in which each man sought his own. Against this error our century has seen a reaction which has sought to overcome the isolation of man from man by imposing upon rebellious individuals a pattern of compulsory and all-embracing state organization, with unlimited power in the hands of the Civil Government. . . . The Christian concept of man is that he is both personal and social. As a person he has rights independent of the state; as a member of society he has social obligations. . . . Freedom has its roots in man's spiritual nature. It does not arise out of any social organization, or any constitution, or any party, but out of the soul of man. Hence to the whole tradition of the Western world, liberty does not come essentially from improved conditions of living either political or economic, but is rather the spring out of which better conditions must flow. A free spirit creates free institutions; a slave spirit permits the creation of tyrannical ones.[131]

A second statement issued in November 1953, "Peter's Chains," returned to the theme of communist persecution of the Church, which the bishops called the "bitterest . . . bloodiest persecution in all history," and a "war against all who believe in God and His Christ, against all who dare claim for man the liberty of the sons of God."[132]

The bishops returned to this theme in their November 1954 statement, "Victory . . . Our Faith,"[133] and in a Christmas 1955 statement issued on behalf of the hierarchy by Archbishop Francis P. Keough of Baltimore, chairman of the NCWC administrative board.[134] The conjunction of the Soviet invasion of Hungary and the Suez crisis in 1956 led to a slightly more developed statement in November 1956:

> It is not mere rhetoric to say that at this juncture the world is poised on the brink of catastrophe: it is grim realism. . . . War in modern terms would be a nightmare of unimaginable horrors. It can only annihilate; it has no power to solve our problems. If, in the ultimate resort, it is the duty of man to resist naked aggression, it is still obvious that every possible means consistent with divine law and human dignity must be employed and exhausted to avoid the final arbitrament of nuclear warfare.

The bishops still affirmed the United Nations: "If there have been mistakes in its decisions and faltering in its procedures, that is no more than a commentary on our human condition. The fact remains that it offers the only present promise we have for sustained peace in our time; peace with any approximation of justice." The bishops also praised the Eisenhower administration for its handling of the crises, and joined with Pius XII in a plea for strengthening international law.[135]

The last in this series of commentaries came three years later in a November 1959 statement, "Freedom and Peace," issued shortly after Soviet Premier Khrushchev's visit to the United States. "Freedom and Peace," although not nearly as refined a document as the statements of the 1940s, hearkens back to several key themes the bishops had developed then.

The bishops argued that the basic issue in the world was not capitalism versus communism; "the choice that men and nations must make today is between freedom and coercion." Freedom was a human birthright; that recognition, and the acknowledgment that human freedom is God-given, are "indelibly woven into the origin and history of the American republic." It was this heritage of freedom that defined the distinctive contribution of America to the rebuilding of the world: "Above and beyond the material aid that we distribute so generously to those around the world in need, we should be equally concerned in sharing our ideals of liberty and justice." Freedom was related to peace, for "peace, as demonstrated by our nation's experience, rests on disciplined freedom with its attendant virtues." This meant a freedom exercised in accordance with moral norms; "not even international organizations and international law, essential as they are for order in the world, can bring about world peace. Fundamentally, that peace depends on the acceptance by men and nations of a fixed, unchangeable, universal moral law."

What were the present obstacles to peace and freedom? The bishops cited world communism; excessive nationalism, not only in the developed world, but among the colonial states just coming into independence; poverty, disease, and hunger; apathy toward refugees and their resettlement. There were obstacles to peace and freedom in America as well: continued racial injustice, a breakdown of the family, and selfishness in economic matters. The bishops taught that "communism is the overriding danger to peace and freedom," but also warned that an emphasis on the threat of communism "should not deter us from seeking to solve other problems that may endanger peace and freedom. . . . The social and economic problems of the world . . . and particularly those of Asia, Africa, and some areas of Latin America pose a two-fold challenge that can be met." Stopgap charitable measures were insufficient; the foundations of a sustained prosperity had to be laid primarily through private sector investment and free trade.

Finally, on the confrontation with communism, "our goal is nothing less than the conversion of the Communist world." The conflict was not a psychological matter or a question of misunderstanding: "It is a delusion to place hope in seeking real understanding when the true problem is a conflict of essential principles, not lack of understanding." But, like the Lord in the Gospels admonishing the Pharisees, the bishops warned that "we cannot live like materialists and expect to convert others to our system of freedom and peace under God." As if to refute later charges that they were merely chaplains to a consumer society, the bishops warned:

> Instead of upholding boldly the principles of peace and freedom under God [in the international war of ideas], we have emphasized the material fruits of our freedom. . . . Instead of proclaiming freedom under God as we did in a more robust time in our history, we have so praised a program of supplying machines and calories and pleasure that these fruits of freedom and peace are made its substitutes. . . . We must convince the world . . . that the grandeur of our heritage and the extent of our contribution to the world is not measured in dollars and machines, but in the spirit of God's freedom and the dignity of the human person. . . . To accomplish this . . . we must be ready to give our country's principles the same unlimited devotion that led to the birth of our nation. Mankind will follow only those who give it a higher cause and the leadership of their dedication. It is up to us to give that leadership to mankind in the cause of God's freedom and peace.[136]

The bishops' concerns over communism in the 1950s paralleled a great domestic debate on the same subject launched by Senator Joseph McCarthy. McCarthy was a Catholic, and the common wisdom on McCarthyism has seen it as a distinctively American Catholic phenomenon. According to the typical portrait of the McCarthy era, the senator was launched on his red-baiting pogrom by a senior Catholic intellectual, Fr. Edmund Walsh, S.J., of the Georgetown University School of Foreign Service; he was aided and abetted throughout his volcanic career by the Catholic hierarchy; and he was supported by the broad mass of American Catholics. The link between McCarthyism and American Catholicism has become such a given in the common understanding of the period that it is often assumed to be true, by Catholic historians and publicists in contemporary debates over war and peace. McCarthyism was something for which American Catholicism should be held responsible, and for which contemporary American Catholics should pay a kind of ideological reparation.

In fact, however, as Donald Crosby, S.J., has shown in a detailed study, *God, Church, and Flag: Senator Joseph R. McCarthy and the Catholic Church, 1950–1957*, virtually all the assumptions of the "McCarthyism = American Catholicism" thesis are false.[137]

Crosby demonstrates that McCarthy made few, if any, efforts to exploit his Catholicism for political purposes: "The blunt fact of the matter is that Catholicism had precious little to do with the senator's hunt for subversives. His designs were political and practical, as his friends William F. Buckley, Jr., and Roy Cohn readily concede."[138] Crosby also casts severe doubt upon the theory that Fr. Walsh "created" McCarthy's crusade over dinner in the Colony restaurant in Washington. "What can one say about the most famous dinner party McCarthy ever attended?," Crosby asks, after reviewing the evidence. "The meeting almost certainly took place, and the participants may well have discussed Communism. Given the senator's growing interest in domestic Communism and Walsh's constant pre-occupation with what he called 'international Marxism,' it is also possible that they had an exchange on the politics of anti-Communism. It is highly improbable, however, that Walsh urged McCarthy to go on a crusade against subversives on the ground that it would bring him instant political stardom."[139]

What, then, about McCarthy's alleged support from the broad mass of American Catholics? In the aftermath of McCarthy's Wheeling, West Virginia, speech denouncing purported communists in the State Department, the Gallup poll found that the pro- and anti-McCarthy figures were virtually identical for Catholics and non-Catholics.[140] Nor was McCarthy's supposed hold on the political loyalties of American Catholics much in evidence during the Senate race in Maryland in 1950, where McCarthy went out of his way to help defeat Senator Millard Tydings. Despite the almost universal press judgment of the time that the "Catholic vote," manipulated by McCarthy in one of the shabbier campaigns in American political history, defeated Tydings, the "reality of the Maryland election was . . . a complex set of voting patterns, party factions, and shifting structures of party loyalty."[141]

McCarthy's vincibility among Catholic voters also showed up in his own 1952 Wisconsin reelection campaign, in which "Catholics generally followed the pattern of the rest of the state, voting against McCarthy when they lived in cities and with him if they lived in farming, back-country areas. His support among Catholics, therefore, was not 'Catholic' as much as geographical and political."[142] Yet the myth of McCarthy's drawing strength among Catholics was so tenacious that veteran Milwaukee Congressman Clement Zablocki thought years later that McCarthy had carried his district; in fact, McCarthy had lost miserably there, drawing some 76,000 votes against his opponent's 136,000.[143]

Joseph R. McCarthy had influential proponents among American Catholics; Cardinal Spellman supported him, as did Joseph P. Kennedy, and a fair portion of the Catholic press, local and national.[144] On the other hand, McCarthy was opposed by *Commonweal*, Bishop Bernard Sheil of Chicago, and Msgr. George Higgins of the NCWC. Crosby believes that this opposition had salutary effects later: "Catholic liberals were especially active during the McCarthy episode, striving to articulate an ethic both of Catholic civil libertarianism and of reasonable anti-Communism."[145] Yet, to this day, the notion that McCarthyism was a distinctively Catholic aberration remains widespread among many of those concerned with the Church's response to the problem of peace and freedom. In the generation after "Tail-Gunner Joe," American Catholicism failed to develop an ethic of sophisticated anticommunism. What did follow in McCarthy's wake (perhaps in an unconscious effort to expiate the sins of his crusade) was a species of anti-anticommunism that would lead to serious dilemmas for the American Church in its response to world affairs.

A second American Catholic senator rose to public prominence in the 1950s: John F. Kennedy of Massachusetts. Kennedy's candidacy for the presidency created yet another episode in the seemingly endless debate over the "compatibility" of Catholicism and American democracy. The details of that argument, prior to and during the 1960 presidential campaign, have been recorded elsewhere.[146] What does bear brief examination here is Kennedy's famous address in September 1960 to the Greater Houston Ministerial Association. If John Tracy Ellis and others are right in claiming that Kennedy's victory marked the political emancipation of American Catholics,[147] and freed the Church in the United States for a more vigorously critical role in the national debate over war and peace, then the Houston speech must stand as a major event, not only in American electoral history but in American Catholic history. The Houston speech, on this analysis, helped create the conditions for the possibility of the 1983 pastoral letter, "The Challenge of Peace," with its sharp critique of U.S. national security policy. But what precisely did the Houston speech, which has long since passed into political folklore, say? And what did it really portend for the evolution of American Catholic thought on the centrality of democracy in the peace of rightly ordered political community?

Analysis of the speech requires stripping away one layer of Kennedy mythology: Arthur Schlesinger's suggestion that Kennedy was a Catholic intellectual manqué, perhaps a secularized version of John Courtney Murray.[148] The evidence for this claim seems rather thin. John F. Kennedy's Catholicism was, like his ethnicity, an inherited attribute. And, while there are surely worse forms of religious commitment, there is virtually no evidence that Kennedy was much interested in Catholic social theory, or sought to engage his mind with the intellectual heritage of Catholic thought on the problems and promise of public life. Recognizing this does not absolve the unfairness, and in fact crude bigotry, to which Kennedy was subjected during the 1960 campaign, from figures as publicly eminent as Norman Vincent Peale.[149] That the Houston speech effectively pulled at least the rhetorical fangs from some nativist bigots was no small accomplishment, and in that minimal sense, the speech was an emancipation for all the candidate's coreligionists.

On the other hand, the Houston speech tried to face down the perennial nativist problem by putting as much distance between Kennedy and the Church as possible.

The distancing began at the outset of the address. Kennedy quite correctly reminded his audience of Protestant ministers that there were "far more critical issues in the 1960 election" than the "so-called religious issue."[150] Communist Cuba, hungry children in West Virginia, the elderly who could not afford medical care, the decline of family farms, "an America with too many slums, with too few schools, and too late to the moon and outer space"—these were the real issues needing debate. Who, in retrospect, could disagree? But in the very next sentence Kennedy argued that these issues, many of which had crucial moral components, were "not religious issues—for war and hunger and ignorance and despair know no religious barrier." The realm of public policy, and the realm of religious commitment, Kennedy seemed to imply, were separated by a wide chasm: what a historian of philosophy would call the fact/value distinction. Facts, we can argue about; values are matters of private concern only.[151]

Kennedy then stated "what kind of America" he believed in, in cadences that have become familiar over the years:

> I believe in an America where the separation of Church and State is absolute—where no Catholic prelate would tell the President (should he be a Catholic) how to act and no Protestant minister would tell his parishioners for whom to vote—where no church or church school is granted any public funds or political preference. . . . I believe in an America that is officially neither Catholic, Protestant, nor Jewish—where no public official either requests or accepts instructions on public policy from the Pope, the National Council of Churches, or any other ecclesiastical source—where no religious body seeks to impose its will directly *or indirectly* upon the general populace or the public acts of its officials—and where religious liberty is so indivisible that an act against one church is treated as an act against all. . . . I believe in a President whose views on religion are his own private affair, neither imposed by him on the nation nor imposed by the nation upon him as a condition to holding that office.

Kennedy's conclusion was eloquent: "But if this election is decided on the basis that 40,000,000 Americans lost their chance to be President on the day they were baptized, then it is the whole nation that will be the loser in the eyes of Catholics and non-Catholics around the world, in the eyes of history, and in the eyes of our own people." But the main burden of his address was to drive a broad wedge between religiously based values—*anyone's* religiously based values—and the American public policy arena. Not only were religious institutions *as such* to eschew any "direct" role in shaping public policy, but also any "indirect" role. There were no "religious issues," which on any serious exegesis must mean that questions of moral value had no place in national debates over public policy.

At Houston, Kennedy appealed to the "American ideal of brotherhood"; but he singularly failed to note that, for the overwhelming majority of Americans of his time (and ours, for that matter), the source of a commitment to brotherhood was the Judeo-Christian tradition's resolute insistence on the inviolable dignity and worth of each individual human being—not an abstract allegiance to Locke or Mill. The classic Catholic notion that politics is a matter of virtue was not only tactically absent from Kennedy's Houston speech; it was absent in principle. On this view, John F. Kennedy was the prophet of what Richard John Neuhaus called, a generation later, the "naked public square."[152]

Whatever the tactical merits of the Houston speech, it hardly stands as a major moment in the evolution of American Catholic thought on the democratic experiment. In order to meet the attack of Protestant nativism, Kennedy adopted, wittingly or unwittingly, the theory of secular nativists: that the First Amendment had ruled religiously based values out of the American public arena. Such a concession could not but have a deteriorating effect on the capacity of religious institutions to help shape the moral dimension of the national debate over war and peace. The gravamen of Kennedy's Houston speech was that that moral dimension did not exist, at least as far as the world of public policy was concerned.[153]

On the brink of Vatican II, then, American Catholicism seemed in a rather mixed relationship to its heritage of moral reflection on war and peace, freedom and democracy. The 1950s marked a highpoint in American Catholicism's reception and extension of the classic Catholic heritage of *tranquillitas ordinis*: this was the decade in which John Courtney Murray claimed the mantle of that heritage and developed it to its most sophisticated and penetrating point. On the other hand, the American bishops, during the 1950s, had failed to maintain the high quality of reflection and public commentary on world events and the American role in international affairs that they had established during World War II. Had the bishops continued along the path laid out during the war—had they developed the kind of sophisticated Catholic internationalism that correctly located the problem of communism as a fundamental threat to the evolution of a democratically based peace of *tranquillitas ordinis*—McCarthyism might have been at least mitigated, if not avoided entirely.

Over a span of 170 years, then, from John Carroll to the opening of the Second Vatican Council, American Catholicism had received the heritage of *tranquillitas ordinis*, and deepened the meaning of that concept by the distinctively American claim that democracy constituted the best available means of "rightly ordering" the peace that was one end of governance. How would that reception and evolution—already showing signs of stress—fare during the Second Vatican Council, the transformative event of twentieth-century Catholicism?

CHAPTER 3

American Catholicism and the Tradition in Transition: The Second Vatican Council

The Second Ecumenical Council of the Vatican was indisputably the most important event in the history of Roman Catholicism since the Reformation. A generation after its close, the meaning of Vatican II is still widely and loudly disputed, within and without the Church. But that the Council was the most significant Catholic event since the Counter-Reformation Council of Trent is agreed by virtually everyone, both supporters and detractors of the changes that the Council wrought in the internal life and public ministry of the Church.

"The changes in the Church" mark the point at which most argument about the Council's meaning gathers: changes in liturgical and penitential practice, in the internal organizational structure of the Church, in its attitude toward the social, economic, and political life of the modern world, in the intellectual methods by which the Church appropriates and expresses its tradition. Like many simplifications, a focus on "the changes" mandated by Vatican II has an element of truth in it. But that truth is distorted if one assumes that "the changes" began at the Council. As with everything else in human history, the Council had important antecedents.

The reign of Pope Pius XII (1939–58) appeared, at the time of his death, as the high point of an ecclesiastical self-understanding and style that had begun with Pope Pius IX. Piux IX's long pontificate (1846–78) not only saw the loss of the Papal States in the reunification of Italy; it also witnessed the most concentrated intellectual assault on Catholic teaching since Western Christianity divided during the Reformation. "Modernity," whose roots Catholic intellectuals traced to Cartesian skepticism, and which seemed to take decisive (and decisively anticlerical) political form during the French Revolution, had to be resisted at all costs. The First Vatican Council's definition of papal infallibility, no matter how constricted in scope it actually was, appeared as a profound statement of defiance: Catholicism, even if alone among Western institutions, would not accommodate to the intellectual challenge mounted by modern historical, literary, sociological, psychological, political, and scientific theory. Modernity's critique of traditional understandings of the origins of Scripture, the meaning of doctrine, and the role of religious authority in personal and civic life would not be met in a spirit of dialogue but of confrontation. Apologetics, the systematic explication of defenses against intellectual attacks on the Church, became a major course of study in Catholic seminaries, colleges, and universities. The truth of the faith, considered immutable not only in its essential meaning but in the forms of expression given it, was to be defended in the manner of a medieval castle: the drawbridge over the moat was raised and the only questions to be settled had to do with the most effective means of reinforcing the walls of the bastion.[1]

The apologetic tradition, which demanded an increasingly centralized Church authority focused on the papacy, continued from Pius IX through Pius XII. But even within Fortress Romana winds of change blew, albeit gently. Pius IX's successor, Pope Leo XIII (1878–1903) encouraged certain of the breezes by his great social encyclical *Rerum Novarum* (1891), which, however implicitly, suggested that the altered conditions of modern life required a development, however modest, in Catholic social theory. In the encyclical *Aeterni Patris* (1879) Leo XIII also encouraged a fresh study of St. Thomas Aquinas, which returned to the original Thomistic sources rather than reading Thomas through the filters imposed by centuries of commentators. Leo XIII even reached a measure of agreement with the secularist French republic, urging French Catholics to participate in the life of their country rather than standing aside and plotting monarchical restoration.[2]

The first pope of the twentieth century, Leo's successor, Pius X (1903–14) is popularly remembered as a man of deep personal spirituality who opened reception of the Eucharist to small children and died of a broken heart at the outbreak of World War I. In fact, Pius X stood firmly in the tradition of Pius IX. Pius X conducted what can only be described as a witchhunt of Catholic intellectuals suspected of the sin of "modernism," particularly those who wished to adopt the methodologies of modern Scriptural exegesis that had evolved in European Protestantism since Hermann Samuel Reimarus and Friedrich Schleiermacher. A frozen carapace subsequently smothered much of public Catholic intellectual life; any exploratory work would be, almost by definition, of an underground nature. And yet even Pius X, the man who imposed the "antimodernist oath" on all Catholic clerics and on all teachers of philosophy and theology in Church-recognized colleges and universities, opened a small window for change in his reform of liturgical music.[3]

Pius X's two immediate successors, Benedict XV (1914–22) and Pius XI (1922–39), were occupied with matters of high politics. Benedict made several fruitless efforts to mediate a settlement of World War I, and virtually bankrupted the Holy See by his generosity to war relief. Pius XI was in continual contest with Italian fascism, German Nazism, and Soviet communism throughout the 1930s. Whatever the loss of the Papal States had been assumed to have meant in the days of Pius IX, it was clear by the time of Pius XI that it did not mean the Holy See's withdrawal from international politics and diplomacy. Whereas Benedict XV is often portrayed as a mildly open man, in the mold of Leo XIII, and Pius XI is thought to have been in the apologetic tradition of his Pian predecessors, it was Pius XI who, amid a rigorous rejection of both communism and socialism, further developed the social theory of the Church in the encyclical *Quadragesimo Anno*, issued on the fortieth anniversary of *Rerum Novarum*.[4]

The Church that Pius XII inherited on the eve of World War II was an institution firmly in control of its internal life and discipline, and restored to prominence in world affairs. Throughout his long reign, Pius XII worked diligently to maintain the image of a Church founded firm on the Rock of Ages, impervious to assaults moral, intellectual, or physical, self-sufficient within its own tradition and within its classic methods of understanding that tradition.

Pius's own personal style, which he cultivated to great effect, was one expression of this self-sufficiency: distant but not remote; paternal rather than fraternal; able to address a multitude of contemporary questions out of the fixed font of traditional wisdom and truth; open, in a personally modest way, to some of the benefices of a technological age (it was thought to be a marvelous thing that the pope used a telephone, a typewriter, and an electric razor). At his death, Pius XII symbolized what most people, within and without the Church, understood when they said "pope" and "Catholic Church."[5] In Pius XII, it seemed, apologetic Catholicism had reached full, perhaps even final, expression.

Yet there were other currents at work, even within the mind of Pius XII who, contrary to much common wisdom on the subject, was the real precursor of Vatican II. Pius XII's encyclical *Mediator Dei* (1947) was a magna carta for Catholic liturgical studies and opened the way for the liturgical reforms of the Council. The encyclical *Divino Afflante Spiritu* (1943) created modern Catholic Scripture studies, which would have a profound impact on the Council's understanding of both the Church and its responsibilities in the world. *Mystici Corporis Christi* (1943) set the stage for the Council's reflection on the Church's self-understanding as a faith community. The extraordinary range of topics addressed by Pius XII in his audiences—ranging from commentaries on modern entertainment to detailed discussions of medical ethics and just-war theory—demonstrated in fact, if not in theory, that the Church's intellectual life was developmental rather than static (even if Pius himself put a firm brake on any serious public dialogue with contemporary intellectual life in the 1950 encyclical *Humani Generis*).[6]

The winds of change blew even more vigorously outside the Vatican. Leo XIII's program of recovering the Thomistic heritage bore fruit in the work of neo-Thomists like Jacques Maritain and Etienne Gilson, and even more adventurously in the transcendental Thomism developed by Joseph Marechal, Karl Rahner, and Bernard Lonergan in explicit dialogue with Kantian critical philosophy.[7] Karl Rahner, his brother Hugo, and the Frenchmen Yves Congar, Marie-Dominique Chenu, and Henri de Lubac, all of whom were under the suspicion of the Vatican Holy Office during the time of Pius XII, also conducted detailed historical studies on the Church's ecclesiological self-understanding, its penitential practice, and its developing understanding of revelation; this work helped build the intellectual foundations of Vatican II.[8] During the 1950s, Catholic liturgical research flourished in Europe and America; it seemed clear that the Tridentine liturgical practice of the Church was not a fixed point in the universe, but a historically conditioned expression of faith and piety that, however beautiful, could, in principle, change.[9]

The reign of Pius XII also saw an explosion of Catholic energy in various forms of social action, nowhere more than in the United States. Young Christian Workers (YCW) and Young Christian Students (YCS) were originally European movements that transplanted fertilely to the United States, where homegrown organs such as the Cana Conference, the Christian Family Movement (CFM) and the Catholic Youth Organization (CYO) flourished in many dioceses—not merely as instruments for personal spiritual development (although that was stressed in each) but as vehicles for Catholic action in business, labor, and politics.[10] Through these agencies, and through more traditional fraternal organizations such as the Knights of Columbus, American Catholicism was being prepared for the notion that active involvement with the social and political issues of the day was an important dimension of a Catholic commitment. The Catholic Association for International Peace (CAIP) was the principal expression of this organizing impulse in the war/peace field; but CAIP differed from organizations such as YCW, YCS, and CFM in being a confraternity of international affairs professionals rather than a mass popular movement.[11]

Those who assumed that apologetic Catholicism had reached a fixed point by the death of Pius XII were wrong. Energetic and exploratory Catholic intellectual life had continued after World War II, despite occasional interventions from the Vatican; it simply happened, in most cases, out of the public eye. The resistance strategy of Pius IX had been abandoned, if not by the official Church, then by many of its most vigorous minds. The question of Catholic theology and modern critical thought (philosophical, historical, sociological, psychological, and scientific) remained open. Moreover, the burgeoning weight and influence of the American Church within both its own society and the universal Church suggested that the whole tenor and theory of the Church's relationship to modern forms of governance, particularly democracy, had to be addressed. A reiteration of Leo XIII's defense of the confessional state would be insufficient, particularly in a time of decolonization and the emergence of the Third World.

At the death of Pius XII, then, the Church had a considerable amount of unfinished business requiring serious attention. What should the Church say about the accomplishments of science, which had developed to the point where human beings were, in the most explicit sense, "co-creators" of the world? How could Catholicism speak timeless truths in an intellectual environment where the historical and social conditioning of sacred texts and ancient ecclesiastical teachings had been established beyond any reasonable doubt—and by men of faith, not by Jacobin anticlericals? If the Church affirmed the value of democracy for civic life, what, then, about its internal life? What about liturgical renewal and reform? And what about the ecumenical movement of the twentieth century, which sought an end to the scandalous division of Christianity, particularly in the face of modern tyrannies and paganisms?

These questions, whether debated quietly in faculty lounges or lived experientially by the Catholic laity, had to be addressed publicly, not privately. If it is too much to suggest that Catholicism in 1958 was a volcano waiting to erupt, it may not be too inappropriate an image to suggest that it was a bottle of sparkling wine waiting to be uncorked.

The wine steward was Pope John XXIII.

American Catholic thought on the moral problem of war and peace would be decisively altered by both the explicit teachings and implicit "spirit" of Vatican II. But it is impossible to understand this particular transition without recollecting John XXIII's vision of the Council's grand agenda.

John XXIII wanted the Second Vatican Council to break Catholicism out of the apologetic mold defined by Pius IX and perfected by Pius XII. John had no doubt that the truth his Church carried was of an eternal nature; but as he put it in his opening address to the Council fathers in October 1962, "The substance of the ancient doctrine of the deposit of faith is one thing, and the way in which it is presented is another."[12] No clearer abandonment of apologetic Catholicism on the Pian model could be imagined within the scope of one sentence. But John wanted the Council to do more than break new intellectual ground in exploring the ancient deposit of faith. The pope's grand agenda was made clear in his opening address, which has now become part of conciliar legend:

> In the daily exercise of our pastoral office, we sometimes have to listen, much to our regret, to the voices of persons who, though burning with zeal, are not endowed with too much sense of discretion or measure. In these modern times they can see nothing but prevarication and ruin. They say that our era, in comparison with past eras, is getting worse, and they behave as though they had learned nothing from history, which is, nonetheless, the teacher of life. They behave as though at the time of former Councils everything was full triumph for the Christian idea and life and for proper religious liberty.
>
> We feel we must disagree with those prophets of gloom, who are always forecasting disaster, as though the end of the world were at hand.
>
> In the present order of things, Divine Providence is leading us to a new order of human relations which, by men's own efforts and even beyond their expectations, are directed toward the fulfillment of God's superior and inscrutable designs. And everything, even human differences, leads to the greater good of the Church.[13]

John XXIII wanted to put his Church firmly on the side of hope, through the experience and teaching of an ecumenical Council. The Church would be a herald of evangelical hope only if it left Fortress Romana and inserted itself into the maelstrom of modern history. John XXIII, so surely an optimist about institutions and history as well as individuals, wished his Church to be optimistic, too. Johannine optimism infused the entire ethos of Vatican II, and had a profound impact on the Council's and the Church's understanding of the problems of war and peace, security and freedom.[14]

PACEM IN TERRIS

Pope John's opening address to the Council suggested a new Catholic attitude toward the modern world: in addition to being *ecclesia docens*, a teaching Church, Catholicism would be *ecclesia discens*, a learning church. The notion that the Church had much to learn from the modern world, as well as much to teach it, was another example of John XXIII's break with apologetic Catholicism. The pope's opening address also introduced the image of a "maturing humanity" into official Catholic thought.

These themes were given full expression less than a year later in the encyclical *Pacem in Terris*, which, though not itself a conciliar text, captured much of the flavor of Vatican II's new openness to modernity. Issued on April 11, 1963, only two months before John's death, the encyclical was the first papal document addressed not only to the hierarchy and faithful of the Church, but to "all men of good will."[15]

While *Pacem in Terris* is often remembered as a decisive break with the Catholic past, a contemporary commentator was accurate in arguing that "the encyclical must not be viewed as an isolated treatise occasioned by a momentary historical situation. It is a highly developed and integral chapter in the total, relatively timeless teaching of the Church on social questions."[16] Nor was *Pacem in Terris* an effort by John XXIII to dismiss or quietly inter the teaching of his predecessor, Pius XII. Almost half of its seventy-three footnotes contained references to the writings of Pius XII. Where John extended his predecessor's thought, events since 1958 seemed responsible.[17]

Yet *Pacem in Terris* did constitute a new moment in Catholic thought about the moral problem of war and peace in the modern world. It made new applications of traditional teaching. But—and this perhaps explains its extraordinary impact both within and without the Church—it also breathed deeply of the spirit of Johannine optimism. Gone were the days of the anathema; John's teaching, though following a classic natural law analysis, aimed at evoking a conversation with modernity, including as partners in dialogue even the leaders of the communist world.

Six themes in the encyclical deserve special attention because of their centrality to the pope's argument and their impact on the subsequent American Catholic debate.

Pacem in Terris was addressed to a world which had entered *a new moment in history*. Workers and women were active participants in public life.[18] Most significantly of all, "the conviction that all men are equal by reason of their natural dignity has been generally accepted."[19] This had led not only to decolonization and the condemnation of racial discrimination but to a situation in which human beings had become "conscious of spiritual values, understanding the meaning and significance of truth, justice, charity, and freedom."[20] This development had political and economic significance, but it also presaged a situation in which men and women could be brought to a better understanding of the true, personal, and transcendent God. They might thereby make "the ties that bind them to God the solid foundation and supreme criterion of their lives, both of that life which they live interiorly in the depths of their own souls and of that in which they are united to other men in society."[21] Human community, Pope John seemed to believe, was on the eve of a new evolution of the spirit, out of which might emerge—indeed, had to emerge—new forms of political organization.

If human community had evolved to a point where interdependence among nations and peoples characterized the modern age,[22] then the classic Catholic social-ethical standard of the common good had to be applied to the global human community as a whole. And thus the second key point of *Pacem in Terris*: that there was a *universal common good* to which nations, their leaders, and individuals were accountable.[23] This evolution of human history carried profound implications for the nation-state system:

> In times past, it seemed that the leaders of nations might be in a position to provide for the universal common good, either through normal diplomatic channels, or through top-level meetings, or through conventions or treaties, by making use of methods and instruments suggested by natural law, the law of nations, or international law. In our time, however, relationships between states have changed greatly. On the one hand, the universal common good poses very serious questions which are difficult and which demand immediate solution especially because they are concerned with safeguarding the security and peace of the whole world. On the other hand, the heads of individual states, inasmuch as they are juridically equal, are not entirely successful no matter how often they meet or how hard they try to find more fitting juridical instruments. This is due not to lack of goodwill and initiative but to lack of adequate power to back up their authority.[24]

The pope did not hesitate to draw the inescapable conclusion of his analysis: "Therefore, under the present circumstances of human society, both the structure and form of governments as well as the power which public authority wields in all the nations of the world, must be considered inadequate to promote the universal common good."[25]

This judgment led directly to the third key theme in *Pacem in Terris: the universal common good required that the peace of political community be established on an international scale.* Morality and right reason demanded a "universal public authority" whose powers were sufficient to tackle the global problems before which individual states were relatively powerless.[26]

The creation of international political structures able to satisfy the requirements of the universal common good would be difficult and perilous. An international political authority would have to be created by consent, not by force; it must be impartial in its actions; it must respect the legitimate claims of national sovereignty.[27] For these reasons, the universal public authority had to take account of the rights of the human person and the principle of subsidiarity:

> Like the common good of individual states, so too the universal common good cannot be determined except by having regard for the human person. Therefore, the public and universal authority, too, must have as its fundamental objective the recognition, respect, safeguarding, and promotion of the rights of the human person; this can be done by direct action when required, or by creating on a world scale an environment in which leaders of the individual countries can suitably maintain their own functions.
>
> Moreover, just as it is necessary in each state that relations which the public authority has with its citizens, families and intermediate associations be controlled and regulated by the principle of subsidiarity, it is equally necessary that the relationships which exist between the worldwide public authority and the public authorities of individual nations be governed by the same principle. This means that the worldwide public authority must tackle and solve problems of an economic, social, political or cultural character which are posed by the universal common good. . . .
>
> The worldwide public authority is not intended to limit the sphere of action of the public authority of the individual state, much less to take its place. On the contrary, its purpose is to create, on a world basis, an environment in which the public authorities of each state, its citizens and intermediate associations, can carry out their tasks, fulfill their duties, and exercise their rights with greater security.[28]

The fourth key theme in *Pacem in Terris* was suggested in the pope's sketch of the boundaries within which a universal public authority must operate: *the protection and promotion of human rights were the fundamental objectives of a universal public authority.* Any human rights scheme worthy of the name would begin with the fact of human personhood.[29] Human rights were not a benefice conferred by government; human rights inhered in persons, prior to their relationship to governments, by the very fact of human nature.

The first human right, according to Pope John, was "the right to life, to bodily integrity and to the means which are suitable for the proper development of life"; the pope referred to food, clothing, shelter, rest, medical care, necessary social services, and social security for those unable to care for themselves.[30]

Next, the pope claimed that "every human being has the right to respect for his person, to his good reputation; the right to freedom in searching for truth and in expressing and communicating his opinions, and in pursuit of art, within the limits laid down by the moral order and the common good; and he has the right to be informed truthfully about public events."[31] Natural law also gave human beings the right to share in culture; this meant a right to basic education, technical or professional training, and higher studies commensurate with an individual's native abilities and acquired talents.[32]

The pope then argued for religious liberty as a right that included the freedom to practice one's faith privately and publicly.[33]

Human rights also included the right to vocational choice, the right to form a family ("with equal rights and duties for man and woman"), and the right to follow a religious vocation.[34] Family rights, including the right of parents to determine their children's education, existed prior to the state, and the state was under a natural law obligation to protect them.[35]

The pope then addressed the question of rights in "the economic sphere," listing a right to an opportunity to work; to work without coercion; to safe working conditions; and to a just wage. The pope also argued that private property was a natural law right that carried social obligations.[36]

The pope defended the rights of assembly and association, and the right of peoples to give their societies the political form they deemed most suitable. The good society would have "a variety of organizations and intermediate groups . . . capable of achieving a goal which an individual cannot effectively attain by himself. These societies and organizations must be considered the indispensable means to safeguard the dignity of the human person and freedom while leaving intact a sense of responsibility."[37]

The pope claimed rights of free movement and emigration for all human beings, because citizenship in one state did not detract from "membership in the human family as a whole, nor from . . . citizenship in the world community."[38]

Finally, the pope argued that human dignity carried the right to active participation in public affairs, whereby individuals made their contributions to the common good.[39] This right, and all the rest, rested on the fundamental human claim "to a juridical protection of . . . rights, a protection that should be efficacious, impartial, and inspired by the true norms of justice."[40] The pope concluded his discussion of human rights with a discourse on those duties that are "inseparably connected, in the very person who is their subject" with the rights just outlined.[41] Here the pope cautioned that any human society built on "relations of force" was inhuman, because it repressed or restricted the personhood of its victims.[42]

The fifth distinctive theme of *Pacem in Terris* was its *approach to the problem of communism*. John XXIII was aware of the dangers of totalitarianism. But, ever optimistic about the human person, he argued that error and the erring person must not be confused. The erring person remained a human being with an inherent human dignity. The human thirst for truth could not be completely extinguished; nor would God be found wanting in opening human minds to enlightenment:[43]

> [It was, therefore] especially to the point to make a clear distinction between false philosophical teachings regarding the nature, origin, and destiny of the universe and

> of man, and movements which have a direct bearing either on economic and social questions, or cultural matters or on the organization of the state, even if these movements owe their origin and tenets to these false doctrines. While the teaching, once it has been clearly set forth, is no longer subject to change, the movements, precisely because they take place in the midst of changing conditions, are readily susceptible to change. Besides, who can deny that those movements, in so far as they conform to the dictates of right reason and are interpreters of the lawful aspirations of the human person, contain elements that are positive and deserving of approval?[44]

This thinly veiled reference to communism effectively repealed Pius XI's proscription on common work with communists; but John XXIII also cautioned that such efforts be approached "with the virtue of prudence."[45]

The sixth distinctive theme in *Pacem in Terris* was its *approach to deterrence and the imperatives of disarmament*. The pope knew that weapons competition was not a question of madness, and he accepted the claim that many arms were meant for deterrence, not aggression.[46] But notwithstanding his existential appreciation of the situation in which governments found themselves, the pope also mourned "the enormous stocks of armaments that have been and still are being made in more economically developed countries, with a vast outlay of intellectual and economic resources."[47] One result of this stockpiling was that "other countries . . . are deprived of the collaboration they need in order to make economic and social progress."[48]

The pope understood the theory of deterrence through balance of power, but claimed that it resulted in a climate of fear which exacerbated the dangers of war through accident or miscalculation.[49] The pope also argued that atmospheric nuclear testing could jeopardize "various kinds of life on earth."[50] John XXIII then pleaded for general and complete disarmament, inspected and enforced:

> Justice, then, right reason and consideration for human dignity and life urgently demand that the arms race should cease; that the stockpiles which exist in various countries should be reduced equally and simultaneously by the parties concerned; that nuclear weapons should be banned; and finally that all come to an agreement on a fitting program of disarmament, employing mutual and effective controls. . . . All must realize that there is no hope of putting an end to the building up of armaments, nor of reducing the present stocks, nor, still less—and this is the main point—of abolishing them altogether, unless the process is complete and thorough and unless it proceeds from inner conviction: unless, that is, everyone sincerely cooperates to banish the fear and anxious expectation of war with which men are oppressed. If this is to come about, the fundamental principle on which our present peace depends must be replaced by another, which declares that the true and solid peace of nations consists not in equality of arms but in mutual trust alone. We believe that this can be brought to pass.[51]

John adduced three reasons for his confidence in the possibilities of disarmament: right reason demanded it; everyone desired it; and its benefits would be felt throughout the world.[52] The pope concluded his remarks on disarmament with the summary judgment that "in an age such as ours, which prides itself on its atomic energy, it is contrary to reason to hold that war is now a suitable way to restore rights which have been violated."[53]

The *Pacem in Terris* Debate

Pacem in Terris evoked more public commentary than any papal statement in history. John Cogley limned the spectrum of debate in a bit of doggerel:

> David Lawrence read it Right
> Lippmann saw a liberal light
> William Buckley sounded coolish
> Pearson's line was mostly foolish
> Courtney Murray wasn't certain
> (We haven't heard from Thomas Merton)
> *Nation*-readers learned to hope
> That J.F.K. would heed his Pope
> Welch saw Red, red, redder than titian
> As Rome fell under Birch suspicion
> *Time* caressed each Lucid text
> While *Playboy* found it undersexed
> *Pravda* praised the portions peacenik
> (No comment on the UN policenik)
> The Dept of State was terribly kind
> The Pope, it said, had *us* in mind
>
> By now we know the simple trick
> Of how to read Pope John's encyc
> To play the game, you choose your snippet
> Of *Peace on Earth* and boldly clip it.[54]

Cogley's sardonic close was not far off the mark; *Pacem in Terris* was highly susceptible to selective citation for political or theological ends—a phenomenon magnified by the stake that all debaters perceived in having the beloved pontiff on their side of the ongoing argument.

In terms of positive appraisal, the editor-in-chief of *America*, Thurston Davis, S.J., saw the encyclical as an expression of Johannine optimism about the world, and the new style of the Church's approach to it:

> One would be blind not to see that the Church here inserts herself into the history of modern man with an uncanny sense of timing and with a spacious understanding of the hopes and dreams that men entertain for a better economic life, for a deeper sense of security, and especially for an organization in the world that will ensure the tranquillity of order and new prospects of unity.

Davis was also taken with the less abstract, more existential style of the encyclical. The old "essentialist" truths were suffused here with a "fresh sense of respect for history, contingency, change, adaptation to change."[55] Davis welcomed the reassertion of the Church's concern for liberty of conscience and its claim that interdependence and international organization were demands of "the meaning and dignity of the human person under God."[56] Although arguing that the pope did not advocate a worldwide dialogue with communists, Davis also saluted the pope's new approach to the Soviet Union, and hoped that "something constructive will come of it all," especially for the persecuted Church in "captive nations."[57]

Commonweal welcomed the pope's "magnificent encyclical" in a lead editorial on

April 26, 1963, and was eager to help put a definitive interpretation on *Pacem in Terris*:

> In the past, what one theologian has called the "studied ambiguity" of papal encyclicals has often served to allow any number of incompatible interpretations. There is certainly some of this ambiguity in *Pacem in Terris*; but its thrust is very clear—and, in many ways, revolutionary. It draws on the best of traditional Catholic thought while at the same time displaying an acute sensitivity to the empirical realities of contemporary life.[58]

Commonweal editor John Cogley was particularly enthusiastic, writing that *Pacem in Terris* was not only among the greatest of modern encyclicals, but could be "one of the historic documents of our time, destined to shape the world to come." The remarkably warm reception given to the encyclical was based on the perception that John XXIII had made "a kind of breakthrough to a new way of thinking about world order," which could have revolutionary effects.[59]

Cogley was especially impressed with the warm humanism of the encyclical:

> Just a month ago, I said in these pages that the question haunting our generation is, Who speaks for Man? *Pacem in Terris* came as a kind of dramatic reply. In it, the Pope stated specifically that he spoke not only as the head of the Church but as one who voiced the longing of all humanity. He addressed himself, consequently, not only to Catholics but to people of all faiths and of none, a precedent-breaking attempt to rally all men of good will to the cause of peace.[60]

Cogley welcomed the fact that the encyclical was not "overly specific" in its attempts to apply principles to historical circumstance, and was grateful that the pope did not identify the advent of peace with the conversion of all souls, but recognized that the emergence of a transnational political authority would have to take account of human beings as they were: as men and women, not angels. *Pacem in Terris* was thus a "soundly realistic document," which recognized that "the world we have to keep from blowing to smithereens is the imperfect world we live in—the only one we have." Pope John's optimism was not based on naivety. Reason and faith were in "exquisite conformity" on the central problem of war and peace; the pope clearly recognized that peace required political action and structural organization.[61]

Another *Commonweal* editor, James O'Gara, saw *Pacem in Terris* as the death knell for American Catholic isolationism; the encyclical taught that "isolationism has become a luxury we cannot afford." O'Gara reminded his readers that modern papal teaching had long been committed to the goal of a juridical organization of the international community that would "permit collective action against unjust aggressors." O'Gara agreed with Cogley that this strand in contemporary papal teaching had always been "intensively realistic," because it stressed "the role of justice in creating peace."[62]

Commonweal's lead editorial for May 24, 1963 welcomed the encyclical's "implicit opening to the left," its abandonment of the anathema tradition in the face of the reality of communism:

> The Pope has made an important, perhaps historical, attempt to change the direction and tenor of the Cold War. He is surely visionary; but that does not mean he is

> unrealistic. . . . Nothing, for instance, could be more realistic than the sharp distinction the Pope draws between "false philosophical teachings regarding the nature, origin, and destiny of the universe and man" and "historical movements that have economic, social, cultural, or political ends." Such movements, he holds, "cannot avoid . . . being subject to change, even of a profound nature."[63]

Several weeks later, in an editorial marking the pope's death, *Commonweal* suggested that this opening to the Left had already borne good fruit: "better conditions for Catholic worshippers in the Russian satellite lands were in the offing at the time of [John's] death."[64]

Michael Novak linked *Pacem in Terris* and the American Catholic experience, seeing in the pope's dialogic approach to the problem of communism a reflection of "the secret of Anglo-American political cooperation." Novak summed up the case for *Pacem in Terris* by writing that Pope John had "launched [the Church] once more on the currents of human history with hope, with courage, with joy, with the exhilaration proper to those who see in the darkness the star of eternal life. So doing, he made it possible for Catholics to speak of good news to their companions who do not see, and to learn from those who do not see the humility of the human situation."[65]

Gratitude for the analysis and prescriptions of *Pacem in Terris* was not limited to Catholic circles. Protestant social ethicist John Coleman Bennett claimed that the encyclical might be "the most powerful healing word" that had been spoken during the Cold War. In some communist nations, Bennett argued, *Pacem in Terris* was being "taken seriously." Moreover, it called Christians in the West to abandon the anticommunism with which they were often "obsessed."[66]

Protestant theologian Alan Geyer agreed. In a symposium on the encyclical in the ecumenical journal *Worldview*, he claimed that the "St. Joseph's Crusade" against communism announced by Pius XI in *Divini Redemptoris* was over:

> *Pacem in Terris* is an encyclical utterly lacking in righteous indignation. Militancy has yielded to mediation. . . . There is no talk of "enemy" or of "victory." The "vast campaign" [called for by Pius XI] is abandoned in an appeal for a new world order which "satisfies the objective requirements of the universal common good." *Pacem in Terris* thus marks a fundamental shift in the stance of the Vatican in world affairs.[67]

Rabbi Everett Gendler went even further in the *Worldview* symposium, announcing that, with *Pacem in Terris*, "that strange (perhaps mythical?) animal, the 'just war,' guided by our contemporary pillars of fire by night and cloud (mushroom variety?) by day, seems to have crept away from the human scene altogether."[68]

Gendler's theme was picked up, and the encyclical taken to its outer hermeneutic limits, by another Catholic commentator, James W. Douglass, whose analysis of *Pacem in Terris* formed a core chapter in his influential book *The Non-Violent Cross.* For Douglass the power of *Pacem in Terris* was nothing other than the power of Gandhian nonviolence, which was as "pervasive as it is undefined" in the encyclical.[69] The encyclical decisively abandoned the Catholic tradition of moderate realism, and represented a "totally non-violent commitment to man."[70] John XXIII, according to Douglass, "hated nationalism . . . [and] would acknowledge no fundamental divisions in the family of man."[71] On this analysis, Gandhian nonviolence was a "natural law imperative" that could move humanity "from an irrational law of fear to a reasonable law of love."[72] The dichotomy between fear and mutual trust defined the choice between war and peace, Douglass argued, and the encyclical implied that this opposition could be overcome only through Gandhian *satyagraha*, "truth force."[73] *Pacem in Terris* was itself an expression of the liberating power of *satyagraha*.[74]

It was precisely on this issue—the relationship of *Pacem in Terris* to the Catholic tradition of moderate realism—that criticism of the encyclical centered.

John Courtney Murray's reaction to *Pacem in Terris* was mixed. Murray welcomed the spirit of the encyclical:

> [The pope had offered] a shining example of everything he means by his own word, *aggiornamento*. He situated himself squarely in the year 1963. There is not the slightest note of nostalgia, nor of lament over the past course of history or over the current situation that history has evoked here on earth. The Pope confronts all the facts of political, social, economic and cultural change that have been the product of the modern era. Generously and ungrudgingly, he accepts those elements of historical progress which can be recognized as such by the application of traditional principles as norms of discernment.[75]

Murray saw *Pacem in Terris* as a development *within* the Catholic tradition of *tranquillitas ordinis*, rather than as a break with that heritage:

> [The pope's] acute sense of the basic need of the age is evident in the word that is so often repeated in the encyclical and that sets its basic theme. I mean the word "order." This does seem to be the contemporary issue. The process of ordering and organizing the world is at the moment going forward. The issue is not whether we shall have order in the world; the contemporary condition of chaos has become intolerable on a worldwide scale, and the insistent demand of the peoples of the world is for order. *The question is, then, on what principles is the world going to be ordered*.[76]

Pope John, Murray argued, developed the constitutionalism of Thomas Aquinas in *Pacem in Terris*. The pope's political ideal was Thomistic: "the free man under a limited government." This ideal was constituted by three principles: human freedom as an inherent right; limited—that is, constitutional—government; and popular participation in governance.[77] Murray welcomed the pope's claim that "the first function of the state and of all its officers is to guarantee . . . the whole order of human rights and duties." He also noted that "in the past, papal pronouncements on political and social order have always been suspended, as it were, from three great words—*truth*, *justice*, and *charity*"—but *Pacem in Terris* had added a fourth element to this traditional triad, the word *freedom*:

> Freedom is a basic principle of political order; it is also *the* political method. The whole burden of the encyclical is that the order for which the postmodern world is looking cannot be an order that is imposed by force, or sustained by coercion, or based on fear. . . . By sharply accenting this theme, the Pope clearly takes sides against movements on the march today that would organize the world and create an order in it on the basis of force and not on the basis of the principle, which we are proud to call American as well as Christian, that the ordering forces in the world must be the forces of "freedom under law."[78]

Pacem in Terris thus affirmed and endorsed the evolution of political freedoms on the Western model, which previous popes had found so troubling because of the anticlerical, secularist currents within that evolution.[79]

Murray thus found much to admire in *Pacem in Terris*; but he also found grounds for concern, even worry. Noting that Pope John's hope for an international public authority was a direct descendent of similar statements by Pius XII, Murray nevertheless conceded that he, like others, could not see how this hope could be "concretely realized" apart from that worldwide moral and political consensus on which an international public authority would have to be built. Still, Murray claimed that John's hope was "not utopian idealism. It is possible of realization. It seems to be sustained, in the final analysis, by the confidence that breathes through the whole encyclical—a confidence in the power of the human person, in association, to 'insure that world events follow a reasonable and human course.' It is therefore a hope that no reasonable man can fail to share, no matter what the difficulties in the way may be."[80]

Yet Murray also warned that the pope might not have taken the full measure of the "fundamental schism" in the modern world.[81] Sophisticated observer of the polemical scene that he was, Murray knew that "on this difficult subject . . . there will be much argument." He wanted the debaters to be clear that the issue was not an accommodation to Marxism, but the degree to which a deterministic view of history had pervaded the West, particularly its elites:

> I think the Pope deeply understands the disastrous extent to which men today are gripped by the myth of history which the Marxists have so diligently inculcated. . . . In this view, man has lost command of his own destiny on this earth; his destiny is determined by the events of history, and he is himself powerless to control these events. The conclusion is that history today is surely and certainly carrying man toward catastrophe with an inevitability against which man is helpless.[82]

Against this determinism, so much a part of the popular and elite war/peace debate in the 1960s (and 1980s), the pope took a firm position on the side of freedom and intentionality: the "principle of fear" that underwrote the world's fragile peace had to be replaced by another principle. Humanity must resist the temptation to despair, the temptation to believe that events were out of human control:

> At least in this respect, the Pope will command the agreement of all men of good will who believe that there are energies in the free human spirit whereby man may fulfill his destiny on earth, which is to be, not God, but the image of God. All men who believe in God are agreed that He is the Master of history. Man, therefore, manifests himself as the image of God chiefly by his intelligent, confident efforts to master the course of historical events and direct it toward the common good of the peoples of the earth.[83]

Murray's colleague in the Catholic Association for International Peace, William V. O'Brien of Georgetown University, was considerably blunter in his criticism of *Pacem in Terris*. Though agreeing that the encyclical reiterated classic teachings of the natural law tradition, O'Brien argued that "when the encyclical treats of the central dilemma of our times, the balancing of the risks of Communist domination with those of nuclear destruction, it appears at first sight to be in rather sharp contrast with the general lines of contemporary American thought and policy on this dilemma."[84]

The pope's optimism about greater understanding with communists, his trust in an ever more powerful U.N., and his belief that peace would ultimately rest on trust were not shared by most American Catholics, who had "steadfastly insisted that the threat of communism must be viewed in terms of a virtually immutable, incompatible, hostile movement which could not, by its internal logic, be at peace with the non-Communist world." This principled anticommunism was the reason why these same Catholics had little faith in the possibilities of "true world government" in a world so divided as this one. Thus American Catholics, O'Brien argued, had "unwillingly accepted the dangers and burdens of defending the Free World through the threat of nuclear deterrence, indeed, by all available means."[85]

This realism was not a peculiarly Catholic phenomenon. It cohered with the consensus of the intellectual and policy community on the nature of the world conflict. This consensus held two points: that the threat posed by communism to the values of *Pacem in Terris* was not accidental, but had to do with the very nature of Leninism; and that the leaders of the communist world were not suicidal, and had therefore decided to conduct their conflict with the West through nonnuclear means.[86] This consensus had led to the theory and practices of deterrence. The Western consensus did not "condone first use of nuclear weapons, even to free the victims of terrible injustice. But the will to implement the deterrent threat is equally clear and, it would seem, taken with a clear conscience."[87]

O'Brien claimed that this policy consensus had not been insensitive to the dangers of deterrence, but those holding the responsibilities of governance had learned the failures, since the 1920s, of "*simpliste* disarmament schemes." The academic and policy communities had therefore tried to make use of "the balance of terror itself as a stabilizing force. Under the 'umbrella' of stable deterrence one can conceive of reduction of armament and of arrangements reducing the risks of accidental war. This is called not disarmament but 'arms control.'"[88]

Events had moved rather faster than theory and policy consensus: changing weapons technologies and "the obscure but palpable changes within the Communist world, all made one's carefully considered position of yesterday more or less obsolete tomorrow." But, O'Brien continued, "until Holy Thursday, 1963, few American Catholics would have guessed that the very ground on which most of them stood in the debate on the nuclear dilemma would be shaken, not by the latest findings of a nuclear scientist, defense analyst or Kremlinologist, but by the Sovereign Pontiff of the Catholic Church himself!"[89]

Was there any possibility of reconciling the pope's optimism and the skepticism of those who believed that deterrence provided the only stable means for defending the free world?[90] O'Brien tried to find some common ground in the pope's assertion that "to proceed gradually is the law of life."[91] But a mutual recognition of the necessity of prudent change "must not be made into an escape valve whereby the tensions between the encyclical and the beliefs and practices of American Catholics and their government are deprecated. The tensions are there and the honest and right thing to do is to acknowledge them."[92] Perhaps, O'Brien suggested, there could be some agreement in the "halfway house of arms control."[93]

Ecumenical and interreligious criticism of the encyclical raised questions similar to O'Brien's. Will Herberg confessed that "even upon a third reading, I am left with a strange and uncomfortable feeling." His uneasiness was not so much with the principles of the encyclical; at a certain level of abstraction, they were unexceptionable. But these principles were then "made to apply—in a simple, unrefracted manner—to *short-range* problems of policy and program. . . . The highly complex problem of disarmament is dealt with as though it were a simple corollary of the principle of peace. But it isn't! Between the principle and the policy there is the distorting and refracting medium of the actual world situation."[94]

Reinhold Niebuhr was impressed by Pope John's achievement in having "co-opted the modern theory of 'natural rights' as an extension of . . . [Roman Catholic] natural law theory."[95] But, like Herberg, O'Brien, and Murray, Niebuhr was not so impressed by Johannine optimism as applied to the concrete realities of the historical situation in 1963:

> The difficulty with this impressive document is that the Church absorbs some of the voluntarism of the social contact theory, which underlies modern liberalism, and speaks as if it were a simple matter to construct and reconstruct communities, not by the organic processes of history but by an application of "the sense of justice and mutual love." The Pope speaks of the "community of mankind" without making clear that this community is in one sense a reality and in another sense an ideal, since mankind is divided by a multitude of languages, customs, traditions, and parochial loyalties. In this sense Augustine's criticism of Cicero's universalism, by calling attention to the confusion of tongues in the community of mankind, is not heeded. The encyclical is thoroughly modern in many ways, but particularly in breathing a Pelagian, rather than an Augustinian, spirit.[96]

The pope's idealism was "a little too easy" on the problems of disarmament and international organization. John's advocacy of disarmament did not adequately take account of the immediate security problems of both superpowers in "the horrible nuclear dilemma." Moreover, the encyclical too easily transcended those often intractable political problems that made the United Nations not so much an embryonic world government as a "minimal bridge of community in a world riven by a cold war."[97]

Niebuhr's conclusion was, well, Niebuhrian:

> *Pacem in Terris* is no doubt a historic document, but it may be heeded more by idealists than by responsible statesmen. Perhaps it will be regarded by the latter as a prod rather than as a guide. It may be heeded as Pope Innocent III heeded Francis of Assisi. This pope of the 13th century was the responsible and powerful head of a unified Christendom. He could savor Francis' pure idealism. Some of it, however, was irrelevant to his responsibility; and some of it may have been contradictory to his power schemes. The wheel of history has turned full cycle. Modern popes have no power but only the responsibility for articulating the ideals of the Church. The men of power and responsibility today are not a pope such as Innocent III but the two K's, Mr. Kennedy and Mr. Khrushchev.[98]

The most thoughtful critique of the encyclical came from Paul Ramsey, who conceded that the encyclical taught some "sound and exceedingly important political truths that were in search of a voice humanly great enough to utter them."[99] But noting that only two sentences referred to human sinfulness, Ramsey argued that John's work was "lacking . . . a proper sense that the human, moral world is also a world of 'indeterminate, irrational equations,' that there is a 'surd' and an 'indispensable negative' or 'minus sign' also located in human nature."[100] The result was ecumenical, if disconcerting:

> The highest tribute to be paid the encyclical is that—both in tone and substance—the pontiff sounds so much like a liberal Protestant parson. This is also a way to express succinctly the chief criticism to be made: the encyclical sounds too much like liberal Protestant statements which, while rightly stressing what positively needs to be done for the attainment of the universal common good, fail to grapple with the problem of power except in *those* terms. We cannot fail to take notice of the degree to which the pontiff's powerful expression of the aspirations of the whole of mankind . . . becomes a nonpolitical statement in the arena of the actual practice of politics.[101]

Like others, Ramsey worried about the pope's confidence in negotiation; moreover, Ramsey believed that the pope had not sufficiently analyzed the "many degrees of *prudential* anti-communism or of *prudential* collaboration" with Marxist governments that might be required in the real world of international politics and economics.[102]

But the central problem of *Pacem in Terris* was its lack of sustained moral analysis of the reality of power in the world. Here, too, Ramsey saw parallels between Johannine optimism and the liberal Protestantism of his day (and ours, for that matter): "It is a striking fact that among Roman Catholics, an unalloyed doctrine of natural law leading mankind on to world justice and, among Protestant liberals, confidence in the power of Christian love pure and undefiled to reconcile every opponent, *lead religiously motivated political diagnosis away from the problem of power.*"[103] Politics without power was like "Macbeth" without murder:

> Every historical moment is a time of waiting and of action without idols. This is . . . why *Pacem in terris* should not have failed to address itself to the problem of power as well as to the need for a just world order; and why the prudence of statesmen needs also to be informed by the moral principles governing the use of force when this is required.[104]

The encyclical's failure to address the problem of power gave an air of unreality to the pope's discussion of the problem of international organization and governance, and the Pope's focus on the enormity of the problems besetting the universal common good led to a mistaken concept of world politics as a cultural activity. Political community was thereby confused with other forms of human community that seek the common good. "Yet political community differs from all other human communities by the fact that its authority exercises a monopoly of power."[105]

There was no way around the problem of power, politically *or morally*. But unlike Niebuhr, Ramsey was not willing to let the matter simply drop there. What was needed was an approach to the problem of peace-as-political-community *through* rather than *around* the central question of power. And for this, the best available moral resource was just-war theory:

> If one wants to make the strongest *political* case for radically changing the nation-state system, he must start and remain with the church's traditional teachings about the just conduct of war. For these criteria are also the principles intrinsic to purposive political action; they specify the moral economy governing the use of force, which must always be employed in politics—domestically within the nations; internationally in the postfeudal, pre-atomic era; and under the public authority of any possible world community that may be established during the nuclear age.[106]

The pursuit of Pope John's vision required hard grappling with the problem of power and political community. Only an international public authority could translate the tendency of the modern papacy to withdraw the right to war into political and moral reality. "These two things are opposite ends of the same see-saw: the moral right to use force cannot go down faster than the public authority and enforcement of a world community is organized."[107]

Ramsey's critique of *Pacem in Terris*, then, was that while it evoked just-war theory to critique arms competition in the nuclear age, it did not apply the political wisdom contained within that theory to the inextricably linked problem of how an international political community meeting those minimum standards of freedom the pope wished to support might be created. Just-war theory had not been rendered obsolete by the advent of nuclear weapons; for its criteria, which "are also the principles intrinsic to purposive political action," are precisely those that would have to guide the morally sound evolution of the kind of political community that, in the pope's vision, could provide an alternative to war or its threat in the resolution of conflict.[108]

An Evaluation

The importance of *Pacem in Terris* in the evolution of Catholic thought on the moral problem of war and peace has not diminished since 1963. The encyclical marked a further evolution of Catholic thought on the meaning of peace with freedom in the modern world. If the critics of the document were right in their claim that Pope John virtually emptied *tranquillitas ordinis* of its Augustinian realism, it is also true that *Pacem in Terris* articulated, in a measured way that its critics (save Ramsey) usually missed, the linkages between the goals of disarmament, international organization and law, human rights, democracy, and economic and social development. This was an important advance over previous Church teaching on peace as dynamic political community. For, while it is true that Pope John gathered many strands of thought previously sketched by Pius XII, it is equally true that he *gathered* them in one powerful statement that was given added weight by the undoubted goodness of its source. Johannine hope, coupled with a rather sophisticated analysis of the minimal requisites of peace in the modern world, made for a compelling vision. After *Pacem in Terris*, Catholic thought on the problem of war and peace, if it wished to be true to the Johannine legacy, would have to make moral and political connections between issues such as peace, human rights, and democratic forms of governance that were often segregated in the past (as well as the present).

On the other hand, several criticisms of the encyclical were sound. Its catalogue of human rights did not address the conceptual and practical problems that are inevitable when the notion of "rights" as protections from the intrusive power of the state is married without distinction to "rights" as entitlements to social, cultural, and economic desiderata.[109]

The encyclical was also murky on the connection between peace and justice, a muddle that would lead to severe problems in the next generation of Catholic thought on these issues. *Pacem in Terris* would have been strengthened has it addressed more directly those *minimal* claims of justice (e.g., essential protections of basic human rights; democratic forms of governance) that had to be satisfied for a morally worthy peace of *tranquillitas ordinis* to be established, especially when the peace of political community was to be international in scope. The encyclical did not, in sum, clarify what the world could agree to disagree on, and still enjoy the peace of a righly ordered political community.

Murray may have been right in arguing that John's distinction between "false philosophical teachings" and "historical movements" had to do with continental liberalism and laicism rather than Marxism, but the net impact of *Pacem in Terris* was to weaken official Catholicism's ideological position vis-à-vis world communism. Pope John's "opening to the Left," if that indeed was his intention, has not had a discernible impact on the condition of religious believers in the communist-dominated world. It is not the fault of *Pacem in Terris* that religious persecution has intensified in the Soviet Union, Eastern Europe, and Cuba since the days of Pope John. But there would seem to be a connection between Johannine optimism about the prospects of a dialogic relationship with world communism and the subsequent emergence of anti-anticommunism in many sectors of the Church, notably in the United States. Pope John need not have reiterated the increasingly fossilized anticommunism of Pius XII. But *Pacem in Terris* would have been a much stronger document had it combined a resolute and principled anticommunism with suggestions for how the religious community might facilitate the change that the pope correctly sensed was underway in the communist world, and help nudge that change in liberalizing, humanizing directions.

But the chief criticism that must be levied against *Pacem in Terris* is on the question of "trust" in international affairs. This is the point at which Johannine optimism lapsed into naivete. As Ramsey acutely observed, trust was not the basis on which political communities formed, at least insofar as trust was understood in the pope's essentially spiritual context. One can agree with Pope John that political community *is* the only available this-worldly moral answer to the problem of war; but such agreement does not require one to conclude that international conflict, particularly between democratic and totalitarian states, is essentially a matter of misunderstanding that can be assuaged by mutual trust. That, for all the encyclical's relatively sophisticated understanding of the conditions for the possibility of world political community, it did not address itself to the central problem of power within that community, but instead opted at the critical moment for a solution in the order of personal spirituality rather than in the order of politics, was the core failure of *Pacem in Terris*.

Pope John's vision was magnificent, and needed. His sense of that toward which humanity should strive—political community on an international scale—was correct. His guidance for morally sound political action across the immense chasm between things as they are and things as they ought to be was, at best, deficient. One result of that deficiency would be a virtual abandonment of the pope's vision and answer within much of American Catholicism.

GAUDIUM ET SPES: VATICAN II ON "THE CHURCH IN THE MODERN WORLD"

There is hermeneutical gamesmanship involved in determining the "most important" document of the Second Vatican Council. Many would argue that the "Dogmatic

Constitution on the Church" (*Lumen Gentium*) should hold pride of place. Theologians often cite the centrality of the "Dogmatic Constitution on Divine Revelation" (*Dei Verbum*), which has had an enormous influence on Catholic biblical studies, the pastoral use of Scripture, and the self-understanding of the Church's theological enterprise. Since a common answer to the question, "What did Vatican II do?" would be "It changed the Mass into English," one would have to acknowledge the importance of the "Constitution on the Sacred Liturgy" (*Sacrosanctum Concilium*). Those who work for the reunification of Christianity often rate the "Decree on Ecumenism" (*Unitatis Redintegratio*) as among the most significant works of Vatican II, whereas both Church historians and bishops see the "Decree on the Bishops' Pastoral Office in the Church" (*Christus Dominus*) as completing the unfinished work of Vatican I by reclaiming the concept of episcopal collegiality. In American Catholicism, the "Decree on the Apostolate of the Laity" (*Apostolicam Actuositatem*) had important effects, and the final counciliar statement, the "Declaration on Religious Freedom" (*Dignitatis Humanae Personae*), so greatly influenced by the work of John Courtney Murray, is widely and accurately remembered as the American Church's particular gift to Vatican II.

But whatever taxonomy of significance one finally decides upon, it is indisputably true that the most important Council document for the Church's address to the moral problem of war and peace was its "Pastoral Constitution on the Church in the Modern World" (*Gaudium et Spes*). This lengthiest of conciliar statements (the Latin text runs to over 23,000 words) was also one of the most controverted documents of Vatican II. It was debated, in one form or another, at all four sessions of the Council, and was promulgated at literally the last moment of the Council, on December 7, 1965, a day before the ceremonial close of Vatican II. But the particular significance of *Gaudium et Spes* lay in its style and tone: unlike other conciliar documents, which followed the traditional expository form of official Church teachings, "The Church in the Modern World" was crafted in a much more dialogic mode. In that sense, *Gaudium et Spes* was *the* quintessentially Johannine document of Vatican II: it drank deeply from the well of Pope John's optimism; its attitude toward the world was fraternal rather than inquisitorial; it explored the pope's call to lower the drawbridge from apologetic Catholicism and engage modernity as the world's servant, not its judge.

Gaudium et Spes took this invitation to dialogue with a vengeance: its conversation with modernity ranges over an extraordinary breadth of topics, from the dignity of the human person and the meaning of Christian humanism, to the achievements of modern science and the contemporary circumstances of family life, on through to issues of social, economic, and political organization in the world. (The document even admonished motorists to observe speed limits.) Nowhere does the image of Pope John as a wine steward uncorking a sparkling vintage seem more apt; the cup of "The Church in the Modern World" indeed runneth over. Nowhere else in the conciliar documents does John's intention to identify his Church with the yearnings of the human family come to fuller expression than in the prologue to *Gaudium et Spes*, with its affirmation that "the joys and the hopes, the griefs and the anxieties of the men of this age, expecially those who are poor or in any way afflicted, these too are the joys and hopes, the griefs and anxieties of the followers of Christ. Indeed, nothing genuinely human fails to raise an echo in their hearts."[110]

Gaudium et Spes was the child of one of the great Council fathers, the Belgian primate, Cardinal Leo-Jozef Suenens. Toward the end of the first session of Vatican II, on December 4, 1962, Suenens, a leader of the Council's liberal wing throughout its four-year history, called on the fathers to "present the Church to the world as the light of the nations . . . that is, the Church in herself, in her nature and in her mission as Mother and Teacher; then the Church in the face of the great problems which trouble today's world, beginning with those concerning the human person and going on to those referring to society with its demands of justice and peace."[111] Cardinal Giovanni Battista Montini (who within six months would be elected Pope Paul VI) seconded Suenens' proposal, arguing that the Church did not exist for itself but as the servant of all humanity.[112] Thus was *Gaudium et Spes* launched with two distinguished sponsors.[113]

Gaudium et Spes was finally structured much as Cardinal Suenens proposed in 1962. The document began with a meditation on "The Situation of Men in the Modern World," which included the famous assertion that the Church must constantly scrutinize the "signs of the times" and "interpret them in the light of the Gospel."[114] Here, the decisive shift in style, tone, and, most importantly, analytic method was established. *Gaudium et Spes* would use essentially *evangelical* criteria for its conversation with the world: "What does the Gospel require?" would be the central question. The document continued with a reflection on the dignity of the human person, and discussed the problem of modern atheism. The Council fathers then reflected on "The Community of Mankind" and "Man's Activity in the World." Part One closed with a discussion of "The Role of the Church in the Modern World," stressing that the Church is a leaven in human society, a community immersed in history even as it understands that history as beginning and ending with Christ, the Alpha and the Omega of the human story.

Part Two of *Gaudium et Spes* took up problems of "special urgency," among which were "Fostering the Nobility of Marriage and the Family," "The Proper Development of Culture," "Socio-Economic Life," and "The Life of the Political Community." Only after all this had been discussed did the document turn to "The Fostering of Peace and the Promotion of a Community of Nations," with which *Gaudium et Spes* ended.

The transition that the pastoral constitution represented in Catholic thought on the problem of war and peace can be summarized in one phrase. After reviewing the threat posed by modern weapons technology, the fathers stated flatly that "all these considerations compel us to undertake an evaluation of war *with an entirely new attitude*."[115] The Council did not intend to abandon the just-war heritage of the church; just-war analysis was used when the pastoral constitution discussed specific problems. But what focused the fathers' immediate attention was the fact that "the whole human family has reached *an hour of supreme crisis* in its advance toward maturity."[116] Nor was the crisis simply one of weapons technology; interdependence had made the problem of war ever more urgent:

> Moving gradually together and everywhere more conscious already of its oneness, this family cannot accomplish its task of constructing for all men everywhere a world more genuinely human unless each person devotes himself with renewed determination to the reality of peace. Thus it happens that the Gospel message, which is in harmony with the loftier strivings and aspirations of the human race, takes on a new luster in our day as it declares that the artisans of peace are blessed, "for they shall be

> called children of God" (Matt. 5:9). Consequently, as it points out the authentic and most noble meaning of peace and condemns the frightfulness of war, this Council fervently desires to summon Christians to cooperate with all men in making secure among themselves a peace based on justice and love, and in setting up agencies of peace. This Christians should do with the help of Christ, the Author of peace.[117]

This paragraph captures the essence of the Council's approach to the problem of war "with an entirely new attitude." Ours is not a time of historically familiar contests for power; ours is a time of "supreme crisis." The problem of war and peace is not solely, perhaps even primarily, a matter for statesmen, scholars, and experts in international affairs; it is a matter for "each person," who is under a moral obligation to be devoted "with renewed dedication to the reality of peace." This is not only a matter of right reasoning from natural law principles; it is a Gospel demand, and peacemakers will be judged by the criteria of the Sermon on the Mount. The Church not only deprecates modern war; it "condemns" it. And the task of the Church is larger than conscience-formation; the Church must "cooperate with all men" in securing a peace that will be based on "justice and love." The transitional character of *Gaudium et Spes* in its address to the moral problem of war and peace, peace and freedom, is nowhere clearer than in these first sentences with which the Council fathers took up the topic.

The fathers then affirmed that "Peace is not merely the absence of war. Nor can it be reduced solely to the maintenance of a balance of power between enemies. Nor is it brought about by dictatorship. Instead, it is rightly and appropriately called an enterprise of justice (Is. 32:7)."[118] Respect for individuals and peoples, and the practice of brotherhood, were essential for peace, which was an expression of the love that transcends justice.[119] Temporal peace "symbolizes and results from the peace of Christ," and for this reason "all Christians are urgently summoned 'to practice the truth in love' (Eph. 4:15) and to join with all true peacemakers in pleading for peace and bringing it about."[120]

In an innovative statement for official Catholicism, the fathers then praised "those who renounce the use of violence in the vindication of their rights." Although such renunciation of violence must be undertaken "without injury to the rights and duties of others or of the community itself," the Council fathers here warmly applauded nonviolent methods of defense.[121] Evangelical criteria were prominent on this question.

The fathers returned to a more traditional just-war analysis in discussing war-fighting and personal conscience. The natural law bound all consciences; actions that violated just-war limits were criminal and could not be excused by blind obedience to civil or military authority. The fathers specifically cited "those actions designed for the methodical extermination of an entire people, nation, or ethnic minority," which were vigorously condemned. Those who refused to participate in such crimes merited "supreme commendation."[122]

A slightly different note was then struck: "As long as the danger of war remains and there is no competent and sufficiently powerful authority at the international level, governments cannot be denied the right to legitimate defense once every means of peaceful settlement has been exhausted."[123] But having affirmed this classic just-war position, the fathers argued that self-defense and aggression were completely different. The fact that one possessed weapons did not legitimate their use, which had to be constrained by classic *in bello* limits.[124] On the other hand, military service was an honorable profession, and soldiers should regard themselves as "agents of security and freedom on behalf of their people." In doing so, they made their own contribution to peace.[125]

The Council fathers then discussed "Total War" and "The Arms Race," asserting that "any act of war aimed indiscriminately at the destruction of entire cities or of extensive areas along with their population is a crime against God and man himself. It merits unequivocal and unhesitating condemnation."[126] This one anathema of Vatican II was based on the bishops' concerns about all weapons of mass destruction, but particularly nuclear weapons; "the unique hazard of modern warfare . . . [was that] it provides those who possess modern scientific weapons with a kind of occasion for perpetrating just such abominations" as the bishops had condemned.[127]

The bishops then acknowledged the reality (and moral possibility) of deterrence:

> Scientific weapons, to be sure, are not amassed solely for use in war. The defensive strength of any nation is considered to be dependent upon its capacity for immediate retaliation against an adversary. Hence this accumulation of arms, which increases each year, also serves, in a way heretofore unknown, as a deterrent to possible enemy attack. Many regard this state of affairs as the most effective way by which peace of a sort can be maintained between nations at the present time.[128]

Yet the arms race, deterrence, and the balance of power were not safe ways to preserve the peace, for they exacerbated the causes of war.[129] Therefore, "it must be said again: the arms race is an utterly treacherous trap for humanity, and one which injures the poor to an intolerable degree."[130]

In the face of these awesome realities, the bishops argued that every possible effort must be made to outlaw war by international consent. This would require a "universal public authority acknowledged as such by all, and endowed with effective power to safeguard, on behalf of all, security, regard for justice, and respect for rights."[131] Here the bishops adopted John XXIII's prescriptions in *Pacem in Terris*, as well as John's notion of "trust" as the foundation of an international peace of *tranquillitas ordinis*:

> Peace must be born of mutual trust between nations rather than imposed on them through fear of one another's weapons. Hence everyone must labor to put an end at last to the arms race, and to make a true beginning of disarmament, not indeed a unilateral disarmament, but one proceeding at an equal pace according to agreement, and backed up by authentic and workable safeguards.[132]

The Council did not mean to deprecate the work on this agenda that was already underway in the world. The good will of statesmen should be acknowledged. "Though burdened by the enormous preoccupations of their high office, these men are nonetheless motivated by the very grave peacemaking task to which they are bound, even if they cannot ignore the complexity of matters as they stand."[133] Still, work for peace could not be left to political leaders alone. All men and women had to "have a change of heart." All should be instructed in "fresh sentiments of peace." The force of public opinion for peace could cause significant change in international politics, the fathers implied.[134]

The bishops made several points that would have a profound influence on subsequent Catholic thought in their discussion of "Building Up the International Community." "If peace is to be established," they claimed, "the primary requisite is to eradicate the causes of dissension between men."[135] War would end, in other words, when the causes of war were eliminated. Chief among these were injustice, "excessive economic inequalities," the "quest for power," and "contempt for personal rights."[136] The deeper causes lay, of course, in human sinfulness.[137]

Without relating the abiding reality of sin to the question of outlawing war through international consent, the bishops then discussed present international institutions, which "stand as the first attempts to lay international foundations under the whole human community for the solving of the critical problems of our age, the promotion of global progress, and the prevention of any kind of war."[138]

Gaudium et Spes then argued that international economic development would require a "profound change" in the practices of the modern business world.[139] The bishops acknowledged the role of investment in promoting development, but warned that "if an economic order is to be created which is genuine and universal, there must be an abolition of excessive desire for profit, nationalistic pretentions, the lust for political domination, militaristic thinking, and intrigues designed to spread and impose ideologies."[140] The bishops recognized that development was not solely a matter of technical or financial assistance, but involved the "complete human development of . . . citizens as the explicit and fixed goal of progress."[141]

Gaudium et Spes ended its discussion of the moral problem of war and peace with an exhortation on the duties of Christians:

> Christians should collaborate willingly and wholeheartedly in establishing an international order involving genuine respect for all freedoms and amicable brotherhood between all men. This objective is all the more pressing since the greater part of the world is still suffering from so much poverty that it is as if Christ himself were crying out in these poor to beg the charity of the disciples.
>
> Some nations with a majority of citizens who are counted as Christians have an abundance of this world's goods, while others are deprived of the necessities of life and are tormented with hunger, disease, and every kind of misery. This situation must not be allowed to continue, to the scandal of humanity.[142]

The pastoral constitution's final proposal was that "some agency of the universal Church be set up for the worldwide promotion of justice for the poor and of Christ's kind of love for them. The role of such an organization will be to stimulate the Catholic community to foster progress in needy regions, and social justice on the international scene."[143] The result of this proposal was the new pontifical commission *Iustitia et Pax.*

The *Gaudium et Spes* Debate

Many commentators welcomed *Gaudium et Spes* as the finest expression of Pope John's intention that the Council should offer the world the medicine of mercy rather than the severity of condemnation. Msgr. George Higgins of the National Catholic Welfare Conference, a theological advisor at the Council, wrote that "the Fathers . . . took Pope John's words to heart and willingly adopted his recommendation to proceed, not deductively—as their own philosophical and theological training might have tempted them to do—but inductively, if you will, carefully striving, under the guidance of the Holy Spirit, to read and to decipher the hidden meaning of the principal signs of the times."[144]

Some observers thought that *Gaudium et Spes* was too optimistic, theologically and politically.[145] Higgins, however, believed that arguments about this or that specific prescription in the pastoral constitution were "less important than the tone or the spirit of the conversation [the document conducted with the world] and the methodology which underlies it."[146] With *Gaudium et Spes*, the detailed explication of Pope John's opening address to the Council, "a whole era in the history of the Church was declared closed, and a new era was ushered in. Some have described this new era as the age of dialogue: dialogue within the Church, dialogue between the Church and other religious bodies, and, last but not least, dialogue between the Church and the modern world."[147] In sum, Higgins welcomed *Gaudium et Spes* as an example of what a Council should really be about.[148]

James Douglass was not only grateful for the tone and spirit of *Gaudium et Spes*; he was enthusiastic about several specific developments he detected in the text. Douglass agreed with and welcomed the Council fathers' awareness that "a spiritual revolution of some kind is necessary" if there were to be peace in the world.[149] Douglass also applauded the Council's tacit rejection of Pius XII's argument that the principal moral defect of nuclear weapons was their alleged "uncontrollability," and the fathers' return to the just-war principle of "discrimination" as the basis on which to judge nuclear weapons and deterrence.[150] But Douglass, a lobbyist at the Council during the drafting and debate over *Gaudium et Spes*, clearly wished that the document had gone further in condemning modern war.

Douglass thought that the Council had failed to recognize the incompatibility of war with the Gospel of reconcilitation;[151] and the Council had been utterly irresponsible in its failure to condemn nuclear deterrence. For Douglass, the possession of nuclear weapons established a conditional intention to use them, which was itself forbidden by the conciliar condemnation of acts of war "aimed indiscriminately at the destruction of entire cities or of extensive areas along with their population." Why, then, did the bishops not draw the inevitable conclusion and condemn deterrence itself, which was not only immoral on the grounds of a conditional intention to commit indiscriminate mass slaughter, but was also the principal expression of that preparation for "total war" that the Council found so appalling a fact of the modern age?[152]

Douglass would have preferred that *Gaudium et Spes* follow the advice of Cardinal Joseph Ritter of St. Louis who, in a written intervention during the debate over the document, called for "an absolute condemnation of the possession of arms which involve the intention or the grave peril of total war."[153] Following the arguments of the English Abbot Christopher Butler, Douglass claimed that Ritter's message could be heeded, even within the terms of *Gaudium et Spes*, by unilateral disarmament.[154]

But the real achievement of *Gaudium et Spes*, Douglass concluded, was "to bring down the curtain on the just-war doctrine":

> To transpose a moral principle traditionally applied to sexual conduct, one can characterize all modern war as a thermonuclear occasion of genocide. Judged by the Council's declaration, modern war itself is a crime against God and man and merits

> unequivocal and unhesitating condemnation. If we wish to take the Council seriously in its central declaration, in spite of the Council's own evident hesitancy to face that declaration squarely throughout its statement, we must declare the just war dead.[155]

To replace the theory he had interred, Douglass proposed that the Church "re-open that scripturally founded tradition of non-violence which has remained largely unexplored since the early age of the Church but which now has Gandhi's, Dolci's, and King's experiments in truth as proof of its untapped power."[156] Beyond an exploration of nonviolent defense, Douglass also called on the Church to reclaim another heritage lost behind the mystifications of just-war theory: the church, on Douglass's reading of *Gaudium et Spes*, had to rediscover "the salvific meaning of accepted crucifixion,"[157] in a return to that pristine, primitive community of love described in the Acts of the Apostles.[158]

What the Council would not do explicitly, James Douglass proposed to do hermeneutically. Another commentator on *Gaudium et Spes*, Paul Ramsey, had a rather different reaction to the meaning and import of "The Church in the Modern World." In Ramsey's mind, *Gaudium et Spes* vindicated just-war theory in its grandest dimensions: the effort to relate power, proportionate use of force, and political community to each other in a morally and rationally sound way.

Ramsey's general judgment on the document was enthusiastic. "The Pastoral Constitution on 'The Church in the Modern World' . . . deserves the admiration of non-Catholics the world over. . . . It may be the greatest achievement of the Council."[159] The document was uneven; some parts were stronger than others. But its overall impact was all to the good. Ramsey was particularly impressed that, on the question of war and peace, *Gaudium et Spes* "did not undertake to say everything that humanity or even all the sons of the church need to hear in this or in any hour, but only that part of what needs to be said that can and may and must be said on the basis of moral values and spiritual truth the church as such is competent to know something about."[160]

Contrary to Douglass's view, Ramsey believed that the constraints on the document's specific prescriptions came, not primarily from compromise within the Council, but from "remembering all the sons of the church and men everywhere in the political and military sectors, in the armed forces of the nations, in huts of poverty because there are arms, in prison for conscientious objection, secretly troubled or not troubled enough in conscience. Above all, the impulse and the limit came from endeavoring to say as fully and adequately as possible what can and may and must be said in Christ's name and only what can possibly be thus said by the whole church to all who bear and do not bear His name."[161]

Ramsey found three "climactic utterances" around which *Gaudium et Spes* was organized.

The first was the famous declaration that "any act of war aimed indiscriminately at the destruction of entire cities or of extensive areas along with their population is a crime against God and man himself. It merits unequivocal and unhesitating condemnation." Here the Council had rightly raised up the just-war principle of discrimination and thereby vindicated the attempt of just-war theory to apply moral reason to the limit case of warfare.[162] "The cardinal point in the declaration," Ramsey argued, "is not a condemnation of any use of nuclears in war. It is rather a call to the citizens and magistrates of all the nations to clarify their consciences in terms of the basic principle governing the use of these or any other 'scientific' weapon."[163] Douglass saw the Council's decision not to explicitly condemn nuclear weapons as a failure of moral nerve; Ramsey found the Council properly interested in an even more fundamental question: establishing the conditions for the possibility of *any* moral discourse about war and peace.

The second crucial utterance was the Council's treatment of deterrence.[164] Here, too, Ramsey failed to perceive the waffle that so exercised Douglass. The Council was not concerned to make a political judgment about the efficacy of deterrence:

> [It allowed these arguments, for and against,] to stand for what they are, or for what responsible people in the political and military sector may judge them to be. Across words that concede the possible necessity and value (such as it is) of deterrence ("*Whatever be* the case with this method of deterrence . . . ") and across words that anyway put deterrence in its place ("peace of a sort"), the Council is concerned rather to direct the sons of the church and mankind in general to the work of political construction needed to alter fundamentally these conditions.[165]

Deterrence, Ramsey implied, provided a necessary regulatory mechanism in world affairs, a stability that set the conditions under which "the work of political construction" could go forward. Ramsey found nothing in *Gaudium et Spes* contrary to this view; the internal logic of the pastoral constitution seemed to support it (even if in the interstices of the bishops' argument).

Deterrence provided a pause, an interlude, in which the work of building *tranquillitas ordinis* could continue.[166] This would require more than mechanics, the fathers understood:

> There must be a real political order which, while it can be brought into existence without conquest or tyranny only by the manifold works of international consent and political construction, still cannot then be broken by mere disconsent. It will be a world order of enforceable law and justice. . . . Thus, the Vatican Council makes it clearer even then *Pacem in terris* that a *single* world political authority is needed. . . . to change this peace of a sort into a steady peace in the world. In expressions drawn from the language and intent of the original Charter of the United Nations (whose realism has been rendered more nugatory by subsequent developments in UN practice), we may attribute to the Council fathers the belief that only a universal public authority with radical and justly used decision-making capabilities, interpositional peace-keeping and interventionary threat-removing powers can safeguard security, justice, and rights on behalf of all; and that only the achievement of just world government will remove from among the burdens and responsibilities of the leaders of nations the right and the duty to maintain a very steady peace. Until then, it will remain *among* the duties of statesmanship (though, of course, not its only responsibility) sometimes to resort to war on behalf of a juster order and a relatively more secure peace.[167]

In Ramsey's judgment, this was not utopian dreaming, for the Council had also addressed the interim steps that would help close the gap between things as they are and things as they ought to be. The conciliar call to the leaders of existing international and regional organizations was not to negotiate for the sake of negotiating, but to alter a structurally defective situation in which all the good will in the world could not guarantee a just peace. The path beyond deterrence required a "manifold work of political intelligence," not expressions of lofty sentiment. "Radical world political reconstruction is a rational requirement in the nuclear age. This is not optional, but mandated."[168] Noninterventionism had to be rejected, because it was "precisely not a principle of world or regional order."[169]

Ramsey worried that the Council's welcome probing to the roots of the problem of war in human sinfulness had not been adequately connected to its vision of international *tranquillitas ordinis:*

> If both injustice and animosities must be removed from among men, if the peace of which we are speaking will evidence a regard for justice and mutual respect for the rights of all, and if world security will finally rest in these achievements and upon the consent of all and not in new political institutions that are established by conquest or maintained by any measure of tyranny, if distrust and pride and other egotistical emotions that rupture the harmony of things must be uprooted, that would seem a reasonable facsimile of the kingdom of God.[170]

Still, Ramsey was essentially sympathetic with the bishops' dilemma:

> Our brothers in Christ who wrote those words, and the Council that adopted the statement, which finally penetrated to the fact that human conflicts are rooted in human sinfulness that is not going to be expunged without redemption, were apparently not unaware of the borderline between this age and another to which simple realism had driven them. In the midst of those somber words they fittingly paused simply to confess our fallen social existence and to describe how the entirety of this human life of ours will appear in the final judgment upon it by the coming age in which their hope was fixed: "Man cannot tolerate so many breakdowns in right order." This is the political realism that is laid bare when man's existence is penetrated and fully revealed in the light of Christ, or in the light of an authentic peace on earth that can only be described as the restoration of the historical socio-political order to harmony.[171]

Ramsey, too, felt the inevitable tension involved in moral thinking about the problem of war and peace, a tension to which the bishops' Christian commitment had brought them. Every Christian, indeed every human being, lived between the Fall and the Kingdom to come, in a dialectic between things as they are and things promised:

> At his peril he ignores either of these dimensions, by failing to take responsibility for the preservation of real political order in the world or by failing to take responsibility for introducing radical changes into the existing world political system. . . . At all times we must decide between the respective claims of preservation and government and of the higher and fuller forms of community toward which God preserves the world by these means. It is better to say that at all times our decisions will be *in between* these respective claims. They will be choices and actions composed by reference to *both* these claims.[172]

Ramsey's final concern was that the Council's theology not be lost amid the attention focused on its political vision. The conciliar understanding of both present circumstances and future requisites "flows naturally enough from bringing all the perspectives of Christian theological ethics to bear upon the problems of politics." Therefore, "unless extreme care is taken in excising the Council's political teaching from its theology, the men of good will who do this are exceedingly apt merely to have their world-historical utopianism confirmed. That, in turn, will only add idealistic fury to the ruptures in the harmony of things that many cannot bear; and this will finally render mankind ungovernable."[173]

An Evaluation

In the retrospect of twenty years, it may appear that the fathers of Vatican II operated less than successfully in the inevitable tension that Ramsey well understood, and with which he rightly and charitably sympathized. Six problems should be noted.

First, *Gaudium et Spes* offered several denotations of peace: a peace of the heart, flowing from our relationship with God in Christ; a peace of restored harmony in the creation (the Isaian vision of *shalom*); the peace of *tranquillitas ordinis*. All of these are legitimate Christian understandings of the richly textured reality of "peace." But the document would have been strengthened by more carefully delineating the relationship among these three meanings of "peace," and by more stringently considering the relationship of the "peace of Christ" and *shalom* to the peace of *tranquillitas ordinis*. That they are related is not in doubt; but the question of *how* is very much part and parcel of moral judgment in this time between Easter and the coming of the Kingdom in fullness.

A parallel problem is the document's too neat suture between pastoral and evangelical encouragement and political analysis and prescription. The degree to which hearts must be changed before political institutions can be created is never carefully assessed by *Gaudium et Spes*. One result in subsequent Catholic argument has been an unfortunate tendency to pose political conflict in psychological categories of misunderstanding. The document itself is not really susceptible to this charge; but others, drawing as they claimed on the inspiration of *Gaudium et Spes*, would be.

The conciliar claim that arms races lead to war is a partial truth that, if absolutized, can lead to serious misperceptions of international reality. An arms race did precede World War I and, without falling into the fallacy of *post hoc ergo propter hoc*, a case can be made that that arms race did help bring about war. But the opposite was true in the 1930s, when failure to arm in the face of Hitler's aggression also created conditions for the possibility of war. The relationship between weapons acquisition and war is much more complicated than the Council allowed, particularly given the deterrent nature of nuclear weapons. *Gaudium et Spes* would have been a wiser statement had it acknowledged this complexity.

The Council was surely right to urge that public opinion be marshaled for the sake of peace. But *Gaudium et Spes* should have noted that the problem of public opinion and peace in totalitarian societies is not a matter of "fresh sentiments," but of structure. One measure of the threat posed by the Soviet Union is the fact that public opinion has little if anything to do with the foreign and military policy of the Soviet leadership, a fact that was clear in 1965. No one doubts the concern of the peoples of the U.S.S.R. for peace; the problem is that that concern has no politically curbing effect on the Soviet leadership, given the structure of the Leninist system. The welcome and admirable concern for human freedom that permeates *Gaudium et Spes* ought to have been brought to bear forthrightly on this problem: not as a polemical exercise, but precisely as a key problem for peacemaking.

Like *Pacem in Terris*, "The Church in the Modern World" understood, correctly, that disarmament could not proceed very far without a parallel evolution in structures of international law, governance, and conflict-resolution; that these structures would require a new measure of political community in the world, based on respect for basic human rights; and that those states primarily interested in change, not the status quo, had to be assisted in their economic and political development. These four minimal goals of a transformed international political system—disarmament, law and governance, political community based on human rights, and development—were all noted in *Gaudium et Spes*, but they are not well related, theoretically or functionally. Nor did the Council seriously address the further problem of how one gains agreement on the pursuit of these goals without further exacerbating the kinds of conflict that lead to war. The conciliar endorsement of nonviolent techniques was a welcome innovation; but a further statement that law was the most widely utilized means for the just resolution of conflict could have been made, and the achievements of democratic governance as a form of nonviolence praised.[174]

Finally, one might raise questions about the "inductive" method so lauded by Msgr. Higgins. To do so is not to yearn and pine for the days when Catholic social theory was a matter of rationalistic deduction from abstract principles. Moral reasoning is under a moral obligation as well as a rational obligation to take serious account of the facts of the matters on which it makes judgments. But there is no *pure* inductive reasoning. The Council did not fashion its approach to the problem of development merely by reading the signs of the times, the poverty and degradation in which millions of people lived, too often in misery and pain, and died. The Council fathers "read" these realities through a set of prior assumptions and judgments about the causes, nature, and cure for poverty. And it was these assumptions and judgments, not the signs of the times, that led to their prescriptions and recommendations. Critical analysis of the preconceptions through which one reads the signs of the times is of the utmost importance. There can be no escape from careful hermeneutics here.

Gaudium et Spes was, in many ways, a remarkable and wonderful document. Its opening sentences are as fine a statement of Christian humanism as can be found in the modern history of the Church. But the ambiguities and weaknesses of "The Church in the Modern World" are also apparent. The document's welcome evocation of the evangelical call to peacemaking was issued at the expense of a failure to develop and extend the Catholic tradition of moderate realism, of peace as *tranquillitas ordinis*. The bishops' admirable intention to read the signs of the times was not combined with reflection on an adequate moral and political hermeneutic for that necessary process. Other spirits, less intelligent, less measured, less noble than those of the fathers of Vatican II would, with both good and ill will, turn these weaknesses to ends that it is difficult to imagine the Council would have endorsed. *Gaudium et Spes* stands, then, as both a high moment in Catholic internationalism according to the moderate-realist tradition, and as the key transitional document in what would be, over the next generation, a process of abandoning that heritage.

THE MINISTRY OF THE LAITY IN THE WORLD, RELIGIOUS FREEDOM, AND A SUMMING UP

Two other Council documents bore on the war/peace debate within American Catholicism.

The "Decree on the Apostolate of the Laity" (*Apostolicam Actuositatem*) gave new prominence to the mission that Catholics who shared in the priesthood of the baptized, but not that of the ordained, exercised in the world. The concept of a specific lay apostolate had existed in Christianity since the New Testament period,[175] but was given higher definition and significance in light of the conciliar images of the Church as the "People of God" and the "Body of Christ"—decisive breaks with apologetic Catholicism's tendency to identify the Church with the hierarchy.[176]

In the "Dogmatic Constitution on the Church," the fathers wrote that the lay apostolate was not an adjunct to the "real" ecclesiastical work of the hierarchy, but was "a participation in the saving mission of the Church itself."[177] The fathers also made clear that that apostolate of the laity had its own specific integrity and purpose: "the laity are called in a special way to make the Church present and operative in those places and circumstances where only through them can she become the salt of the earth. Thus, every layman, by virtue of the very gifts bestowed upon him, is at the same time a witness and a living instrument of the Church herself, 'according to the measure of Christ's bestowal' (Eph. 4:7)."[178] This theme was made even more explicit in *Apostolicam Actuositatem*, which taught that the renewal of the world was the "special obligation" of the laity. Christian social action was an "outstanding" form of the lay apostolate, which the Council wanted extended to touch the full range of public life.[179]

As the Council fathers taught:

> A vast field for the [lay] apostolate has opened up on the national and international levels *where most of all the laity are called upon to be stewards of Christian wisdom.* In loyalty to their country and in faithful fulfillment of their civic obligations, Catholics should feel themselves obliged to promote the true common good. . . . Catholics skilled in public affairs and adequately enlightened in faith and Christian doctrine should not refuse to administer public affairs, since by performing this office in a worthy manner they can simultaneously advance the common good and prepare the way for the Gospel. . . .
>
> Among the signs of our times, the irresistibly increasing sense of solidarity among all peoples is especially noteworthy. It is a function of the lay apostolate to promote this awareness zealously and to transform it into a sincere and genuine sense of brotherhood. Furthermore, the laity should be informed about the international field and about the questions and solutions, theoretical as well as practical, which arise in this field, especially with respect to developing nations.[180]

The Council's ecclesiology of the laity was thoroughly transformationalist (or "conversionist," in H. Richard Niebuhr's typology:[181] the Church was not a sect set over against the world, but a leaven within it. This mission of leavening the affairs of the world with the values of the Gospel was specifically and properly the mission of the laity, according to the Council. Theirs was the right to this mission by reason of baptism and confirmation; and theirs were the specific competencies required so that the leaven may bring forth appropriate results. The apostolate of the laity in the world was not a bequest from the hierarchy but a right derived from incorporation into the Body of Christ in baptism. In this sense, the laity was "the Church in the modern world" in a decisive way. Finally, the laity, in the exercise of their worldly ministry, had a claim on the support, wisdom, and counsel of the hierarchy. This was not to be understood, in the Council's mind, on a model of control—the laity carrying out the instructions of the more illuminated hierarchy—but on a model of fraternal dialogue. The laity ought to look to the hierarchy for counsel on the values of the Gospel and the teaching of the Church; the obverse, implied by the documents, was that the hierarchy ought to look primarily to the laity for counsel on the application, through the virtue of prudence (a particularly lay virtue, in conciliar ecclesiology), of Christian principles in the affairs of the world.

The implications of this teaching for an American Catholicism deeply engaged with the moral problems of peace and freedom are profound. The Council gave its virtually unqualified blessing to the specific competence of the laity in the affairs of this world, which surely include political life. The Council envisaged a partnership between the hierarchy and the laity in the transformation of the "temporal order:" but it is equally clear that the primary responsibility for this transformation falls on the laity—again, not as a bequest, but as a responsibility specific to their baptism and their particular knowledge of the world's affairs. Clericalism, the assumption that wisdom on prudential matters in the public order is conferred by priestly or episcopal ordination, was decisively rejected by Vatican II. The lay apostolate was a legitimate component of the ministry of the Church, with its own integrity and sphere of competence.[182]

The "Declaration on Religious Freedom" (*Dignitatis Humanae Personae*) was the special gift of American Catholicism to the universal Church at Vatican II. The principal architect of the declaration, through five drafts, was the American Jesuit John Courtney Murray. The declaration not only vindicated Murray's formulation of Catholic church/state theory; it also vindicated the American experience of Catholicism, and the Catholic experience of America.[183]

The central affirmation of the declaration is well known:

> This Vatican Synod declares that the human person has a right to religious freedom. This freedom means that all men are to be immune from coercion on the part of individuals or of social groups and of any human power, in such wise that in matters religious no one is to be forced to act in a manner contrary to his own beliefs. Nor is anyone to be restrained from acting in accordance with his own beliefs, whether privately or publicly, whether alone or in association with others, within due limits. The Synod further declares that the right to religious freedom has its foundation in the very dignity of the human person, as this dignity is known through the revealed Word of God and by reason itself. This right of the human person to religious freedom is to be recognized in the constitutional law whereby society is governed. Thus it is to become a civil right.[184]

The declaration was not without significance for Catholic thought on the linked problems of peace and freedom. While it is true that the Council did not issue any formal condemnation of communist totalitarianism, it is hard to imagine a more antitotalitarian statement than the "Declaration on Religious Freedom." If the essence of totalitarianism is to deny that there is any sphere of privacy in the human person; if totalitarians insist that all matters, even of conscience, are public matters; then totalitarianism is flatly rejected by the declaration. The Council's insistence that there remains within each person a sanctuary of privileged privacy that cannot be desecrated by "any human power" is a fundamental challenge to totalitarian claims. The Council thus implicitly asserted that the right of religious freedom was the most fundamental of human rights, and that any human rights scheme that failed to acknowledge this (or that qualified it, as in the Soviet Constitution), was spurious. Moreover, the Council asserted that this right was not a matter of religious conviction through revelation, but was in principle knowable by human reason. No one, therefore, had any excuse for violating the right of religious freedom, which was a fundamental part of the patrimony of humanity itself.

The "Declaration on Religious Freedom" can also be taken as a tacit, but critical, affirmation of democracy and democratic pluralism. Not only was the confessional state envisioned by apologetic Catholicism abandoned; the declaration clearly implied that democratic pluralism, under the conditions of the modern world, was the most appropriate embodiment of Catholic social theory. The Council did not specifically apply this to the question of war and peace, but a further implication can be drawn: that democratic pluralism is the most appropriate expression of *tranquillitas ordinis* in the modern world. Murray's work at Vatican II thus bore on questions far beyond the specific issue at stake in the "Declaration on Religious Freedom."

The shift from apologetic Catholicism to Catholicism in dialogue with the modern world had been anticipated in the experience of the American Church. Yet the evangelical tone and style by which the Council wished to conduct this dialogue raised its own hermeneutical questions. Granted that the Church in the modern world was to be *ecclesia discens* as well as *ecclesia docens* (a welcome understanding with deep roots in classic Thomism), one still had to determine the principles by which that learning from the world was organized and appropriated. This is particularly crucial in the Church's reflection on the moral problem of peace and freedom in a world of weapons of mass destruction. It is a perversion of the intention and teaching of Vatican II to suggest that the Council ever meant to imply that "the demands of the Gospel" led, without the mediation of moral reasoning, to specific policies and practices. Whatever one's disagreements with this or that formulation of *Gaudium et Spes*, its authors intended to reinforce, not mitigate, the claims of moral reason as a mediating instrument between Gospel values and the exigencies of the age. Here, Paul Ramsey was surely right.

Such a tradition of mediating reason had informed the American Catholic appropriation of the Catholic heritage of *tranquillitas ordinis* and the Catholic response to the American democratic experiment since John Carroll. Given the conciliar affirmation of this style of moral reasoning on these issues, it seems appropriate to examine next the work of the man who, in all of American Catholic history, best exemplifies the Catholic tradition of moral reasoning on the great issues of public life, John Courtney Murray.

CHAPTER 4

The John Courtney Murray Project

John Courtney Murray, S.J., the preeminent American Catholic theologian of the politics of peace and freedom, was not a publicly-notorious figure in his lifetime. The basic outline of Murray's biography is strikingly simple: born in New York City on September 12, 1904, of Catholic parents; boyhood in Queens; joined the Jesuits in 1920; educated at Weston College (B.A., 1926) and Boston College (M.A., 1927); regency in the Philippines, teaching Latin and English; preliminary theological studies at Woodstock College in Maryland, culminating in priestly ordination in 1933; graduate studies in theology at the Pontifical Gregorian University in Rome, with doctorate in 1937; professor of theology at Woodstock until his death in 1967. During his teaching career, Murray served as editor of *Theological Studies*, and worked as the religion editor of *America* magazine in the mid 1940s.

As a friend once reminisced:

> In his most influential days, Father John Courtney Murray, SJ, was by most popular yardsticks a relatively anonymous person. His writing appeared for the most part in esoteric or specialized journals of limited circulation. His campus presence, except for a brief period as a visiting professor of philosophy at Yale, was confined pretty much to a seminary in rural Maryland. He moved in the councils of the mighty—the intellectually mighty—but not conspicuously so far as publicity and renown were concerned. Even at Vatican Council II, an event which he helped give its decisive shape, John Courtney Murray moved behind the scenes, a non-voting figure shunning the spotlight of public debate and the platform of propagandists. . . . As an actor on the stage of history, John Courtney Murray needed no periodic excursions into the limelight, no plaudits, no curtain calls. The performance was what counted; more particularly, the results.[1]

The homily preached at Murray's funeral by a longtime Woodstock colleague, Fr. Walter Burghardt, S.J., puts flesh onto the bones of Murray's curriculum vitae, and suggests the scope of his accomplishment:

> How do you bring to life a man who taught with distinction in the Ivy League and on the banks of the Patapsco; who served country and Church in Washington and Rome; who graced the platform of so many American campuses and was honored with degrees by nineteen; who researched theology and law, philosophy and war; who was consulted "from the top" on the humanities and national defense, on Christian unity and the new atheism, on democratic institutions and social justice; whose name is synonymous with Catholic intellectualism and the freedom of man; whose mind could soar to outer space without leaving our shabby earth; whose life was a living symbol of faith, of hope, of love?

How does one recapture John Courtney Murray? . . .

I remember a mind. Few men have wedded such broad knowledge with such deep insight. Few scholars can rival Father Murray's possession of a total tradition and his ability to tune it in on the contemporary experience. For whether immersed in Trinitarian theology or the rights of man, he reflected the concerns of one of his heroes, the first remarkable Christian thinker, the third-century Origen. He realized with a rare perceptiveness that for a man to grow into an intelligent Christianity, intelligence itself must grow in him. And so his intellectual life reproduced the four stages he found in Origen.

First, recognition of the rights of reason, awareness of the thrilling fact that the Word did not become flesh to destroy what was human but to perfect it. Second, the acquisition of knowledge, a sweeping vast knowledge, the sheer materials for his contemplation, for his ultimate vision of the real. Third, the indispensable task that is Christian criticism: to confront the old with the new, to link the highest flights of reason to God's self-disclosure, to communicate the insight of Clement of Alexandria that Father Murray loved so dearly: "There is but one river of truth, but many streams fall into it on this side and that." And fourth, an intelligent love: love of truth wherever it is to be found, and a burning yearning to include all the scattered fragments of discovered truth under the one God and His Christ.

The results . . . were quite astonishing. Not in an ivory tower, but in the blood and bone of human living. Unborn millions will never know how much their freedom is tied to this man whose pen was a powerful protest, a dramatic march against injustice and inequality, whose research sparked and terminated in the ringing affirmation of an ecumenical Council: "The right to religious freedom has its foundation," not in the Church, not in society or state, not even in objective truth, but "in the very dignity of the human person." Unborn millions will never know how much the civilized dialogue they take for granted between Christian and Christian, between Christian and Jew, between Christian and unbeliever, was made possible by this man whose life was a civilized conversation. . . .

With the mind went the manner. What John Murray said or did, he said or did with "style." I mean, the how was perfectly proportioned to the what. . . .

Each of you has his or her private memory of the Murray manner. How your heart leaped when he smiled at you; how your thoughts took wing when he lectured to you; how good the "little people" felt when *he* spoke to *you*. How natural it all sounded when *he* ordered a "Beefeater martini desperately dry." How uplifted you felt when he left you with, "Courage, Walter! It's far more important than intelligence."

Each of you has his or her memory of the Murray manner. How aloof he seemed, when he was really only shy—terribly shy. How sensitive to your hurt, how careful not to wound—with his paradoxical belief, "A gentleman is never rude, save intentionally." . . . How open he was, to men and ideas, as only "the man who lives with wisdom" can be open. How stubborn and unbending, once the demands of truth or justice or love or freedom were transparent. . . . How delighted he could be with the paradoxes of life—as when the Unitarians honored this professional Trinitarian. . . . And how confident he looked as he predicted that the post-conciliar experience of the Church would parallel the experience of the bishops in council: we will begin with a good deal of uncertainty and confusion, must therefore pass through a period of crisis and tension, but can expect to end with a certain measure of light and joy. . . .

The captivating thing is, the manner was the man. As the mind was the man.

Here was no pose, no sheerly academic exercise. Here was a man. In his professional, academic, intellectual life, he lived the famous paragraph of Aquinas: "There are two ways of desiring knowledge. One way is to desire it as a perfection of oneself; and that is the way philosophers desire it. The other way of desiring knowledge is to desire it not simply as a perfection of oneself, but because through this knowledge the one we love becomes present to us; and this is the ways saints desire it." Through Father Murray's knowledge, the persons he loved, a triune God and a host of men, became present to him. . . .

John Courtney Murray was the embodiment of the Christian humanist, in whom an aristocracy of the mind was wedded to a democracy of love. Whoever we are—Christian or non-Christian, believer or atheist—this tall man has made it difficult for any of us who loved him to ever again be small, to ever again make the world and human persons revolve around our selfish selves. We have been privileged indeed: we have known and loved *the* Christian man, the "man who lives with wisdom."[2]

John Courtney Murray's life-project can be thought of as three concentric circles. The innermost circle, for which Murray is best remembered because of his work at Vatican II, was the issue of religious freedom. Murray sought to develop a theory of religious liberty that cohered with classic Catholic thought on the rights of conscience and on the relationship between Church and state, but extended that tradition in light of contemporary experience and circumstances. The conciliar "Declaration on Religious Freedom" (so aptly styled in its Latin title, *Dignitatis Humanae Personae*) vindicated Murray's pathfinding exploration of this issue.[3]

The middle circle of Murray's project was the issue of American Catholicism and American democratic pluralism. Murray took responsibility for demonstrating the compatibility of Catholic social theory with the American experiment. Here we may locate Murray's controversy with the "new nativism" of Paul Blanshard, and his work within the nascent Catholic/Protestant ecumenical dialogue. Murray's ecumenism was endorsed by Vatican II; and his forthright defense of Catholicism in American society and politics helped prepare the ground for the election of John Fitzgerald Kennedy as president of the United States in 1960.

The third and most comprehensive circle of the Murray Project reversed the roles of defendant and judge in the second circle. Rather than defending Catholicism at the bar of American democratic theory, Murray asked a larger, more fundamental question: What were the moral roots of *any* democratic experiment, and especially the American experiment? With Lincoln, Murray believed that America was a "proposition" to be tested, not a given to be assumed. The testing of that proposition was more problematic for modern Americans because the "self-evidence" of the truths on which the experiment was based had ceased to be apparent. Murray's thought on the moral problem of war and peace should be located within this most comprehensive circle of the Murray Project. War and peace involved, Murray knew, basic questions of national identity and purpose: How were we to defend (and promote) the values and understandings that had given birth to the American experiment?

Three great issues thus shaped the "civilized conversation" of Murray's life and work. Murray's thought on religious freedom was rooted in Thomistic constitutionalism and in Murray's own hermeneutic of the church/state teaching of Leo XIII; but it was decisively influenced by his experience as an *American* Catholic. Here, religious freedom was not an abstract theorem to be scholastically defended or attacked, but a lived experience to be appropriated and understood.

Religious freedom was also deeply enmeshed with the problem of the new anti-Catholic nativism, a nativism perhaps more insidious because of its secularist roots. How could the American theory and practice of religious freedom be maintained if the experience of religious freedom was foreshortened for a major religious community in America?

Finally, there was the question of building the moral foundation of any democratic experiment. Could an experiment "so conceived and so dedicated" long survive if its public discourse was systematically stripped of those religiously based values and understandings that had contributed to its Founding? How could those values and understandings help form a sustaining public consensus on the bases of the American experiment, across denominational divisions, and across the divide between believers and unbelievers? Murray was, rightly, dubious about the prospects for democracy's survival absent such a foundational consensus. The collapse of reasonable debate about the great issues of war and peace exemplified the deterioration of public discourse that would necessarily result from a lack of agreement on moral coordinates for the democratic experiment. There was little real argument here, Murray feared, only cacophony. Various polemicists shouted past each other, ignorant of the source of their disagreement.

Murray's great project was left incomplete at his untimely death in 1967. The "civilized conversation" he wished to inaugurate within the Church, and between the Church and American society, became increasingly uncivilized within half a decade of Walter Burghardt's funeral homily. Murray's death, and a new incivility in Catholic and public argument, were related. Understanding how they were related requires a detailed review of the Murray Project itself, both its great accomplishments and its weaknesses.

THE MURRAY PROJECT: A NEW HERMENEUTIC

A generation after his death, John Courtney Murray stands as the great theological synthesist of the American Catholic experience: the man who most creatively extended American Catholicism's development of the tradition of "moderate realism" in its moral assessment of the American experiment, and in its understanding of how that experiment bore on the contemporary problem of peace with freedom.

The popular hermeneutic of the Murray Project holds that Murray made it possible for American Catholics to participate fully in our national political life by "explaining" Catholicism to Protestant America in a palatable way.[4] Here, the "inner circle" of the Murray Project, the issue of religious freedom, touched the second circle, the defense of American Catholicism against traditional nativism and the new nativism of Paul Blanshard.

This popular view of Murray's importance is true as far as it goes; but it unnecessarily limits the scope of Murray's intention and accomplishment. So, too, does the suggestion that Murray's work on religious freedom, as well as his efforts to meet the challenge of nativisms old and new, dissolved the distinctiveness of Catholicism under the pressures of the American religion of "civility";[5] here, Murray is reduced to being the codifier of Catholic acculturation.[6]

This sociological diminution of Murray has been reinforced in critiques mounted by several key figures in the post-Murray generation of Catholic social ethicists who find Murray theologically deficient in his fondness for the American democratic experiment. David Hollenbach, S.J., for example, thinks that post-Murray American Catholic theologians should undertake "a critical re-evaluation of the American civil liberties tradition which Murray brought into creative contact with the Catholic social tradition."[7] Hollenbach admits that the establishment of moral norms for underpinning the American democratic experiment was the task that truly engaged Murray's mind, and that the religious liberty issue must be understood in that context.[8] But Hollenbach also argues that "one can no longer suppose, as Murray did, that the American public philosophy is rooted in and supported by the broad theological tradition of Christian history . . . [for] the missing element in the public ethos of America is the sense of the sacred in history and in society and human interaction."[9]

Whereas Murray wanted to reestablish a "civilized conversation" within America on the basis of natural law philosophy, Hollenbach wants to take up the same task of reconstruction by means of the explicit use of religious categories and biblical images in a new "public theology"[10] that would not be "misunderstood by the left as a defense of the American *status quo* and as insensitive to the realities of oppression and systemic exploitation which are such central concerns of Latin American liberation theology." Nor would it be "co-opted by the right as a legitimation for an uncritical merging of Christian faith with the American civil liberties tradition."[11] Hollenbach does not suggest that such concerns should have exercised Murray in the 1950s. He implies, however, that the large-scale Murray Project was basically flawed because of its uncritical and specifically Catholic affirmation of the basic character of the American democratic experiment: "questions have been raised about Murray's optimistic judgment on the fundamental compatibility of the Christian vision of a just society with the accepted principles which govern moral discourse about social policy in the United States."[12]

John Coleman, S.J., has similar worries. He, too, wishes to reclaim a public role for biblical imagery in a North American version of liberation theology, for the lack of such imagery in John Courtney Murray "skews [his] writings on public issues too strongly in the direction of liberal individualism, despite his own intentions."[13] This led to three weaknesses in Murray's grand project:

> [a] bias toward liberty at the expense of justice in the American public-philosophy tradition and [a] concomitant individualistic tone . . . a failure to admit that his own [i.e., Murray's] theory of natural law rests on particularistic Catholic theological principles and theories which do not command widespread allegiance . . . [and an inability] to evoke the rich, polyvalent power of religious symbolism, a power which can command commitments of emotional depth.[14]

Although several of the conclusions of these post-Murray commentators seem unwarranted, their understanding of the breadth of the Murray Project is more accurate than either the popular or sociological reductions of Murray as the man who made Catholics "acceptable" to American Protestants and religious liberty "acceptable" to the Vatican. Certainly one main intention of Murray's lifework was to challenge, on the basis of a genuine development of doctrine evolving from reflection on the American Catholic experience, the traditional "thesis" that a confessional state was most congruent with Catholic social teaching (in light of which the American paradigm of religious pluralism in a confessionally neutral state was only a tolerable "hypothesis"). But Murray's essential purpose was never to make Catholicism "acceptable" within American political culture according to the criteria of that culture.

Murray's ambitions were considerably larger: to provide the moral-theoretical underpinnings for a liberal-democratic experiment in America. Murray's grand project was to bring into explicit philosophical and theological understanding American Catholicism's lived experience of *tranquillitas ordinis*, and to do so in a way that would redound to the common good of the entire experiment. This would not be a simple christening of Americanism; it would be a *critical* exercise in both the popular and technical meanings of the term.

AMERICA AS PROPOSITION

John Courtney Murray's answer to Crevecoeur's classic question, "What, then, is the American, this new man?" was Lincolnian: the American is the bearer of a proposition.[15] The American proposition had both theological and practical dimensions. It had to be defended intellectually, and worked out historically in patterns of governance. The Founders and Framers "signally succeeded" in both asserting and giving structure to the American proposition.[16] But there was nothing in the Founders' and Framers' success that carried any guarantee for the present; America is *in principle* never finished. The American proposition is tested anew in each generation of Americans. All Americans, whatever the accident of their date of birth, are Founders; all are Framers:

> Neither as a doctrine nor as a project is the American Proposition a finished thing. Its demonstration is never done once for all; and the Proposition itself requires development on penalty of decadence. Its historical success is never to be taken for granted, nor can it come to some absolute term; and any given measure of success demands enlargement on penalty of instant decline. In a moment of national crisis Lincoln asserted the imperilled part of the theorem and gave impetus to the impeded part of the project in the noble utterance, at once declaratory and imperative: "All men are created equal." Today, when civil war has become the basic fact of world society, there is no element of the theorem that is not menaced by active negation, and no thrust of the project that does not meet powerful opposition. Today, therefore, thoughtful men among us are saying that America must be more clearly conscious of what it proposes, more articulate in proposing, more purposeful in the realization of the project proposed.[17]

The American proposition was not to be confused with political mechanics. Against those who proposed the Declaration of Independence, and particularly the Constitution, as utilitarian exercises in the management of interests, Murray saw the Founders and Framers, then and now, as engaged with matters of truth and value:[18]

> The epistemology of the American Proposition was made clear by the Declaration of Independence in the famous phrase: "We hold these truths to be self-evident . . ."

> Today, when the serene, and often naive, certainties of the eighteenth century have crumbled, the self-evidence of the truths may legitimately be questioned. What ought not to be questioned, however, is that the American Proposition rests on the forthright assertion of a realist epistemology. The sense of the famous phrase is simply this: "There are truths, and we hold them, and we here lay them down as the basis and inspiration of the American project, this constitutional commonwealth."[19]

Because it rested on an assertion of truth, the American proposition was not based on philosophical pragmatism, although it utilized practical reason to construct institutions of governance. But unless the truth of the proposition was understood to be more than utilitarian, there could be "no hope of founding a true City, in which men may dwell in dignity, peace, unity, justice, well-being, freedom."[20] The aim of the American proposition, then, was nothing less that the building of a true City, *tranquillitas ordinis*, guided by the conviction that a politics of virtue was possible, even for fallen human beings in that most temptation-filled of enterprises, the exercise of power. However it had been traduced by pragmatists and utilitarians, degraded by the ambitious, misunderstood by secularized elites, or reduced to offensive slogans by crude popularizers, the American proposition was founded on a notion of truth that had to be reclaimed by each successive generation of Founders and Framers.

And therein lay the contemporary problematic: the self-evidence of those truths was no longer self-evident. The very notion of a publicly knowable truth, one that was in principle available to all and could thus form the basis of a constitutional commonwealth, was under systematic assault. Murray wrote:

> What is at stake [no less truly of his time than ours], is America's understanding of itself. Self-understanding is the necessary condition of a sense of self-identity and self-confidence, whether in the case of an individual or in the case of a people. If the American people can no longer base this sense on naive assumptions of self-evidence, it is imperative that they find other more reasoned grounds for their essential affirmation that they are uniquely a people, a free society. Otherwise the peril is great. The complete loss of one's identity is, with all propriety of theological definition, hell. In diminished forms it is insanity. And it would not be well for the American giant to go lumbering about the world today, lost and mad.[21]

When American self-understanding was not only a matter effecting our own lives, but had profound effects on the unfolding of the human story, the truths to be asserted within the American proposition needed reclamation and restatement.

Murray's primary concern was not legislative solutions to particular problems of public policy; the issue was more basic. It was no less than American society's "calendar of values,"[22] what Murray termed the country's "constitutional consensus, whereby the people acquires its identity as a people and the society is endowed with its vital form, its entelechy, its sense of purpose as a collectivity organized for action in history."[23] This "consensus" was not registered in electoral exit polls; nor was it a matter of psychological rationalization or economic interests. It could not be measured by pragmatic criteria alone.[24] Rather:

> [The consensus was] an ensemble of substantive truths, a structure of basic knowledge, *an order of elementary affirmations* that reflect realities inherent in the order of existence. . . . This consensus is the intuitional *a priori* of all the rationalizations and technicalities of constitutional and statutory law. It furnishes the premises of the

> people's action in history and defines the larger aims which that action seeks in internal affairs and in external relations.[25]

The consensus was not a matter of argument over issues; it was the very condition for the possibility of argument. Lacking consensus, there was no argument, only barbarous discord. Agreement on society's elementary affirmations, its fundamental calendar of values, was the precondition to the civil discourse that marked a genuine *polis*, and was the true City's most distinctive characteristic.[26]

Why do we hold to "these truths"? Initially, Murray wrote, "because they are a patrimony." We have received them from history, "through whose dark and bloody pages there runs like a silver thread the tradition of civility." Because the consensus is an "intellectual heritage," which can be forgotten or distorted, it must be kept alive by "high argument," conducted civilly and in public.[27] But the consensus is more than a patriotic heritage, which could be susceptible to sentimentality: "We hold these truths *because they are true*. They have been found in the structure of reality by that dialectic of observation and reflection which is called philosophy." But here, too, the consensus could never be taken for granted: its vitality must be continually reexamined in a dialectical reflection that mediates between first principles and political experience. Examination against experience protects the consensus from being simply a "structure of prejudice"; examination against principle is a safeguard against utilitarianism. Public argument on specific issues thus had to be related to first principles, and vice versa.[28]

This never-ending, always-expanding process of public argument was a particular problem because of the pluralist character of American society. American pluralism was unique because it was the country's original condition.[29] This pluralism, today as in 1776 and 1787, and continued to provoke public argument over society's calendar of values. The old antagonisms—Christians against Jews, Protestants against Catholics—contributed to the contemporary debate. But in Murray's judgment, these had been overshadowed by the emergence of those secularists who questioned the very idea of a principled experiment, if the principles concerned had any connection with religious conviction. Amid the many arguments of American public life, one thing seemed clear to John Courtney Murray: the time for reclamation of the American proposition was pressingly at hand. For "the fact is that among us civility—or civic unity or civic amity, as you will—is a thing of the surface. It is quite easy to break through it. And when you do, you catch a glimpse of the factual reality of the pluralist society . . . [which is] honestly viewed under abdication of all false gentility . . . a pattern of interacting conspiracies."[30]

Murray took the word "conspiracy" in its classical sense of "unison, concord, unanimity in opinion and feeling, a 'breathing together.'"[31] In this sense, American society was a plurality of conspiracies, a perilous condition for a *polis* based on ideas and values. Perhaps, then, the contemporary problem was this:

> [To make the] great conspiracies among us conspire into one conspiracy that will be American society—civil, just, free, peaceful, one. . . . We cannot hope to make American society the perfect conspiracy based on a unanimous consensus. But we could at least do two things. We could limit the warfare, and we could enlarge the dialogue. We could lay down our arms (at least the more barbarous kind of arms!), and we could take up argument. . . . [Then] amid the pluralism a unity would be

> discernible—the unity of an orderly conversation. . . . Thus we might present to a "candid world" the spectacle of a civil society.[32]

The Foundations of the Experiment

Reconstructing an "order of elementary affirmations" that would set the ground for reasoned public argument required an understanding of "the essential contents of the American consensus, whereby we are made *e pluribus unum*, one society subsisting amid multiple pluralisms."[33] Murray argued that one claim within the American proposition was absolutely fundamental: that the American experiment was "a nation under God," which meant under judgment, not merely as individuals, but as a community. The experiment itself, as experiment, was under transcendent judgment, and was unanswerable to transcendent norms:

> The first truth to which the American Proposition makes appeal is stated in that landmark of Western political theory, the Declaration of Independence. It is a truth that lies beyond politics; it imparts to politics a fundamental human meaning. I mean the sovereignty of God over nations as well as over individual men. This is the principle that radically distinguishes the conservative Christian tradition of America from the Jacobin laicist tradition of Continental Europe. The Jacobin tradition proclaimed the autonomous reason of man to be the first and the sole principle of political organization. In contrast, the first principle of the American political faith is that the political community, as a form of free and ordered human life, looks to the sovereignty of God as to the first principle of its organization. . . . The affirmation in Lincoln's famous phrase, "this nation under God," sets the American proposition in fundamental continuity with the central political tradition of the West. . . . By reason of this fact, the American Revolution was "less a revolution than a conservation." It conserved, by giving newly vital form to, the liberal tradition of politics, whose ruin in Continental Europe was about to be consummated by the first great modern essay in totalitarianism.[34]

The American experiment was under transcendent judgment; from this foundational agreement, a consensus with four special features had emerged.

First:

> [The] consensus was political, that is, it embraced a whole constellation of principles bearing upon the origin and nature of society, the function of the state as the legal order of society, and the scope and limitations of government. "Free government"—perhaps this typically American shorthand phrase sums up the consensus. "A free people under a limited government" puts the matter more exactly. It is a phrase that would have satisfied the first Whig, St. Thomas Aquinas.[35]

The first building block of the American consensus was a fundamental distinction, with roots in medieval Catholic thought, between society and the state. Society, the natural human habitat, exists prior to the state. The state is at the service of society, not vice versa. This implied a limited function for the state, and meant that government was "not simply the power to coerce, though this power was taken as integral to government." Rather, "government, properly speaking, was the right to command. It was authority. And its authority derived from law. By the same token its authority was limited by law."[36] Thomistic constitutionalism came to contemporary flower in the first building block of the American experiment: "Constitutionalism, the rule of law, the notion of sovereignty as purely political and therefore limited by law, the concept of government as an empire of laws and not of men—these were ancient ideas, deeply implanted in the British tradition at its origin in medieval times."[37]

The American experiment had extended this ancient tradition by devising the written constitution. The American Constitution was not the benevolent grant of a monarch. The distinctively American devices of a constitutional convention followed by popular ratification meant that the American Constitution was an act of the people: "By the Constitution the people define the areas where authority is legitimate and the areas where liberty is lawful. The Constitution is therefore at once a charter of freedom and a plan for political order."[38]

Here, then, was the second building block of the American consensus, the principle of the consent of the governed. Murray found medieval roots for this principle, too:

> The American consensus reaffirmed this principle, at the same time that it carried the principle to newly logical lengths. Americans agreed that they would consent to none other than their own legislation, as framed by their representatives, who would be responsible to them. In other words, the principle of consent was wed to the equally ancient principle of popular participation in rule. But, since this latter principle was given an amplitude of meaning never before known in history, the result was a new synthesis, whose formula is the phrase of Lincoln, "government by the people."[39]

The result was that Americans had "limited" government in a new way. Government was limited by the people's will, expressed not only in the Constitution, but in the statutory law written by freely elected legislators and in the frequent changing of administrations. "The people are governed because they consent to be governed; and they consent to be governed because in a true sense they govern themselves."[40]

This was, to be sure, "a great act of faith in the capacity of the people." It was not an act of faith in each citizen's expertise in public policy; it was not an act of faith that the consensus that sustained popular governance would (or should) ever go far beyond first principles; it was surely not an act of faith in the perfectibility of human beings. But it was an act of faith in "the premise of medieval society, that there is a sense of justice inherent in the people, in virtue of which they are empowered, as the medieval phrase had it, to 'judge, direct, and correct' the processes of government."[41]

This medieval notion of a sense of justice "inherent in the people" was another expression of that Thomistic insistence that society was a "natural" phenomenon. Against the grosser claims of individualism—and against the interpretation of the American experiment as exclusively Lockean, now oft repeated by liberation theologians critical of the very premises of the experiment—Murray understood that the American experiment was about an organic community, and that this was the basis on which early American freedoms, such as the institutions of free speech and a free press, grew. Murray argued:

> In the American concept of them, these institutions do not rest on the thin theory proper to eighteenth-century individualistic rationalism, that a man has a right to say what he thinks merely because he thinks it. . . . The proper premise of these freedoms lay in the fact that they were social necessities. . . . They were regarded as conditions essential to the conduct of free, representative, and responsible government. People who are called upon to obey have the right first to be heard. People who are to bear

> burdens and make sacrifices have the right first to pronounce on the purposes which their sacrifices serve. People who are summoned to contribute to the common good have the right first to pass their own judgment on the question, whether the good proposed be truly good, the people's good, the common good.[42]

The third element in the foundational consensus of the American experiment also hearkened back to classic and medieval political theory: the "profound conviction that only a virtuous people can be free." Americans did not assert that free government was inevitable, only that it was possible. Its possibility rested on the people's agreement with Acton's postulate, that freedom is "not the power of doing what we like, but the right of being able to do what we ought."[43] Freedom and virtue were inseparable. The American experiment in democracy could not be construed merely as a matter of political mechanics, a comfortable compromise across the ebb and flow of narrow interests. The Marxist explanation, that "bourgeois democracy" was simply an epiphenomenal expression of economic forces, was equally deficient.

Here, the religious affirmation that all human communities are under judgment met the particular American affirmation that this democratic experiment is answerable to norms that transcend it. Here, we come to understand that democracy is not just an experiment in politics, but an exercise in human spirituality and morality. A successful democracy required a virtuous people; "men who would be politically free must discipline themselves."[44]

The American proposition would stand or fall, Murray believed, not on the measure of its gross national product, important as that might be, but on the measure of its civic virtue: "Institutions which would pretend to be free with a human freedom must in their workings be governed from within and made to serve the ends of virtue."[45] The American experiment was, in sum, an experiment in *tranquillitas ordinis*, as Catholic theory had come to understand the full meaning of that term.

The fourth constitutive element of the American consensus was the claim that human rights were the rights of human beings *as human beings*, prior to, and not dependent upon, their condition as citizens:

> [Incarnated in the Bill of Rights, this assertion] was also tributary to the tradition of natural law, to the idea that man has certain original responsibilities precisely as man, antecedent to his status as citizen. These responsibilities are creative of rights which inhere in man antecedent to any act of government; therefore they are not granted by government and they cannot be surrendered to government. . . . Their proximate source is in nature, and in history insofar as history bears witness to the nature of man; their ultimate source, as the Declaration of Independence states, is in God, the Creator of nature and the Master of history.[46]

The Bill of Rights was fundamentally a matter of virtue, of the common moral apprehension of the people. It was effective, not because it was written down, but because it was lived: because "the rights it proclaims had already been engraved by history on the conscience of a people." Lockean theory and distinctively American religious leaders like Roger Williams had helped shape this conscience; but its roots also reached back to the Christian medieval theory of man and society, and to Thomistic constitutionalism. The Bill of Rights was surely influenced by the Enlightenment; but it was also "the product of Christian history," behind which one could discern "the older philosophy that had been the matrix of the common law. The 'man' whose rights are guaranteed in the face of law and government is, whether he knows it or not, the Christian man, who had learned to know his own personal dignity in the school of Christian faith."[47]

Did this fourfold consensus, these elementary affirmations of the American proposition, still seize the imaginations of Americans? Murray was skeptical. Some legal scholars and practitioners showed an interest in the "revival of this great tradition," and in the theory of natural law. But "one would not talk of reviving the tradition, if it in fact were vigorously alive." The American people might still hold to it; Murray found "cheerful" Clinton Rossiter's claim that "the people, who occasionally prove themselves wiser than their philosophers, will go on thinking about the political community in terms of inalienable rights, popular sovereignty, consent, constitutionalism, separation of powers, morality, and limited government."[48] But if this were the case, it was with no thanks to America's intellectual elites. Murray mused:

> Perhaps the American people have not taken the advice of their advanced philosophers. Perhaps they are wiser than their philosophers. Perhaps they still refuse to think of politics and law as their philosophers think—in purely positivist and pragmatic terms. The fact remains that this is the way the philosophers think. . . . The American university long since bade a quiet goodbye to the whole notion of an American consensus, as implying that there are truths that we hold in common, and a natural law that makes known to all of us the structure of the moral universe in such wise that all of us are bound by it in a common obedience.[49]

Still, if the elementary affirmations of the American proposition and experiment—derived from and related to public agreement that a discernible trajectory in the human experience placed moral demands on each of us—had quietly expired in the elite universities, Murray saw one constituent component of the American political community where the original American consensus still endured: where America as a project of civic virtue still held fast against America as a mechanical arrangement for managing competing claims and interests. That was in the American Catholic community.

The Catholic Moment and the Reconstruction of a Public Philosophy

Thus did Murray, who must have savored the delicious irony, turn the tables on nativists old and new. Where, amid the complex pluralism of American society, would one be likely to find the bearers of the originating tradition? Who had now assumed (if unintentionally, and almost certainly unknowingly) the basic tasks of culture formation, the forging of that "order of elementary affirmations" on which everything else stood? None other than the despised Catholics. Some 180 years after John Adams wondered "how Luther ever broke the spell," it was the allegedly bewitched Catholics whose moment it was, in the great task of reclaiming and reconstructing the foundational consensus of the American proposition.

Why had this come about? Murray paused briefly over the "paradox . . . that a nation which has (rightly or wrongly) thought of its own genius in Protestant terms should have owed its origins and the stability of its political structure to a tradition whose genius is alien to current intellectualized versions of the Protestant religion, and even to certain individualistic exigencies of Protestant religiosity."[50] Murray insisted:

> Catholic participation in the American consensus has been full and free, unreserved and unembarrassed, because the contents of this consensus—the ethical and political principles drawn from the tradition of natural law—approve themselves to the Catholic intelligence and conscience. Where this kind of language is talked, the Catholic joins the conversation with complete ease. It is his language. The ideas expressed are native to his universe of discourse. Even the accent, being American, suits his tongue.[51]

This "Catholic moment" was hardly a time for sectarian self-congratulation, though; the issues at stake were too grave. The consensus was in such disarray that its final dissolution was a real possibility. One day, Murray worried, "the noble many-storeyed mansion of democracy [may] be dismantled, levelled to the dimensions of a flat majoritarianism, which is no mansion but a barn, perhaps even a tool shed in which the weapons of tyranny may be forged." Should it happen, it would be because of public amnesia over the true roots of the experiment, and its native condition of self-consciously standing under transcendent judgment. Murray argued:

> If that evil day should come, the results would introduce one more paradox into history. The Catholic community would still be speaking in the ethical and political idiom familiar to them as it was to their fathers, both the Fathers of the Church and the Fathers of the American Republic. The guardianship of the original American consensus, based on the Western heritage, would have passed to the Catholic community. . . . And it would be for others, not Catholics, to ask themselves whether they still shared the consensus which first fashioned the American people into a body politic and determined the structure of its fundamental law.[52]

The paradoxes would be greater than Murray had imagined, as things worked out. Many Catholic intellectuals in the post-Murray generation would set themselves the task of delegitimating the American proposition—although many understood themselves to be constructing a prophetic critique of U.S. foreign policy—with an enthusiasm, indeed a passion, that far outstripped the efforts made by any department of philosophy in any American university at the time Murray wrote *We Hold These Truths*. The delegitimators would not win, precisely. But they would help create a climate in which Murray's call that American Catholics assume the task of culture-formation, of reclaiming and restating the "order of elementary affirmations" undergirding the experiment, could not be heard, much less acted upon, because it seemed to involve an ignoble acquiescence to the "American status quo." Understandings of the Church's distinctive civic role shifted dramatically in the years immediately after Murray's death: the primary task was not culture-formation, but judgment. The tone of interaction would not meet Murray's standard for the climate of the true City: "cool and dry, with the coolness and dryness that characterize good argument among informed and responsible men." Rather, the general temper would be what Murray once called "hot and humid, like the climate of the animal kingdom."[53]

This abandonment of a culture-forming mission would have sorely disappointed Murray, whose sense of the need for a reconstituted American public philosophy was as acute as his sense of the breakdown of the prior consensus. This need expressed itself not merely in the order of ideas, as a matter of intellectual aesthetics, but just as pressingly in the order of public affairs. America was not doing well at a critical moment in human history; this was the fact from which a new public philosophy should begin. There had been other times of crisis—the Civil War, Pearl Harbor. But goals were clearer then: capture Richmond, or Tokyo, or Berlin. There were no such symbolic reference points for "victory" today, because the contemporary crisis of American insecurity had not so much to do with a physical enemy as with the fundamental matter of national purpose:

> So baffling has the problem of our national purpose become that it is now the fashion to say that our purpose is simply "survival." The statement, I think, indicates the depth of our political bankruptcy. This is not a purpose worthy of the world's most powerful nation. It utterly fails to measure the meaning of the historical moment or to estimate the opportunities for greatness inherent in the moment. Worst of all, if we pursue only the small-souled purpose of survival, we shall not even achieve survival.[54]

The construction of a public philosophy faithful to the originating traditions of the American experiment and capable of illuminating the American task in the contemporary world was thus an issue of the first importance. Given the inescapable fact of American pluralism, such a public philosophy had to be constructed in language and modes of thought that could reach across the various "conspiracies" of American life, and appeal to common understandings and values. For this task, only a natural law theory would suffice, Murray argued.

Murray was eager to strip from the notion of "natural law" the cobwebs that often smothered it: not only the refusal of modern academic philosophy to take natural law theory seriously, but what a later generation of Catholic scholars would call "decadent scholasticism"—a natural law theory devoid of a sense of human historicity. Natural law theory, for Murray, was not a matter of "Roman Catholic presuppositions." Rather, the "only presupposition" of natural law "is threefold: that man is intelligent; that reality is intelligible; and that reality, grasped by intelligence, imposes on the will the obligation that it be obeyed in its demands for action or abstention." And even these basic postulates "are not properly 'presuppositions,' since they are susceptible of verification."[55]

Nor was natural law theory an exercise in bloodless abstraction. Moral imagination and intelligence were built into the structure of human knowing, "as reason itself emerges from the darkness of infant animalism"; but moral imagination and intelligence develop through experience. First, "after some elementary experience of the basic situations of human life, and upon some simple reflection on the meaning of terms, intelligence can grasp the meaning of 'good' and 'evil' in these situations and therefore know what is to be done or avoided in them."[56]

As the experiences of life unfold, as human relationships are developed and reflected upon, further principles of right behavior can be apprehended, and known to be matters of obligation. There are no guarantees; "man is not an animal, ruled by unerring instinct," who will always choose the good and avoid what is evil. Murray was fully aware of the abiding effects of sin on the human condition. Natural law did not suppose that human beings were angels, only that they were possessed of intelligence, and inhabited a universe that was not, at bottom, absurd, but intelligible. But natural law theory also held out the prospect of human advancement, of a growth in moral knowledge that would approach wisdom. Natural law involved "particular principles which represent the requirements of rational human nature in more complex human relationships and amid the institutional developments that accompany the progress of civilization."[57] Prudent judgment, and not simply knowledge of facts and principles, is required here. "Little reflection on experience is needed to know the principle of justice, *suum cuique* ("to each what is his")," Murray would write. "But an extensive scientific analysis of the functioning of economic cooperation is needed to know what a just settlement of a wage-dispute might be."[58]

Natural law reasoning was not, then, a simple matter of comprehending right principles and applying them mechanically to each and every situation of life. History had its claims in the calculus of natural law reasoning. Creativity altered the human condition; the future was open, not closed. Murray was a strict Thomist here: "'The nature of man is susceptible of change,' St. Thomas repeatedly states. History continually changes the community of mankind and alters the modes of communication between man and man, as these take form through external acts." In this sense, human nature changes in history, for better or for worse, "at the same time that the fundamental structure of human nature, and the essential destinies of the human person, remain untouched and intact. As all this happens, continually new problems are being put to the wisdom of the wise; at the same time, the same old problems are being put to every man, wise or not."[59]

Natural law theory did not promise more than it could deliver: "It does not show the individual the way to sainthood, but only to manhood."[60] Natural law methods of moral reasoning do not directly promote personal religious conversion. But natural law reasoning suggests:

> [It is a] mistake . . . to imagine that the invitation, "Come, follow me," is a summons somehow to forsake the universe of human nature, somehow to vault above it, somehow to leave law and obligation behind, somehow to enter the half-world of an individualist subjectivist "freedom" which pretends to know no other norm save "love" . . . [for] the law of nature, which prescribes humanity, still exists at the interior of the Gospel invitation, which summons to perfection. What the follower of Christ chooses to perfect is, and can only be, a humanity. And the lines of human perfection are already laid down in the structure of man's nature. Where else could they be found?[61]

Such a method of natural law reasoning was essential today, Murray argued, not only because it is true, but because of the structure of the problem of human liberty in the world. Contemporary human beings yearned for liberty in more than a formal sense; they sought "liberty with a positive content within an order of liberty of rational design. Rousseau's 'man everywhere in chains' is still too largely a fact. Our problem is still that of human freedom, or in juridical terms, human rights. It is a problem of the definition of freedom, and then, more importantly, its institutionalization."[62]

Natural law theory was the only way to address the problem of freedom in an age in which freedom was no longer a utopian ideal, but a matter of habits, associations, and institutions.[63] The problem of reclaiming the American proposition, and the problem of human freedom in a world too often in chains, were similar in structure. Both required a reclamation and extension of natural law principles if the foundations of their solution were to be properly laid.

Catholicism, Murray believed, could offer important insights to an effort to reclaim and extend the American proposition. Against an "eschatological humanism," which counseled withdrawal from the task of culture-formation, Murray offered Catholic "incarnational humanism." Incarnational humanism asserts that "the end of man . . . is indeed transcendent, supernatural; but it is an end of *man* and in its achievement man truly finds the perfection of his nature. Grace perfects nature, does not destroy it. . . . The heavens and the earth are not destined for an eternal dust-heap, but for a transformation. There will be a new heaven and a new earth; and those who knew them once will recognize them, for all their newness."[64]

This perspective on the human condition and prospect had profound implications for political life here and now. Because human historicity was not set over against transcendence, but was the matter in which the transcendent could be apprehended, known, acted on, and, ultimately, loved, human beings could be freed from "a Greek bondage to history and its eternal cyclic returns."[65] Catholic incarnational humanism taught the lessons of innate human dignity, the true meaning of equality "under God," and a non-sentimental understanding of the unity of the human race.[66] These lessons were foundations for the creation of a true City, in the peace of rightly ordered, dynamic political community. This was not a utopian delusion; one proof of that was the American experiment, understood in terms of its originating proposition and the elementary affirmations that gave that proposition content in history.[67]

Incarnational humanism was the best answer to what Romano Guardini called "the interior disloyalty of modern times": a rejection, not merely of a particular state, or even of religious faith, but of the very structure of human being in the world. This "interior disloyalty" of modernity had reached its dreadfully logical expression in communism, which was "political modernity carried to its logical conclusion,"[68] the contemporary expression of that Jacobin/laicist revolutionary impulse against which the American proposition and experiment stood as a "conservation" of the authentic political heritage of the West. The nature of the contemporary world conflict was clarified at this point; here, Murray's comprehensive project had to address more explicitly the American role in world affairs.

THE MORAL PROBLEM OF WAR AND PEACE

The gravamen of Murray's approach to the moral problem of war and peace was summed up in his query to a friend puzzled by the debate over "morality and foreign policy." When his friend asked what foreign policy had to do with the Sermon on the Mount, Murray quickly answered, "What makes you think that morality is identical with the Sermon on the Mount?"[69] Moral reasoning, Murray insisted, was *not* simply a matter of repeating Scriptural injunctions, particularly in so dense a field as foreign policy. Unlike those who saw the problem of "morality and foreign policy" as weighted on the side of "foreign policy" (the assumptions being that all disputants understood what was meant by "morality," and that the only remaining questions involved the application of that "morality" to the policy agenda), Murray thought that the central question was the nature of moral reasoning itself. The basic confusion in the debate over morality and foreign policy lay in inadequate answers to one straight-forward question: "What is morality?"[70]

To answer that question it was necessary to recognize "the shortcomings and falsities of an older American morality that dominated the nineteenth century and still held sway into the twentieth." This older American morality, or *moralism*, had a distinctive style, particular sources, a discernible mood, and a dominant spirit:

> Its style was voluntarist. It sought the constitution of the moral order in the will of God. The good is good because God commands it; the evil is evil because God forbids it. The notion that certain acts are intrinsically evil or good, and therefore forbidden or commanded by God, was rejected. Rejected too was the older intellectualist tradition of ethics and its equation of morality with right reason. Reason is the dupe of interest and passion. . . . In the search for moral principles and solutions reason can have no place. . . .
>
> In its sources the older morality was scriptural in a fundamentalist sense. In order to find the will of God for man it went directly to the Bible. There alone the divine precepts and prohibitions are stated. They are stated in so many words, and the words are to be taken at their immediate face value without further exegetical ado. . . .
>
> In its mood the old morality was subjectivist. Technically it would be called a "morality of intention." It set primary and controlling value on a sincerity of interior motive; what matters is not what you do but why you do it. And it was strong on the point that an act is moral only when its motive is altruistic—concretely, when the motive is love. If any element of self-interest creeps in, the act is corrupt and sinful.
>
> Finally, in its whole spirit the old morality was individualistic. Not only did it reject the idea of a moral authority external to the individual conscience. It also set its single focus on the individual existence and on the moral problems that arise in interpersonal relationships. As for society, it believed in a direct transference of personal values into social life. . . . Its highest assertion was there would be no moral problems in society, if only all men loved their neighbor.[71]

It was against this moral*ism*, which provided "no resources for discriminating moral judgment" amid the complexities of world affairs, that the realist school reacted in the 1940s. Hans Morgenthau, Dean Acheson, and George Kennan were prime exponents of the realist position. They joined with Murray's friend Reinhold Niebuhr in warning that general principles should be held suspect in the conduct of America's business with the world, and that moral principles were the worst sort of general principles because they compounded the dangers of unwarranted universalism with zealotry and utopianism. Muddying the already murky waters of foreign policy with moral principles, the realists argued, led to a naive belief that the flashier forms of diplomacy—conferences, verbal proclamations of peace, and the like—could be substituted for altered facts on the ground.

Niebuhr in particular insisted that diplomats, above all others, must take human beings as they were, not as one might like them to be. Diplomats must recognize, and act on their recognition of, the fact that the peculiar perversity of sin, as it is manifested in politics, is its capacity to mask egotistical drives with universalist pretensions. The realists taught that war was the norm, and peace the welcome deviation, in the human condition. Those who failed to understand this and conduct their policy accordingly would only worsen the misery of an already miserable world.[72]

The realist case against the application of moral norms in the design and conduct of foreign policy must be understood in its historical context. Wilsonian idealism (so typical of the "older morality" sketched by Murray) had been broken on the wheel of twentieth-century tyrannies, the worst the world had ever seen. A morality of intentions, let loose in the League of Nations palace in Geneva, had led to such pieces of fluff as the Kellogg-Briand Pact outlawing war. More cruelly, it had helped create the circumstances in which the West seemed paralyzed in the face of Hitler. Given a tyranny even worse than Hitler's, this time armed with nuclear weapons, the case for a diplomacy based on realpolitik was understandable.

John Courtney Murray understood this point of view; even sympathized with it. But he could not, in the final analysis, accept it, because realism as Morgenthau in particular defined it smacked of the schizophrenic: it split the human universe in two, with the moral order and the order of political reality and reason separated by an unbridgeable chasm. Murray would have been grateful for Charles Frankel's reaction to Morgenthau's claim that "to know with despair that the political act is inevitably evil, and to act nonetheless, is moral courage." This was, Frankel asserted, simply "moral melodramatics."[73] Murray would also have agreed with Frankel that "thinking about morals and thinking about power may interact on each other."[74] The use of power could be guided by moral principles even while the casuistry of their application was refined through their intersection with the realities and possibilities of historical circumstance.

What the realists rejected, in other words, was the moralism that Murray also decried. Moralism set up what Murray called three "pseudo problems" that had to be dissected before a morality based on reason could address itself, through the virtue of prudence, to the tangled web of foreign policy.

The first pseudo problem (and here Murray would part company with Reinhold Niebuhr) was the alleged "gulf between the morality of the individual and collective man." Posed in these terms, the question becomes how to close the gap between "personal" and "social" ethics. When the gap appears uncloseable, one "is driven back upon the simplist category of 'ambiguity,'" or "sadly admits an unresolvable dichotomy between moral man and immoral society."[75]

This pseudo problem did not exist "within the tradition of reason—or, if you will, in the ethic of natural law." In this tradition, society and the state were "natural" institutions, with their own "relatively autonomous" purposes. Inasmuch as these purposes were public, not private, the obligations of society and the state were "strictly limited," and were "not coextensive with the wider and higher range of obligations that rest upon the human person (not to speak of the Christian)." The "ought" in political and social ethics reflected the distinctive ends and obligations of society and the state, as moral reason grappled with the five obligatory ends of politics:

> Justice, freedom, security, the general welfare, and civil unity or peace. . . . [Therefore] the morality proper to the life and action of society and the state is not univocally the morality of personal life, or even of familial life. . . . The effort to bring the organized action of politics and the practical art of statecraft under the control of the Christian values that govern personal and familial life is inherently fallacious. It makes wreckage not only of public policy but also of morality itself.[76]

The second pseudo problem was self-interest. The definition of the national interest was always under moral scrutiny; self-interest, national interest, should not be considered a matter of *raison d'ètat*, a concept based on an unacceptable absolutization of state sovereignty.[77] Rather:

> The tradition of reason requires, with particular stringency today, that national interest, remaining always valid and omnipresent as a *motive*, be given only a relative and proximate status as an *end* of national action. Political action stands always under the imperative to realize, at least in some minimal human measure, the fivefold structure of obligatory political ends. Political action by the nation-state projected in the form of foreign policy today stands with historical clarity (as it always stood with theoretical clarity in the tradition of reason) under the imperative to realize this structure of political ends in the international community, within the limits—narrow but real—of the possible. Today, in fact as in theory, the national interest must be related to this international realization, which stands higher and more ultimate in political value than itself. *The national interest, rightly understood, is successfully achieved only at the interior, as it were, of the growing international order to which the pursuit of national interest can and must contribute.*[78]

The answer to moralism was not realpolitik; it was moral reasoning, properly understood. Once that had been settled, and agreement reached on the moral horizon against which the "national interest" could be judged, the serious work of statecraft could begin.

The third pseudo problem was power. The older moralism was one in which "a cold breath of evil more than faintly emanates from the very words 'power' and 'force.'"[79] Americans seemed to have thought that they alone, among the great nations, would be absolved from grappling with the dilemmas of power. But it was not to be; history determined that the question could not be avoided. Realpolitik, with its schizophrenic view of the relationship between the orders of morality and politics, was of little help in addressing the inescapable issue of power. The tradition of reason, however, could help resolve the resident confusions. For although natural law theory "rejects the cynical dictum of Lenin that 'the state is a club,'" it did not propose that the state be fashioned "in the image of an Eastern-seaboard 'liberal' who at once abhors power and adores it (since by him, emergent from the matrix of American Protestant culture, power is unconsciously regarded as satanic). The traditional ethic starts with the assumption that, as there is no law without force to vindicate it, so there is no politics without power to promote it. All politics is power politics—up to a point."[80]

That crucial point was determined by "multiple criteria." The most basic, standard-setting distinction was between "force" and "violence." "Force is the measure of power necessary and sufficient to uphold the valid purposes both of law and of politics." Anything exceeding this was violence, which destroyed both politics and law. Instrumentally, force was "morally neutral in itself." The essential moral calculus had to do with its "aptitude or ineptitude" for accomplishing the obligatory purposes of public life. Murray knew full well that here, too, the moral calculus was "endlessly difficult, especially when the moralist's refusal to sanction too much force clashes with the soldier's classic reluctance to use too little force. In any case, the theory is clear enough. The same criterion which governs the state in its use of coercive law for the public purposes also governs the state in its use of force, again for the public purposes."[81]

Indeed the calculus would be "endlessly difficult," particularly in a nuclear age.

Still, a reclamation of the tradition of reason not only dissolved the pseudo problems that corroded both moral argument and public policy; it could set the ground on which civil argument about the "endlessly difficult" casuistry could commence. Such a reclamation would address both the potentially paralyzing problem of "ambiguity" and its cousin, the problem of "complexity." Murray would hardly deny that complexity was the hallmark of moral reasoning on issues of war and peace. But complexity need not lead to a paraplegia of undifferentiated anxiety: "It is as if a surgeon in the midst of a gastroenterostomy were to say that the highly complex situation in front of him is so full of paradox ('The patient is at once receiving blood and losing it'), and irony ('Half a stomach will be better than a whole one') and dilemmas ('Not too much, nor too little, anesthesia') that all surgical solutions are necessarily ambiguous."

The answer to this paralysis, Murray believed, was to face our problems frankly in the full measure of their messiness, and then get on with the task of moral reasoning and prudential judgment: "Complicated situations, surgical or moral, are merely complicated. It is for the statesman, as for the surgeon, to master the complications and minister as best he can to the health of the body, politic or physical."[82] The body politic was "neither a choir of angels nor a pack of wolves"; it was the human community as it built a margin of safety between itself and barbarism. Maintaining that margin involved moral reason, prudent law and policy, and the discriminate use of force. The tradition of reason did not promise complete success; it did not "expect that man's historical success in installing reason in its rightful rule will be much more than marginal. But the margin makes the difference."[83] Understanding the nature of moral reasoning, and thereby rejecting moralism would, Murray argued, strengthen that civilizing margin of difference—even at the limit case of war.

Murray's Just-War Theory

Murray began his discussion in "War as a Moral Problem"[84] by noting three typical entry points for the argument. One could focus on weaponry, and the devastation that modern war would cause; one could take such a single-mindedly anticommunist stance as to make some version of the crusade the only possible choice; or one could start with the world's faltering, yet important, efforts to create international organizations capable of resolving conflict without mass violence.[85] Each of these starting points raised important questions. But "if . . . one adopts a single standpoint of argument and adheres to its narrowly, one will not find one's way to an integral and morally defensible position on the problem of war." There was, however, "an order among these questions."[86]

The first order of business was to determine the "exact nature of the conflict that is the very definition of international life today."[87] Communism was one expression of that conflict; but Murray knew that "the exact nature of the international conflict is not easily and simply defined. The line of rupture is not in the first instance geographic but spiritual and moral; and it runs through the West as well as between East and West."[88] The first question was, essentially, one of values. What are the values we wish to defend, because we perceive them to be at risk in the contemporary world? What is the precise nature and extent of the danger in which those values stand? Is the menace we face simply military, or is it also "posed by forms of force that are more subtle?"[89] Only when these questions—which fall within the traditional *ius ad bellum* category of "just cause"—had been explored could one develop a morally satisfactory approach to the problem of war.

The second question had to do with means: What instruments could defend the primary values that we believe to be threatened? Here, too, Murray warned against an exclusive focus on military hardware; conflict could be prosecuted through many instruments, and unless the full range of these means was considered, war would have no meaning as the *ultima ratio*. "Moreover, the value of the use of force, even as *ultima ratio*, will be either overestimated or underestimated, in proportion as too much or too little value is attached to other means of sustaining and pressing the international conflict."[90]

Only when these questions had been addressed could one reasonably take up the question of the *ultima ratio* itself, "the arbitrament of arms as the last resort."[91] Contemporary circumstances indeed raised novel issues here. Murray agreed with Pius XII that "never has human history known a more gigantic disorder,"[92] and that the "gigantic disorder" reflected fundamentally clashing visions of the human person. The conflict between those visions was being waged by many means, not all military, and not simply between nation-states; international organizations had become arenas for the great conflict. Still, "the most immediately striking uniqueness comes to view when one considers the weapons for warmaking that are now in hand or within grasp."[93] Moral reasoning had to take account of actual or potential developments in both defensive and offensive weapons. It also had to consider the various strategic doctrines within which the possible use of such weapons, for deterrence or for actual combat, was located. All of this was part of the moral problem of war.

It is important to underscore the structure of the argument as Murray saw it. Weapons were not the primary issue, nor was weapons-use the best starting point for moral analysis. First, one had to attend to the nature of the conflict in the world, to the issue of just cause. Then one had to think through the plurality of means available for prosecuting that conflict; "proportionate and discriminating means" included more than military hardware (although this was surely included as well). Only after all the rest had at least been opened up and explored did one get down to the questions of strategies and weapons technologies. If moral reasoning was to function responsibly on the question of war and peace, it could *not* begin where so many wished to start: with questions of hardware and strategy. Cause, and a multiplicity of means, were prior issues of moral analysis. Unless they were addressed first, the moral argument would inevitably be skewed and distorted.

Murray's review of the state of the Catholic position in 1959, as it had been sketched in various statements of Pius XII, illustrated how the Church tried to speak to the "gigantic disorder" of modernity. All "wars of aggression" were banned. Pius XII had never defined precisely the notoriously slippery term "aggression," but some points had been clarified. The pope's ban on "aggressive war" was an attempt to curb claims that the state possessed absolute sovereignty to make *ad bellum* decisions without reference to any moral categories. In a world increasingly governed by realpolitik, Pius was urging that the debate return to the fundamental issue of just cause: for here was the essential link between the use of armed force and the moral order. This entire question, Murray thought, needed more public clarification through debate.[94]

That debate would have to address the question of why aggression was bad.

Aggression could lead to disproportionate acts of violence, as the history of the century had shown time and again. But Pius, and Murray, were also concerned that aggression blocked the way to the evolution of a well-functioning international legal and political order that could adjudicate competing claims of justice without resort to armed force. "To continue to admit the right of war, as an [absolute] attribute of national sovereignty, would," Murray wrote, "seriously block the progress of the international community to that mode of juridical organization which Pius XII regarded as the single means for the outlawry of all war, even defensive war."[95]

Murray was a realist about the immediate possibilities of accomplishing Pius's vision. He noted that the pope had not specified how this moral imperative could be translated into historical reality—a problem that "has hitherto been insoluble."[96] But its insolubility to date did not render Pius's "solution" to the problem of war chimerical; it could stand as an horizonal concept in moral analysis and political/strategic thought—identifying the preferred future to be worked toward, and thus casting light on decisions taken in the here and now.

The second point in the Catholic theory as of 1959 was the obverse of the proscription of aggressive war: "A defensive war to repress injustice is morally admissible both in principle and in fact."[97] Catholic theory had always admitted such a possibility. It was not "a contradiction of the basic Christian will to peace . . . [but] the strongest possible affirmation of this will."[98] The classic Catholic insistence that the human universe was always and everywhere a moral universe was not mute when facing that limit case of the human experience, organized mass violence. Even here, the theory claimed, moral reason had its claims. Primary among these was that the resort to war had to have, as its final object, the restoration of peace—a concept with origins in Augustine. But in the contemporary context, Murray noted:

> The reiteration of the right of defensive war derives directly from an understanding of the [basic] conflict [in the world], and from a realization that nonviolent means of solution may fail. The Church is obliged to confront the dreadful alternative: "the absolute necessity of self-defense against a very grave injustice that touches the community, that cannot be impeded by other means, that nevertheless must be impeded on pain of giving free field in international relations to brutal violence and lack of conscience."[99]

Even here, classic Catholic theory sought to put limits around the resort to armed force in the vindication of violated rights. Nuclear weapons, as well as conventional weapons, had to meet the traditional tests of the just-war theory: that there be a grave injustice at issue; that the use of armed force be the last resort, all other means of redress having been tried and having failed; that proportion *ad bellum* be observed (". . . consideration must be given to the proportion between the damage suffered in consequence of the perpetration of a grave injustice, and the damages that would be let loose by a war to repress the injustice";[100] that proportion *in bello* be rigorously followed; and that the limiting principle of *discriminate* use of force be always observed.[101]

This architectonic structure of moral reasoning did not yield easy or simple answers, Murray admitted. But it was a "*Grenzmoral*, an effort to establish on a minimal basis of reason a form of human action, the making of war, that remains always fundamentally irrational."[102] Without such a moral framework, irrationality—

the unrestricted and absolute resort to violence—would hold sway. This, Murray believed, was the problem with pacifism, as he understood the term; pacifism in Murray's experience was simply the obverse of "bellicism." The pacifist, too, denied the capacity of moral reason to order human judgment and action at the limit case of war. What must be defended, first and foremost, were the prerogatives and potentialities of right reason.

Murray concluded his survey of Catholic teaching prior to Vatican II by noting that Pius XII had allowed the "legitimacy of defense preparations on the part of individual states," or deterrence. The legitimacy of deterrence rested on two empirical conditions: the absence of "a constituted international authority possessing a monopoly of the use of armed force in international affairs," and "the threat of 'brutal violence and lack of conscience.'" Pius's acceptance of deterrence was not "open-ended." The principle the pope articulated was extremely general, Murray conceded, and did not address "the morality of this or that configuration of the defense establishment of a given nation. The statement does not morally validate everything that goes on at Cape Canaveral or at Los Alamos."[103]

What was this traditional Catholic theory good for, in the face of the pressures and seemingly desperate circumstances of the day? Murray thought it performed three important functions.

First, it resolved false dilemmas. By establishing a moral position between "soft sentimental pacifism" and "cynical hard realism," it eliminated "the false antinomy between war and morality" that both the sentimental pacifist and the cynical realist propounded in different ways.[104] Such a mediating moral position could greatly improve the climate of public debate on issues of war and peace. The falsely dichotomous choice between sentimentality and cynicism had helped to create a "public climate of miasma that . . . tends to smother the public conscience."[105] Only the traditional theory of moral reason, only just-war theory with its insistence that the moral order and the political order were interpenetrable, could drain the miasmal swamps in which moralism wrestled with realpolitik.

Secondly, the tradition of moral reason could also expose the fallacy implicit in the phrase "Better red than dead." Such slogans posed a false choice. "The Catholic mind, schooled in the traditional doctrine of war and peace, rejects the dangerous fallacy involved in this casting up of desperate alternatives. Hidden beneath the fallacy is an abdication of the moral reason and a craven submission to some manner of technological or historical determinism."[106] It was much better, of course, to be neither Leninized nor vaporized. Moral reason and imagination could, Murray insisted, point a path beyond the alternatives of desperation.

Thirdly, the classic theory could set the right terms for public debate about specific policy choices. If there was to be no schizophrenic division between the orders of morality and politics, if just-war limits had to be observed, even in the great conflict of the day, then moral reason had to address the problem of creating strategic and military circumstances in which just—that is, *limited*—wars could be prosecuted as a last resort. If limited wars may be a necessity, Murray argued, then the possibility of fighting them according to just-war norms "must be created."[107] Murray was addressing himself here to the specific issue of limited nuclear war, which had been raised by Henry Kissinger, among others.

Murray's claim that the possibility of just and limited nuclear war had to "be created" did not spring from mindless anticommunism. Still less did he take pleasure at the thought of the nuclear battlefield. Murray's theory of limited war followed logically from his insistence that the moral and political orders not be sundered. War and peace, to Murray, were not antinomies, but points on a continuum of human action. The resort to war did not mean that moral norms were suspended by political exigencies; neither did the condition of peace absolve us from the responsibility of thinking about the morally legitimate use of armed force. Even in a world constructed according to the vision of Pius XII (and John XXIII), in which an international authority held a monopoly on the legitimate use of armed force, the moral legitimacy of the resort to that force would have to be faced. There was no escape, in other words, from the rigors of the traditional theory of moral reason. To mount such an escape would not only be to abjure morality; it would be a revolt against the human.[108]

Murray concluded his essay "War as a Moral Problem" with a call for a new "master strategic concept," which reflected, from within his larger project, his analysis of the need for a reclamation of American national purpose in the world. Technology would not, of itself, provide such a sense of purpose and direction; "technology tends toward the exploitation of scientific possibilities simply because they are possibilities."[109] What was needed was a master strategic concept that wed power to moral principle. Murray argued:

> Power can be invested with a sense of direction only by moral principles. It is the function of morality to command the use of power, to forbid it, to limit it, or, more in general, to define the ends for which power may or must be used and to judge the circumstances of its use. But moral principles cannot effectively impart this sense of direction to power until they have first, as it were, passed through the order of politics; that is, until they have first become incarnate in public policy. It is public policy in all its varied concretions that must be "moralized" (to use an abused word in its good sense). This is the primary need of the moment.[110]

That is surely as true of our own time as it was of Murray's. The fundamental debate to be conducted is a moral debate: about identity, purpose, values, and appropriate means to public ends. Here, the grand Murray Project and Murray's specific address to the moral problem of war are, at bottom, one effort. They cannot be separated without grave damage to both.

The Problem of World Order

Murray's caution about Pius XII's vision of world political community—that it involved "hitherto insoluble" technical problems—should not be taken to mean that Murray never turned his mind to the question of international organization. In a signed editorial in the Fordham University quarterly *Thought*, December 1944,[111] Murray outlined the question in a way strikingly parallel to the American bishops' discussion in their World War II pastoral letters.

"Peace," Murray began, "does not lie in the triumph of force, but in the permanent, ever-renewed triumph of law over force, in the triumph of political intelligence operating under the guise of moral sense."[112] At the height of World War II, this truth was "beginning to be grasped." There was a "mounting conviction that the peace of the world will not be assured unless the community of nations is somehow juridically organized."[113] Neither a return to prewar isolationism, nor the "imperialist isolationism" that might tempt America in the future would suffice. Public opinion in the United States and indeed throughout the world must "grasp the fact that the juridical organization of the international community is an inescapable demand of social justice, a true and genuine moral imperative, laid upon the collective conscience of States and peoples by the moral law and sanctioned by the sovereignty of God."[114]

What kind of international organization was required by this moral imperative? It would not be "a World State, a supranational government, under which national sovereignties would be more or less destroyed. Such a political conception is unrealizable; it has no basis in Catholic political and moral science; it looks, in fact, like a projection on the international plane of the individualistic liberal illusion that has contributed to the wreck of national society itself."[115]

A supranational world state would conflict with the classic Catholic social-ethical principle of subsidiarity, which defended the prerogatives of mediating institutions between the individual and the state. "It would deny the right to prosecute by institutional, organized action the intermediate special goods of particular groups—Labor, Industry, the Church, particular sovereign States."[116]

The world state would not do; but neither would the moral imperative of the moment be met by some sort of ongoing negotiating process or forum. This would quickly dissipate into the impotence characteristic of the League of Nations. Thus, "our moral imperative . . . falls on a mode of international organization that stands between the minimalist Continuous International Conference and the maximalist World State."[117]

Murray saw no necessarily irresolvable theoretical problem between the facts of national sovereignties and the "moral imperative" of international organization:

> Moral law demands that the international community be organized in terms of its natural political units (sovereign States); and it demands that the international community be truly organized. *Its essential demand is for juridical institutions in the international field that will control, not destroy, national sovereignties.* Demanding these institutions, it demands that there should exist in the international scene a coercive power adequate to protect the juridical order and to vindicate it in a case of violation—a coercive power which would be at the service of an international institution, to be employed in the interests of the common good.[118]

This was as far as the demands of moral law could carry us, Murray believed. Beyond this, "political prudence, guided by moral law," would have to determine the structure of this institution, and the precise way in which it would use armed force for the common good.[119] Moral reason, in other words, could suggest only the general outline of what was required. But even this would "take us further than men of inferior vision and excessive timidity—that is, the 'political realists'—are willing to go."[120] Nor was the outline without substantive implications:

> A limit must be placed on the right to arm without limit. . . . The moral and juridical process which is peace must . . . evolve to the point where national States will be able to trust themselves to the international protection of their rights. . . . The international community must move toward a state of civilization in which material force will be replaced by moral force, in which armed force will be employed only at the instance of an international institution and under its control.[121]

Murray was well aware of the multiple obstacles that stood in the way of a juridical organization of the international community that would meet the test of, say, the "elementary affirmations" of the American proposition. Yet he also contended:

> If the so-called practical objection is raised that such a system will not work, then the really practical answer must be given: "Then no lasting peace is possible." In this matter the shallow "realists" are prophets of despair. If we despair of organizing a world order based on justice, we despair of the moral nature of man. . . . There are only two ways to world peace. The first—the system of power—has not worked. The second—the system of juridical organization—has not been tried. If we do not make it work, we must resign ourselves to the barbarism of a Third World War. *It must be made to work.* The system must be organized as soon as possible. It will not be at once ideally perfect. But the machinery must be kept going, no matter how it creaks and groans; and we must improve it in the light of experience. . . . World order is a demand of moral law: and *we are bound to do what is possible at the moment, however remote the ideal may be.*[122]

For Murray, the moral structure of the problem of international organization was parallel to the moral structure of the problem of the legitimate use of armed force: what was necessary *must be created.* Not overnight, to be sure; but unless that horizon of possibility was kept open, and immediate policy decisions oriented to it, the demands of moral imagination and moral reason would not be met.

Murray on the Soviet Union

If there were problems with the imperative of world order in the moral imagination of "realists," Murray also knew that there were severe political problems for world order because of the international role of the Soviet Union. Murray's lucid outline of the nature of the Soviet threat was part and parcel of his approach to the moral problem of war and peace. Failure to comprehend the exact nature of this threat posed problems for the reclamation of the American proposition, as well as for the immediate demands of foreign policy.

In Murray's view, there were four unique dimensions to the Soviet empire.

First, the U.S.S.R. was unique as a *state.* It controlled the mass of Eurasia. It was a police state of great efficiency which had achieved a distinctive centralization of political power based on a rigorous party discipline. Its concept of "legality" was perverse, in that it had no concern for justice or human rights as these had been understood in the classic tradition of the West. And it had a completely socialized economy, which had created a military power of the first rank.[123]

Secondly, the U.S.S.R. was unique "as an empire, as a manner and method of rule, as an *imperium.*" It was organized, and its policy was led, by a revolutionary doctrine in which atheistic materialism was a first principle. Moreover, its view of historical change made it an "inherently aggressive" force in world affairs.[124]

Thirdly, the Soviet Union was unique as an *imperialism.* It was "essentially an empire, not a country. Nearly half [its] subjects should be considered 'colonial peoples.' Many of the 'sister republics' are no more part of Russia than India was of Great Britain." Within this empire, "patriotism" was not a traditional matter of "blood and soil but of mind and spirit." It may be true that the U.S.S.R. lacked, in practical terms, a "finished imperial design," but it had something "far better for its purposes, which are inherently dark. It has a revolutionary vision." That vision might not be of "world domination," as the phrase was usually understood; but the intention within the vision was clear: "the imperialism of the World Revolution."[125]

Finally, the U.S.S.R., although "the inheritor both of Tsarist imperialism and of mystical panslavist messianism," represented a fundamental break with the political tradition of the West. Murray flatly rejected the "pernicious fallacy" that communism was a "Christian heresy." Rather, communism represented "a conscious apostasy from the West. It may indeed be said that Jacobinism was its forerunner; but Jacobinism was itself an apostasy from the liberal tradition of the West, as well as from Christianity, by its cardinal tenet . . . that there are no bounds to the juridical omnipotence of government, since the power of the state is not under the law, much less under God."[126] Communism had taken up the task of Jacobinism, Murray argued—"putting an end to the history of the West."[127]

If the U.S.S.R. were simply another great power, the kind of conflict in which it was engaged with the United States would not have ensued. In fact, it was precisely on the central issues of political doctrine (for Murray, of course, always moral in content) that the nature and extent of the Soviet threat was made clear. It was irrelevant whether or not that doctrine was "true" (as Murray was sure it was not), as long as it provided the *modus operandi* for the leadership of the Soviet imperial state.

All of this had implications for U.S. policy, which seemed to Murray, in the late 1950s, muddled at best and wrongheaded at worst. Our domestic approach to the internal problem of communism had been a "fiasco,"[128] and our international approach to the Soviet threat seemed incapable of meeting the challenge posed, which was fundamentally in the order of values. The root of both the domestic fiasco and the international fumbling was a failure of understanding. We did not understand the essentially ideological character of the contest in the world, and we had compounded this failure by reducing our alternative ideological position—democracy—to "an ensemble of procedures."[129] Once again, Murray's grand project and the moral problem of war and peace in the modern world touched.

What, then, ought we to do? The first requisite was a more adequate understanding of the adversary, and particularly of his view of the use of force. Murray believed that Soviet doctrine required a foreign policy combining a maximum of security with a minimum of risk.[130] Soviet foreign policy would continue to probe for Western weaknesses, exploiting them whenever and wherever possible. Therefore, U.S. policy should be guided by the principle of "a minimum of security and a maximum of risk." America should "seize and retain the initiative in world affairs," rather than merely react to Soviet probes on an ad hoc and inevitably piecemeal basis.

Murray argued that American policymakers wasted too much time in worrying about a bolt-from-the-blue first-strike, and invested too little imagination in considering how the Soviet method of probe-and-push might be met. What was needed, in other words, was an American policy that operated simultaneously on two fronts. The United States should conduct the ideological battle with the Soviet Union forthrightly, aggressively, and in a manner that regularly seized the moral initiative. Complementarily, in the face of the Soviet probe-and-push tactics and in light of the Soviet belief that military power yielded political power, America needed a public doctrine on the legitimate use of limited military force to meet the threat of totalitarian aggression.

Murray believed that the concept of the cold war had been "overmilitarized and therefore superficialized." This had "tended to obscure or even discredit the validity of the very concept of the cold war," which was "lamentable, because the concept is fully valid, if it is interpreted in the light of the full reality of the Soviet empire in its fourfold uniqueness."[131]

The muddle had been compounded by the prevalence of the category of "survival" in the public debates. Murray wrote:

> We may be quite sure that the Communist mind, with its realistic and strategic habits of thought, has carefully separated the problem of the "survival" of the Communist Revolution from the problem of war. The Communist leadership has no slightest intention of making "survival" the issue to be settled by force of arms. In fact, it is prepared to abandon resort to arms, as soon as the issue of "survival" is raised. Survival is the one thing it is not willing to risk. In contrast, America is not prepared to resort to arms until the issue of "survival" is raised. Survival is the only thing it is willing to risk. Not the least irony in the current situation is the fact that the West has surrendered to the East its own traditional doctrine, that "survival" is not, and never should be allowed to become, the issue at stake in war.[132]

The crucial problem for American policy at the moment, Murray concluded, was one the Soviets had already solved: how to use armed force for achievable political ends while keeping the apocalyptic prospect of a "war for survival" off the agenda of superpower conflict. The development of the "correlation of forces" in the world between the Cuban missile crisis and the early 1980s makes it hard to argue with Murray's critique of American analytic and policy failures. The issues he raised concerning the nature and source of the Soviet threat are as pressing today as they were in the late 1950s, and his critique of American incapacities in the face of that threat—especially American incapacities in the nonmilitary means of conducting ideological conflict—remain equally germane.

Selective Conscientious Objection

Murray's fears about the fragility of civil argument in America were borne out in the anarchy of the Vietnam debates. His sense that the American proposition needed fundamental reclamation, and that a newly struck agreement on the moral coordinates for national security policy would have a central place in that project, was truly (and cruelly) prescient. It was thus appropriate that Murray's last major public statement, on selective conscientious objection, involved the moral problem of war and peace. The commencement address at Western Maryland College on June 4, 1967, stands as a summary of his just-war thought, located well within the larger designs of his grand project. "Summary" is apt biographically as well as intellectually; John Courtney Murray died some two months later.

Murray argued that selective conscientious objection could be understood only as an issue involving "the whole relation of the person to society."[133] The issue was further complicated, Murray believed, by a typical American oscillation between "absolute pacifism in peacetime and extremes of ferocity in wartime." The pacifist pole of this oscillation claims that "no nation has the *ius ad bellum*," ever. Then, when historical circumstances compel a resort to war, "no ethic governs its conduct. There are no moral criteria operative to control the uses of force. There is no *ius in bello*.

One may pursue hostilities to the military objective of unconditional surrender. . . . [The result was] a paroxysm of violence of which Hiroshima and Nagasaki are forever symbols."[134]

Such extreme options were no longer tolerable. A "discriminating doctrine—moral, political, and military—on the uses of force" had to be developed. The debate over selective conscientious objection might, Murray hoped, make the development of such a doctrine more possible: because the debate over selective conscientious objection could "contribute to a revival of the traditional doctrine of the just war."[135]

The Vietnam debate had reinforced the civic importance of just-war theory. Just-war theory was publicly accessible; it was not a "sectarian doctrine." Its principles were open to the understanding of all men and women of good will. Moreover, just-war theory insisted that "military decisions are a species of political decisions," and that "political decisions must be viewed, not simply in the perspectives of politics as an exercise of power, but of morality and theology in some valid sense."[136] Unless these norms were understood within society, the result would be "the degradation of those who make" political decisions "and the destruction of the human community."[137]

Just-war theory did not, Murray repeated, provide a simple template for the conduct of national security policy; "the issue of war can never be portrayed in black and white." American failure to understand this could only lead to "a polarization of opinion that makes communication among citizens difficult or even impossible."[138] How impossible would appear only in the public chaos over Vietnam in the late 1960s, a spectacle that Murray, perhaps mercifully, did not live to witness. In the absence of agreed moral coordinates, "the inevitable tension between the person and the community," which was, properly understood, "a tension of the moral order," could only "degenerate into a mere power struggle between arbitrary authority and an aggregate of individuals, each of whom claims to be the final arbiter of right and wrong."[139]

Murray concluded his address with a plea for "a manifold work of moral and political intelligence"[140] that would re-create the conditions for the possibility of an orderly, civilized debate on the relationship between private conscience and public responsibility. That work, it seems safe to say, remains to be done.

THE INCOMPLETE PROJECT

In the work of John Courtney Murray, American Catholic thought on the great heritage of *tranquillitas ordinis* was honed to an especially fine edge. Murray not only helped vindicate the place of American Catholics within the American experiment; he did so in a way that strengthened the experiment itself by clarifying the contemporary nature of its self-definitional task. Murray's impact extended to the life of the universal Church as well: the Vatican Council's "Declaration on Religious Freedom" is the most obvious monument to Murray's influence, but the Church's blessing on democratic values and institutions (in *Pacem in Terris* and *Gaudium et Spes*) was also at least indirectly related to the Murray Project in its largest dimensions.[141] That Murray's call for a reconstituted American public philosophy has not been taken up within our body politic is undeniable; but the idea is increasingly recognized as valid.[142] Murray's project, in its grandest design, was left unfinished. But the project itself remains of the utmost importance, not only for Catholics, but for the future of the American experiment.

On the question of the moral problem of war and peace—which Murray understood to be at the core of the larger project—several themes in Murray's thought retain great contemporary significance. Murray's insistence that the main line of Catholic thought was characterized by "incarnational humanism" still sets the appropriate theological and moral context for the ongoing war/peace debate within the American Catholic community. Incarnational humanism and its "tradition of reason" gives the American Catholic community ground on which to stand between what Max Weber called an "ethic of absolute ends" and an "ethic of responsibility." The tradition of reason affirms both the reality of transcendent moral norms and the capacity of human moral imagination and intelligence to apply those norms to the immense complexities of human affairs, even at the limit case of war. We need *not* choose between an absolutist ethics and a morality of accommodation, Murray insisted. Incarnational humanism, tempered by Augustinian realism yet infused with a Thomistic sense of human possibility, made it possible to conceive of a more organic, "natural," textured relationship between the order of morals and the order of politics than Weber allowed. Murray undoubtedly accepted the cautions against moralism that Weber raised. But Murray's refusal to absolutize Weber's distinction was both faithful to the mainstream of the Catholic heritage and appropriate to the requirements of the age.[143]

Murray's retrieval and modernization of what he termed the "Gelasian tradition" in Catholicism is sometimes attacked today. In 494 A.D., Pope Gelasius I, writing to the Byzantine Emperor Anastasius I, affirmed that "two there are, august Emperor, by which this world is ruled on title of original and sovereign right—the consecrated authority of the priesthood and the royal power."[144] Murray saw here the basis for a Christian theological rejection of any monistic view of human society, of which totalitarianism was the modern and most dangerous form. Gelasius's assertion of the independence of the Church (which necessarily implied, *mutatis mutandis*, the independence of civil authority) not only protected the Church against the possible encroachments of state power. It also created the possibility of a desacralized public order—which meant a public order open to the possibility of constitutional, and ultimately democratic, governance.

Catholic monists have never accepted this comfortably (although the Gelasian tradition was vindicated at Vatican II). The traditional example of such Catholic monism was often seen to be Franco's Spain. But there are new monists at work in the contemporary Catholic debate over the right ordering of society. In their view, Murray's Gelasianism was too "dualistic," and to be a "dualist" in postconciliar Catholicism is to be a very bad thing, at least insofar as many American Catholic intellectual and activist circles (mis)understand the term.

The charge that Murray was a "dualist" because of his reappropriation of the Gelasian tradition is unfounded. The central purpose of Murray's insistence on the "tradition of reason" was to reinforce our understanding that the political order and the moral order were dimensions of one human reality. Murray's development of just-war theory was similarly meant to confound the dualistic view that, in the limit case of war, moral reasoning was irrelevant. If monism, in either its Francoist or Sandinista form, is a classic Catholic temptation, then we need *more* of Murray's modernization of the Gelasian tradition, not less.[145]

Murray is also criticized today for his dependence on natural law categories of analysis. The charge is that such thinking is ahistorical. But in fact Murray insisted on the historical character of the human nature he wished to analyze. As James Hennesey once put it, Murray's "great contribution was to alert American Roman Catholic theologians to the role of historical sensitivity in their discipline."[146] The vehicle for this was Murray's demonstration of a genuine development of doctrine in the Catholic theory of Church and state. A further charge against Murray's natural law ethic is that it is an inappropriate methodology for the reclamation of the American proposition that was Murray's grand project. Yet other thinkers, not operating from the "Catholic presuppositions" with which Murray's natural law analysis is charged, have recently made similar claims.[147]

It is no service to Murray's memory and his remarkable corpus of work to claim for it a completeness that it did not possess. The large-scale Murray Project was left radically open-ended: the need for a public philosophy was asserted, but Murray never took up the task of providing it, beyond the outlines discernible in *We Hold These Truths*. It is a signal failure of contemporary American Catholicism that this work has not been taken up by Murray's successors in the tradition of American Catholic social theory.

On the specific question of the morality of war and peace, Murray also left significant holes to be filled. His approach to pacifism was inadequate at best. In Murray's own time, efforts were underway to relate the pacifist conscience to the real-world tasks of conflict-resolution and governance; Murray seems to have been unaware of these, or at least uninterested in them. This lack of interest has had grave effects. Murray's failure to be in active intellectual dialogue with nascent American Catholic pacifism helped create the circumstances in which the "sentimentality" Murray most feared among pacifists would be given full rein, not merely among lay activists, but at the highest levels of the American Church's official leadership.

Murray also failed to develop the thought on international organization as an important dimension of *tranquillitas ordinis* in the modern world that he so suggestively sketched in 1944. This failure was widely shared by the entire Catholic Association for International Peace. The result was a vacuum in which there was no sophisticated, developed American Catholic understanding of what the positive content of "peace" might be in the contemporary world. Murray's instinct in 1944, that moral reason demanded something other than the balance of power, *even if that "something other" be a horizonal concept rather than an immediately achieveable objective*, was correct. It met not only a requirement of moral reason but a profound, common, human yearning for a peace that was something better than that of an armed and surrounded camp. In the absence of a reasonable faith about the possibilities and instrumentalities of peace and freedom, unreasonableness and sentiment would win the day. The tradition of *tranquillitas ordinis* would fade into obscurity, and all manner of passionately held concepts of "peace" emerge—up to and including the survivalism that Murray rightly dreaded.

Murray's cautions on *Pacem in Terris*, as well as on the vision of Pius XII from which John XXIII built, remain important. But how much better off American

Catholicism would have been if Murray had had the time (and the inclination) to take up Paul Ramsey's challenge to social ethicists after *Pacem in Terris*: relating the morality of power to the creation of the international public authority that Pope John (and Murray in 1944) believed essential to resolving the moral problem of war in a thermonuclear age of interdependence, where human freedom remained at risk and the human prospect remained to be ensured.

These lacunae in the Murray Project do not invalidate the essential rightness of the work that John Courtney Murray set for himself, nor the basic structure of thought with which he sought to carry out his task. Murray's defense of democratic pluralism as an expression of the peace of dynamic political community remains sound. That he did not (could not, perhaps would not) extend his thought to bring these "elementary affirmations" of the American proposition to bear explicitly on the problem of war and international political community defines the task remaining for others to complete.

CHAPTER 5

The Promise and the Vulnerability of the Heritage

At the close of the Second Vatican Council in 1965 (which was, fatefully, the very moment when debate over the war in Vietnam first reached a boiling point), American Catholicism was the bearer of an ancient heritage of moral thought about the problem of war and peace. It was a heritage in transition. But even under the impact of the Council, the heritage remained a comprehensive and coherent body of thought (as expressed, for example, in Murray's work). The modes of analysis and expression had evolved since the days of Augustine and Aquinas, but the essential stance endured: the Catholic answer to the moral problem of war was *tranquillitas ordinis*, the dynamic peace of rightly ordered political community.

This was an answer not only in that it suggested an alternative to war in prosecuting and resolving conflict. It was also an answer in that Catholic moral reasoning on *tranquillitas ordinis* had shaped the Church's approach to the question of just conduct within war. War and peace did not exist in different universes of moral discourse: this was a central, defining claim of the Catholic theory. Questions about the legitimate use of armed force occurred in any serious analysis of the peace of political community, as well as in the narrower questions of just military operations. John Courtney Murray's phrase, "incarnational humanism," captured the basis of this core conviction. The human beings who, as the images of God, built and governed political communities were the same human beings who took up the burden of arms when the vindication of violated rights demanded it.

This heritage, by the end of Vatican II, seemed full of promise. It also, perhaps more clearly in retrospect, bore a weight of vulnerability.

THE PROMISE OF THE HERITAGE

The promise of the heritage lay in a rich set of intellectual resources which gave hope that Catholicism, and perhaps especially American Catholicism, might develop a body of teaching capable of meeting Camus's challenge that the men and women of modernity be neither victims nor executioners.

The first promising element of the heritage was its view of the human condition. Whatever Jansenist currents may have moved through popular American Catholicism in the days before Vatican II, the main line of Catholic theological anthropology was set, and clear. The extreme options of Pelagius and Hobbes had been rejected. Man was the image of God in history. Even marked by sin (Catholic theological wags used to joke that the doctrine of original sin was the only dogma for which there was irrefutable empirical evidence), human beings remained open to a Godly word in history, and capable of fulfilling their charge to be the stewards of creation. Human reason, even after the Fall, could grasp the requirements for living justly in community.

This anthropology was the basis on which the post-Augustinian development of *tranquillitas ordinis* as a positive, rather than punitive, concept was built. It was essential to Aquinas's constitutionalism. It grounded the teaching of the American bishops on the moral foundations of democracy. In a modern age where the threat of war and the threat of totalitarianism were inextricably intertwined, Catholicism's measured, realistically optimistic view of the human condition was a precious resource. It spoke to both the human quest for freedom and the human experience of tyranny and violent conflict. It suggested the possibility of a politics beyond barbarism, even as it understood that the barbarian temptation remained within every human breast. Given the realities of sin, the margin between barbarism and civility might be narrow indeed. But as John Courtney Murray would insist, the margin made the difference.

The second important element in the Catholic heritage as it had evolved through Vatican II was the universality of the Church itself. At the precise moment when, under the impact of interdependence, we might speak meaningfully of a "world history," Catholicism was an intellectually integrated transnational institution carrying a common tradition of thought on the moral problem of political community, war, and peace.

Traditional ecclesiology had always considered universality one of the four distinguishing marks of the Church. Now, under the conditions of modernity, universality took on a more richly-textured historical meaning. Among other things, it might mean that there was a multiracial, multiethnic transnational institution which brought a delicately crafted, politically sophisticated ethic to the central problem on the common human agenda—the problem of the continuation of the human story, in freedom and security. The Church's common ethic could address that problem in its manifold dimensions. For Catholicism had had powerful experiences of governance under both premodern and modern conditions; it had fought, and suffered under, wars and rumors of wars; it continued to bear the burden of tyrannical persecution. In short, it carried within its mind and spirit both the scars of the modern age, as inflicted by war and totalitarianism, and a moral and political vision of a better future—not only for itself, but for all men and women of good will. Nothing quite like it existed amid the vast plurality of institutions and communities in the world.

The American Church added a third precious resource to these elements in the common Catholic heritage: its own experience of democratic pluralism. In the civil rights revolution (in Murray's term, "conservation") of the late 1950s and early 1960s, American Catholicism had tested and found still supple the capacity of American democracy to make legally meaningful the promises of its foundational proposition. The process had been both painful and uplifting. But as the Council closed, American Catholicism seemed to believe that the gold of democratic pluralism had been tried in fire, and had come out sound—in fact, sounder.

In addition, the American experiment in democratic constitutionalism had been tacitly, yet genuinely, endorsed by the universal Church: in *Pacem in Terris* (with its affirmation of political community as alternative to war, and its claim that human rights formed the essential basis of a morally worthy City), and in the conciliar "Pastoral Constitution on the Church in the Modern World" and "Declaration on Religious Freedom." The universal Church, under American leadership, had abandoned its fondness for the unitary confessional state. American Catholicism had opened up, for the Church as a whole, the possibility of addressing the global problem of political community from a stance that combined evangelical religious commitment and a lived experience of democratic pluralism. In American Catholicism, then, the tradition of "incarnational humanism" as applied to the Catholic heritage of peace as political community had reached a fine edge of existential sophistication. The Murray Project, even in outline, showed that this experience could be brought to a high pitch of intellectual sophistication as well.

The American Church, in its relationship to American society, was in a strong position to contribute to the development of the heritage. By the mid-1960s, American Catholics had been freed from the necessity to demonstrate their loyalty to the American experiment through narrow forms of nationalism and patriotism. While the classic historiography of the American hierarchy's relationship to U.S. national security policy exaggerated the degree of this narrowness, the atmospherics for lay American Catholics within American society had surely changed—and for the better. The American Church was well positioned to focus its received heritage of moral reason through the prism of contemporary issues in ways that might contribute to a revitalized consensus on America's right role in world affairs. By the end of Vatican II, American Catholicism was in a position to apply the heritage of peace as political community to the exigencies of U.S. foreign policy without, as it were, continually looking over its shoulder for fear of nativist misinterpretation. In fact, American Catholicism carried, by the mid-1960s and through the efforts of scholars like Murray, a more sophisticated framework for moral analysis of the linked problems of peace, security, and freedom than any other religious body in the nation.

It was the comprehensiveness of the Catholic heritage that made this assertion possible. American Catholicism bore a tradition that resolutely declined narrow or partisan perspectives. It did not view the problem of war and peace solely in terms of weapons capabilities. It did not read all international events, dilemmas, and possibilities through the single prism of the communist threat; yet it was quite clear on the nature and the dangers of totalitarianism. It did not view international peace in terms of religious conversion, but knew it to be a problem of political and moral imagination and effort—one that was, in principle, comprehendible by all men and women of good will. Several key themes in this comprehensive view of both the problem and its solution may be reviewed briefly.

On the question of *human nature*, Catholic incarnational humanism saw human beings as neither demonic nor angelic, but only as themselves: fallen and weak, but still the images of God in history. If the demons within us made war likely, so did the better angels of our nature create political communities where conflict could be resolved through law and governance.

Catholicism understood that the meanings of peace were many, and intertwined. Peace was surely a question of personal spiritual interiority, arising from a right relationship with God. Peace also stood before us as a Kingdom vision of *shalom*, the time when all tears would be wiped away, the time of the eternal banquet in justice and righteousness. The *shalom* vision was eschatological, a reality to be brought about in God's time, by God's action in the completion of human history; the vision was also a horizon against which to measure the brokenness of the present, and perhaps bring to that brokenness a measure of healing. But peace also had a distinctive political denotation: peace was *tranquillitas ordinis* in freedom, charity, justice, and truth. Peace as a governed political community, a true City, *was* a this-worldly possibility. The creation of that City was not an option, but a moral imperative. Within this vision, the Catholic heritage had a realistic understanding of the inevitability of conflict: conflict was a given of the human condition, the political meaning of original sin. But conflict need not lead to mass violence if a rightly ordered peace of political community had been established. There was, then, no ineluctable slippery slope from conflict to war, just as there was no necessary disjunction between political conflict and *tranquillitas ordinis*. The American experience of democratic pluralism bore heavily on this understanding.

Catholics had discovered interdependence long before it became a tag word in academic and journalistic circles. For the Catholic heritage, the issue of *intervention* had long been settled. This world was one world, the Church, itself a transnational reality, affirmed. Now, under modern conditions of economic, social, and political interdependence, there could be no nonintervention policy, especially for the world's principal democratic power. The question the Catholic heritage posed on the matter of intervention was a question of *how*, not *if*: Toward what ends, by what legitimate means, through which institutional matrices? These were the questions of moment, not the old arguments of interventionism vs. isolationism.

The Catholic heritage fully understood the reality of *military force* in the world, and was most concerned with how it might be used justly to defend and advance the peace of political community. Time and again, from Augustine through John Courtney Murray, Catholic thinkers insisted that the moral questions of military force and peace were inextricably related, not disjunct. The primary issue was not whether the use of military force might ever be countenanced, but how it might contribute (along with a host of other instrumentalities) to the building of *tranquillitas ordinis*. This had become an acutely pressing, world-historical problem in the nuclear age, and under the threat of totalitarianism. But the basic principles, the heritage argued, remained sound and applicable.

The heritage saw both the reality of the nation-state and the insufficiency of the *present international system*. Structures for the realization of peace with freedom were required on the international level; yet the nation-state would remain, for the foreseeable future, the basic unit of international society and the primary object of meaningful political loyalty. The question was how these two facts of life might be addressed simultaneously. The Catholic heritage did not lend itself to the vision of a world state; but it was profoundly dissatisfied with the reigning condition of international anarchy. Therefore, it proposed an approach to the problem of international political community that allowed international institutions to control the bellicosity of nation-states without requiring their abandonment or abolition. The heritage may have erred on the side of optimism about contemporary international institutions. But it persistently argued that the peace of political community was a necessity on the international level, and thus had to be created. The issue was not one of time—that is, whether this would happen in the next year, decade, or century. The issue was the maintenance of a horizon of possibility to guide present policy.

The heritage understood the complexity of that task, and therein lay one of its primary strengths. It called for disarmament, balanced and verifiable. But it knew that disarmament could not take place without the parallel rise of international legal and political institutions credibly and effectively able to do the necessary job that war (or its threat) now did: resolve the argument, older than Thucydides, over who rules. International legal and political institutions, moreover, required the evolution of at least a minimal sense of political community across national borders. The Church, itself a transnational reality with hundreds of millions of members in the formerly colonized world, also understood the linkage between economic, social, and political development in the Third World and the problem of *tranquillitas ordinis* on the international level. Means must be found to meet the legitimate concerns of those for whom change, not the status quo, was the first requirement.

The Catholic heritage had, by the time of Vatican II, evolved a measured understanding of the *boundaries of political obligation.* It recognized and blessed patriotism; it also called its members to a sense of political obligation that extended beyond their own borders. The Church's own missionary activity, regularly recommended to the charity of American Catholics, helped make that call something other than a vague assertion of "global citizenship." The heritage knew that one could only be a citizen of a *polis*, and that political community in that sense did not yet exist on an international scale. But the transnational nature of the Church itself made it an important instrument for the development of a sense of moral and political obligation that did not stop at the water's edge.

The Catholic heritage, even at the close of Vatican II, had a sophisticated, realistic, and detailed understanding of the threat of communist totalitarianism, especially as expressed in the U.S.S.R. This was not a McCarthyite perspective, all charges to the contrary notwithstanding. There did not seem much purpose in debating whether or not the U.S.S.R. was a threat to the West; that answer was clear. The point, as Murray and others argued, was to craft policies that challenged the threat rather than simply react to it. Even as Vatican II opened up the intellectual possibility of a Christian-Marxist dialogue, the basic stance on the historical danger posed by Soviet power and intentions remained intact. The issue was how that threat might be met, and not solely by military means of deterrence.

These themes had been gathered together in an approach to the perennial argument over *morality and foreign policy* that mediated between the unacceptable alternatives of moralism and realpolitik. Here, what Murray termed the "tradition of reason" was most evident; the insistence of Catholic theory on the unity of human experience gave it ground on which to argue that the undeniable complexities of international relations did not abrogate the claims of moral principles, but in fact made them more important in the pursuit of peace with freedom and security. The Catholic development of just-war theory had given this claim meaning in history, and not only in theory: a scheme of sophisticated casuistry was available for those who wished to combine both statesmanship and moral conviction.

For all that it had suffered under the slings and arrows of nativist suspicion and persecution, American Catholicism saw the truth of this comprehensive view of the problem of conflict and political community borne out by its own experience. The

Catholic experience in America was an experience of the solution of the moral problem of war. A morally worthy peace of public order in dynamic political community *could* be built in this world, amid plural peoples gathered from across the globe. Democratic institutions of law and governance *could* provide a this-worldly alternative to mass violence in the resolution of social, economic, racial, ethnic, sectarian, or political conflict. The American experience had demonstrated that the problem of war was, fundamentally, not a problem of weapons, but a problem of political community: its creation according to morally sound "elementary affirmations," and its maintenance by common commitment to a republic of virtue. Here was the crucial importance of Murray's grand project for the specific moral problem of war and peace, security and freedom: his attempt to provide a moral rationale for democratic pluralism, and his sketch of the moral standards by which such an experiment might be carried out, defined, in microcosm, the contours of the Catholic response to the moral problem of war, and did so in reference to an experience of this-worldly governance, rather than simply in theory.

The heritage, then, seemed full of promise, and the future full of a Johannine optimism about the prospects of that promise. But the heritage also carried a weight of vulnerability, which would, over the next generation, prove more decisive than the promise.

THE VULNERABILITY OF THE HERITAGE

None of the constitutive elements of this comprehensive view of the moral problem of war and peace was so firmly fixed in the Catholic moral imagination as to be unchallengeable. Questions were in the air. The heritage itself was not without fault lines. Tremors and quakes would soon rumble through them.

On the question of *human nature*, Johannine optimism about the human prospect could become a modern Pelagianism: too detached from the realities of sin, too sanguine about the capacity of the better angels of our nature to win through over their demonic cousins. One could ask, of documents such as *Pacem in Terris* and *Gaudium et Spes*, whether they had sufficiently measured the world's continuing suffering, particularly under modern totalitarianism. Would postconciliar American Catholicism, eager to distance itself from the Church's previous rejection of the Promethean siren songs sung by modern prophets of human perfectibility, prove vulnerable to those very temptations?

The question of the meaning of peace was also unsettled. Both *Pacem in Terris* and *Gaudium et Spes* spoke of the multivalent meanings of peace for Christian understanding. Would the intellectual and political tensions among this welcome richness of understandings be maintained? Or would American Catholicism reduce the plural meanings of peace to categories of personal spirituality (and psychology), or to the creation of the *shalom* Kingdom in this world? If so, what would happen to the concept of peace as dynamic, rightly ordered political community, the element of the Catholic heritage that was most existentially applicable to the moral problem of war?

Classic answers to the question of *intervention* and the *present international system* were particularly vulnerable to distortion through simplification. Could American Catholicism think through the dangerous and irreducible complexities of fundamental change in international politics while remaining committed to the defense and advance of democracy in the world (of which the United States was the principal defender)? Under the pressures of the age, both nuclear and totalitarian, would the insistence of the heritage on moral and political linkage among the questions of disarmament, world order, international political community, and development hold firm? Or, as both theoreticians and diplomats continued to divide these issues into so many discrete puzzles, would American Catholicism follow suit? Would, for example, the question of disarmament be uncoupled from the problem of political community and human rights, and made an independent variable? Would approaches to the problem of development evolve from a genuine dialogue between the emphasis of the heritage on integral human development and the exigencies of life in the Third World, or would Third World definitions of both the nature and solution of the problem (increasingly statist, as the Bandung conference had shown) be seen as primary?

The heritage was also vulnerable on the question of the *boundaries of political obligation*. The Church's marvelous experience of transnationalism was both a blessing and a potential solvent eroding a careful balancing of the claims of national loyalty and transnational obligation. Paul Ramsey's claim that the peace of political community in the world would be established *through*, not *around*, the nation-state was not firmly established in American Catholicism. Would Pope John's plea that "trust" be the basis of international life become, particularly under the nuclear sword of Damocles and the nerve-wracking experience of a seemingly endless contest with totalitarianism, a prayer whose net effect would be to minimize the claims of existing political community in the name of a nascent international political community?

Would the "opening to the Left" that so seized the imaginations of several commentators on *Pacem in Terris* become the occasion to lose touch with the reality of totalitarianism? The fact that Nikita Khrushchev had embarked on a new campaign of religious persecution in the U.S.S.R. at precisely the time the universal Church was moving away from the rigorous, public anticommunism of Pius XII should have been a warning signal here. But would it be heeded? Did the universal Church, or the American Church, still understand the essentially ideological nature of the great conflict in the world? Could it combine a firm, principled, sophisticated anticommunism with a casuistry that emphasized other-than-military means of changing the Leninist nature of Marxist totalitarian societies? Would vigorous defense of the persecuted Church still be seen as essential to the slow growth of the peace of public order, as the American bishops had taught in their World War II pastoral letters? Would the Christian/Marxist dialogue, a potentially important advance for scholarship, come to mean a more easily accommodating approach to the historical problem of a Leninist superpower armed with nuclear weapons? These were all open questions as Vatican II ended.

Another key theoretical weakness in the heritage should have been apparent. The Council's affirmation of the moral claims of those who rejected personal participation in mass violence was overdue, and welcome. But the Church had no developed theory of pacifism—neither a theology of pacifism nor a means of relating pacifist personal convictions to the requirements of governance in a world that would always remain, in the main, nonpacifist. The twin threats of nuclear weapons and totalitarianism seemed to suggest the necessity of developing nonmilitary means to prosecute conflict. The American experience of democratic pluralism commended institutions of law and governance as the most widely used forms of nonviolent conflict-resolution in the world. Would the Church become an agent for the development of a politically and morally sophisticated pacifist ethic, or would the sentimentality that Murray feared come to infect American Catholic pacifism?

Catholicism understood that more than cold rationality was needed in building the peace of dynamic political community in the world. *Cor ad cor loquitur* ("Heart speaks to heart."), read the great Cardinal Newman's coat of arms. How could those passionate truths of the heart be wedded to the tradition of reason so that both just-war theory and pacifism were enriched? The disinclination of a just-war classicist like Murray to open this argument was one dimension of the vulnerability of American Catholicism. So, too, was the absence of an intellectually rigorous American Catholic pacifist ethic. Both these flaws boded ill for the future, particularly as the American Church took up the Council's challenge to reclaim its Scriptural heritage. A pacifist/just-war dialogue, in all openness and honesty, was essential. Challenges to the "growing end," as Murray would have put it, of the heritage were necessary if the danger of complacency was to be avoided. Such challenges were inevitable, in any event: the question was whether they would contribute to the growth, or the abandonment, of the heritage. But there was no experience on which a keen, intellectually rigorous dialogue might be built. This was surely a hole of large proportions in the heritage of the American Church.

Intellectual vulnerabilities were compounded by institutional weaknesses. Murray was almost certainly too sanguine about the degree to which the heritage of *tranquillitas ordinis* and its just-war ethic had been appropriated by the leadership elites of American Catholicism. One looks in vain through the curricula of Catholic seminaries of the 1950s and early 1960s for course materials that challenged the future ordained leadership of the Church to make its own the rich, complex tapestry of reflection reviewed above. So, too, does one search in vain for active theological argument on these issues in Catholic institutions of higher education. Catechesis of Catholic social teaching was marginal, at best, in Catholic elementary and secondary schools, as well as in Catholic colleges. The Catholic Association for International Peace remained a small band of theological and political professionals. It did not (given its numbers, could not) become the vehicle through which the richness of the heritage unfolded within the American Church as a corporate community. That the heritage existed, cannot be doubted; that the heritage had been appropriated—had become part of the intellectual and moral bone and spirit of the Church in the United States—was another matter indeed, and altogether uncertain.

In retrospect, then, we see a heritage at once immensely rich and terribly vulnerable. The failure of American Catholic intellectuals to take up Paul Ramsey's challenge in the wake of *Pacem in Terris* and *Gaudium et Spes*—to think through the relationship between the realities of power and the creation of the peace of political community on an international scale—helped create an existential as well as theoretical vacuum. What did it mean for an American Catholic to be "for peace"? The war in Vietnam would soon demand an answer. What did it mean to conceive of the peace of political community as a moral necessity in the modern world, and yet not minimize the multiple obstacles that stood between things as they were and things as they ought to be? No satisfactory answer to this larger theoretical issue had been offered, nor had the analytic framework for developing it become a firm template in the American

Catholic moral imagination. Churches, no less than anything else in nature, abhor a vacuum. The vacuum would be filled, one way or another. The times demanded answers.

Viewed through the prism of its promise, the Catholic heritage seemed, by 1965, to be poised on the verge of a renaissance of potentially world-historical importance. Its teaching about *tranquillitas ordinis* as a this-worldly, achievable peace had been forged in the fires of Augustinian realism. But the heritage had been developed, and reset in a more measured theological anthropology, by St. Thomas Aquinas. One result of that development had been an opening to democratic constitutionalism—an opening that had been fully exploited by American Catholicism. Within the American church, the tradition of *tranquillitas ordinis* had been refined in reflection on the American national experience of democratic pluralism, and under the pressures of two world wars. In the work of John Courtney Murray, these two centuries of Catholic experience in America had been brought to a fine pitch of theological and political sophistication, and the main lines of that distinctively American appropriation of the Catholic heritage had been validated by the judgment of a general council of the Church. By the end of the Council, American Catholicism seemed poised to make the kind of contribution to the revitalization of the American experiment that Murray saw as a never-ending task—and to do so in ways that would advance the international peace of political community in freedom, charity, justice, and truth. American Catholicism stood prepared, by reason of its heritage, to make a signal contribution to meeting the challenge set by Camus: that of being neither victim nor executioner, in a world marked by the twin terrors of modern war and the totalitarian state.

Thus the heritage stood, in all its promise. Within a decade, though, that heritage would be virtually abandoned, and the task of culture-formation set by Murray viewed as distasteful, if not actually iniquitous. Elements of the heritage would remain; its truth was too profound for it to be utterly swamped by the flood of countercurrents that would be set loose in the debates over Vietnam, over nuclear strategy, over whether there was any positive role for America in the world. The instruments of the abandonment would be those who might have been expected to make the Murray Project their own: the religious leadership of the Church, and most especially its intellectual and activist elites. How, and why, that happened is the tale and the analysis to which we must now turn. Set against the richness of the heritage, incomplete as it was, it is a painful exploration. But it must be undertaken. For only if the way in which the heritage was abandoned is understood can there be any possibility of reclaiming and developing it in the future.

INTERLUDE: FIVE PORTRAITS

Dorothy Day
Gordon Zahn
Thomas Merton
Daniel and Philip Berrigan
James Douglass

"Ideas," Lenin is reported to have said, "are more fatal than guns." On this point, at least, Lenin was right. But the history of ideas does not unfold in an existential vacuum. Ideas are borne in history by persons. The human qualities of the bearers of ideas can have much to do with the power of their thought, especially in religious communities.

The five portraits that follow illustrate this point. The ideas borne into American Catholic reflection on the moral problem of war and peace by Dorothy Day, Gordon Zahn, Thomas Merton, Daniel and Philip Berrigan, and James Douglass contributed to a radical transformation of that reflection. Whether that transformation was a genuine development in understanding or an abandonment of the heritage received and developed by the American Church up through the Second Vatican Council and the work of John Courtney Murray is a matter of great dispute. For me, abandonment is an accurate description of what transpired through these lives, and many others.

But, whatever my judgment on the merit of the arguments that follow, the power of these ideas surely had something to do with the quality of the persons involved. Each of these individuals was, in a distinctive way, a person of deep faith and deep commitment to Catholic Christianity. Each of them touched and shaped innumerable lives and spirits. American Catholicism is a "coat of many colors," and that garment would be the poorer were it dyed in one particularist hue. The task here, then, is not condemnation but understanding. We owe each other that much, precisely because of the profundity of our differences.

DOROTHY DAY AND THE CATHOLIC WORKER MOVEMENT: CATHOLIC ANARCHISM

At her death in 1980, historian David O'Brien wrote that Dorothy Day, the co-founder of the Catholic Worker movement, "was the most significant, interesting, and influential person in the history of American Catholicism."[1] At the very least, Dorothy Day was the single greatest influence on the transformation of American Catholic thought on the moral problem of war and peace since Vatican II. That influence can be traced organizationally, and through a network of interlocking biographies: the Catholic Worker was the seedbed from which grew various American Catholic peace organizations, such as "Pax" and the Catholic Peace Fellowship. The history of the Catholic Worker movement and Pax Christi is also densely (and delicately) intertwined.[2] But Dorothy Day's most profound impact was in the order of ideas, not organizations.

Sorting through those ideas is no mean task; as former Worker John Cort put it, the Worker's ideas "floated around the movement like chunks of meat, potatoes and carrots in a typical C.W. stew."[3] One looks in vain, in the intellectual history of the Catholic Worker and in the thought of Dorothy Day, for the kind of precise analysis that marked the writings of John Courtney Murray. Worker ideology reminded Cort of "the remark made by Oscar Peterson, the great jazz pianist. When some admirer, knowing that Peterson had studied classical music as a boy, asked if he hadn't detected a trace of Mozart in a number he had just played, Peterson replied, 'I was going so fast I don't know what I was playing.'"[4] But throughout a complex history, certain themes reemerge time and again. Whatever their relationship to each other, these themes had an immense impact on American Catholicism, at whose margins, but firmly within whose nest, the Catholic Worker movement has lived for well over half a century.

Five Worker themes were of paramount importance.

In the first instance, the thought of Dorothy Day and the Catholic Worker represented *a fundamental break with a reigning myth of modern life*: what Worker historian William Miller called *"the acceptable humanism of progress . . . located in the dogmas of liberalism."*[5] Not for the Catholic Worker Murray's modern Thomistic analysis. If St. Thomas represented one current of the medieval heritage and St. Francis of Assisi another, then Dorothy Day and the Catholic Worker movement stood decisively with the Franciscan religious intuition and existential impulse, breaking with the mainstream tradition of American Catholicism and its view of the American experiment that began with Archbishop John Carroll, and ran, in a reasonably straight line, through to John Courtney Murray.

The roots of this division lay in the second principal theme in Catholic Worker thought: the idea of *personalism*, as it had been expounded by Dorothy Day's mentor and Worker co-founder, the French peasant-publicist Peter Maurin. Maurin, who was influenced by Emmanuel Mounier and the early Jacques Maritain, believed that modernity had broken faith with the essential integrity (and thus sanctity) of the individual human person. The great mega-structures of modern life, in governance and political economy, were fundamentally inhuman and must be rejected. Maurin's thought hearkened back to the agrarian world of medieval Christendom, not to the world of the medieval universities. Like Evelyn Waugh's character Scott-King, Maurin believed that "it would be very wicked indeed to do anything to fit a boy for the modern world."[6] Modernity was unfit for *persons*, boys and others.

The personalism of Peter Maurin, Dorothy Day, and the Catholic Worker was set in a decisively Christian context, which argued that Christian love should be brought from the margins to the center of public affairs. Redemption began with the individual human being next to you, not with the "process" of history. Touching that neighbor was the meaning of St. Paul's injunction to put on Christ. True faith demanded a liberating break with the bourgeois world of institutions, manners, and values, which was a hollow shell.[7]

Worker personalism led to *Catholic anarchism*. Peter Maurin once told Ammon Hennacy, recently come to the Worker from the activist world of American anarchism, "Sure, I am an anarchist; all thinking people are anarchists."[8] Maurin preferred the name "personalist." But "anarchist" was not a title he, or Dorothy Day, ever explicitly rejected. Their Catholic anarchism was not a revolt against all organization; but it rejected all concentrations of power, especially economic power, as incompatible with the personalist ideal.

Catholic Worker anarchism had some affinities with the traditional Catholic social-ethical principle of subsidiarity, which taught that decision-making should be left to the lowest possible level in any hierarchy, commensurate with the common good. But Dorothy Day's anarchism focused on two targets that were not forthrightly condemned by mainstream Catholic social ethics: nationalism and capitalism. Both nationalism and capitalism encouraged the struggle for power in the world, Peter Maurin and Dorothy Day believed, and were obstacles to the fulfillment of the ethic of love that was the heart of Catholic Worker personalism. Classic Catholic social ethics taught the limits of the claims of nationalism and capitalism; the Catholic Worker taught their rejection.

Personalism and anarchism, in Worker thought, were influenced by a *radically eschatological view of history*, with apocalyptic overtones. Dorothy Day, according to Miller, "had a vision of the radicalism of the Gospels so profound that it aimed at ending time itself."[9] Dorothy Day and Peter Maurin were both influenced by Nikolai Berdyaev's apocalyptic view of the historical present, and Day was particularly drawn to the vision of Fyodor Dostoevsky. But the source of the Worker's eschatology was its distinctive Christology. In Christ, history has been brought to an end. The individual must now "put on Christ" in a radical act of personal freedom through voluntary poverty—an abandonment of the world's time, and an anticipatory entry into the Time beyond time. Even when the darkness seemed to close around us, the times "when one must live by blind and naked faith," Dorothy Day believed that "God sends intimations of immortality. We believe that if the will is right, God will take us by the hair of the head, as he did Habakkuk, who brought food to Daniel in the lions' den, and will restore us to the Way and no matter what our wandering, we can still say, 'All is grace.'"[10]

Catholic Worker *pacifism* flowed naturally from Worker personalism, anarchism, and eschatology. Dorothy Day had been a pacifist before her conversion to Catholicism, and carried her pacifist commitment into the Church. The movement split over the issue of pacifism during World War II. Key Catholic Worker leaders could not countenance the prospect of a world controlled by Adolf Hitler and joined the armed forces of the United States.[11] Dorothy Day, however, remained committed to the pacifist conscience throughout her life.

Her eschatology was especially infuential here. Like the monk Father Zossima in *The Brothers Karamozov*, Day believed that "the radicalism of love ignores time," and thus the world's demands in history—even under Hitler. Dorothy Day did not flinch from the implications of this radically eschatological view of our present responsibilities in this world. This "most unsentimental of saints"[12] joined Father Zossima in the belief that "love in action is a harsh and dreadful thing compared to love in dreams. . . . Just when you see with horror that in spite of all your efforts you are getting further from your goal instead of nearer to it—at that very moment you will reach and behold clearly the miraculous power of the Lord who has been all the time loving and mysteriously guiding you."[13] Others in the Catholic Worker movement could not accept this, perhaps believing that the miraculous power of the Lord worked through human agency, even in the limit case of war, and especially in the face of a Hitler. But to this view, Dorothy Day would hold fast all the days of her life.

To her pacifism, Dorothy Day added a commitment to nonviolent resistance and social change. Worker pacifism was not a species of quietism; it demanded an *activist* stance toward the principalities and powers of this world. The Catholic Worker movement was not a refuge from the trials and tribulations of this time between Christ's first and second coming; before and after her conversion, Dorothy Day was a committed political activist. Under her leadership, the movement took up a multitude of causes: trade unionism, civil rights, resistance to American military and civil defense programs, resistance to the war in Vietnam (particularly draft resistance), the farmworkers' movement, the old antinuclear activism of "ban the bomb" days, and the new antinuclear activism of the 1980s.

These five key themes in Dorothy Day's thought—personalism, anarchism, a radically eschatological view of history, pacifism, and political activism, all subsumed under a profound religious sensibility that decisively rejected the modern myth of progress—worked themselves out in a variety of secondary teachings, which were even more influential in the evolution of American Catholic thought on war and peace than the primary themes themselves.

Despite intermittent charges to the contrary, the Catholic Worker movement was not a Marxist enterprise. Dorothy Day was aware of the materialism and atheism at the root of communist ideology. But, determined to combat the distorting influence of bourgeois culture (also perceived as materialist and atheist) on the Church and the Gospel it proclaimed, the Catholic Worker movement came to espouse an anti-anticommunism that would become a key theme in American Catholic activist and intellectual circles during and after the Vietnam War. Dorothy Day would say, of various communist regimes, what she said of the Third Reich in 1939: "Let us realize that we are responsible as much as Hitler."[14]

Dorothy Day did not have in mind here Churchill's analysis, that World War II should be called the "Unnecessary War," because Hitler could have been stopped early on by the minimal application of will through armed force. Day's 1939 judgment on Hitler, although shaped by historical interpretations of the First World War that emphasized the role of the "merchants of death," was most profoundly influenced by her eschatological view of history, the eschatological meaning of the Christ event, and the apocalyptic nature of the present. When politics is judged against an apocalyptic horizon, everything is leveled: all the works of our hands are under judgment. Understanding *that* is the first requisite for the person of faith. Taxonomies of relative evil are essentially delusory. It is not as important to tell bad from worse as it is to drink deeply from the well of eschatological love, to confront the fact that history is at an end, and to prepare to suffer the consequences of this radical act of faith.

The desperate nature of the threat posed by Hitler eventually led Dorothy Day to understand something of the falsity of her 1939 analysis; this both deepened her commitment to the active pursuit of nonviolence and strengthened her belief in the spiritual dimension of the Catholic Worker commitment. Over the course of a series of Worker retreats during World War II, Day came "to see Christ not primarily as a social reformer but as the exemplar of all-sufficient love."[15] But the Catholic Worker approach to the problem of communism remained distorted by the apocalyptic horizon and its failure to distinguish relative evils. Dorothy Day claimed that she objected to Soviet nuclear weaponry and nuclear testing as much as she did to American weapons programs, "but the personalist way was not in name-calling."[16]

What were the roots of Dorothy Day's anti-anticommunism, which led to a mirror-image understanding of the superpowers' role in the world? Day's early history in the American Left played its part. But the deepest source of this perspective was not political; it was religious. Dorothy Day truly believed that "history had already ended. All was the present and what mattered was bringing the spirit into the world."[17] The depth and profundity of this belief is not in question. But the results were not a balance in judgment. Over the years, the *Catholic Worker*, the movement's penny newspaper, would dedicate considerably more space and energy to condemning American "militarism" than to a similar critique of the Soviet role in the world.

The problem of moral judgment in public life is a perennial one for the Christian conscience. Given the Weberian choice between an "ethics of responsibility" and an "ethics of absolute ends," Dorothy Day unhesitatingly chose the latter, the ethic of a harsh and dreadful love, even within a world where "all was grace." But Dorothy Day and the Catholic Worker movement did not heed Weber's advice to eschew politics. The movement may have rejected "politics as a vocation," but it eagerly embraced politics as an avocation. Dorothy Day and the *Catholic Worker* maintained a running commentary on public affairs for over half a century; that commentary led many others into the political arena, carrying the primary and secondary themes of Catholic Worker ideology. The result would not be an American Catholicism fully in accord with Worker personalism, anarchism, pacifism, and eschatology. It would be a Church increasingly comfortable with Worker anti-anticommunism; with a tendency to mirror-image analysis of the United States and the Soviet Union; and with a politics that focused on resistance to American military programs and national security policy rather than on a comprehensive critique of militarism throughout the world, or on the establishment of the peace of political community on an international scale.

It can be argued that the Catholic Worker's lack of a positive program of social change, including a positive vision of an achievable, this-worldly peace, was not a flaw; its function, and Dorothy Day's charism, was to witness to truth and to love. No one can doubt that the witness of the movement, and of Dorothy Day herself, had a profound impact in putting the issue of war and peace on the American Catholic agenda. But *how* the issue was posed is as important as its having been posed. For the movement was not an impartial witness above politics, the arena of compromise and prudential judgment. Dorothy Day and the Catholic Worker made many judgments: about the character of the American experiment, about the nature of conflict in the world, about the meaning and threat of totalitarianism, about the relationship between individual conscience and civic responsibility—and about the implications of all these for U.S. policy in the world. These judgments are not validated by the intensity of the religious intuition that gave birth to, sustained, and still nurtures the Catholic

Worker movement; they are not validated by the personal sanctity of Dorothy Day. The witness of Dorothy Day and of the Catholic Worker movement in its self-giving love and radical commitment to the corporal and spiritual works of mercy sets a standard against which American Catholics might measure their own lives. But affirming this does not require an acceptance of Catholic Worker theology. Nor does it require agreement with the political judgments and prescriptions that were made by Dorothy Day, and are still being made by her disciples.

Dorothy Day's life and witness remains a powerful sign in modern American Catholicism, and the power of that sign derives in part from the pain she suffered out of commitment and love. Dorothy Day's autobiography was appropriately entitled *The Long Loneliness*. "I wonder how many people realize the loneliness of the convert," she would muse on her seventy-fifth birthday. She had given up everything she owned—"a whole society of friends and fellow workers"—in order to follow that Master who was "all-sufficient love," whose life assures us that "all is grace."[18] He had promised the cross before glory, and Dorothy Day had come to understand this, too.

Sometimes, she would admit, she was "almost on the verge of weeping" from the demands on her goodness. "The burden gets too heavy, there are too many of them, my heart is too small." But once chosen, the path could not be abandoned; the burden, once assumed, could never be laid down, save in the Time beyond time. "If you will to love someone," Dorothy Day would say, "you soon do. You will to love this cranky old man, and someday you do. It all depends on how hard you try."[19] Of this "great witness to . . . profound Gospel simplifications,"[20] this pioneer of lay initiative in the Church, this defender of pacifism and nonviolence in American Catholicism, it can truly be said that no one ever tried harder. The enduring truth of Dorothy Day's life rests, though, not in her political judgments, but in her faith that "all is grace."

GORDON ZAHN: "TRADITIONALIST CATHOLIC PACIFISM" AND THE REFORM OF THE CHURCH

The parable of the mustard seed is often interpreted in Christian spirituality as a reminder that large consequences may follow small beginnings. Such seems to have been the case with that small band of American Catholic conscientious objectors during World War II, and particularly of one of their number, Dr. Gordon Zahn. During the war in Vietnam and since, the Catholic COs of World War II came to be seen as pioneers of a new Catholic sensibility on the moral question of war and peace. No one did more to articulate that sensibility than Gordon Zahn.

As Zahn's doctoral dissertation on his fellow COs demonstrated, Catholic conscientious objectors during World War II were a very mixed bag.[21] Generally mature in years (average age, 26) and urban in origin (the largest numbers coming from New York, Pennsylvania, and Illinois), they tended to be well educated, significantly beyond the national norm. But there the generalities stopped. Zahn identified five different ideological and theological positions at work among the Catholic conscientious objectors of World War II.

There were the followers of Father Charles Coughlin, who believed that the United States was on the wrong side in the war.

There were those who based their conscientious objection on philosophical grounds, and who referred to humanitarian rather than distinctively religious ideals in defending their position.

Then there were those with identifiably Catholic positions. Zahn defines 42 percent of this small band of 135 men as "evangelicals:" those for whom the war was a break in the Mystical Body of Christ, in the brotherhood of man under the Fatherhood of God. Then there were "perfectionists," men who believed that Catholicism obliged them to a literal living of the counsels of perfection, and who held themselves accountable to a supernatural ethic that forbade personal participation in violence (those who had been touched by the Catholic Worker movement tended to fall into this category). Finally there were the "traditionalists," who either believed that World War II was unjust under traditional just-war criteria, or who held that all modern war was banned according to those standards.

World War II conscientious objectors were put in Civilian Public Service camps run and financially supported by three traditional peace churches—the Friends, the Mennonites, and the Brethren, through their respective service committees—and by the Association of Catholic Conscientious Objectors (ACCO), which was created by the Catholic Worker movement and depended upon the Workers for financial support.[22] The Catholic camp was originally set up in Stoddard, New Hampshire, and was called Camp Simon in honor of "the Cyrene who was 'conscripted' to help Christ carry his cross to Calvary."[23] The camp was later moved to Warner, New Hampshire, where it remained until closed in March 1943. "Special service units" were also set up at the Alexian Brothers Hospital in Chicago and at the Rosewood State Training School for the mentally retarded at Catonsville, near Baltimore (where Zahn worked). Among all the Civilian Public Service camps, the Catholic camps were the most rebellious. Many of the Catholic COs saw Civilian Public Service as "nothing more than a program of slave labor offered by the State as an alternative to outright imprisonment."[24]

The social and political thought of World War II Catholic COs was diffuse, although in some instances it paralleled Catholic Worker thought. The COs opposed, of course, the use of violence for political ends, even the defense of violated rights. Some COs came out of the mainstream tradition of Catholic Action; others favored the Catholic anarchism associated with the Catholic Worker movement. The COs were uniformly critical of the acquiescence of their church leadership to governmental claims in the matter of conscription. Catholic COs were concerned about racism and antisemitism. Some of them came out of trade union backgrounds, and wanted to see management/labor relations governed by a Christian ethic of love. Christian personalism was a thread connecting these diverse interests and issues. But "the one theme which brings these all together . . . is the absolute insistence upon judging the behavior of men in society according to the standards of individual conduct required by Christian moral teachings."[25]

World War II Catholic COs would have agreed with John Courtney Murray's refusal to dichotomize "moral man and immoral society." They would have vigorously disagreed with Murray's insistence that the morality of politics not be confused with the ethics of interpersonal relationships, because social ethics had its own integrity, built on the Catholic view of the state as a natural institution. Catholic anarchist COs, and those with other theological or ideological positions in the CO camps and hospitals, joined in rejecting this classic theme in the Catholic heritage.

Zahn's experience was not atypical of the World War II Catholic CO:

> I came into Civilian Public Service as a conscientious objector, feeling that I would be almost alone as a Catholic in the program. I hadn't even heard of the *Catholic Worker*, for example, until after I had had my first tussle with the draft board. In other words I'd gone, applied as a Catholic for this conscientious objector's classification, and was told by the local Catholic pastor, who I believe was chairman of the draft board at that time—or at least sat on it—that it was absolutely impossible for a Catholic to be a conscientious objector. I imagine largely because of his position on this, my first classification was turned down and I had to go to appeals. And it was while this appeal was in process that somebody pointed out to me there was a group of Catholics that took the same position and were publishing a paper in New York. And that was the first I had *heard* of other Catholics who held this position.
>
> And so when I got into Civilian Public Service, in the Catholic camp, my position [was] the one I've described as integralist. It was a humanitarian type of Catholicism; you had on the one hand the nature of war, the spirit of war, the genius of war, if you would call it that, and on the other hand the genius or spirit or nature of Christianity. And I just made a judgment—largely emotional, part intellectual, I suppose—that the two were irreconcilable. Then when I was in camp I *met* educated Catholics who began pointing out that I was holding an untenable position in their eyes, and they instructed me in the traditions of the just war, and quite converted me at this point into a traditionalist Catholic pacifist.[26]

Gordon Zahn has spent the better part of his life, as a Catholic activist and an academic, in putting flesh on the bones of his "traditionalist Catholic pacifism." Unlike others for whom conscientious objection during Vietnam was a rebellion against the Church as well as against U.S. foreign policy, Zahn has remained firmly and persistently inside the Church. He has not identified his "traditionalist Catholic pacifism" with every item on the social and political agenda of the Left, taking, for example, a principled stance against abortion in 1971. Zahn is quite willing to be viewed as an "absolutist" where the right to life is concerned. To that charge, he wrote:

> I am ready to plead guilty. At a time when moral absolutes of any kind are suspect and the fashions in theological and ethical discourse seem to have moved from situationalism to relativism and now to something approaching indifferentism, it strikes me as not only proper but imperative that we proclaim the value of every human life as well as the obligation to respect that life wherever it exists—if not for what it is at any given moment (a newly fertilized ovum; a convicted criminal; the habitual sinner), at least for what it may yet, with God's grace, become. It is not just a matter of consistency; in a very real sense the choice is between integrity and hypocrisy. No one who publicly mourns the senseless burning of a napalmed child should be indifferent to the intentional killing of a living fetus in the womb. By the same token, the Catholic, be he bishop or layman, who somehow finds it possible to maintain an olympian silence in the face of government policies which contemplate the destruction of human life on a massive scale, has no right to issue indignant protests when the same basic disregard for human life is given expression in government policies permitting or encouraging abortion.[27]

At least a decade before it was popularized by Cardinal Joseph Bernardin, Gordon Zahn was weaving his "traditionalist Catholic pacifism" into a "seamless garment" approach to the "life" issues of nuclear war, abortion, and capital punishment.

Zahn has been a tireless organizer and leader of Catholic peace organizations, most recently Pax Christi–USA, but this important work is of less interest here than his ideas about Catholic peace activism. Zahn's approach is captured in his reflections on the Catholic peace movement in the Vietnam era. In a December 1973 essay, Zahn praised and criticized the "Catholic peace-and-resistance movement" and its accomplishments over the previous decade. Zahn called for immediate and unconditional amnesty for draft resisters, exiles, and other "prisoners of conscience;" freedom for the political prisoners "held by the tyrannical regime our nation's leaders have seen fit to install and maintain in South Vietnam;" and "reparations" in the form of reconstruction aid to North Vietnam. Zahn also wanted a fundamental examination of the "lessons of Vietnam" for U.S. foreign policy, to prevent "other Vietnams" around the world, especially in Latin America.[28]

On the positive side of Catholic activism during Vietnam, Zahn believed that sheer numbers told an important tale. Whether in draft resistance, conscientious objection, protests, lobbying, tax resistance, or civil disobedience against the war, the "great Catholic peace conspiracy [was] a phenomenon unique in modern Church history."[29] Although unsure of the depth of commitment the movement represented, Zahn yet argued that it had enjoyed "truly significant victories," particularly the hung jury in the Berrigan conspiracy trial at Harrisburg and the acquittals in the Camden trial of Catholic "resistance" leaders. Most importantly, Zahn believed that the movement had made a historic dent on the official leadership of the Catholic church, which had been made visible in the bishops' support for both general and selective conscientious objection, and in their "public call, overly qualified though it may have been, for amnesty toward anti-war exiles and deserters."[30]

Zahn also saw problems that had to be addressed if the movement was to maintain its momentum. The tendency toward diffusing political energy among a host of causes had to be checked, and the focus maintained on the priority of "peace" issues. Here Zahn demonstrated his conviction that a "traditionalist Catholic pacifism" need not adopt wholesale the Left's entire social and political agenda. Furthermore, the Catholic peace movement should be "recognizably Catholic." Zahn worried that many in the movement had been "inclined to play down or ignore—in some cases consciously divest themselves of—a Catholic identity and the behavioral limitations it might impose upon them. . . . If it means anything at all, a Catholic identity would seem to require that the movement and its members, at the very least, present themselves as loyal and committed members of the institutional Church performing an important, though sadly neglected, part of her mission."[31] This was not only a matter of moral consistency. It was prudent tactics: "A clearly and explicitly Catholic peace movement is the only kind which can hope to reach what has to be its primary target audience, that great majority of Catholics, lay and clergy, who are not yet alert to the pacifist implications of their religious tradition."[32] Zahn went so far as to ask that "individuals who come into positions of prominence [in the Catholic peace movement] recognize that this might impose special obligations of self-discipline, and, if they no longer feel willing or able to meet the standard behavioral criteria for (one is almost embarrassed to use the term) 'the good Catholic,' as defined by those we are trying to persuade, they should lower their profile as symbols or spokesmen for a movement that identifies itself as Catholic."[33]

Gordon Zahn took his own advice to heart in a decade of writing and activism after Vietnam. More and more, though, his thought seemed to turn to the internal reform of the Church's teaching according to his criterion of "traditionalist Catholic pacifism." Here Zahn was influenced by his scholarly research on the acquiescence of German Catholicism under Hitler.[34] Believing that the nuclear threat posed dangers no less profound than those of Nazism, Zahn urged the American bishops to abandon the just-war tradition, or at the very least consider that the application of traditional just-war criteria required that certain weapons systems be placed "under formal interdict," forbidding any Catholic involvement with them.[35]

Zahn's courage in criticizing aspects of Catholic activism should not be gainsaid. But his analysis of the teaching and political methodology of the Vietnam-era peace movement was too taken up with tactical issues, and insufficient in its critical appraisal of the ideological themes that came to dominate the "Catholic resistance" leadership. Did the movement correctly identify the source of the war, and take the full measure of North Vietnamese militarism, which has been made all too plain in the aftermath of the fall of Saigon? Whatever the depredations of the Thieu government, why did they not result in hundreds of thousands of refugees fleeing into the sea in small boats? Was the only possible response to the torture of Vietnam to call for American withdrawal? Why did the killing continue after that objective had been realized? Were the postulates that had led America into Vietnam so mistaken in their assessment of the nature of that conflict (no matter how cruelly or ineptly they were expressed in actual policy)? Gordon Zahn did not address these questions in 1973.

Nor did Zahn take up the moral cudgels against those American Catholic "peace-and-resistance" activists who, during the Vietnam era, joined with others in spelling "America" as "Amerika," thus linking the United States, in its policy and its public morality, with Hitler. Gordon Zahn was the biographer of Franz Jaegerstaetter, an Austrian martyr who paid with his life for his resistance to the Nazi regime. Why did Zahn not condemn the grotesque identification of the United States, under whose laws he was free to oppose the war in Vietnam, with the Nazi regime? Gordon Zahn himself would reject the identification of the United States with the Third Reich. But could he not have made this clear in a way that strengthened the impact of his pacifist objection to the Vietnam War? For whatever reason, Zahn fell short of the full pacifist witness of which he was capable in his critique of Vietnam-era Catholic activism.

Gordon Zahn's measured, polite, traditionalist approach to the Catholic conscience and the question of war and peace has often been a refreshing oasis in the desert of activist polemics. Zahn's quiet yet firm insistence that Catholic pacifism remain within the orbit of the Church, that it not become simply another species of the American Left, makes an important contribution to the ongoing Catholic discussion. But Gordon Zahn has never answered the question of how pacifism avoids that sentimentality (with its attendant weakness of political judgment) against which John Courtney Murray warned; and he has not done so for two reasons.

First, he does not seem to have moved beyond the core conviction he identified among the Catholic COs of World War II—that is, "the absolute insistence upon judging the behavior of men in society according to the standards of individual conduct required by Christian moral teachings." There is an important grain of truth in this classic pacifist assertion. To abandon moral responsibility for individual human beings in the name of abstractions—even such grand abstraction as peace, or freedom, or justice—is to do a terrible thing.[36] But this element of truth in the pacifist conscience is not the only norm relevant to moral reasoning in the face of war and tyranny. Nor is Zahn's standard capable of guiding the evolution of legal and political alternatives to war in the prosecution of human conflict.

And this suggests the second flaw in Zahn's formulation of "traditionalist Catholic pacifism." He has never attempted, in a systematic fashion, to connect the pacifist conscience to the concept of the peace of dynamic political community as it had evolved in American Catholicism through Vatican II and the work of Murray. Other issues, primarily resistance to American foreign and military policy, have seemed to him more urgent than the question of how *tranquillitas ordinis* might be constructed by pacifists and nonpacifists alike in a world that will always be marked by conflict, and in which totalitarianism, not just nuclear weapons, poses a radical present danger. This has been a loss for all concerned with the health of the American Catholic debate.

THOMAS MERTON AND CATHOLICISM AT THE APOCALYPSE

The life of Thomas Merton, disguised in fictional form, might well be rejected as too fantastic by even the most credulous readership. Edward Rice, one of Merton's many biographers, told of two encounters he had had while writing *The Man in the Sycamore Tree*:

> An Eastern lady wanted to know what I had been working on so diligently. I replied that I was writing a book about an Englishman who became a Communist, then a Catholic, later a Trappist monk, and finally a Buddhist, at which point, his life having been fulfilled, he died. An American businessman asked the same question. I replied that I was writing a biography of a fellow who had made a million or two as a writer but gave the money to a religious order. "Must have been crazy," said the businessman.[37]

Rice exaggerated, but not by all that much.[38]

Merton was, of course, much more than a political commentator and demi-activist. Almost single-handedly, he opened the tradition of monastic spirituality to American Catholics: first, in his best-selling autobiography, *The Seven Storey Mountain*; later in an extensive series of books on virtually every aspect of the spiritual life. Merton also built important bridges between the Western and Eastern traditions of mysticism, was an accomplished poet, and actively participated in his Cistercian community at the Abbey of Our Lady of Gethsemane in Kentucky, where he served as novice master before setting up his hermitage.

But it is Merton the activist-monk-theologian who concerns us here. There are interesting parallels with the life and work of Dorothy Day. Both came out of literary-political backgrounds and a youthful involvement with the old American Left. Both were converts who wished to change their Church's approach to the contemporary problems of race and war, while remaining firmly within the Catholic family. Both were profoundly influenced by the eschatological dimension of Catholicism.

Merton's apocalyptic view of history profoundly shaped his thought on war and peace. The autobiography of conversion may have been influential here. The early Merton, so eager to leave the world behind at the gates of the Abbey of Gethsemane, would use as the text of his ordination card, "He walked with God and was seen no more because God took him."[39] But Merton's apocalypticism evolved far beyond the youthful romance and conviction of his first years with the Trappists.

Merton believed, and taught, that we now lived in a "post-Christian era." Ours were apocalyptic times; the history of salvation had come to a "final and decisive crisis." Whether or not modern men and women could understand the term "the end of the world," they had to grasp the central fact of their era: that the human future was literally a matter to be decided by the present generation and "its immediate descendents, if any."[40]

The gracious outcome of history was assured, Merton believed, by the Resurrection of Jesus Christ. Those who had entered into the fullness of the Christ-event were thus freed to seek truth rather than victory in life. This bore on the moral problem of war and peace. The Christian approach to the problem was fundamentally eschatological, according to Merton. "The Christian does not need to fight and indeed it is better that he should not fight, for in so far as he imitates his Lord and Master, he proclaims that the Messianic kingdom has come and bears witness to the presence of the *Kyrios Pantocrator* in mystery, even in the midst of the conflicts and turmoil of the world."[41] Merton looked to the Book of Revelation for the imagery capable of conveying this message of the Church's spiritual combat with the powers of this world.

Merton's apocalypticism also colored his reading of the contemporary American scene. Despite his general approbation of democracy, Merton believed and wrote that he was living in "Magog's country."[42] This judgment, perhaps a flash of poetic excess at first, hardened throughout the 1960s. Continuing racial conflict intensified Merton's view of America as Magog, and Vietnam confirmed the truth of the imagery. The curse of Magog was violence. Asked to comment on the contemporary American scene by the Kerner Commission (the National Commission on the Causes and Prevention of Violence, set up in the wake of the first urban riots of the 1960s), Merton wrote that "the real focus of American violence is not in esoteric groups but in the very culture itself, its mass media, its extreme individualism and competitiveness, its inflated myths of virility and toughness, and its overwhelming preoccupation with the power of nuclear, chemical, bacteriological, and psychological overkill."[43]

Merton knew that Gog was also loose in the world. Having flirted with the Communist Party in his student days at Columbia, he was not enthralled with communism as political system or as ideology. Merton understood that the Soviet Union was not a benign power. He was repelled by militant Soviet atheism, and unlike a later generation of monk-activists, he did not indulge himself in a romantic view of Third World revolutions.[44]

But totalitarianism was not the facet of modern life that seized and held Merton's attention. What was more important was to refuse to acquiesce in the "'great illusion' that the United States was a paradigm of virtue, the lover of peace, and always right while the Communists were the embodiment of everything evil and base."[45] Merton feared that a black-and-white moralistic approach to the problem of world communism would lead only to an emotional, thoughtless anticommunism of little use in the world.[46] No doubt this was a possible danger; Murray, too, wanted to ground anticommunism in a sophisticated moral and political analysis. But the result of Merton's warnings against moralism was not a wise, disciplined, and rational anticommunism. Whatever Merton's personal intentions, his work served to buttress anti-anticommunism in American Catholic elite circles.

Merton could hardly have been surprised by this, given the mirror-image analysis by which he regularly lashed out at both West and East. The Western world was beginning to have more in common with the communist world than with the "professedly Christian" societies of its past. Social pathologies scarred both sides of the Iron Curtain. The root of these pathologies was, in both cases, materialism. Both East and West had a deterministic view of morality that left little or no room for human initiative and responsibility. Both forms of political economy were marked by a "demonic activism, a frenzy of the most varied, versatile, complex, and even utterly brilliant technological improvisations, following one upon the other with an ever more bewildering and uncontrollable proliferation." On both sides, politicians maintained the illusion that they were in control of events; in fact, no one was.[47]

One need not hold that all is well with the spiritual condition of the West, or deny that profound spiritual insights come from behind the Iron Curtain, to understand that Merton's mirror-imaging was, to use his term, "overkill." Western materialism is not of the same genus as Soviet materialism; nor can these be conceived as two parallel expressions of the same fundamentally flawed project, modernity. As with Dorothy Day, the apocalyptic horizon of historical judgment had, in Thomas Merton, leveled all present human experience to the point where one could not really distinguish (or, perhaps better, saw no public point in distinguishing) between bad and worse, or between imperfect systems and fundamentally flawed, even evil, systems.

Merton's tendency to mirror-image analyses of East and West was also influenced by his conviction that nuclear warheads were the "eschatological weapon." Nuclear weaponry had made final destruction not only possible, but probable: the political leadership of both the U.S. and the U.S.S.R. was committed to the threat of extermination rather than surrender.[48] All this was "madness";[49] and madness defined the climate "in which all Christians are facing (or refusing to face) the most crucial moral and religious problem in twenty centuries of history."[50]

Merton was not sanguine about the possibilities of Christian response to this apocalyptic situation. Most Christians in America did not understand the fundamentally religious nature of the nuclear weapons issue; "the average priest and minister seems to react in much the same way as the average agnostic or atheist." American Christians confused the interests of the free world, NATO, and the Church. "And the possibility of defending the West with a nuclear first strike on Russia is accepted without too much hesitation as 'necessary' and a 'lesser evil.'"[51]

Yet the Merton who made this unsubstantiated charge was also capable of more measured analysis. He was not a sectarian; he understood that "moral and political problems are inextricable from one another, and it is only by sane political action that we can fully satisfy the moral requirements that face us today as Christians."[52] Merton also knew that the nuclear issue could not be disentangled from the general problem of war, and the moral and political necessity of political community:

> The clarification of the basic moral issue of nuclear war is an all-important first step, but there is much more to be done after that. What faces us all, Christians and non-Christians alike, is the titanic labor of trying to change the world from a camp of warring barbarians into a peaceful international community from which war has been perpetually banned. Chances of success in this task seem almost ludicrously impossible. Yet if we fail to face this responsibility we will certainly lose everything.[53]

Thomas Merton knew that there were better and worse ways to approach that task. He once wrote that "the conventional 'pacifist' position is inadequate, if by this we mean a peace movement of individuals associated in a protest based on personal conscience." The task of peace was international in scope, not merely a matter of clarifying individual consciences. Neither protest, nor conscientious objection, nor withdrawal from society were adequate responses. Nor did Merton support unilateral disarmament, which was "naive, and would do more harm than good. . . . [It] might precipitate a war."[54]

Merton was interested, however, in "*unilateral initiatives* in *gradual* steps toward disarmament." These had to be linked, he thought, to "*alternatives* to military defense."[55] The tradition of *tranquillitas ordinis* ought to have suggested itself to Merton at this point. But no; the alternatives he proposed involved "a transformation of all the attitudes and methods that now govern our thought and action in politics." Such a transformation could take place only when we see "the spiritual and psychological evils in our own current social situation as a vitally important part of the problem: boredom, tension, affluence, individualism, and irresponsibility." Beyond personal and social conversion, then, would lie the path of nonviolent defense.[56]

Merton was as responsible as anyone for introducing Gandhian thought to the American Catholic debate on war and peace. Unlike some others, Merton remained a strict Gandhian. He affirmed the necessity of the rule of law, and was critical of those who, in the name of nonviolent protest, engaged in violence against property and persons. Merton was not, for example, an exponent of the civil disobedience style and theory developed by the Berrigans at Catonsville. Nonviolence had to do with truth, not power, he wrote. Its immediate aim was not political results, but the epiphany of an important truth. "Nonviolence is not primarily the language of efficacy, but the language of *kairos*. It does not say 'We shall overcome,' so much as 'This is the day of the Lord, and whatever may happen to us, *He* shall overcome.'"[57] Like Gandhi's, Merton's was a nonviolence of the strong. Like Gandhi, Merton preferred violent resistance to cowardly surrender when fundamental rights were at stake.[58] Merton was also aware of the temptation to self-righteousness that always threatened a nonviolence not firmly grounded in commitments to truth.

Thomas Merton was not a systematic theorist of the ethics of war and peace; different currents ran, rapidly, through his thought. The same Merton essay could combine rhetorical overkill with measured analysis and prescription. The multiple currents of Merton's peace concern, as well as the mixed quality of his reflection, were all displayed in his writings on Vietnam.

Apocalyptic themes provided the entry point here, as with nuclear weapons. "The war in Vietnam," Merton wrote in his preface to the Vietnamese edition of *No Man is an Island*, "is a bell tolling for the whole world, warning the whole world that war may spread everywhere, and violent death may sweep over the entire earth."[59] Apocalyptic also colored Merton's view of America in Vietnam; Vietnam hardened Merton's view of the United States as a warfare state dominated by the military-industrial complex in collusion with pusillanimous politicians.[60] The "overwhelming atrocity of Vietnam" also led to the familiar mirror-image:

> The people we are "liberating" in Vietnam are caught between two different kinds of terrorism, and the future presents them with nothing but a more and more bleak and hopeless prospect of unnatural and alienated existence. From their point of view, it doesn't matter much who wins. Either way it is going to be awful: but at least if the war can stop before everything is destroyed, and if they can somehow manage their own destiny, they will settle for that.[61]

Yet Merton could not bring himself to endorse the kind of protest of which the Berrigan draft-file burning at Catonsville had become a movement ikon. Merton worried that the peace movement might be poised at the brink of violence. Classic Gandhian theory emphasized respect for just law in order to highlight the injustice of the law being protested. Nonviolence was not "a sort of free-floating psychological threat"; it should be directed clearly at a target that even the adversary had to admit was unjust. "But if nonviolence merely says in a very loud voice, '*I don't like this damn law*,' it does not do much to make the adversary admit that the law is wrong. On the contrary, what he sees is an apparently arbitrary attack on law and order, dictated by emotion or caprice—or fanaticism of some sort."[62]

Thomas Merton brought the authority of the country's premier spiritual author to his writing on war and peace. That authority, rather than the precision or depth of his moral-political analysis, most accounts for Merton's influence on the transition in the American Catholic war/peace debate during the mid- and late-1960s. That it was *Thomas Merton* who said these things was of paramount importance, as Gordon Zahn concedes.[63]

But Merton also contributed to a positive evolution of the Catholic heritage, as the broker, into American Catholic thought, of principled Gandhian nonviolence. Unlike those who saw Gandhi as a weak-minded fakir, Merton knew that Gandhi's political thought could be sinewy and tough, because it did not burke the question of power. What Gandhi urged was that power not be identified exclusively with violence or its threat. There were other means by which to wield power and force change, means that could be deliberately related to a politics of virtue (in Gandhian terms, a politics of truth). Merton took this dimension of Gandhi seriously, and refracted his thought faithfully into the American Catholic scene.

Thus Merton, himself not a pacifist in principle (if he was, on virtually all matters of policy, a practical pacifist), offered American Catholic pacifism a means to grapple with its central weakness: the sentimentality that made the question of power seem dirty, unworthy, unclean. Power, no less than some human beings, was not untouchable. This element of Gandhi's insight could have been a decisive influence in averting the corruption of pacifist witness that would take place during and after Vietnam.

That it was not, may also have been due to Merton's influence. Merton's apocalypticism, his sense of the eschatological immediacies of the present, captures and reflects an important Christian experience and sensibility—the Easter faith that the gracious outcome of history has been assured in Christ. But the heightened apocalyptic dimension in Merton's view of history would, as with Dorothy Day, become a great leveler of his vision. It created the horizon against which Gog and Magog, the Soviet Union and the United States, could be portrayed as two expressions of the same disease.

Merton's apocalyptic leveling of present reality was problematic in another sense as well. Apocalypticism can paralyze a Christian call to political responsibility. How does one respect the law, as Merton-the-principled-Gandhian urged, if the laws are the laws of Magog, as Merton-the-apocalyptic told his readers? If we are really living in Magog, a contemporary Babylon, why should we seek to make that poisonous environment even marginally better? Is not all intercourse with the principalities and powers of Magog polluting?

Merton's apocalyptic view of nuclear weapons and deterrence would also lead to difficulties. Nuclear apocalypticism can result in the most grotesque forms of fear-mongering, which in turn lead to an amorality of survivalism. The imperative of survival transvalues all values, by reducing them to the nothingness of utter relativity in the face of the "death of death," as Jonathan Schell would put it in *The Fate of the Earth.* Merton's apocalyptic rhetoric and imagery was intended to put the nuclear issue before American Catholicism, and in fact did just that. But it posed the issue in a way that made the Church vulnerable to a shallow survivalism that destroyed the very connection between morality and politics that Merton affirmed.

The same apocalypticism led to similarly unhappy results in Merton's analysis of Vietnam. In this case, the net result was not an opening to survivalism, but a definition of the key issue as American withdrawal for the sake of an end to the killing. America withdrew, and the killing continued. But in a larger sense, America has never withdrawn from Vietnam. And Merton's failure to criticize, sharply, the "Amerika" currents in the Catholic peace effort would, like Gordon Zahn's, leave open the festering wound, and allow further falsifications of reality a decade and a half later in the American Catholic debate on Central America.

Merton does not seem to have been in serious dialogue with the sophisticated form of the tradition of *tranquillitas ordinis* as it had been developed by Murray. Why, is difficult to understand. Why did the Merton who would reclaim, reexpress, and thus renew the monastic tradition of spirituality for millions of Catholics prove unable to provide the same kind of translation for the tradition of peace as dynamic, rightly ordered political community? His offense at what he perceived to be the corruptions of bourgeois American life—the life he had fled, both in his student radical days and in the Cistercians—may have been at play here, since it is inconceivable that Merton, in many respects a brilliant man, was unaware that this tradition existed. Whatever the cause, the fact that Merton was not in critical dialogue with the tradition received up to and through Murray was a major loss, especially in terms of the needed encounter between just-war theory and principled Gandhian nonviolence.

Thomas Merton's legacy to the contemporary war/peace debate is thus mixed. His mantle of principled nonviolence continues to be borne by occasional figures in the current Catholic peace activist world. But the same figures often bear the grimmer side of the Merton legacy, particularly in their understandings of the American experiment. In this sense, Merton opened the door, by his immense authority as well as by his thought, to the very self-righteousness and simplemindedness that he feared, and contributed to the abandonment of the heritage that, in his advocacy of principled nonviolence, he had opened to the possibility of a positive and needed development.

DANIEL AND PHILIP BERRIGAN: THE CATHOLIC NEW LEFT

If there has indeed been an abandonment of the heritage of *tranquillitas ordinis* as it had been received and developed in American Catholicism through the Second Vatican Council, then no two men share a larger portion of the responsibility than Daniel and Philip Berrigan. To them, of course, whatever "abandonment" there has been has been all to the good. What had been received was of no use, particularly in the face of Vietnam. The heritage was not a set of moral coordinates for reflection and judgment. It was an albatross, a perversion of the Christian heritage, a selling-out to the principalities and powers of this world that had blinded Church and country to those overwhelming facts of evil that were the very stuff of American public life.

Daniel and Philip Berrigan were, during and immediately after the war in Vietnam, prolific writers and publicists. In his preactivist days, Daniel Berrigan once won the distinguished Lamont Poetry Prize. It seems best, then, both for accuracy and fairness, to let them speak first in their own words. Over the past decade, the timbre of their language may have faded from our mind's ear. It deserves to be recalled before attempting any judgment on what the Berrigans wrought.

A Berrigan Anthology

On the Burning of Draft Records at Catonsville, Maryland, May 17, 1968

We destroy these draft records not only because they exploit our young men, but because they represent misplaced power concentrated in the ruling class of America.[64]

On America

I arrived so American, such an idiot [*Daniel Berrigan, reflecting on his time in France in 1953, a period in which he was taken up with the French worker-priest movement*].[65]

Do you honestly expect that we could so abuse our black citizens for three hundred and forty years, so resist their moral and democratic rights, so mistreat, exploit, starve, terrorize, rape, and murder them without all this showing itself in our foreign policy? Is it possible for us to be vicious, brutal, immoral, and violent at home and be fair, judicious, benificent, and idealistic abroad? [*Philip Berrigan before the Newburgh, New York, Community Affairs Council, in 1965*].[66]

Your honor, I think that we would be less than honest with you if we did not state our attitude. Simply, we have lost confidence in the institutions of this country, including our own churches. . . . We have come to our conclusion slowly and painfully. We have lost confidence, because we do not believe any longer that these institutions are reformable. . . . I am saying merely this: We see no evidence that the institutions of this country, including our own churches, are able to provide the type of change that justice calls for, not only in this country, but around the world. We believe that this has occurred because law is no longer serving the needs of the people [*Philip Berrigan's statement to Judge Thomsen, from* The Trial of the Catonsville Nine].[67]

Judge: "Father Berrigan, you made your points on the stand, very persuasively. I admire you as a poet. But I think you simply do not understand the function of a court.

Daniel Berrigan: I am sure that is true [*from* The Trial of the Catonsville Nine].[68]

In a word, I believe in revolution, and I hope to make a non-violent contribution to it. In my view, we are not going to save this country and mankind without it [*Philip Berrigan, in a 1967 letter to James Forest, resigning from his co-chairmanship of the Catholic Peace Fellowship*].[69]

Do the American Bishops accept the implications of their country's control over one-half the world's productive capacity and finance? Do they realize that, despite our affluence, we have institutionalized poverty for perhaps one-quarter of our own people, plus millions in the developing world? Will they admit that these appalling realities are not an accident but a cold calculation; that they follow the logic of profit and policy? [*Philip Berrigan, writing to Bishop William W. Baum, who had requested Berrigan's reflections on the forthcoming 1971 international Synod of Bishops, which would discuss "justice in the world"*].[70]

I think especially that a pure political solution in electoral politics is a dead end. That has no future and that really suffocates the independence of the Gospel [*Daniel Berrigan, in an interview after his release from Danbury Prison, where he served his Catonsville term*].[71]

On November 25, 13 of us post-americans entered the White House with the usual throng of tourists, but with a slightly different idea in mind. We wanted to dramatize the current nuclear death race in which our government is Number One Downhill Runner. We carried banners and costumes concealed on our persons. . . . Some of us unfurled our messages . . . others of us threw aside jackets and stood revealed, funereal and celebratory; four in specter of death black, one in a resplendent Uncle Sam outfit. Sam appeared to be positively jocose; a successful life detergent and death hunter. He embraced Death with fervor. His business is death and death is good business. . . . As I reflect on it, our action was a message, looking toward Christmas. A child is born in the mad inn of the world. . . . His slaughter, in the person of millions of children of the world, is contemplated calmly by the mad prestidigitators who have made us part of their obscene act—digging like rodents. Something in us would like to say, Thank you, criminal sirs, but no thank you [*Daniel Berrigan's 1975 Christmas greetings to friends*].[72]

All this government needs to lead the world to nuclear ruin is an irrelevant vote every four years; a sizable slice of our income (for war), and silence [*Philip Berrigan, in 1978*].[73]

On the Church and Moral Witness

Of course it [the Catonsville raid] had not been a useful act, a political act. Too many people were hung up on usefulness these days. If you're useful, you know, you become disposable. Who wants to end up in the trash can with Godot? How useful were the acts of the Martyrs? How many martyrs ever had any practical programs for reforming society? Since politics weren't working anyway, one had to find an act beyond politics: a religious act, a liturgical act, an act of witness. If only a small number of men could offer this kind of witness, it would purify the world. Wasn't there a time in England when every Quaker was in jail? What a great scene that must have been! Perhaps that's where all Christians should be today [*Daniel Berrigan, at a rally before the trial of the Catonsville Nine, as reconstructed by Francine du Plessix Gray*].[74]

I am convinced that our Church has helped to write our modern liturgies of death because with all its talk of sacraments of peace its faith has lain in war, in the continuation of original sin. How different from our early Church, in which the liturgy of baptism was an induction into a community of nonviolence! In their rite of baptism, the early Christians laid aside the imperialist state and all its social demands. . . . Oh I don't care if it was only that way for two hundred years. There, in that early Church, is that pure model of nonviolence we must return to [*Daniel Berrigan, at the Cornell University chapel in 1968*].[75]

That, after all, was in good part what Catonsville was about: the Church's total misunderstanding of the Gospel. And what was the Gospel about? Oh well, the Gospel was about revolution, and if you didn't understand that, forget Christianity, for the Gospel was the greatest revolutionary document of all times [*Philip Berrigan in conversation in the Baltimore County jail, reconstructed by Francine du Plessix Gray*].[76]

The American Church has been one of the staunchest allies of nationalism and the natural armor of weaponry and militarism that nationalism demands; of the imperception and witch-hunting relative to communism; of the trivia and baubles of American bourgeois living. . . . The American Church, in regard to Vietnam, has already reached the measure of default of the German Church under Hitler and a position far less defensible, since in speaking of the immorality of Hitler's aggressive wars the German Church had to confront a totalitarian regime, and we do not [*Philip Berrigan, in a 1965 lecture*].[77]

The kids and I felt that for the first time in a thousand years we were building community around the altar [*Daniel Berrigan, reflecting on his ministry at LeMoyne College in the early 1960s*].[78]

On Nonviolence, Just War Theory, and Civil Disobedience

We invite our friends in the peace and freedom movements to continue moving with us from dissent to resistance [*Philip Berrigan, statement to the media at the first draft board raid, in Baltimore, October 17, 1967*].[79]

Some beautiful polarization! [*Daniel Berrigan, on the impact of the Catonsville Nine raid, reconstructed by Francine du Plessix Gray*].[80]

Well, I just don't like the term [nonviolence] because I think it is kind of ambiguous. I would prefer to say that our part in the resistance has never had violence attached to it. We were non-violently attacking idolatrous property in order to raise what we felt was a crucial distinction between militarized possessions on the one hand and human life on the other [*Daniel Berrigan, after his release from Danbury Federal Prison*].[81]

I think that Phil and myself sort of sweated through the last vestiges of the just war discussions represented by Courtney Murray in the early '50s. Even at that time we found ourselves without a very good formulation of our own and yet with a desperate sense that his [Murray's] language and his thinking were already outmoded by the whole nuclear invasion of weaponry and technology. Then it took us some time, and some thought and writing, to come up with an understanding that the [just-war] theory as such was no longer useful. It was a great help, I think, to us to realize that the Vatican Council was willing to go so much further into the roots of Catholic resistance and emphasize once again the radical and explicit statements of Christ with regard to violence and to put the Gospel, rather than later philosophical formulations, at the center [*Daniel Berrigan, in 1972*].[82]

The prophets Isaiah and Micah summon us to beat swords into plowshares. Therefore, eight of us from the Atlantic Life Community come to the [General Electric] King of Prussia, Pa. [Re-Entry Division] plant to expose the criminality of nuclear weaponry and corporate piracy. We represent resistance communities along the East Coast; each of us has a long history of non-violent resistance to war [*Daniel and Philip Berrigan in a statement after they, and six others, had taken hammers to Mark 12A reentry vehicles on September 9, 1980*].[83]

On the Apocalyptic Present

But what of us? How shall we lead our lives? Everything from Vietnam to Lewisburg [Federal Penitentiary, where Philip Berrigan was imprisoned] suggests to me that men who hope at this point, for other directions than further repression, further wars, more jailings of resisters, are whistling into the prevailing winds. To expect the worst is the only realism worth talking about. For we are going, downhill and pell-mell, into a dark age, a progress led by neanderthals armed to the teeth. What lends a sinister despair to their flight plan is the simple knowledge that they face, as do all who misuse the world, simple extinction. The subhumans are struggling against death, the humans are struggling toward birth. Our lifetime sees the conflict joined. We must expect bloodshed, agony, prison, exile, psychic and physical injury, separation, the rupture of relationships, the underground; these are the symptoms and circumstances that precede a new age, a new mankind [*Daniel Berrigan, from underground, in 1970*].[84]

The Berrigan Legacy

The tremendous impact of Daniel and Philip Berrigan on the organization of American Catholic activism during and after the Vietnam War has been documented by others.[85] The question to explore here is this: What did the Berrigans *teach* American Catholicism? For those teachings have had an influence far beyond what the Berrigans achieved as movement tacticians.

Themes in the Berrigans' teaching can be traced back into the Catholic Worker movement and Thomas Merton. With Dorothy Day and Merton, the Berrigans shared an apocalyptic view of history, a profound distrust of American institutions and their capacity to effect needed social change, a species of Catholic pacifism, an anti-anticommunist view of the problem of totalitarianism, and a passion that American Catholicism become a "Peace Church," as they defined the term. But the impact of the Berrigans' teachings came from the fact that Daniel and Philip Berrigan recast these themes in the social and political ideology of the Vietnam-era American New Left.

The "New Left" must be distinguished from the old American Left. The New Left did not, in the main, share the fondness of the Old Left for Marxism-Leninism as an all-encompassing politico-philosophical worldview. Nor did the New Left pose the Soviet Union as the model of a humane and decent society to which all would wish to conform themselves. New Left models were states like Cuba, Tanzania, and pre–Cultural Revolution China. Finally, the New Left did not have an organized expression like the American Communist Party. In short, the New Left tended to be more philosophically eclectic, politically diffuse, and open to countercultural currents of thought and lifestyle than its Old Left predecessors of the 1930s.

Although the New Left was a variegated phenomenon, certain key themes consistently shaped its politics and activism.[86] These themes were absorbed by the Berrigans, and formed the matrix out of which they taught American Catholicism during and after the Vietnam War.

The core of New Left ideology was a vulgarized Marxism. Power in America was corrupt, because it resided in an unholy alliance among profit-obsessed corporations, a virulently anticommunist military establishment, and politicians in thrall to the corporate and military sectors. Corporate economic interest was a primary determinant of American policy, domestic as well as foreign. The Cold War was not a conflict between freedom and tyranny, but a device behind which American corporate capitalism advanced its interests in a contemporary variant on the Open Door policy.[87] This concatenation of forces had resulted in a racist, imperialist (later, sexist) society, bent on maintaining a permanent underclass at home and quasi-imperial hegemony abroad. This led, ineluctably, to Vietnam. The moral and political tasks of the age were clear: a radical reform of American society through a redistribution of economic and social power, and a withdrawal of American influence from the contest for power in the world.[88]

These basic New Left themes were fundamental determinants of the Berrigans' teachings. America was a corrupt society, and its corruption should be understood in terms of class analysis. The main fault lines of American society, Daniel Berrigan argued, were "a matter more, of, say, middle class or upper middle class versus . . . some form of social involvement."[89] The chief expressions of that corruption (poverty, racism, militarism) were linked together by the violence at the heart of the American darkness.[90] Liberal politics had failed: "The same old kettle of fish, stinking worse than ever in the boiling juices of change," as Daniel Berrigan put it.[91] America, particularly in its foreign policy, and especially in its role in Vietnam, should be approached as one would approach Hitlerite Germany in the dock at Nuremberg (a favorite theme of Old/New Left lawyer William Kunstler at the trial of the Catonsville Nine).[92]

In such circumstances, revolutionary politics was morally and politically essential. By Catonsville the Berrigans had come to believe that "men don't grow by . . . gentle escalation."[93] Radical apocalypticism played its role here. As their biographer Francine du Plessix Gray put it, "Like the Old Testament prophets who wrote in moments of historical crisis, the Berrigans feel that purification is immediate and necessary. Like the early Christians on whom they model their vocations, the Berrigans see the Second Coming—either man's perfectibility or his destruction—as imminent in their own lifetimes. It is a view in which there is no time to atone, to reform, to do penance."[94] Polarization was the order of the day. The times were too "inexpressibly evil" to allow for the gentler political arts of persuasion, civil debate, compromise, and mutual agreement. Politics was about utility, and what was needed was not utility, but cleansing.

Classic Gandhian nonviolence—insisting on the prerogatives of just law, believing in the convertibility of the adversary who is also a seeker of truth, willing on principle to take the legal consequences of one's acts of civil disobedience for the sake of the integrity of the law—was rejected. Destruction of property did not abrogate nonviolence. Some property, one of the Catonsville Nine argued at their trial, did not have a right to exist.[95] Nor should one countenance the legal strictures of "Big Bro Justice," as Daniel Berrigan would call the American legal system.[96] "Why play mouse, even sacred mouse, to their cat game?"[97] The net result, over time, would be an unwillingness to make judgments about others' violence. As Daniel Berrigan said of the Weathermen, "I believe that their violent rhythm was induced by the violence of the society itself. I don't think we can expect . . . passionate young people to be indefinitely nonviolent when every pressure put on them is one of violence, which I think describes the insanity of our society, and I can excuse the violence of those people as a temporary thing."[98] Similar ideas colored Daniel Berrigan's teaching about social change in Latin America. Although he might hope to be, in such a situation, a force for nonviolence, it was not for him to pass judgment on others' choices.[99]

The Berrigans also taught American Catholicism about their Church. A twisted piece of steel, in the abstract form of a fish, was "Jesus Christ after the institutional church got through with him."[100] There is no doubt of Daniel and Philip Berrigan's sense of Catholic Christian commitment, but it was a commitment marked by extraordinary amounts of anger, self-righteousness, and arrogance. Daniel Berrigan's claim that, with his LeMoyne students, he was "for the first time in a thousand years . . . building community around the altar," whatever allowances one makes for poetic excess, stands as one of the most remarkable statements of self-congratulation in the contemporary history of the Church.

The Berrigans taught an ecclesiology of repristination. Only the early Church, the Church of martyrs and nonviolence, was authentic. For those who wished to take up the task of institutional reform, that was the goal to be sought: the Church of the catacombs, a classically sectarian view. But even here, New Left politics obtruded. The classic, sectarian Peace Church was a Church that had eschewed the task of culture-formation in society; that was not the Berrigan view. The sectarian Church was to bring society to judgment, by its own witness and by building cadres for revolutionary social change.

Even at the height of their fame during Vietnam, the Berrigans were not without thoughtful critics. Novelist Walker Percy, like Dorothy Day and Thomas Merton a convert to Catholicism, registered his disagreement in a 1970 letter to *Commonweal*: "God save us from the moral zealot who places himself above the law and who is willing to burn my house down, providing he feels he is sufficiently right and I sufficiently wrong."[101]

Similar criticism came from Jewish pacifist Robert Pickus, who was just as opposed as the Berrigans to the Vietnam War. Pickus rejected the Berrigans' concept of violence in politics, their understanding of the war, and their approach to issues of law and conscience in a democracy. To the first, "You are either opposed to the resolution of political questions by self-authorized violence or you are not. It is clear to me that using fire as your mode of expression . . . is a violent act." Pickus condemned the "fashionable newspeak" that broadened the term "violence" to include all types of oppression, because blurring the meaning of violence weakened inhibitions against it. The Berrigans could not have it both ways. They could not claim the mantle of pacifism and then engage in clearly violent acts.

Pickus also charged the Berrigans with selective outrage over Vietnam, another falsification of their claim to a principled pacifism. Where was the Berrigan condemnation of Viet Cong and North Vietnamese atrocities? But the gravamen of Pickus's critique had to do with the Berrigans' deprecation of democracy. One need not always obey the law, even a just law. But one could not replace an absolutism of law with an absolutism of individual conscience. "To treat the law lightly; to believe that my *feeling* alone is warrant for any act is to abandon the attempt at moral choice. Such a stance forgets the centrality of intelligence to morality." It also forgot that law in a democratic society was an effective way to achieve needed social change without violence. The Berrigans, Pickus concluded, had contributed to the collapse of a genuine pacifist tradition in America, which had been "swamped in a politics of hate that locates all the world's evil in the structure of American society and the evil motivations of an American establishment." This was a particular tragedy for those who, like Pickus, wished to wed pacifist insights to a politically responsible approach to the problem of war.[102]

Pickus's challenge to the Berrigans sets much of the framework of the argument that would burn through American Catholicism between the Second Vatican Council and the 1983 pastoral letter "The Challenge of Peace." That argument—in terms of the main themes that eventually shaped the official church's address to the moral problem of war, to the American role in the world, and to the very nature of the American experiment—was more a victory for the Berrigans' perspective than it was for Pickus's. The Church would not adopt the higher forms of Berrigan rhetoric or ideology (although some in the Church, not uninfluential, did). Yet the legacy of Daniel and Philip Berrigan, men of deep religious faith and equally deep political convictions, would be a debate shaped more by their teachings than many of its participants (or the Berrigans themselves) seemed aware. By pushing the argument so far in one direction, the Berrigans created a new center of gravity for American Catholicism's grappling with the moral problem of war and peace, security and freedom. That center would look, over time, much more like the American New Left than it would resemble John Courtney Murray.

JAMES DOUGLASS AND COUNTERCULTURAL CATHOLICISM

It was a wet, cold morning in January 1982 as about one hundred of us crossed Puget Sound on a Washington State ferry to attend a day-long retreat at the Ground Zero Center, just beside the Trident submarine base in Bangor, Washington. We had been invited to participate by the Justice and Peace Center of the archdiocese of Seattle. Our retreat would be led by Jim and Shelley Douglass, founders of the Ground Zero Center. On arriving at Ground Zero, we trudged through ankle-deep mud to a semifinished wooden geodesic dome, under which the retreat would be conducted. Once inside the dome, after our eyes adjusted to the dim light provided by a creaking electric generator outside the door, we could see a gigantic golden Buddha, before whom was laid a basket of oranges, and in whose honor joss sticks burned, adding to the dimness and fragrance of our surround. We were led in an opening prayer by the archbishop of Seattle, and then told by the Douglasses that the purpose of our time together was not to think analytically about the arms race, or the varieties of Catholic moral response to it, but, rather, to "get in touch with our feelings."

The retreat, which was over about 4 P.M., featured a series of presentations (on the Trident base, on the nuclear arms race) that involved a catechesis of the Douglasses' geopolitical, strategic, and ethical views. On the ferry ride back, I noticed, hunched over a table with two large plastic glasses of beer in front of him, a leading Catholic educator who had listened quietly to the Douglasses. Staring into his beer, he mused on how the Church had probably invested half a million dollars in his education—and now he was being told not to think, but to "get in touch with his feelings."

It had not begun this way for James Douglass. In the early and mid-1960s, Douglass was a rising young Catholic academic, a lobbyist for pacifism at the Second Vatican Council with access to many of the Council fathers involved in drafting the "Pastoral Constitution on the Church in the Modern World." Douglass's first book, *The Non-Violent Cross*[103] was well received by reviewers, and made a serious attempt to criticize just-war theory, particularly as formulated by Paul Ramsey, from the vantage point of an evangelical Christian pacifism enriched by the Gandhian theory and practice of nonviolence. In the book, Douglass acknowledged his debts to Dorothy Day, Thomas Merton, Gordon Zahn, and the Berrigan brothers.[104]

Douglass became involved with anti-Vietnam War "resistance" while teaching at the University of Hawaii, and engaged in civil disobedience—pouring blood over Air Force files at Hickam Field. In the late 1960s, Douglass became interested in Zen thought through the influence of Merton's writings on the Zen master D. T. Suzuki. The volatile mixture of pacifism, Gandhian thought (in a traduced form similar to that of the Berrigans, with respect to violence against property), Zen, and revolutionary ideology led Douglass to an appreciation of Fidel Castro and Che Guevara ("lives . . . in which belief was made flesh in action and in a struggle with the powers of oppression to the point of suffering love and death."[105]

This judgment was foreshadowed in the quotation from Frantz Fanon with which Douglass opened *The Non-Violent Cross*: "What counts today, the question which is looming on the horizon, is the need for a redistribution of wealth. Humanity must reply to this question, or be shaken to pieces by it." Douglass saw a connection between Fanon, an avowed exponent of revolutionary terror and violence, and *Gaudium et Spes*, for immediately after the Fanon citation Douglass memorialized the Council fathers' assertion that "we must undertake an evaluation of war with an entirely new attitude."[106] The bridge between these perceptions and Vietnam was soon built; Douglass wrote that the Viet Cong were "strong in poverty, community, and the truth."[107] Douglass did not endorse Viet Cong violence; throughout his career, Douglass has maintained a rigid personal moral code of nonviolence. But the key point here is not his undoubted commitment to nonviolence; it is his assertion that the Viet Cong were "strong in . . . the truth."

How could this be? How does a man deeply devoted to nonviolence come to speak of the "truth" in regard to an implacably murderous force? In Jim Douglass's case, the path lay through countercultural interests, particularly Eastern religious and mystical thought. In his second book, *Resistance and Contemplation*, Douglass posed the two nouns of his title as the "yin and yang" of a nonviolent life: realities in tension, two mutually supportive but also mutually challenging dimensions of the same human condition. The task was not to challenge the truth one found in the revolutionary politics of Castro, Guevara, Fanon, or Camilo Torres; these were men who had taken up the cause of the poor, who had identified themselves with the wretched of the earth. The task was to purify their politics of violence with a baptism in Gandhian thought and Christianity, which would stress the resurrecting power of nonviolent, self-sacrificial love.

Jim Douglass and his wife Shelley added feminism to their ideological armamentarium in the 1970s. During the time I knew them in Seattle and watched their work at Ground Zero, the notion that "patriarchy" was the root evil of American society, the explanatory principle behind its violent commitment to such engines of Armageddon as the Trident submarine beside whose first base they lived, became prominent. But it was precisely at this time that Jim Douglass achieved his greatest impact on American Catholicism. For through his work at Ground Zero he met, and had an enormous influence on, Archbishop Raymond G. Hunthausen, who became archbishop of Seattle in 1975 after thirteen years as the bishop of his native Helena, Montana. Douglass's tutelage led Archbishop Hunthausen to become one of the most prominent episcopal critics of U.S. deterrence strategy, a stance that helped pave the way for the 1983 pastoral letter, "The Challenge of Peace."

In a June 1981 address to the Pacific Northwest Synod of the Lutheran Church in America, Archbishop Hunthausen claimed, following Jim Douglass's analysis, that American "willingness to destroy life everywhere on this earth, for the sake of our security as Americans, is at the root of many other terrible events in our country." The Archbishop felt a particular responsibility to speak against the Tridents, which were based in his diocese. "And when crimes are being prepared in our name, we must speak plainly. I say with a deep consciousness of these words that Trident is the Auschwitz of Puget Sound."

Hunthausen's prescription was also couched in Douglass's language:

> As followers of Christ, we need to take up our cross in the nuclear age. I believe that one obvious meaning of the cross is unilateral disarmament. Jesus' acceptance of the cross rather than the sword raised in his defense is the Gospel's statement of unilateral disarmament. We are called to follow. Our security as people of faith lies not in demonic weapons which threaten all life on earth. Our security is in a loving, caring God. We must dismantle our weapons of terror and place our reliance on God.

Like Jim Douglass, Archbishop Hunthausen could not see the value of deterrence, even in the face of Soviet power and purpose. There were those who argued that unilateral disarmament in the face of Leninism was insane, the archbishop conceded, but his own view was that "nuclear armament by anyone is itself atheistic, and anything but sane." What Douglass could not accomplish at Vatican II, a flat rejection of deterrence theory and practice, he continued to espouse almost a generation later through the words of an American archbishop, himself among the youngest of the Council fathers in 1962.

Jim Douglass contributed more than analysis and prescription to Archbishop Hunthausen's seminal speech; he also contributed a Vietnam-era resistance strategy. Archbishop Hunthausen argued that some form of nonviolent resistance was demanded by the prospect of nuclear holocaust, and he endorsed tax resistance:

> We have to refuse to give incense—in our day, tax dollars—to our nuclear idol. On April 15 we can vote for unilateral disarmament with our lives. Form 1040 is the place where the Pentagon enters all our lives, and asks our unthinking cooperation

> with the idol of nuclear destruction. I think the teaching of Jesus tells us to render to a nuclear-armed Caesar what that Caesar deserves—tax resistance. . . . Some would call what I am urging "civil disobedience." I prefer to see it as obedience to God.[108]

Six months later, Archbishop Hunthausen followed his own, and Jim Douglass's, vision. In a pastoral letter to his archdiocese, dated January 28, 1982, Hunthausen wrote that "after much prayer, thought, and personal struggle, I have decided to withhold fifty percent of my income taxes as a means of protesting our nation's continuing involvement in the race for nuclear supremacy." He did not enjoin such an action on his congregants; rather, his action was meant to awake those who remained somnambulant in the face of global destruction, and to spur those who disagreed with him to find a better path to peace than through deterrence.[109]

This is, by now, a familiar theme: that a particular act of civil disobedience is meant to put an issue on the agenda. Yet Archbishop Hunthausen's action, as the Berrigans' over a decade before, was full of ideological and political content, much of it shaped by Jim Douglass. In a speech at Notre Dame University the day after the pastoral letter announcing his tax-resistance, the archbishop argued that "In considering a Christian response to nuclear arms, I think we have to begin by recognizing that our country's overwhelming array of nuclear arms has a very precise purpose: it is meant to protect our wealth. The United States is not illogical in amassing the most destructive weapons in history. We need them. We are the richest people in history."

Frantz Fanon, via classic New Left themes espoused by the Berrigans, through James Douglass, to the archbishop of Seattle, speaking to an adulatory audience at the nation's premier Catholic university: the circle had come complete. "Our nuclear war preparations are the global crucifixion of Jesus," the archbishop warned, in words almost certainly drawn from Jim Douglass's pen; "You and I, friends, are paying for that crucifixion, at least until we become tax resisters to our nuclear idol."[110]

James Douglass continues to lead the Ground Zero Center at Bangor, Washington. Its "action focus" in 1984 was the "death train" on which nuclear warheads were supposed to be carried into the Trident base. Douglass, his wife, and their large family continue to live in voluntary poverty, an influential fact in measuring their impact on Archbishop Hunthausen, and on their neighbors in Kitsap County, Washington: some look on the Douglasses with horror, but others have been profoundly moved by the Douglasses' witness to their understanding of the truth. The depth of James Douglass's commitment to that truth is not at issue; what is important is to understand what he taught (and is teaching) American Catholicism.

That Douglass's teaching has made an impression is clear not only from Archbishop Hunthausen's statements in 1981 and 1982, but from the review of a decade and a half of argument on the morality of war and peace in American Catholicism that follows. There it will be clear that Lenin was indeed right in his claim that "ideas are more fatal than guns." Douglass's ideas, drawing on and in a sense summing up themes taught earlier by Dorothy Day, Gordon Zahn, Thomas Merton, and Daniel and Philip Berrigan, would, in conjunction with those of his "resistance movement" colleagues, result in an abandonment of the heritage of Catholic thought on war and peace as it had been received and developed in the American Church through the Second Vatican Council. At the end of that process of abandonment, Archbishop Hunthausen's would no longer be a lonely voice among the official leadership of the Church in the United States.

PART TWO

The Heritage Abandoned

Many of the younger Jesuits present, veterans of the peace movement, needed no convincing of the bankruptcy of American society and the evil consequences of its influence abroad. One soft-spoken seminarian, recently released from prison where he had served twenty-two months for draft resistance activities, quietly asked if there was any name for American influence abroad other than murder or genocide. Another peace veteran took a more historical perspective. Was not oppression the American tradition—not simply Asians and Latin Americans abroad, but blacks and Indians at home, and before them, the immigrant waves from Europe? There did not seem too much left to what John Courtney Murray liked to call "the American Proposition."

Joseph G. O'Hare, S. J., "The Summer of '72," *America*

CHAPTER 6

Themes for the Abandonment of the Heritage

Camus's challenge—to create a politics in the modern world in which human beings need not choose between being victims or executioners—was anticipated by classic Catholic moral theory as it evolved from Augustine through the Second Vatican Council. The Catholic heritage never claimed that we would be absolved from moral choice; never promised a utopian world free from conflict; never believed that the Kingdom of God, present in human history in an anticipatory way in the Resurrection of Jesus Christ, would be completely realized in the temporal order. What the heritage *did* claim was that moral choice was a function of reason, not sentimentality; that conflict need not lead to mass violence; and that dynamic, rightly ordered political community could provide morally sound means for resolving conflict without resort to the mass violence of war.

Catholic theory understood that the ability to achieve common purpose was a central reality in all human communities, especially political communities. And it had evolved a sophisticated casuistry to guide moral decisions about the use and ends of power. Whatever its weaknesses, this heritage seemed, by the end of Vatican II, well positioned to contribute to a thoughtful grappling with Camus's challenge. This was particularly true of American Catholicism, which had led the common heritage of the Church to an appreciation of democratic values and institutions in building *tranquillitas ordinis* in the modern world.

Thus, the situation in 1965. Twenty years later, the landscape had dramatically changed. The nature and extent of the change in the American Catholic debate on war and peace, security and freedom, was so fundamental, and involved such a profound alteration of understandings, as to constitute an abandonment of the heritage as it had evolved through Vatican II and the work of Murray. That is a hard judgment, but it is accurate. The American Catholic debate on war and peace lost its bearings over that twenty-year period, to the point where Nietzsche's complaint, "I have forgotten why I ever began," would be an exaggerated, but not essentially mistaken, encapsulation of what had taken place.

Many things were not forgotten, of course, within American Catholicism. The fundamental proposition that politics was a moral enterprise would remain, as would the forms of the just-war theory. But the moral *content* of the American experiment would be construed quite differently. And the result would be a dramatically different set of strategic and policy prescriptions.

Explorations of these changes tend to focus on this last level of the transition: the new prudential judgments and policy recommendations regularly proposed by the American bishops, or by their public policy agency, the United States Catholic Conference. This is precisely the wrong place to begin analyzing what the past twenty years have wrought. Why, has to do with the intellectual structure of the war/peace debate.

THE STRUCTURE OF THE WAR/PEACE DEBATE

Even at the elite levels of government, most arguments over national security issues are framed in terms of binary options: Are you for or against a particular weapon, a certain military tactic, a given diplomatic strategy? To take examples from this morning's newspaper: Are you for or against the MX missile? The development and deployment of strategic defense systems? Disinvestment in South Africa? Aid to the anti-Sandinista forces in Nicaragua? Cultural and scientific exchange programs between the United States and the Soviet Union? Applying human rights criteria in determining America's relationship to South Korea, the Philippines, Chile, Pakistan, or the Peoples Republic of China? (And if so, what human rights criteria are you for or against?) The list of binary arguments is virtually endless, a litany of the dilemmas of statecraft in the modern world.

There is a point, of course, at which binary choices must be made. This is particularly true in the Congress, but also in the executive branch of government. Try as the Department of State might to "preserve the president's options," at certain moments the green or red light will be given. Vacuums, even for contemplation, are no more tolerable in governance than in nature. Somebody has to decide, and something has to be decided (even if the decision is *not* to decide for a while). All this is granted.

But a primary focus on yes/no, for/against decisions can obscure the fact that the binary option chosen is but the tip of the decision-making iceberg. Policy conclusions are heavily influenced by a complex array of presuppositions, assumptions, and prior judgments. A U.S. senator's judgment to vote for or against the MX missile is not a simple matter. It involves certain political calculations (in the narrow sense of the term, i.e., relationships with constituents and party leadership, vote-trading across issues, etc.). But it also involves a set of previously determined answers to prior questions: about the nature and purpose of nuclear weaponry; about the deterrence system, and possible alternatives to it; about the present course and future prospects of the U.S./Soviet relationship; about the legitimacy of the threatened use of armed force (nuclear or otherwise); about the political and moral responsibilities of the United States in the modern world—at bottom, about human nature, and what we can and cannot expect of human beings in a broken world. These questions, and the answers to them, may be more tacit than explicit in the mind of our imaginary senator. But they shape, and often determine, his or her binary choice.

The body of the policy iceberg, often below the surface of consciousness as well as of public debate, is, over the long haul, of even more significance than the specific choices to which it leads at one or another given moment. We may call the "body of the iceberg" the *context* out of which political judgments, which are concurrently moral judgments, are made. Context is of greater long-term importance than binary choice because context creates the possibility of coherent policy over time. Moreover, *it is at the contextual level of the argument that moral imagination and moral reasoning make their primary claims for attention.*

If moral choice is not simply a matter of establishing the self-evident, one-to-one correspondence of eternal principles and specific policy prescriptions, then moral claims are weakest at the level of prudential, concrete choice. These choices are not "less moral"; but moral clarity is more difficult to achieve at the level of prudential judgment because empirical complexities, the possibility of unintended consequences, and the ever-present factor of ignorance make moral certitude less likely here. Only a fool would argue that specific, prudential choices are without moral dimensions; only a greater fool would argue that recognizing those moral dimensions provides transparently obvious policy clarification. Both types of fools are depressingly ubiquitous in American public life today. But the Catholic heritage (in fact, the entire Western tradition of moral discourse) warns us against following their fool's counsel.

Context is more fundamental than binary choice. Moral imagination and moral reasoning make their primary claims, and offer their most salient counsel, at the contextual level of the argument. Grant this, and one must still determine the contours of the context. Identifying the primary questions that, answered, shape our approach to a given binary choice (or our formulation of the available binary choices) can be an exercise rather like searching for the core of an onion. Visible layers of argumentation are stripped away; more opaque, less publicly perceptible layers are found and then peeled off in the search for the gravamen of it all—and one is suddenly and sadly left with a pile of layers. There is no one central, all-determining *quaestio*, the answer to which shapes everything else we think about war and peace, security and freedom (although the question of human nature comes as close to an irreducible core as we are likely to get).

This being the case, one can schematize the basic questions in several ways. There are two criteria for evaluating such schemes of contextual questions. First, does this set of questions, answered, provide a rationally satisfactory explanation of why our imaginary senator (or bishop, or columnist) takes his or her particular positions on binary choices? Put another way: Does this set of questions, this context, explain how our imaginary decision-maker or commentator formulates what the binary choices are, or comes to understand that what may appear the only options are in fact too limited a pair of choices?

Secondly, does this set of questions, answered, provide a rationally satisfactory explanation for our senator's (or bishop's or commentator's) positions on issues over time? Is it, in other words, predictive of what might happen as well as explanatory of what has taken place? Positive answers to these two questions suggest that a useful, and essentially (if not exhaustively) truthful definition of the context of the war/peace debate has been identified. Argument about the validity, internal consistency, and moral worthiness of typical answers to these contextual questions can then proceed in an orderly and, just possibly, enlightening fashion.

Two brief illustrations may make the case here. In the years before World War II, Lend-Lease to Great Britain would have been opposed by a pacifist sympathetic to the Oxford peace pledge; by a farmer steeped in Midwest progressivist isolationism; by a German-American Bundist; and by a disciple of Father Charles Coughlin. Any serious moral analysis of the binary choice made by these four hypothetical figures (no Lend-Lease) would have to take account of the contextual underpinnings of what appear, on the surface, to be four similar prescriptions.

To take a second example. Disinvestment in South Africa today might be supported by black Americans eager to apply the lessons of the Montgomery bus boycott to the dissolution of apartheid; by a Marxist eager to foment revolutionary violence in a strategically sensitive region; by an American bishop morally revolted by racial discrimination; by a human rights activist who believes that the United States should have no truck with oppressive regimes; and by a politician who claims that disinvestment is the only way to force the South African government to change its racial policies. Disinvestment might be opposed by a political realist who sees South Africa as a bastion of stability in a region holding crucial strategic mineral assets; by a trade unionist supporting the nascent black labor movement in South Africa; by admirers of the *Inkatha* movement of the Zulu chief Gatsha Buthelezi; by corporate executives convinced that rigorous application of the Sullivan principles is the best way they can aid in the breakdown of apartheid; by a bishop who takes seriously the antidisinvestment counsel of anti-apartheid South African novelist Alan Paton; and by a politician who believes that U.S. involvement in the South African economy provides a crucial point of political leverage for changing South African racial policies.

Again, it would seem impossible to make a morally responsible analysis of the binary choice of principled investment/disinvestment unless one had explored the terrain of moral and political judgment lying just beneath that choice: an exploration that would involve the large question of ends (What kind of South African society is our goal?) as well as specific questions of means (What are the morally acceptable ways to achieve needed change, without making matters more oppressive, more bloody, less humane?). Understanding the contexts that shape binary choices is important for political activists as well as for moral theorists. Should the bishop who supports disinvestment because of his moral repugnance over apartheid, but who favors peaceful change in South Africa, make common cause with a Marxist whose purpose is to create a violent and bloody revolution?

There are several possible lists of the key questions that, answered, combine to form a "context" in the war/peace debate.[1] The list used here has eight fundamental questions, or key choice-points, the answers to which have proven both explanatory and predictive of binary choices made by politicians, commentators, publicists, and religious leaders. The list includes both moral and strategic/political questions:

1. The question of *human nature*. Is war inevitable, given human nature as we know it?
2. The meaning of *peace*. Do we seek interior spiritual tranquillity before God, a *shalom* kingdom of righteousness and justice, a rightly ordered and dynamic community of *tranquillitas ordinis*? What are the morally and politically important relationships among these three models?
3. The question of *intervention*. Should the United States actively seek to influence world politics and economics?
4. The question of *military force*. Is the resort to it ever justified? If so, under what circumstances and constrained by what limits? If not, why not?
5. The question of the *present international system*. Is the current situation,

essentially one of anarchy among independent sovereign states, viable? If so, what instruments of stability can be built into the system? If not, what should replace it, and how can that replacement evolve without further deteriorations in peace, freedom, security, and justice?

6. The question of *the boundaries of political obligation.* Do they stop at the borders of the nation-state of which one is a citizen? Can there be morally and politically responsible obligation across national borders without an international *polis* to which one can be responsible, and of which one can meaningfully speak of "citizenship?"
7. The question of the *Soviet Union.* What is the nature of the Soviet regime and ideological system, and their relationship to the Soviet role in the world? Can the Soviet Union change? If so, how? If not, how should it be treated as a principal actor in world affairs?
8. The question of *America.* What is the moral worthiness of the American experiment in democratic pluralism? What relevance does that national experience have for U.S. foreign policy? Is American power—ideological, cultural, moral, political, economic and military—a force for good or evil in the world, on balance and considering the alternatives?[2]

This matrix of questions has proven useful (and truthful) in understanding the evolution of the American Catholic debate on the moral problem of war and peace, security and freedom. It helped take the intellectual temperature of the Catholic heritage at the close of the Second Vatican Council, suggesting both the great strengths, and the troublesome weaknesses, in that heritage as it had evolved in American Catholicism. Answers to these contextual questions changed drastically in the intellectual, and ultimately official, centers of the Church in the United States from 1965 to 1985, as American Catholic leaders sought to give content to the "entirely new attitude" toward the problem of war enjoined by Vatican II. Understanding the contours and content of this radical contextual change is the indispensable precondition to understanding the public policy positions taken by the American bishops and the United States Catholic Conference over the past decade.

One final introductory word: the contextual debate over war and peace takes place in what Gabriel Almond once called the "attentive public."[3] The mass public sets the mood of the debate. Specialists formulate the binary choices that must be made by government officials. The "attentive public"—typically without specialist knowledge, yet seeking information on world affairs beyond that available in local newspapers and network television news, and actively engaging in public debate and political activism on foreign policy issues—sets the *themes* of the war/peace debate. This is where contexts for judgment are formed. Catholic intellectuals, commentators, publicists, and religious leaders fall within the ambit of this "attentive public." They are the men and women who shaped the context of the Catholic debate, the level where moral reasoning makes its most persuasive claims and has its greatest public impact. It is in this Catholic attentive public that the abandonment of the heritage as received and developed through Vatican II and the teaching of John Courtney Murray, is most evident.

The American Catholic contextual argument from 1965 to 1985 was not conducted primarily in specialist journals of theology, but in those periodicals aimed specifically at the Catholic attentive public: the lay journal *Commonweal*, the Jesuit magazine *America*, and a host of other publications, which, to name but the most prominent, would include the weekly newspapers the *National Catholic Reporter* and the *National Catholic Register*, and the ecumenical journals *Worldview, The Christian Century*, and *Christianity and Crisis*. It was in these sources, amplified by argument in the major theological quarterlies and by occasional statements by both the national hierarchy and individual bishops, that the contextual debate was shaped, and a new moral horizon expressing the content of an "entirely new attitude" to the moral problem of war and peace formed.

THEMES IN TRANSITION

Human Nature and War

Writing at the turn of the decade, in December 1969, the editors of *America* sensed an "apocalyptic mood" abroad in the land, one that had led Richard Rovere to "an appreciation for the appositeness in a friend's description of the 1960s as a 'sewer of a decade.'" The Jesuits at America House saw some truth in this. Human ingenuity had led to the despoiling of the natural environment. Wars and rumors of wars echoed around the globe. The youth culture, indulged by adults, had only led to a wider generation gap. The religious optimism of Vatican II and the ecumenical movement had given way to unprecedented turmoil within the churches.

All this added up to a somber time of self-appraisal. Yet some things remained constant. "In that magic moment between 1969 and 1970 we will discover in the mirror a strange creature, half monster and half angel."[4] It is, perhaps, a measure of the temper of those times that the editors cited the demonic elements of human nature before the angelic. But, on this point at least, the classic Catholic heritage had remained largely intact. We were neither demons nor angels, but only human beings, still the image of God in history, still capable, for all the swinishness of the years just past, of hearing a Godly word in history.

Catholic moral theory, rooted in Catholic theological anthropology, has never been quite comfortable with the notion that war is an inevitable result of the flaws in human beings. War was not "in the nature of things"; the better angels of ourselves could, the heritage insisted, provide a measure of the peace of political community for prosecuting conflict.

That conflict was a constant of the human condition, the heritage did not doubt. Augustinian realism, even modified by a Thomistic sense of human possibility, had always understood conflict as part and parcel of the human experience, the political meaning of the doctrine of original sin. But the heritage insisted on a distinction between "conflict" and "war." Conflict was inevitable: it was built into the very structure of human being-in-the-world after the Fall. But to conclude from this realism that war was an ineluctable constant of the human condition seemed too despairing a view. Whatever its other deficiencies of analysis and prescription, *Pacem in Terris* expressed this tempered confidence about the human prospect. Pope John did not look forward to a day when strife and conflict would be banished from the earth; he anticipated the day when war, as a means for coping with conflict, would be no more.

This balanced view of the human condition remained largely intact while other key themes in the Catholic heritage underwent profound transitions during and after the Vietnam era. Perhaps one reason for this relative stability lies in the fact that the immediate post-Vatican II period was a time of deep and serious reflection on Christian anthropology in Catholic theological circles throughout the world. A renewed Christian humanism seemed aborning.[5] That new humanism, at least in the hands of philosophical and theological masters, remained in a pattern of continuity with what had gone before. But certain themes, undoubtedly influential in pressing the moral issue of war and peace onto the Catholic agenda, were stressed. Human *historicity* was one such theme. Not only was there no escape from history as the arena of human responsibility; history was pregnant with possibilities for human growth and development. Not everyone shared Daniel Berrigan's apocalyptic view that the times were "inexpressibly evil." But many agreed that the Christian mission in the world was to "redeem the times."[6]

The *evolutionary* character of the human condition and prospect was another important theme in postconciliar Catholic theological anthropology. Arguments over Darwin aside, the main line of Catholic teaching affirmed, as did Murray, that human nature was changing, a view congruent with the teaching of Aquinas. "Evolutionary humanity" suggested that there might come a time, through human intention and action, when wars and rumors of wars were less a staple of international public life. But evolutionary themes could also suggest too unilinear a view of history, too sanguine a confidence that the path toward inevitable "progress" was foreordained. An oversimplified, sentimental view of the human prospect could blind perceptions of evolutionary regression. Totalitarianism comes immediately to mind here. Such things were just not supposed to happen. Might the reports of them be a bit exaggerated?

A third theme in postconciliar Catholic theological anthropology emphasized *freedom* and *intentionality in history*. Classic Catholic theory was vigorously antideterminist in its view of history, politics, and technology. Murray's strictures against determinism, and his concern that the West had been deeply affected by Marxist thematics on the predetermined course of human affairs, exemplified this current. Classic Catholic theory insisted that the world remained a locus of free moral choice and resolutely resisted any reduction of the human prospect to the vagaries of chance, or to impersonal economic, political, or technological forces. Even under the conditions of the Fall, human beings retained the capacity for choice and, under grace, the capacity to choose the good: to shape the future in ways more congruent with our status as the image of God, to "redeem the times." Murray's cautions against technological determinism in thinking through the moral implications of particular weapons illustrated this consistent theme of human freedom in history, which held a prominent place in the theological anthropology of key Catholic thinkers during the conciliar period.

Catholic understandings of sin evolved considerably in the years after Vatican II, and these affected, in different ways, evaluations of the moral problem of war. The concept of "structural evil," central to the analysis of liberation theologies, became prominent. One did not look for evil first in human hearts, but in political and economic institutions and structures. Liberation theologians did not deny the reality of personal sin; but their orienting emphasis was on the evil of structures.[7]

This resulted in a curious inversion. On the one hand, theorists of "sinful social structures" seemed to suggest that the *shalom* kingdom (or a reasonable facsimile) could indeed be built on this earth; wars and rumors of war would thereby end. Yet the instruments for achieving that peace of full justice included, for some liberation theologians, revolutionary violence. In virtually all theologies of liberation, *shalom* required drastic alterations in the structure of power in the United States, source of much of the oppression of the wretched of the earth. If Augustine claimed that "we go to war to gain peace," some liberation theologians seemed to argue that "we go to war to gain justice," to inaugurate in history the Kingdom of *shalom*. At one and the same time, virtual human perfectibility, and utter depravity under the sinful social structures of late capitalism, were being taught.[8]

Liberation theology, under the influence of Paulo Freire, also introduced the concept of "consciousness-raising" to the Catholic debate on war and peace. Consciousness-raisers taught that the poor were largely unaware of their oppression and its causes; they could be brought to a correct understanding by the "pedagogy of the oppressed" Freire prescribed in an influential book by that title. "Consciousness-raising" was, by the late 1970s, a ubiquitous term in Catholic activist circles. Its adherents taught a concept of the limitations of human perception that ran across the grain of the classic notion of the people's "sense of justice." They combined this with notions of the malleability of human consciousness that were incongruent with the mainstream tradition's insistence that the angelic and the demonic remained the fixed poles of the human experience this side of the Kingdom.

Salvation, personal or political, through the raising of one's consciousness is not a notion that finds much resonance in the classic heritage as it had developed through Vatican II. The Catholic heritage knew much about conversion, an act of the intellect, will, and spirit; it knew as well of the demonic capacities of modern politics. But the ideological content of many theorists of "sinful social structures," and the epistemological assumptions implicit in Freire's theory of conscientization, were in contradistinction to the classic theme of human freedom in history that dominated the theological anthropology of Vatican II.[9]

That war was not an inevitable by-product of human sinfulness remained a theme in the Catholic debate on war and peace after Vatican II. Such an understanding was in concert with what Paul Johnson has called the "Promethean" anthropology of Pope John Paul II.[10] But there the agreements would end. The shift in some quarters toward placing a priority on the "structural" dimensions of sin presaged fundamental shifts in understanding on virtually every other aspect of political ethics.

The Meaning of Peace

The multivalent meanings of peace were recognized in classic Catholic theory. At its most profound level, peace was a condition of right relationship between the individual and God, a right ordering of the relationship between creature and Creator made possible in history through grace. Peace also had an eschatological meaning: the completion of the creation in a *shalom* Kingdom of justice and righteousness where the nations streamed to the mountaintop of the Lord, swords were beaten into plowshares, and every tear was wiped away. Eschatological peace was not a matter of human action this side of God's completion of human history. But confidence in that gracious action of God could be a horizon against which to measure, and heal, some of the brokenness of the present. The trajectory of salvation history had been definitively revealed in the Resurrection of Christ; this ought not lead to quietism and withdrawal, but to active service to the world.

On the specific question of peace as the antonym of war, the tradition offered a third understanding: peace was *tranquillitas ordinis*, a dynamic political community rightly ordered in truth, charity, freedom, and justice. The experience of American Catholicism, as reflected in the Council's "Declaration on Religious Freedom," had helped bring the triad of truth, charity, and justice to completion with a fourth indispensable component, freedom.

The image of an "abandoned heritage" is particularly appropriate on this question of the meaning of peace. Between the close of Vatican II and the adoption of the 1983 pastoral letter "The Challenge of Peace," the concept of peace as dynamic, rightly ordered political community virtually disappeared from the American Catholic debate. In its place was substituted a host of confusions.

The unraveling of the heritage on this question occurred during the fierce ecclesiastical and civic debate over the Vietnam War. As early as June 1966, William V. Kennedy was warning that "law and order—'peace'—have never been achieved in any given community or in any group of communities by a working agreement between those who sought to live in peace and those who sought to advance their personal or collective ambitions by murder, terror, and pillage."[11] Kennedy knew that Hanoi fit into the latter camp. And therefore, in his judgment, peace in Vietnam was primarily a question of rational judgment and action, not emotion and wishfulness.[12]

Kennedy recalled John XXIII's proposal for an international political authority capable of resolving conflicts without war, and thought that the obstacles to such a transition in human affairs should be bluntly recognized. But he also argued that we ought to begin by equipping ourselves with whatever knowledge could help start the process of building the pope's vision.[13] This had to include a sophisticated understanding of modern war. But American Catholicism (as other churches) was not very well informed on the topic. It was susceptible to "stereotypes" about the military that had been reinforced by films like *Dr. Strangelove, Fail-Safe,* and *Seven Days in May*, each of which stressed the alleged mendacity, stupidity, or arrogance of the American officer corps. "If religion is to participate effectively in the formulation of a national and international policy that will lead to a genuine world peace, it must have done with this sort of thing," Kennedy warned. "Otherwise, it is doomed to go on wringing its hands on the sidelines."[14]

Kennedy did not want the Church out of the public debate on war and peace; he wanted the Church in the debate, intelligently. But he was not optimistic, concluding his essay with a warning that would go unheeded:

> Within recent weeks, the World Council of Churches on one day admitted the need for lay consultation and study, and then virtually on the next went on to issue a statement on the war in Vietnam that has no more relationship to the resolution of that conflict than to a conflict on some distant planet. Let us hope we can all do better the next time at bat.[15]

The editors of *America* had similar concerns. They argued in October 1966 that Pope Paul VI was not "asking for peace at any price" in Vietnam. War was not the only evil to be avoided.[16]

Bishop John J. Wright of Pittsburgh, a member of the liberal leadership of the American hierarchy at Vatican II, was also worried about the drift of the debate over Vietnam. Addressing the thirty-ninth annual convention of the Catholic Association for International Peace in October 1966, Wright tried to reclaim the tradition of moderate realism, which he apparently thought was being rapidly abandoned by antiwar activists. Wright argued that American Catholicism lacked a genuine theology of peace, which would stress the importance of the heritage of *tranquillitas ordinis* as applied to international institutions. Wright worried that Catholics had shown little effective support for either the League of Nations or the United Nations, and believed the Church was at least partially responsible because of its failure to "develop among Christians a 'sense of organic human community.'" It was not enough, Wright concluded, to settle for a "sentimental hatred of war."[17] The problem of war was a problem of political community, not psychology.

Wright's worries were well founded. Within three years, a representative essay in *America* argued that peace "is no one's responsibility" in the United States government; that "as a nation . . . we are not much interested in peace"; and that the solution lay in the creation of a National Peace Academy that would "diligently and scientifically" research the meaning of peace, which had never been attempted.[18] This essay did not mention the tradition of *tranquillitas ordinis*. Its authors seemed unaware of the sophisticated understanding of peace, its meaning and the minimal requirements for building and sustaining it, that was the heritage of the American Church. Peace was rather a mystery; the mystery might be solved if "the billions of dollars currently directed to researching and developing weapons of war" were "matched in researching and developing weapons of peace."[19]

Where this unraveling of the understanding of peace as *tranquillitas ordinis* could lead was illustrated by one of the first "peace pastorals," issued at Christmas 1971 by Bishop Carroll T. Dozier of Memphis, Tennessee—"Peace: Gift and Task."

Bishop Dozier argued that true peace was *shalom*, a harmony in community that "vibrates with God's blessings" and sets the ground for human growth. Peace, on this ancient Hebraic understanding, was far more than absence of war. The golden age foreseen by the prophets of Israel "began with the Kingdom of God which Jesus preached, a rule of God over the hearts of men." The peace of *shalom* was thus a possibility in history, symbolized by the birth of the Christ Child.

Some would be skeptical of his claim, Bishop Dozier knew, because wars and rumors of wars still surrounded us. Did Christ really bring peace on earth, we are forced to wonder? The temptation to cynicism or despair must be avoided. The peace of Christ that believers feel in their hearts was the basis on which peace in the world could be built. The full peace of Christ meant justice, and justice required both a change in attitudes and a call to action. Justice required a "significant change" in the lives of Memphis Catholics, who had to learn how far our society was from the ideals taught by Christ.

Justice also required religious conversion, a spirit of "self-restraint, our own turning of the cheek, walking two miles with those who force us to walk one with them," the bishop contended. Men and women became peacemakers when they accepted the cross, which was where "Christ's nonviolent life" had led. Early Christianity had understood nonviolence, which was why Christians refused military service in the Roman Empire. The law of love called them to peace, not war.

That witness had been corrupted by the just-war tradition, which evolved when the Church became excessively entangled with the politics of medieval Christendom, Bishop Dozier maintained. The medieval counterpoint to this corruption was the nonviolent Francis of Assisi. Yet even his example had not really stanched the corruption that just-war theory had wrought in the Church. Modern Catholicism had all too frequently yielded its moral principles to the demands of nationalism. "Christian sensitivity . . . was almost completely extinguished during World War II. . . . The atomic bombing of Hiroshima and Nagasaki was the final atrocity," the net result of the numbing of the Christian conscience.

Bishop Dozier cited Vatican II's injunction that war be evaluated "with an entirely new attitude" and then drew his own conclusion: "We must now squarely face the fact that war is no longer tolerable for a Christian. We must speak out loudly and clearly, and repudiate war as an instrument of national policy." This would be an important act of ecclesiastical repristination, based on the fact that we were no longer citizens of one country alone, but of the "international community of mankind." True peace had nothing to do with deterrence, but was based on trust. The only effective human power came from love, "not out of the barrel of a gun." This was not an impractical approach to peace, as Gandhi and King had shown.

Dozier then turned to Vietnam, arguing that "We must stop the war." Vietnam was more than a mistake and more than a tragedy; it was a "sinful situation which swallowed us up." The war was not just, either by the classic criteria or "by the need to bring peace and justice to all men." There was only one moral course open: immediate withdrawal of all U.S. armed forces, abandonment of the South Vietnamese, and war reparations to rebuild the shattered country. Dozier pledged to support any draftee who conscientiously objected to the war in Vietnam, announced the establishment of a draft counselling center in the diocese, urged his priests to make their pulpits "beacons of peace," and called on parents to buy "toys of peace instead of toys of war" for their children.[20]

Bishop Dozier's pastoral letter graphically illustrates how the concept of peace as *tranquillitas ordinis* was abandoned in key American Catholic intellectual and leadership circles less than a decade after the close of Vatican II. The concept of peace as dynamic, rightly ordered political community is simply absent from "Peace: Gift and Task." In its place, Bishop Dozier proposed an amalgam comprised of personal conversion (the "peace of Christ") and elements from the notion of peace-as-*shalom*. The relationship between these connotations of peace was not clarified. What relevance they had to a pluralistic world, let alone a pluralistic country like the United States, was also unstated, as was their relationship to either domestic or international politics. The beginning of peacemaking, for Bishop Dozier, was a shift of attitude in his congregants' minds and hearts; the problem, the bishop implied, was psychological rather than political.

This confusion led to other unhappy results in the Dozier pastoral. The American experience of *tranquillitas ordinis* was demeaned, because our society was to be measured by another standard, "the ideals that Christ taught." Set against these criteria, we would "realize how far away we and our society are from these ideals."

This led Dozier to a teaching reminiscent of the thought of James Douglass: that peace, untimately, was a matter of "the cross [which] was the outcome of Christ's nonviolent life." Such a simplification of the biblical witness led to a romantic view of the early Church. The early Christians' refusal to participate in the military life of the Roman Empire was a logical outgrowth of their temporal proximity to the Sermon on the Mount. That Christian resistance to Roman military service had much to do with the Roman soldier's obligation to engage in idolatrous emperor-worship was not discussed.[21]

Such a tendentious view of early Christian pacifism set the stage for Bishop Dozier's dismissal of just-war theory as an unworthy accommodation to medieval political pressures. Dozier did not mention *tranquillitas ordinis* in his discussion of Augustine, an extraordinary omission, and suggested that Aquinas, had he lived under contemporary conditions of military capability, would have fundamentally altered his approach to the moral problem of war and peace. Bishop Dozier posed St. Francis of Assisi as a more faithful medieval model of Christian response to war and peace, but did not mention Francis's preaching of the crusade.

Dozier adopted, without qualification, the classic historiographic claim that the American Church "until recently gave almost complete support to the public authorities in time of war." The American bishops' pastorals during World War II were ignored. Similar gaps marred Bishop Dozier's exegesis of Vatican II: the injunction to an "entirely new attitude" was stressed, but no mention was made of the Council's teaching on the legitimacy of national self-defense, its tempered approval of deterrence as a regulatory mechanism in world affairs, or its adoption of Pope John's concept of international political authority as a this-worldly alternative to war. What Bishop Dozier took from Pope John was his stress on "trust"—another psychological category—as the foundation of peace.

Bishop Dozier's view of the Vietnam War was severely limited. He did not discuss, at any point, the role of North Vietnam or the Viet Cong in that grim conflict; but the South Vietnamese were "Indochinese mercenaries" hired by Americans. The *only* morally acceptable course was complete American withdrawal, abandonment of our allies, and an American program of war reparations. How this was supposed to lead to peace, freedom, or justice was left unstated. That Bishop Dozier did not consider the possibility of a political settlement in Vietnam, one that would end the killing *and* set the conditions for peaceful resolution of the conflict, seems a telling point, and illustrates the effect on moral imagination and political judgment of abandoning the concept of peace as political community.

The Dozier pastoral was noteworthy for its dependence on New Left analyses of the roots of conflict in the world. After discussing Vietnam, Dozier asked about the causes of immense suffering in the Third World. The causes were in the First World:

> Some are forced to remain poor and are deprived of the means of bettering themselves by an unfair economic system which governs their lives. The conditions under which they live are a breeding ground for violence, for more Viet Nams at home and abroad. . . . Our task, then, is wider than just stopping the war in Indochina, urgent as that may be. We are faced with the need to work for justice among peoples everywhere—in Viet Nam, Pakistan, the Middle East, Northern Ireland, Latin America, Africa, here at home, everywhere—in order that true peace may spread throughout the world.

Conditions of gross deprivation are surely an obstacle to the peace of *tranquillitas ordinis*; but it does not seem accidental that the bishop's list does not include the Soviet Union, Eastern Europe, the Peoples Republic of China, or Cuba. Bishop Dozier evidently believed that justice, understood primarily in redistributionist terms, was the essential precondition to peace. This latter theme became dominant in the American Catholic debate in the twenty years after Vatican II, as did the tendency to minimize (or, as in this case, ignore) deprivation and depredations in the communist world. The fact that Bishop Dozier did not cite the problem of totalitarianism *once* in his "peace pastoral" was an important indicator of shifts in the wind. So, too, was the bishop's claim that his people had to be "educated in peace," on the assumption that the laity was not aware of the problem of war. Notions of consciousness-raising seem to have been at work here.

That Bishop Carroll T. Dozier was genuinely conscience-stricken over the problem of modern war in general, and the Vietnam War in particular, is not in question here. Nor should it be argued that consciences should not have been stricken over these grim realities, which the bishop believed, correctly, the Church had a responsibility to confront. That the evangelical mission of the Church was to call the world to peace is also not in dispute. What can and must be argued about is the bishop's understanding of the meaning of peace and his analysis of the causes of conflict in the world. Bishop Dozier was not engaged merely in religious exhortation or theological speculation; he was teaching his diocese and (through the reprinting of his pastoral letter in *Commonweal*) the entire American Church specific concepts of crucial importance. The bishop's personal intentions are not at issue here. But the theological and political soundness of his teaching, and its consequences on the Catholic debate, are surely open to questioning and critique.

Teachings similar to Bishop Dozier's continued to set the terms of the Catholic discussion on the meaning of peace after Vietnam. In December 1973, for example, *America*'s Christmas editorial argued that "peace is indivisible. If we allow it into our hearts and nurture it by prayer and contemplation, it will begin to spread to others in ever-widening circles." How, was not suggested. What captured the editors' attention was the notion of distributive justice as precondition to peace: although a more just distribution of the world's goods would not eliminate violence automatically, it would set the only possible foundation for peace. This editorial was written at a time when the OPEC oil embargo severely threatened international stability; but the embargo was not addressed as a threat to peace. Rather, the "obvious lesson of the current energy crisis" was that "we use far more than our fair share of the world's resources."[22]

One final illustration demonstrates the degree to which the tradition of *tranquillitas ordinis* was routed in the generation after Vatican II. In a 1978 essay supporting the establishment of a National Peace Academy, Eugene Bianchi argued that Americans still equated national security with arms. Why did we do this? "The answer in the end is not complicated: because we 'think war.'"[23] A federally funded Peace Academy would demonstrate "a visible American commitment toward 'thinking peace.'" Conflict-resolution was an "established science" that could relate the skills involved in domestic counseling to problems of international politics and economics. The knowledge of peace and how to make it was available, or could be developed in tax-supported research; all that was required was the will to peacemaking. In forming that will, and in conducting peace research, "critical perspectives of American minorities and third-world people should have major prominence."

That a National Peace Academy might focus some of its attention on peace as public order in dynamic political community—examining, for example, the role of democracy as a peacemaking instrument in the world—did not appear important to Eugene Bianchi. That this should be so baldly missed says much about the transition in American Catholic thought on the meaning of peace.

The Question of Intervention

American Catholicism might have justifiably been charged, in the early 1960s, with too optimistic a sense of what America might accomplish for the world—a sense shared (and to some extent shaped) on the New Frontier. But the argument over "intervention" seemed settled: America had a responsibility to actively intervene in the world. The means of engagement might be controverted (as in debates over nuclear strategic policy, or on the question of change in the communist bloc and the potential Western role in facilitating such change). But the necessity of engagement was accepted as a first principle. Rare indeed was the voice, like Philip Berrigan's, that talked of an American imperialism whose paradigm was to be found in racial relations here at home. Intervention was an argument settled at Pearl Harbor, and by the continuing crisis of crises known as the Cold War.

This pattern was reversed in the decade after Vatican II. The Catholic debate over the means of intervention was replaced by a debate over the very possibility of any morally worthy American engagement with the world. By the mid-1970s, "intervention" would be a term of opprobrium in American Catholic intellectual, activist, and leadership circles. The consensus on America's responsibility to actively shape world politics and economics was replaced by various forms of neo-isolationism. A hard Catholic neo-isolationism sought the virtual abandonment of an American world role (what U.S. responsibilities remained involved dismantling previously supported regimes in Asia, Africa, and, especially, Latin America). A softer neo-isolationism posited a modest role for the United States in world affairs: as a human rights advocate, or as the bankroller of institutions for multilateral economic aid to the Third World. Both hard and soft neo-isolationism agreed that American *military* intervention—the preferred connotation of "intervention"—was to be avoided in almost any conceivable circumstances.

This drastic alteration in the contextual debate took place with stunning rapidity.

As early as May 1965, William V. Shannon was worried about U.S. intervention in the Dominican Republic, although his concern tended to focus on means and allies. When the Dominican crisis broke, no one argued about intervention, according to Shannon; that was assumed. In fact the entire Alliance for Progress structure anticipated "massive though peaceful" U.S. intervention in support of Latin American reformers. But there were, Shannon argued, "degrees and styles of intervention," and the Johnson administration had contrived to put the United States on the side, not of reform, but of the status quo in the Dominican Republic. But such tactical errors did not invalidate the interventionist case. If the U.S. could not gather itself to act for reform in Latin America, "then the Alliance for Progress and the liberal dream in

Latin America are doomed."[24] The classic position on intervention was holding here, but just barely.

Elements in the abandonment of that position were displayed nine months later in Ronald Steel's *Commonweal* analysis of "Cold War Myths."[25] American intervention in world affairs through alliances reflected a "pactomania of the fifties," itself rooted in "an obsession with aggression in areas where it was often unlikely and frequently impossible."[26] The results were clear in Southeast Asia, where U.S. military intervention in a "civil war among Vietnamese" had polarized the situation, given Ho Chi Minh the mantle of nationalism, and made the United States vulnerable to the charge of imperialism.[27] The postwar world, "dominated by those twin symbols of intransigence, Dulles and Stalin," had been eroded. The Chinese/Soviet split; the rise of nationalism in communist countries; the fact that "NATO has been made largely irrelevant by its own success"—all these required cleansing the governmental Augean stables of Cold War mythology, and reconsidering America's interventionary responsibilities in the world.[28]

Steel's essay was a symbolic benchmark in the transition of the American Catholic debate on the question of intervention: a halfway house between the interventionist consensus *ante* and the triumph of neo-isolationism. Steel did not dwell on the evil of the American system (and its ineluctably bad impact on the world) so much as on American governmental ineptitude. The world was changing, and old "myths" stood in the way of realizing that the guiding horizon of U.S. foreign policy had to change. Steel was not optimistic about the prospects for such a change; but his 1966 pessimism had more to do with empirical judgments on world realities than with a basic attack on America as an actor in world affairs.

The themes of *that* attack emerged a few years later, and were illustrated in an *America* essay reporting on a Jesuit meeting called to consider the war/peace question in 1970.[29] The basic conviction of the participants was that "the contemporary American's easy identification of freedom with his own style of material prosperity . . . is inconsistent with his professed desire for world peace and economic development."[30] Therefore, the first requisite for peace and justice was that "the individual American citizen must recognize that he has no moral choice but to renounce his quest for maximum material prosperity."[31]

The meeting had begun traditionally, Andrew Christiansen noted, with a paper stressing original sin and the human need for redemption. Such considerations were soon discarded, however, and the participants led to an understanding of "the probable incompatibility of American economic structures with goals of international justice and peace" by Richard Barnet of the Institute for Policy Studies. Barnet taught the Jesuits that the root of the world *problématique* was American consumerism.[32] A classic New Left analysis, in other words, turned the tide of the meeting. Theology subsequently took a back seat to a social, political, and economic analysis emphasizing the "incompatibility" of American society and political economy with international justice and peace.

Fr. Joseph Fitzpatrick of Fordham University provided Jesuit reinforcement for Barnet's teaching, arguing that American economic assistance to the Third World was structured so that Third World development would "not interfere with our ability as six per cent of the people of the world to enjoy over fifty per cent of its resources."

These themes struck home; the meeting proposed that, to meet "the threat to peace arising from American economic, political, and military domination in foreign countries," a new research center be established.[33] The result was the Washington, D.C.-based Center of Concern. From its intellectual genesis, it seemed likely that the Center would be (as it has been) a voice for neo-isolationism, and that its neo-isolationist stance would rest on teachings about the evil structures of American society. This was less than five years after the close of Vatican II, and only a decade after the publication of *We Hold These Truths*.

From this point on, neo-isolationist themes dominated the contextual debate on war and peace in the American Catholic attentive public. William Pfaff's description of the decline of "American imperial ambition" was commonplace, as was Pfaff's argument that policy ought to have been redirected during the Kennedy adminstration. But an "ambitious reassertion of American authority" then had paved the way for "our present calamitous situation." Thus in Vietnam, "we continue to populate another foreign graveyard, one which is already close to the moral engulfment of our nation."[34]

On the death of Lyndon Johnson in 1973, *America* editorially portrayed the former president as an embodiment of the myth of the archetypical American which, for a while, had worked to the benefit of many of our citizens. "But beyond our shores its writ did not run. In the end, in the steaming jungles of Southeast Asia, the myth finally betrayed the man who embodied it."[35]

Michael Klare, writing two years later in *Commonweal*, was not so sure that the lesson had been learned, He worried that armed intervention in the Middle East was being planned by those "national security managers" who insisted that "America's privileged world position compels it to prepare for future wars abroad." The key word was "privileged"; the sources of U.S. foreign policy were to be found in the consumerist character of American society.[36]

Klare's concerns were shared by the editors of *Commonweal*, who wrote in April 1975 that "the United States has become the preeminent contender for the title of world champion death merchant" because of its military assistance policies. The lesson of Vietnam was the rigorous avoidance of regional foreign entanglements. U.S. arms aid to Pakistan and various Persian Gulf states suggested that these lessons had not been learned.[37] So did the *Mayaguez* affair, which had been eagerly greeted by the Ford administration and the Pentagon as an opportunity to demonstrate American will. "So the bully flexed his muscles and tingled to the exertion. The elephant-gun was got out to zap the gnat. The brute triumphed. But it was a cowardly performance. Most such exercises are." The administration was not the only problem; it was frightening when the Congress could get swept away "in a patriotic emotionalism born of a lunatic military aggression."[38]

Neo-isolationism shaped commentary on virtually every foreign policy issue. Draft registration would reopen Pandora's box. "If the draft is restored," Murray Pollner wrote in *America*, the United States would "certainly return to its pre-Vietnam policy as global policeman and its willingness to intervene abroad."[39]

America's editors worried that the Carter doctrine on the Persian Gulf was rhetorically excessive and lacked sufficient flexibility. True, the Soviets had posed a "challenge" in their invasion of Afghanistan. But what was most worrisome was that "public militancy" over the invasion would lead to a resurgence of interventionist sentiment.[40]

Commonweal's editors were equally anxious about the effects of Afghanistan on U.S. intervention in the world. Citing George F. Kennan, the editors claimed that the postinvasion "militarization of thought" and argument in Washington was unprecedented since World War II, and reflected the rekindled power of "veteran Cold Warriors." *Commonweal* condemned the Soviet invasion and distanced itself from those who called the Soviet move "defensive." But the editors' principal concern was clear: "the public's mood of frustration makes it a ripe target for the *simplificateurs* of the right," for whom "the world is always on the brink of another Munich."[41]

Such was the state to which American Catholic debate on the question of intervention had sunk by 1980. Debate on intervention in light of the Catholic heritage of peace as dynamic political community had virtually disappeared. Neo-isolationisms ruled the day, despite the persistent claims of *Commonweal's* editors that U.S. military adventurism was the real cause of whatever "isolationist" currents existed in the country.[42] New Left themes, first preached by the Berrigans in the 1960s, had become the basis of the discussion. The key issues were fundamental alterations in the structure of power in American society and a healthy respect for the "complexities" of the world. In extreme form, this neo-isolationism could lead to a foreign policy largely unconcerned about the Soviet invasion of Afghanistan because that meant the U.S.S.R. was "stuck to a tarbaby" in the Third World.[43]

More often, though, neo-isolationism emphasized the great difficulty, if not sheer impossibility, of any useful (much less morally worthy) U.S. "intervention" in the world. Change was in the wind: the forces of history were blowing through bellows not guided by American hands. One might accommodate to those forces gracefully or spitefully; but that accommodation was in order was not much disputed. John Courtney Murray's warnings against historical determinism had been ignored or rejected as incompatible with the new realities of the world scene. This rout took less than ten years to complete.

The Question of Military Force

The classic Catholic heritage agreed with Clausewitz in a limited sense: it sought to link war and politics so that the resort to proportionate and discriminate armed force would serve the ends of *tranquillitas ordinis*. Just-war theory provided the moral calculus for thinking through the hard questions of precisely when (the *ius ad bellum*) and how (the *ius in bello*) military means might be ordered to that end. The deliquescence of just-war theory will be treated later in this study. The immediate concern here is to see how shifts in the Catholic contextual debate over the use of military force were applied to several illustrative cases in U.S. foreign policy.

That profound shifts on this question were underway in the generation after Vatican II is readily demonstrated by two essays written sixteen years apart. In May 1965, *Commonweal* columnist William V. Shannon defended the preemptive use of military force to preclude the Peoples Republic of China (PRC) from gaining a nuclear capability. Shannon did not wish to wage any kind of nuclear war with the PRC.[44] But he feared that a Chinese nuclear arsenal would make war between the

United States and China much more probable in the future. In those circumstances, Shannon did not understand "why we have to sit idly by and do nothing while the notoriously hostile Chinese Communists develop the capacity to blow up the world."[45]

Shannon argued that the 1963 nuclear test ban treaty provided "all the moral sanction for prompt action that anyone could desire," inasmuch as it represented the overwhelming judgment of world opinion that nuclear proliferation had to end.[46] It was regrettable that there were no effective international means to deal with the threat of a nuclear-armed PRC. "If the United Nations were what it should be—a world government ruling the world community under a code of law—it would enforce the test ban treaty by sending its air force to bomb the Red Chinese nuclear plants." But the lack of effective international instrumentalities neither minimized the danger nor lessened the need that something be done to halt nuclear proliferation. Therefore, Shannon proposed "that the US send its own planes to bomb—with conventional bombs—the Chinese nuclear plants."[47]

"By old-fashioned standards," Shannon admitted, "that would be an act of war." But those were not the relevant standards:

> By the new world community standards that the nuclear-and-rocket age require, it would be an act of elementary peacekeeping. It should be no more controversial than the police in any city today disarming underworld gangsters. Notwithstanding all the pro forma protests that would occur, the entire world, including Russia, would experience a profound new sense of security and confidence.[48]

Shannon's call for preemptive military action to support world order goals was controversial in 1965. But the rationale for Shannon's proposal (whatever its strategic and tactical merits or defects) suggests that the heritage of *tranquillitas ordinis* was alive in the Catholic debate at that point. International order required the discreet use of military force, not for national self-aggrandizement, but for "elementary peacekeeping."

The degree to which that understanding was abandoned is illustrated by *America*'s 1981 editorial comment when Israel, operating according to Shannon's standards and methods, preemptively destroyed the Osirak nuclear reactor being built in Baghdad by the government of Iraq (a regime that surely fit Shannon's image of a "notoriously hostile" adversary).

America's editors granted that, behind the almost universal public outcry against Israel there was, in some capitals, quiet satisfaction that Iraq's volatile Saddam Hussein had been denied nuclear technology.[49] Such satisfaction was not shared at America House, though, for the editors worried that the Osirak raid proved "that Israel will never hesitate to act unilaterally when it judges its own security to be threatened." The editors were also concerned that Israeli leaders continued to believe that national security was better protected by armed force than by diplomacy.[50] *America* was not buying the argument that Israel's action was defensive, and designed to prevent the use of a nuclear weapons against Tel Aviv. "What is painfully clear . . . is that Israel, already in possession of its own nuclear capability, intends to maintain a nuclear monopoly in the Middle East."[51]

All military actions carry a heavy weight of ambiguity. But that *America*'s editors would not frankly admit that the Israeli action had removed a potentially disastrous tool from the hands of one of the Middle East's most dangerous leaders; that the editors would charge Israel with a lack of concern for international law, in the face of Saddam Hussein's refusal to even recognize the existence of the state of Israel, erected by the United Nations in 1948; that the editors seemed more offended by the alleged Israeli nuclear monopoly in the Middle East than by Iraqi attempts to break that monopoly and French collusion in the Iraqi effort—we are far indeed from William V. Shannon in 1965. If the editors of *America* could not bring themselves to admit that the Israeli raid on the Osirak reactor came as close as one is likely to get to a use of military force that met the classic *in bello* and *ad bellum* criteria of the just-war theory, especially in the tangled mess of the Middle East, it is hard to imagine what possible use of military force—precisely for the sake of strengthening what fragile peace there is in the world—they would countenance. The contextual debate on the use of military force had not only changed; it had been virtually reversed.

America was similarly angered by the U.S. action in Grenada two years later. Its editors charged the Reagan administration with a "cheap and cynical bit of exploitation" in referring to the possibility of hostage-taking, as in Tehran. Moreover, "restoring freedom and democracy to Grenada . . . seemed implausible from the start, since the crisis had been triggered by conflict between one faction of a Marxist movement and another."[52] The editors wondered "how persuasive would the [administration's] case be when and if the captured documents supposedly outlining this Soviet-Cuban threat were made public?"[53] The upshot of it all was a mirror-image: "To descend on a small island in the Caribbean with such overwhelming firepower in order to make a point elsewhere in the world only surrenders more of that sacred ground that separates the United States from its totalitarian rivals."[54]

Commonweal took a similar line. Peace and security still hinged on the "taint" that attached to unwarranted military action. Just cause "was not present in the case of Grenada."[55] The Reagan administration "wanted this invasion."[56] The editors noted that the U.S. action had been compared, around the world, to the Soviet invasion of Afghanistan, a comparison made "with some justification." There were differences of scale, of course. But there was an important parallel between "America's neat, little exercise in intervention" and the Soviets' "big messy ones": both great powers were determined to exercise a veto over changes in regimes within their "self-defined spheres of influence," an exercise they both conducted with the most "transparent falsifications." America had been given a dose of the climate of falsehood that characterized totalitarian societies, *Commonweal* concluded. "In invading Grenada, we have invaded ourselves—with mendacity and sham."[57]

These reactions paralleled those of the Rev. J. Bryan Hehir, the American bishops' chief foreign policy advisor. The Grenada invasion, Hehir wrote, was a "teachable moment which should be used to help U.S. Catholics understand the interventionist history of U.S. foreign policy."[58] The key analogy was not Grenada/Afghanistan but Grenada/Poland. The precedent for armed Soviet intervention in Poland, "especially if invited in by key actors," had been strengthened by the U.S. action in the Carribean.[59] But what really exercised Father Hehir was the possibility that the Grenada invasion was practice for a similar operation in Central America. "Those of us with doubts about or decisive opposition to the Grenada action should understand where the focus of attention should be," Hehir warned, and that was "establishing a barrier against future interventions, not winning the debate about the last one."[60]

The Israeli action in the skies over Baghdad and the U.S. action in Grenada were not morally unambiguous. No military action ever is, a reality of the soldier's craft recognized by its most thoughtful practitioners and historians. Perhaps there were nonmilitary ways of denying nuclear weapons to Saddam Hussein of Iraq; perhaps there were ways to rid the Grenadian people of the curse of being ruled by the likes of Bernard Coard and Hudson Austin (an exercise in liberation for which the Grenadians seem overwhelmingly grateful). Those are points worth exploring, although the burden of proof lies with those who would argue for the potential success of such efforts. But what seems so disproportionate is the tenor of the reactions to these two episodes. The grotesque parallelisms suggested or overtly drawn between Grenada and Afghanistan—one a country now ruled by a democratic government, the other an abatoir in which systematic policies of genocide are pursued—are only the most egregious examples of what the deterioration of American Catholic thought on the questions of intervention and the use of military force had led to by the early 1980s.

The debate had come far from that "cool and dry" climate of moral reasoning enjoined by John Courtney Murray. The "coolness and dryness that characterize good argument among informed and responsible men"[61] were singularly absent in these examples of recent American Catholic debate on the use of military force. That absence did not signify an improvement in the quality or moral health of the argument.

The Present International System

The heritage of peace as public order in dynamic political community offered Catholic moral theory the basis on which to critique the post-Westphalian organization of the international system, and the assertions of virtually absolute national sovereignty that characterized it. As it had developed through the Second Vatican Council, the heritage emphasized the necessity of new levels of international organization to meet the demands of the common good, now considered international in scope. The American bishops' World War II pastoral letters were one benchmark in this evolving body of Catholic thought, particularly in their emphasis on the centrality of democratic values to any international institutions that might provide a morally satisfactory alternative to war in resolving international conflict.

Given forty years of experience with the United Nations and its related agencies, the Catholic vision of *tranquillitas ordinis* through international organization may appear hopelessly grand. For that reason, it is important to recall that the heritage argued a *minimalist* approach to international organization. It did not claim that international organization could solve every conceivable problem in the world (in fact, the Catholic principle of subsidiarity would warn against any such efforts). The heritage, insisting that conflict and war need not be synonymous, looked to international organization primarily as a means for addressing the dilemma of peace and freedom, which, because of modern weaponry, interdependence, and the proliferation of independent sovereignties in the world, seemed incapable of resolution in an anarchic international system.

The heritage emphasized four, interrelated, minimalist objectives which, if achieved, would contribute to the peace of political community on an international scale. Disarmament was one objective; but disarmament could not proceed faster than international legal and political means for resolving conflict evolved. Thus, the heritage linked progress in arms reduction to progress in the creation of effective international legal and political institutions. These institutions would have to be based on a minimal sense of political community across national borders, a community that had to be built around agreement on the defense of basic human rights. Finally, the heritage linked economic, social, and political development in the formerly colonized world to these other three objectives of disarmament, law, and community: recognizing, in its tradition of moderate realism, that stability had to be joined to peaceful change for those on the underside of the contemporary world.

This fourfold structure of objectives to be sought in the pursuit of peace served several useful functions. First, it was a comprehensive view of international problems, suggesting linkages that had to be made were the world to advance from anarchy to community. Secondly, it provided a distinctive angle of vision on the problem of communism in the international system (as had the American bishops in their insistence that post–World War II planning take account of the principles of the Atlantic Charter). Communism was a major problem, not merely because it was wrong in its world-historical view, but because the internal practices and international behavior of communist states (especially the U.S.S.R.) were threats to peace. Thus, the heritage looked toward the transformation of communist states as a necessary part of the strategy of peace, a vision most optimistically affirmed in *Pacem in Terris*.

Concerns for international order and the transformation of the international system remained part of the American Catholic debate on war and peace in the generation after Vatican II. But each issue in this debate would be framed differently. A fractionation of the four minimalist objectives set in, paralleling similar developments in the worlds of political science and diplomacy. The American Catholic debate became less rigorous in its insistence on the linkage, for example, between disarmament and human rights. The very objective of disarmament was largely abandoned, to be replaced by the "halfway house of arms control."[62] The U.N. system was increasingly viewed from the perspective of the Group of 77 caucus at Turtle Bay, as were issues of economic and social development. The North Atlantic human rights tradition, emphasizing civil liberties and political freedoms, was still affirmed; but the notion that "human rights" included economic, social, and cultural desiderata received increasing (and at times overwhelming) emphasis.

Each of these transitions should be explored in turn.

The fundamental problem of the present international system, it was argued at the time of Vatican II, was that national security was impossible. Humanity had reached the point where a tripling of defense budgets could not offer safety against attack.[63] The basic task was thus to achieve a system of mutual security in which one "covets security for others . . . just as we covet it for ourselves."[64] Through the U.N. system, the world was slowly moving toward conflict-resolution through democratic institutions. Such efforts should be intensified, especially in the face of those who argued for the "irresponsibility of absolute national sovereignty."[65]

Such themes were rare in the Catholic debate less than a decade later. Saving world peace "is too essential a task to be left merely to statesmen," observed *America* editorially at the 1969 closing of the first U.N. Development Decade.[66] The basic threats to national security and international order did not involve what Solzhenitsyn would call a "world split apart" by a basic rift over the question of human nature, but by the gap between the developed and the developing countries.[67] Psychological dysfunction was more crucial than political conflicts. As Patricia Mische, author of the influential *Toward a Human World Order* wrote:

> We have entered a new era of global interdependence. But our spiritual and moral sensitivities and our systems have lagged behind. It is now time to close the gap. . . . We are coming to the end of the dichotomies between "us" and "them." "We"—all of us—are affected by decisions and activities in other areas of the world. . . . Let us learn to weave together with the power that comes from a clearer image of the human and from a clear vision of our own power to be a creative force in history. Let us not be afraid to weave new works and new designs, to be participants in shaping a more human world order.[68]

Where this was heading seemed clear from an *America* essay on the twentieth anniversary of *Pacem in Terris*, which did not mention civil rights and political freedoms as essential components of *tranquillitas ordinis*,[69] and a *Commonweal* article attacking the notion of a National Endowment for Democracy to support the evolution of democratic institutions in the Third World[70].

These were not the only themes in the debate. *America* acknowledged the international character of terrorism and the need to control it through international agreement.[71] Francis X. Winters, in an essay reaffirming the classic view that "the state system is indeed compatible with world order,"[72] criticized those "world order" modeling projects that assumed the disappearance of the nation-state. But voices like Patricia Mische's were more representative of the themes in the Catholic debate after 1965.

The evolution of the Catholic debate on arms control and disarmament will be discussed in detail in Chapter 9, below. For now, we may briefly note the transition between an approach that stressed "disarmament, American style (that is, under a rule of law) . . . [with] arms reductions in stages, paralleling agreement on adequate inspection and control,"[73] and approaches that focused solely on Western responsibilities for conventional arms proliferation[74] or that celebrated the 1982 Central Park anti-U.S. weapons policy demonstration as "the emotional peak and focus of public concern for peace."[75]

America essays in 1970 on the twenty-fifth anniversary of the U.N. reviewed the American bishops' commentary on the postwar planning of the Roosevelt and Truman administrations and recalled the consistent support of the Holy See for international organization as an alternative to war, thus hearkening back to the tradition of *tranquillitas ordinis* as it had evolved through Vatican II.[76] On the same anniversary, *Commonweal* editorially endorsed U.S. ratification of the Genocide Convention and repeal of the Connally Reservation to U.S. participation in the International Court of Justice.[77] But a U.N. twenty-fifth anniversary editorial in *America* argued that "in the coming decades the key to peace in the world will depend in large part on the attitudes, and the economic strategy, of the wealthy nations of the world toward the less fortunate developing nations—a hint that the comprehensive approach that linked disarmanent, law, community, and development was breaking down, with development issues (understood largely in terms of "the gap" between rich and poor) taking priority.[78]

Both *Commonweal* and *America* editorially defended the view that the trouble with the U.N. at age twenty-five was the attitude of its member states. *Commonweal* urged "radical changes in the dispositions" of all, but primarily the major powers, which should be willing to "forego veto privileges and be guided more by the will of the majority."[79] Both journals supported U.N. Charter reform, but focused more on limiting the role of the major powers at the U.N. than on a General Assembly voting system that gave equal weight to the votes of micronations and the United States. Neither journal discussed the relationship between the failures of Third World tyrannies and the sad state of U.N. affairs on the organization's twenty-fifth anniversary.

Commitment to U.N. reform remained relatively constant in the American Catholic debate during the 1970s, with both essayists and editorial writers criticizing the U.N. failure to come to grips with international terrorism, and the increasingly anti-Israeli tone of the General Assembly's debates and resolutions.[80] But *America* argued that "U.S. indignation in the face of the new U.N. majority has been superficial and dishonest," citing the "unwise suspension of South Africa" as an "expression of frustration at the repeated refusal of the United States and others to cooperate with legitimate efforts to bring justice to southern Africa, a refusal that contradicts the spirit and sometimes the letter of many an Assembly consensus."[81]

The U.N. Third World caucus seemed rather benign at America House; "the world's have-nots" had brought the Assembly's attention to bear on the issues of most concern to them. Neither peace nor justice was possible without recognizing the legitimacy of the Group of 77 perspective.[82] Though generally supportive of the new visibility given the U.N. through the ambassadorship of Daniel Patrick Moynihan, the editors wondered whether "Mr. Moynihan's evaluation of the American record in assisting economic progress and the protection of civil liberties" was not "somewhat facile." "U.S. defense of civil liberties [will not] make a credible platform of opposition if attention is directed only at Uganda, Algeria, and the Soviet Union, while American dollars continue to flow to equally abusive regimes in Chile and South Africa."[83] (Moynihan's point, of course, was not that attention be directed solely at Uganda, Algeria, and the U.S.S.R. but that it was not being directed there at all, in the usual pattern of General Assembly debate.)

Several months later, when the General Assembly passed its resolution identifying Zionism as a "form of racism," Thomas Powers, although mildly critical of the resolution, described subsequent press criticism of it as a "three-day riot of recrimination, a nearly unanimous din of vituperation, name-calling, and outright slander. Hysterical is not too strong a word for it. You would think Martin Bormann had entered the delegates' lounge with the Russian Ambassador on one arm, and Arafat on the other."[84]

America on the other hand, opposed another project of the U.N. Third World caucus, the New World Information Order, even while arguing the basic case of the Group of 77—namely, that "cultural differences are real, and there is a real imbalance in the systems of world information."[85] Similar adoption of a Third World–caucus perspective could be found in a lengthy *Commonweal* essay on the Law of the Sea Treaty, which argued that the central issue on the international agenda was the gap between rich and poor nations.[86]

In the generation after Vatican II, then, a commitment to international legal and political institutions remained relatively constant in the American Catholic contextual debate over war and peace. But the standards that had once guided that debate—commitment to Atlantic Charter democratic priniciples, and a comprehensive approach to disarmament, law, human rights, and development within a vision of *tranquillitas ordinis* on an international scale—changed. In their place arose a more fractionated approach to problems of international organization, and a marked tendency to adopt Group of 77 perspectives on both the nature of the international problematic and the functioning the U.N. system.

The classic heritage had taught American Catholics about the necessity of political community beyond national borders as a precondition to international organization. Human rights, primarily understood in terms of the North Atlantic tradition of civil liberties and political freedoms, were taken as the standards by which transnational political community ought to form. Political community, the classic heritage knew, was not simply a matter of structure but of values. The connection between peace and human rights continued throughout the postconciliar period. But the content of the notion of "human rights" changed. Economic, social, and cultural benefices took on a new significance in the American Catholic debate over human rights.

A 1974 *America* essay, for example, reported enthusiastically that the second World Conference of Religion for Peace, held in Louvain, had transcended denominational and sectarian boundaries by agreeing that "global justice" was "the foundation for peace."[87] "The demands of the rich for scarce resources to support wasteful and indulgent patterns of consumption and production" was a global justice issue, because the consumption of the wealthy was "impairing access by the poor to the supplies needed to meet their most basic development needs." The answer lay in the New International Economic Order proposed by the sixth Special Session of the U.N. General Assembly.[88] Toward this end, and echoing the teaching of Paulo Freire, the conference participants urged "all religions to deepen the 'conscientization' of mankind in order to build a public opinion that leads to political action."[89]

David Hollenbach's book *Claims in Conflict: Retrieving and Renewing the Catholic Human Rights Tradition* marked a watershed in the postconciliar human rights debate in American Catholicism. Msgr. George Higgins, reviewing Hollenbach's book, welcomed its call "for a new ideology that will transcend both liberalism and Communism and make its central issue the indivisible rights of man."[90] Hollenbach's study implied that the liberal human rights tradition and the economic and social "rights" claims of Marxist theory were somehow equivalent, in the sense that they each spoke to important dimensions of the one human person and his or her rights and duties, and because both were deficient in lacking a comprehensive approach to the definition and protection of human rights.

The key criterion for any human rights scheme, according to Hollenbach's argument, was its ability to create "greater equality in situations characterized by inequalities of material well-being, freedom, and power."[91] Although the priority given to greater economic equality seems significant, Higgins relayed Hollenbach's confidence that progress could be made, "not by a series of trade-offs between civil and political rights on the one hand and social and economic rights on the other, but by a new strategy that will harmonize the two."[92]

Hollenbach's new "strategic moral perspective" on human rights was built around three principles: "1. The needs of the poor take priority over the wants of the rich; 2. The freedom of the dominated takes priority over the liberty of the powerful; 3. The participation of marginalized groups takes priority over the preservation of an order that excludes them."[93]

Hollenbach's willingness to give equal weight to the economic and social "rights" of Marxist theory, and his tendency to diminish the relationship between political freedom and economic development, presaged a Catholic discussion that ignored or minimized the fatuity of any notion of "rights" in totalitarian societies, and that increasingly saw economic justice (usually posed in egalitarian terms) as the precondition to the evolution of democratic forms of governance.

Hollenbach's book reflected the admixture of rights found in *Pacem in Terris*. In the 1970s, then, New Left sensibilities were married to the universal Church's traditional unease with Liberal ideology to yield a softer form of the human rights discussion that had become common at the United Nations by this time. Empirical evidence on the congruence among democratic forms of governance, basic protections of the individual against arbitrary state power, and economic development was not a staple feature of this discussion.

These were not the only voices in the debate. Thomas Melady argued forcefully against "selective outrage" on human rights issues, and was particularly critical of the double standard on human rights violations regularly applied at the United Nations.[94] *America* argued in 1977 that the "U.S. Government should surrender neither of its ideals, protection of human rights and vigorous pursuit of balanced disarmament."[95] Msgr. George Higgins applauded the forthright Western condemnation of Soviet and East European human rights violations at the Belgrade Review Conference on the Helsinki Accords. A vocal U.S. posture at the Belgrade conference was not a matter of ideological recalcitrance but an affirmation that "real peace between East and West must be based on genuine acceptance and implementation of the Helsinki accords," a process that would be advanced by Western support for the various Helsinki monitoring groups that had sprung up in the communist world.[96]

Yet the basic thrust of the postconciliar discussion was well summed up for *America* readers by associate editor James Brockman's 1980 assertion:

> Not everyone in the world views human rights in the same way. . . . The liberal tradition flowing from the French and American revolutions sees human rights as principles for defending the individual from the state. The Marxist tradition so influential in the Third World views human rights as means to defend the weak classes of society from their exploitation. The liberal is concerned with the rule of law, due process, and free expression of ideas; the Marxist, with bread, shelter, health, and literacy. The two, it would seem, should not exclude each other, and recent Catholic thought supports both approaches and tries to synthesize them.[97]

The fact that Marxist-influenced regimes in the Third World consistently failed to deliver on their economic and social "rights" agendas; the fact that any "rights" in totalitarian societies were chimerical, a cynical adoption of Western language to mask the all-encompassing claims of state and party; the fact that Western societies were not only concerned with "bread, shelter, health, and literacy," but were providing them—these realities were not prominent in the American Catholic debate on human rights theory and practice in the generation after Vatican II. That debate was marked by a considerable ahistoricity, and an extremely abstract view of the human rights issue in which *theoria* counted far more than *praxis*, despite claims to the contrary.

Thus, in a notable paradox, the conciliar admonition that the Church read the signs of the times was not heeded. A distinctive human rights ideology—influenced by New Left concepts of the locus of power in American society and the international system, a sentimental fondness for Third World perspectives, and a particular reading of the weakness in the Catholic human rights tradition—took precedence over a rigorous examination of the circumstances in which human beings were not forced to make the false choice between bread and freedom.

Issues of Third World development had long been part of the Catholic contextual debate over the present international system. In the early 1960s, *America* and *Commonweal* were among the most vigorous American editorial advocates of foreign aid.[98] Foreign aid had multiple rationales and purposes. Humanitarian concerns required it. So, too, did the integration of the formerly colonized countries into a nascent international system capable of resolving conflict without war; hunger, disease, and poverty were incompatible with a morally worthy peace of political community on the international level. Foreign aid was also important in an East/West context: programs like the Alliance for Progress were essential both for meeting human needs and for providing prophylaxis against communism. Arguments about development assistance centered on questions of means. That development was an essential, minimal objective in building dynamic political community in the world was not doubted, nor was the American capacity to act for good ends in the Third World (even if some commentators wondered if the national will could be mustered for the task).

All of this changed dramatically in the years immediately following Vatican II. The transition in American Catholic thought on these issues forms the essential background to Chapter 10, below. Here we may note that the transition was encapsulated as early as 1966 by Denis Goulet, writing in *America* of Brazilian Archbishop Dom Hèlder Câmara's critique of the Alliance for Progress. Goulet agreed with Câmara's critique that an "unworthy form of materialism" was being forced on Latin America in the name of development.[99] Anticommunism was "an irrelevant term and a sterile policy; the real issue was poverty, which denied the Latin American underclass their personal and economic freedom."[100] The first requisite, therefore, was a revolution in American consciousness: "a total conversion from a value system based on acquisitiveness and superfluous wealth—with its corollary, foreign aid, as the instrument of defense of one's economic interests and politico-military strategy—to a value system based on world solidarity in the battle against massive dehumanizing need."[101]

Here were the chief themes by which the development debate in American Catholicism was turned: the fundamental problems were not in Latin America, but in North America. Latin American poverty was caused by North American greed and acquisitiveness. Foreign aid was a sop to guilty consciences at best, and an imperial instrument of control at worst. A radical transformation of the economic system of the North was the precondition to any geniune progress. The key economic issue was distribution, not production. North American development models were either culturally irrelevant, or, because of their moral flaws, unworthy of consideration.

These themes recurred time and again: in a 1967 essay praising Julius Nyerere's nationalization of industry and agriculture in Tanzania;[102] in a 1971 *Commonweal* editorial supporting multilateral aid institutions;[103] in an essay which marked the fortieth anniversary of *Quadragesimo Anno* by asserting that "Americans clearly enjoy a grandiose style of life, like the nobility and gentry of the medieval court, while most of the world's population lives on the level of serfs. And not unlike the old monarchical order, the castle is abundantly protected by its palace guards and knights and soldiers";[104] in an essay arguing that the first concern of development should be "the liberation of man from the oppressive social structures to which he is indentured rather than the alleviating of his sufferings . . . [which] was a Marxist concern before it became a Christian concern";[105] in a disposition to accept at face value the "limits of growth" analysis of the Club of Rome;[106] in a 1974 celebration of Nyerere's collectivizing Ujamaa experiment;[107] in strong support for the U.N. New International Economic Order and for the 1980 report of the Brandt Commission.[108]

These analyses were bound together by two primary teachings. First, "the gap" between the First and Third Worlds defined the gravamen of the development issue. Secondly, market-oriented approaches to development were to be held highly suspect. These dispositions were heavily influenced by key teachings of the theologies of liberation.

Thus, although the chief claim of the classic heritage remained—the present international system had to be altered in order to achieve the peace of dynamic political community in the world—the *contents* of that alteration, and the perceived *obstacles* to it, changed dramatically. The problem of totalitariansim virtually disappeared from the Catholic debate over world order. The primary obstacle to a "world more fully human" lay in the organization of power in American society, and in that society's political, cultural, and economic effects on the world. Change in the present international system would begin at home. According to the main themes in the Catholic debate between 1965 and 1985, it would have to be a drastic change indeed.

The Boundaries of Political Obligation

The classic Catholic heritage, as it had evolved through Vatican II, lent itself to a tempered internationalism. This internationalism has been described above in some detail. Its tempering came from a recognition that, whatever the limitations of nation-states, they would remain the basic building blocks of international public life for the foreseeable future. American Catholicism blessed both national patriotism, rooted in a moral obligation to present political community, and a sense of moral and political responsibility that transcended national boundaries.

This transnational sense of obligation not only reflected the extension of the dynamic concept of *tranquillitas ordinis* to the international level; it also involved the Church's position as the world's oldest and largest transnational institution. American Catholics, through the time of Vatican II (and, for that matter, ever since), thought of themselves as members of a global community. The enormous financial investment of the American Church in missionary activity was one expression of this sense of obligation to communities lying "beyond the water's edge."

The tempered internationalism of the Church was a barrier against a sentimental view of "global citizenship." With Hannah Arendt, and in concert with classic political philosophy, the Church taught that citizenship was a function of a *polis*. Where there was no *polis*, there could be no citizenship in any morally meaningful sense of the term. This understanding began to shift in *Pacem in Terris*, with its emphasis on the "universal common good." Action toward the realization of the international common good was clearly incumbent on Catholics of all nationalities. But that obligation did not minimize one's responsibilities toward existing political communities, nor did it constitute "global citizenship." Even in *Pacem in Terris*, the path to transnational political community, and the international legal and political institutions that would be its expression, was *through*, not *around*, existing national communities.

In the post–Vatican II debate, the argument about the boundaries of political obligation revolved around the extent of one's obligation to existing political community (in, for example, the debate over conscientious objection during the Vietnam War), and the role of the Church in American public life. The question of conscientious objection will be taken up below; here, the arguments over the role of the Church in civil society, and how it might marry its commitment to transcendent moral norms with its responsibilities to American society, will be highlighted.

The *status quo ante* for this debate was summarized in a 1966 *America* editorial commenting on the National Inter-religious Conference on Peace, called to protest U.S. policy in Southeast Asia. *America*'s editors worried that the meeting "gave the impression that religion had nothing better to offer the country than the echo of a university teach-in. The churches were tagging along like tails on a kite."[109] A distinctive voice against such a drift was raised by Bishop John J. Wright, who argued that the particular role of the churches was to "create a moral climate that gives the chance of organic life to the otherwise dead, purely mechanical structures of such peace as diplomats and statesmen can organize." *America* agreed, and its editors warned against the temptation to put foreign policy micromanagement ahead of culture-formation when the Church entered the public policy arena.[110]

Vietnam was one key determinant in the shift from the classic position, which emphasized the Church's primary role as moral teacher at the contextual, rather than policy-prescriptive, level of the war/peace debate. Six years after Wright's speech, *Commonweal* editorially welcomed the fact that "the younger generation of Catholics is not hung up on rigid anti-Communism, on authority, or on the need to 'prove their patriotism'; its members are determined to transform the traditional Christian reconciling role from unobserved theory into nitty-gritty practice," a trend "symbolized by the Berrigans."[111]

Commonweal also took the lead in urging the Church toward "social responsibility" in its investment practices, citing the National Council of Churches' Corporate Information Center as a prod to the Catholic conscience on this issue. *America* joined in urging a reexamination of Church investment practices, in an editorial that acknowledged some of the difficult questions that would have to be faced in this process.[112] The issue was joined on the question of investment in defense-contracting firms.

Gordan Zahn urged a "witness" model on American Catholicism during and after Vietnam. Writing in 1973, Zahn argued that there was little comfort in the fact that the American bishops had not matched the "nationalistic excesses of their German counterparts" under the Nazi regime, for the bishops had been too late in confronting

U.S. war policies and the "inhumanity" of U.S. tactics in Vietnam. What had been missing was an "unambiguous episcopal call" to resistance against the war.[113]

The kind of witness Zahn proposed would not reflect a sectarian ecclesiology, standing aside from the corrupting circumstances of the political world with its inevitable compromises. Catholic witness in the Vietnam and post-Vietnam periods was cast in an activist mold, aimed not merely at shaping the moral dimension of the war/peace debate, but committed to determining specific policy outcomes. A primary example of this was the organization Network, an agency founded to give women religious a lobbying voice in government.[114] Network looked to Saul Alinksy, Daniel and Philip Berrigan, and Ivan Illich as its ideological and tactical role models. [115] In its politics, economics, and activist strategy, Network (and its parallel organization, the Catholic Committee on Urban Ministry) was an expression of "the heart of the Catholic Left—inner-city priests and progressive women religious."[116] Success was to be measured legislatively rather than evangelically. Congress, and state and local legislative bodies, were its chosen arenas for the witness of political activism.[117]

This prophetic witness-through-political-activism model eventually influenced the Catholic bishops of the United States, although their initial entry point into the political arena would be the debate over abortion. In November 1975 the bishops issued a "Pastoral Plan for Pro-Life Activities," which argued that "it is absolutely necessary to encourage the development in each Congressional district of an identifiable, tightly-knit, and well organized pro-life unit . . . which can be described as a public interest group or a citizens' lobby."[118] Paul Weber commented that the bishops seemed to have been influenced by the explosion of "public-interest partisan blocs."[119] If Saul Alinksy was the grandfather of such efforts, it was Ralph Nader whose success the bishops sought to emulate. Nader's influence had legitimated religiously based lobbies in the nation's capital.[120] Weber understood that such advocacy did not jeopardize the constitutional separation of Church and state. Yet he worried whether legislative political clout was really "how the church wants to preach the Good News":

> The relation between the love of God and political power has always been tenuous. While it is true that [James] Madison accepted political activity by religious groups, he also worried that they would degenerate into political factions. What assurances are there that this will not happen to the Catholic episcopacy? When the abortion issue is resolved, one way or the other, what will be the next target? What are the critieria of choice, and who will make the decisions?[121]

John Coleman, S.J., raised similar questions in a reflection on American Catholicism during the national bicentennial. Rejecting a simple assimilationist historiography of the American Catholic experience, Coleman argued that "it seems fair to conclude that, in the post-Vatican II period of drift, Catholics, with few exceptions, have been more shaped by their American environment than they have reciprocally influenced it."[122] Coleman was particularly concerned that "the church seems to have suffered pastoral bankruptcy in dealing with a specifically *religious* agenda at a time when a kind of religious revival of interiority is occurring outside the church."[123] Yet Coleman, who looked forward to a revival of the principle of subsidiarity through a heightened Catholic concern for society's "small platoons"—family, neighborhood, parochial school[124]—seemed influenced, in his analysis of American society, by themes dominant at one distant pole of the public policy debate. Catholics should, for example, be "especially sensitive to the complaints of the various liberation movements among women and blacks that attack the imperial claims of male or upper-class white language."[125] Thus, the crisis of the American Church had to be understood in terms of its "failure to generate critical alternative models either to the cold war or the smug American consumer and status-oriented society."[126]

Coleman also offered mild criticisms of the witness/activist model in a 1977 retrospective on the five-year history of the Jesuit-launched Center of Concern. But his worries were primarily at the level of tactics, not ideology. Coleman warned, for example, that "it is easier to denounce . . . intolerable injustices and conjure up imaginative models of the convivial society than it is to chart the hesitant, cautious, and usually debatable course of step-by-step replacement (by either evolution or revolution) of the 'established disorder'." Yet Coleman's summary on the Center was that its "instincts and program are correct, its dedication and commitments inspiring."[127] The witness/activist model and the ideological freight it carried had become the new status quo. The chief expression of the "established disorder" was the organization of power in American society. America's claims on its citizens' sense of moral obligation were thus attenuated by the "unjust social structures" that dominated public life.

The Question of the Soviet Union

Classic Catholic anticommunism had multiple roots. Marxist-Leninist ideology was antithetical to Catholic incarnational humanism on at least three counts: its militant atheism, its antimetaphysical materialism, and its world-historical view. John Courtney Murray's view of the threat posed by the Soviet Union contained all three of these elements. To Murray, the U.S.S.R., and the ideology central to its domestic and international enterprise, were not warped variants of Christianity, but the antitheses of the Western project in political morality since the Greeks.

But American Catholic anticommunism, and the view of the Soviet Union that shaped the American Catholic contextual debate on war and peace through Vatican II, was not just ideological. American Catholics, many of them the descendents of Central and East European immigrants, were painfully (and in not a few cases, personally) aware of the persecution of the Church in the U.S.S.R., and behind what everyone called the Iron Curtain. American Catholic anticommunism also reflected the Church's commitment to peace. The Soviet Union, in its purposes and power, threatened the evolution of an international political community capable of resolving conflict without the mass violence of war. This dimension of Catholic theory shaped the American bishops' World War II pastoral letters. *Tranquillitas ordinis* was not, in the American Catholic heritage, a matter of mere mechanics; it was fundamentally a question of values. The bishops, during and after the war, were deeply concerned that the abandonment of Atlantic Charter principles in the formation of the postwar world order would jeopardize whatever fragile peace had been won and make a mockery of the sacrifices that the war had demanded. Given Stalin's record, it is hard to argue that the bishops were far off the mark.

Finally, American Catholic anticommunism was not as simplistically McCarthyite in the 1950s as is often assumed. Murray thought that the domestic anticommunist campaign had been a "fiasco." American Catholic support or rejection of Joseph R. McCarthy tended to follow national patterns, and was not as univocal as is often charged. Both Murray and Father John Cronin, S.S., another prominent American Catholic anticommunist theorist of the postwar period, sharply opposed the John Birch Society's definition of the issues and style of political action. In Cronin's case, this reflected his long experience with anticommunist battles in the American trade union movement.[128]

What was missing from this sophisticated anticommunism was serious thought on how the Western democracies might act to *change* the course of Soviet purpose and power. Military means of deterrence were regularly supported, well into the Vietnam era. But the discussion rarely moved beyond containment strategies in which weapons served as the dike behind which the threat was kept at bay. Murray's call for a policy of "minimum security and maximum risk," emphasizing the host of other-than-military means available to conduct the great contest in the world (which, anticipating Yuri Andropov, Murray understood to be a contest for the hearts and minds of billions of human beings around the world) was not explored in any depth.

Through this vacuum of thought blew winds of rapid change. The atmosphere for the transition was set by the Johannine optimism that marked *Pacem in Terris*. Rather than marrying a flexible, multidimensional strategic armamentarium—based on the maintenance of stable deterrence but concurrently addressing the issue of nonmilitary approaches to the war of ideas in the world—to the accurately realistic analysis of Soviet ideology and policy sketched by Murray, the postconciliar debate centered on the abandonment of Murray's realistic analysis. The threat posed by the Soviet Union was dismissed, or, more typically, drastically minimized. The "complexity" of world affairs was stressed, in ways that did not much illuminate the undoubted complexity of an interdependent world, but did serve to reduce perceptions of the centrality of the U.S./Soviet contest. Deterrence came under severe attack, and was often posed as a reflection of American paranoia. The superpowers were regularly portrayed as mirror-images of each other. The persecution of the Church in communist countries became a relatively minor theme in the debate. And the issue of other-than-military approaches to changing the course of Soviet policy remained virtually unexplored.

Vietnam provided the occasion for these transitions. In early 1967, Wilson Carey McWilliams wrote of the "moldiness of the premises" that had guided the Cold War, and warned that "both the great powers are too wedded to their own idolatrous universalities" to seek a common understanding of the world's need for a new ideal.[129]

Similar themes were struck by Georges Morel:

> In writing of the defects of communism, I do not mean to say that everything is black in the Eastern nations and everything white in the West. Each country has to fight against its own selfishness, and the reconciliation for which we are working does not mean that all the nations must take the same path. . . . Today nobody is yet free; but everybody has to become free, or more exactly, we have to become free together. As Karl Marx said, even if only one man is a slave, everybody is in chains; if only one nation is a slave, the others are not yet liberated.[130]

On the fiftieth anniversary of the Bolshevik revolution, William Pfaff argued that the "most obvious" thing about Russia was the gap between "Lenin's vision" and the contemporary realities of Soviet life. What would happen when the Soviet peoples realized what had happened to them?[131] Pfaff suggested that severe problems would follow for Lenin's heirs. After the 1968 invasion of Czechoslovakia, Pfaff noted "that inner stress within Russian society which the literary controversies and arrests of intellectuals in recent years have dramatized." But he did not take the next step and think through what such restiveness within Soviet elites might mean in terms of Western policy. Pfaff's primary concern was that "the more belligerent of the Hawks in this country" not displace their anger at Moscow onto Hanoi and the Viet Cong.[132]

These shifts in perspective accelerated in the 1970s. In 1970, *America*'s readers were told that Salvador Allende was a democrat.[133] Two years later, *America*'s editors rang down the curtain on the Cold War:

> The recognition that political differences are no longer (if indeed they ever were) the most serious obstacles to peace has its corollary in the need to redefine the struggle for human freedom. Human liberation is increasingly seen as the struggle against economic and social oppression rather than a rival political ideology. Few would be comfortable today with the *simpliste* designation of the noncommunist nations as the "free world." Greater sensitivity to the inequities within American society has chastened such confidence. Such inequities mirror the really crucial division in the world, not between Communist and non-Communist nations, but between the rich and the poor, between the developed, that is, economically dominant, nations, and the developing, that is, economically dependent, nations.[134]

Commonweal columnist Frank Getlein thought that Watergate was "the grand climax, the supreme act of *hubris,* and the ultimate bankruptcy of anti-Communism in America." No one else in the American elite seemed to grasp this. Anticommunism had been an "unrelieved disaster" for America since World War II. It had led to U.S. support for dictators around the globe; it had made us oppose democratically elected regimes; it had resulted in "something close to an economic military dictatorship in our own country;" and it had brought nuclear overkill, inflation, and social corruption. Moreover, American anticommunism had not "rolled back" Stalin's postwar conquests; and it had been purblind to the fact that the Soviets "can't take Communist Yugoslavia, let alone the world." If Watergate led Americans to reexamine the anticommunist basis of U.S. foreign policy, it would have been "a very cheap price for an absolutely essential educational leap."[135]

Getlein's extremist formulation was not replicated by all participants in the Catholic debate over world communism and the U.S.S.R. But his themes, in rhetorically modified form, were pervasive.

In 1975, for example, Tom Dorris argued against Western criticism of the accommodation of the Russian Orthodox Moscow patriarchate to the leadership of the U.S.S.R., asking "Who are we to condemn?"[136] A year later, two *Commonweal* essayists, writing on "The Myth of Soviet Aggression," claimed that the "extreme caution" and self-destructive results of Soviet foreign policy were "well known;" that "Stalin stayed out of Greece" [the Truman Doctrine thereby being fraudulent, a key theme in revisionist historiography]; that Cuba had not tried to export its revolution in Latin America; and that "American inflexibility" prevented Castro's distancing himself from Moscow. The authors concluded that the "timid and conservative" U.S.S.R., whose main problem was internal social control, was not an inherently aggressive force in the world. Soviet foreign policy did not require increases in U.S. defense spending. (The authors wrote "defense" in quotation marks.) It was more important to rethink an American policy that had become self-defeating and futile.[137]

This 1976 essay exemplified an important part of the shift in the Catholic debate after Vatican II. The authors did not defend Soviet internal repression. They portrayed the U.S.S.R. as a clumsy, defensive power, hamstrung by bureaucratic arthritis. The authors were not pro-Soviet as, for example, the 1930s American Left tended to be. Rather, they were anti-anticommunists. Their primary concern was to combat anticommunism, not communism, which was irrevelant, stolid, or anachronistic. They abandoned Murray's analysis of Soviet power and purposes. But they did not take up the question of how the West might effect positive change in the communist world. To do so would violate a canon of anti-anticommunism.

Anti-anticommunism led to bizarre, as well as troublesome, results in the Catholic debate. Frank Getlein, for example, described Solzhenitsyn's Harvard commencement address as "reactionary claptrap" marking the "reemergence of . . . 19th-century Russian moonshine."[138] Michael Harrington, in 1978, suggested that Karl Marx was "the greatest thinker the world has ever known," for "Marx's method, values, and dedication to freedom represent the highest point in human thought"—comments reported without criticism by *Commonweal* essayist John C. Cort, who also informed his readers that "James Will . . . pointed out [to a Christian-Marxist dialogue at Rosemont College] that several thousand years before Marx the Hebrew prophets had provided us with a very sharp class analysis, which Christ and the Fathers of the Church had further developed."[139]

Closer to the mainstream discussion was a 1978 *America* editorial arguing that "to resist Soviet advances wherever they take place would be a real mistake."[140] Again, the failure to develop a body of thought on multiple means to counter Soviet push-and-probe tactics resulted in the binary choice of military response or no response. Thomas Powers struck an increasingly familiar note in late 1980 by arguing that U.S. policymakers could not know Soviet global intentions, and that the superpower arms rivalry was motored by an action-reaction cycle in which Soviet policy was determined in response to American action, rather than by its own dynamics.[141]

America's editorial on the death of Leonid Brezhnev neither mentioned nor discussed the increased persecution of churches in the U.S.S.R. that has been underway for the last three years of the Brezhnev regime.[142] For analysis of Soviet policy and purpose, Fr. J. Bryan Hehir recommended George F. Kennan, by 1983 firmly in his antinuclear phase, to readers of *Commonweal* as "the norm against which" others should be measured.[143] Endorsing the idea of an exchange of four thousand Soviet and American students each year, *America* argued that such an effort would "move past stereotyped perceptions and develop common bonds of interest."[144] How this could be protected from Soviet active measures was not discussed.

What this psychologized approach to the important instrument of exchange could mean was amply illustrated when John MacDougall wrote that, although "for an American [visiting the U.S.S.R.] the lack of freedom is hard to take," what an exchange visitor has to realize is that "the situation is more complex than one of mere repression." MacDougall claimed that Moscow was sensitive to non-Russian ethnic concerns, and concluded by reminding his readers that "the Soviet people do not expect the freedoms we take for granted."[145]

The new nuclear debate of the early 1980s also affected the American Catholic argument over Soviet policy and purpose, as was illustrated by reactions to the Soviet destruction of Korean Air Lines flight 007 in 1983. *Commonweal*'s editors complained that George F. Will missed the central point by arguing that no one should be shocked by Soviet brutishness. *Commonweal* was not surprised. Besides, America was also willing to "sacrifice innocent lives in pursuit of its interests." Moreover, the shoot-down of KAL 007 had shown that the Soviets were essentially defensive, rigid in their judgments, and fallible in their perceptions and policies. The U.S. government, too, should recognize that "the entire planet carries a lot of innocent passengers."[146] John Garvey, describing the KAL shoot-down as "the latest tragedy in the history of nation-states versus human beings," worried that the Reagan administration would use the incident "to beef up its MX and Pershing deployment plans, to push for increased defense spending, and to make our own intransigence on defense issues seem reasonable. . . . The holy quiver in Ronald Reagan's voice might really be heartfelt. That scares me more than the possibility that it is not."[147]

The post-1965 American Catholic debate on the U.S.S.R. and the problem of world communism was not univocally anti-anticommunist. Some attention was paid to the persecution of the Church in the U.S.S.R., thanks in part to the remarkable *samizdat* "Chronicles of the Catholic Church in Lithuania." As a *Commonweal* essayist wrote, in a welcome return to the tradition of *tranquillitas ordinis*, "The truth is that the rule of law is the only real safeguard against the nightmare of a nuclear war or an oppressive world order which subverts civil liberties and the human spirit. It is against this backdrop that American Catholics should lend their voice to the cause of the Catholic civil rights movement in Lithuania."[148] Russian Orthodox dissidents like Fathers Dmitri Dudko and Gleb Yakunin received occasional notice. The repression of the Solidarity trade union movement in Poland and the invasion of Afghanistan, like the shoot-down of KAL 007, provided occasions for more sober views of Soviet purpose and policy. Aleksandr Solzhenitsyn was a controversial figure, but his views were usually given a hearing.

But the main drift of the debate was clear, these crosscurrents notwithstanding. Anti-anticommunism oriented the discussion. Détente, if not called precisely that, was the preferred model for policy analysis and prescription. The notion that U.S./Soviet conflict was really psychological—that it was largely a matter of misunderstanding and misperception, rather than a genuine political conflict based on accurately perceived and basic differences—became widespread. The equivalent responsibility of the superpowers for the dangerous condition of the world was a commonplace theme, even if usually hedged by a nod to Soviet brutishness as revealed in Czechoslovakia, Poland, or Afghanistan. Other-than-military approaches to the problem of the U.S.S.R. and peace were occasionally bruited; but they were usually divorced from the realist context that would have given them meaning and credibility.

By the mid-1980s, this component of the Catholic contextual debate over war and peace was in singular disarray. What was sound in the heritage, its realism, had been abandoned. And the weaknesses in the heritage could not be addressed by prescriptions resting on the shaky foundations of anti-anticommunism and détente. American Catholicism's elites, by reinforcing Soviet perceptions of Western vacillation and incapacity, had actually become obstacles to change in the U.S.S.R. and in Soviet international policy.

The Question of America

On May 13, 1964, Msgr. Josiah G. Chatham arose at the annual alumni meeting of Rome's North American College. Chatham, pastor of a Jackson, Mississippi, parish and a World War II military chaplain, urged his fellow priest-alumni to support the civil rights movement, and then proposed a toast:

> Through sin and blood and tears, Divine Providence has assembled representatives of every segment of the human family here in America, and us among them, to give glory to God and hope to man. America is the place where Christians are being asked by God and man whether they will accept God's word concerning Christ's Mystical Body at face value. We, Roman priests, heirs of American freedom, are being given the chance to be progenitors. The world waits for our voices. We are heralds. God wills it. I give you America.[149]

Such confidence in the American experiment, even when facing its failures, became increasingly rare in the post-1965 American Catholic debate on war and peace, security and freedom. Here, the abandonment of the heritage would be decisive. For if, as Murray's project suggested, the American experiment was an important prism through which to think about the peace of dynamic political community in the world, a soured view of the experiment would inevitably lead to altered understandings of both the problem of war and America's possible contribution to its resolution.

An opening wedge for the radical transition in Catholic thought on the American experiment was the claim that the United States had become but another *imperium* on the world stage. Ronald Steel admitted, in 1967, that the "American empire" was not the product of a grand imperial design, and had been maintained by a "sense of benevolence." Yet the very fact of American imperialism seemed to mitigate against notions of benevolence. The origins of the American empire were ideological: "struggling against communism, we tried to build a counter-empire of anti-communism." And although Americans had not exploited their accidental empire, we had been driven to "use imperial methods—military garrisons around the globe, subsidizing client governments and politicians, the application of economic sanctions and even military force against recalcitrant states, and a veritable army of colonial administrators working through such organizations as the State Department, the Agency for International Development, the US Information Agency, and the Central Intelligence Agency."[150]

Things looked less "benevolent" to the editors of *Commonweal* six months later, when they wrote of the "military-industrial-government complex that controls American power abroad and increasingly at home," and wondered what this meant America had become.[151] David Burrell answered that question less than a year later when, in an essay on the flower children, he argued that "the shape of their rejection helps us recognize more clearly the violence latent in the ideals of upper-middle-class America, that class which possesses the substance of what earlier Americans hoped for."[152] Later in that dreadful year, 1968, another *Commonweal* essayist, Jeane Strouse, described the Democratic Convention in Chicago as "a taste of fascism, reminiscent of Stalin, Hitler, Prague, and a warning that it *can* happen here."[153] By "fascism" Strouse did not mean those who called their political opponents "pigs"; the new Hitlers and Stalins were to be found in the mayor's office and police department in Chicago.

America's capacity to deal with its undeniable problems in the late 1960s did not seem very promising. *Commonweal* readers were told that "official American responses to conflict for about the past decade have had underlying patterns of hysteria and disproportionate response to risk, either ignoring problems until they become high pathologies or taking minor problems and blowing them up by compulsive intervention until they become nearly uncontrollable." Nuclear proliferation, combined with America's inept, business-as-usual approach to the world, made even physical survival "highly problematic."[154]

Such ineptitude in problem-solving was hardly surprising in a country where "flag-worship" had become a "blatant form of public idolatry which should have the Christian churches roaring in protest." "To be honored," Edward Cain continued in *Commonweal*, "a country must be honorable. Those who do not find this honor should at least be allowed their silence or their symbolic protest." Cain was referring to a protest in which Rochester Institute of Technology students had photographed a nude woman, wrapped in an American flag, for the cover of a student publication called "GI Joe Meets Wonder Woman."[155]

These arguments were buttressed by revisionist histories of the Cold War. Non-revisionist analyses of the late 1940s were described as "the government-sponsored interpretation of the present and the recent past," reflective of a "Cold War paranoia" that was intent on "a suppression of those Third World peoples struggling for a form of national self-determination at variance with Washington's interpretation of what their future should be." Arthur Schlesinger's *The Vital Center* and the American Committee for Cultural Freedom had done more than the FBI or SDS to stifle intellectual freedom on American campuses.[156]

Richard Miller glowingly reviewed revisionist treatises on the origins of the Cold War for *Commonweal*, in a lengthy essay claiming that John Foster Dulles's successors had "given us the Bay of Pigs, Brazil, the Dominican Republic, Bolivia, Laos, Cambodia, and Thailand—along with more than a half-million American dead, wounded, and addicted, and countless millions of nameless foreigners who have received the final blessing of freedom through the muzzle of an M-16 or the bomb bay doors of a B-52." But Miller's critique went far beyond policies with which he disagreed; the real issue was America:

> We have tried to make the world safe for America by becoming the Titan of Counter-Revolution. Fearing radicalism in any form we have joined the Russians in promoting tyrants abroad and the tyranny of docile conformity at home. The Bundy and Rostow brothers remain in positions of power and prestige, while the Berrigan brothers remain in jail. This is what the Cold War has done to the American character.[157]

This anger and despair was widespread in the American Catholic debate during the Vietnam years. It led *America*'s literary editor, Fr. Philip Rule, to sympathize with the flower children, who probably thought America was not worth growing up in.[158]

One prominent symbol of American corruption was Lt. William Calley and the My Lai massacre in Vietnam. Wes Barthelmes connected Calley to Charles Manson, arguing that "a theme of extraordinarily grisly violence links the two—terrorizing and abusing a selected group of human beings and, afterward, killing them. Calley's misdeed, twenty-two murders financed from the public treasury, occurred in an Asian land we are systematically destroying under color of blind anti-Communism, arrogant idealism, and Occidental face-saving."[159] Raymond Schroth believed that Calley symbolized an America off course, intoxicated by power, and debased by a sense of manifest destiny that had made it abandon its respect for human life.[160]

Such views of the American character and prospect led many to support George McGovern's 1972 call for America to "come home again." James Finn, writing after the 1972 election, tried to strike a more balanced note:

> We *do* need a vision of America that starts from where we are, neither all good—which no one, I think, claims any longer—nor all bad—which many people assert. It will do no good to tell us, however it is phrased, to "come home again." We are home, however confused we may be at the moment. Our task as Americans is to make this a better home and ourselves better neighbors.[161]

But other themes influenced by the New Left remained predominant. The alleged role of the American corporate sector as the guiding force of U.S. foreign policy was a regular target. Two *Commonweal* essayists wrote that anticommunist ideology and the myth of the "free world" served American corporate interests more than any other sector of American society. The ideology had permitted interventions that, examined individually, looked irrational. But the "militarist-interventionist" offspring of American anticommunism had worked to the economic advantage of American oligarchs.[162]

A more apocalyptic note was struck by Episcopal theologian William Stringfellow, cited approvingly by John Deedy in *Commonweal.* In an Advent sermon preached at New York's Cathedral of St. John the Divine, Stringfellow said:

> It is the radical dehumanization of society by war, by apartheid and by persistent official aggressions against the rights of human beings that justifies comparison of the United States with first-century Rome, and which prompts Christians to discern in the incumbent American regime the same spirit of the Antichrist which Christians in the primitive church exposed and opposed in the Roman state.[163]

The passage from America as "benevolent" empire, flawed more in policy than in character, to America as Antichrist had taken less than six years.

Stringfellow was an extreme, but not unique, case. The urge to turn virtually every issue and occasion into a condemnation of this or that flaw in the American experiment was often irresistible. Andrei Sakharov's dissidence in the Soviet Union was, for Sakharov-admirer Peter Steinfels, an occasion to wonder "how we would react . . . were the Soviet intellectuals our gadflies instead of Moscow's?" Sakharov, Solzhenitsyn, and the other Soviet dissidents were men whom America admired more at a distance than it would up close.[164] The debate over domestic social policy prompted Father William J. Byron, S.J. (who would be appointed president of the Catholic University of America in 1982) to observe:

> [Americans] are at a self-enclosed point of preoccupation with skin care, clothes care, lawn care, floor care, appliance care, car care, and pet care. We are weight-watchers, clock-watchers, girl-watchers. We neglect our old and spoil our young. We pollute our air, poison our wells, and devour our nonrenewable natural resources. If challenged on any of these points, we justify ourselves as taxpayers, property-owners, and free citizens in the richest and most powerful nation on earth.[165]

The national bicentennial was another occasion for mourning the American present. *America*'s editors called for a soft neo-isolationism, resting on a "more mature faith" born of "the experience of disillusionment." Such a chastened faith could restore a proper focus to the national experiment: "not a manifest destiny based on wealth and power nor a screen of rhetoric about the righteousness of our leaders and their policies, but the idea that the human person is important, that he possesses rights given to him by his Maker, that, as a result, the task of justice is never finished because the full promise of the human person is never realized."[166] Another editorial in the same issue described Daniel Patrick Moynihan as "the *enfant terrible* of the diplomatic set," and jailed American draft resisters and deserters as "our own 'prisoners of war.'" Yet a third editorial charged (incorrectly) that most of the world's military expenditure went to support the threat of nuclear war.[167]

A year later, reflecting back on the bicentennial and forward on the impending Carter administration, *America*'s editors argued that three of the "many myths" about America that needed recasting were "the myth of the most powerful nation in the world," "the myth of the champion of free nations," and "the myth of the abundant life." These and nameless other fictions had "ceased to reflect the planetary realities of the last quarter of the twentieth century."[168]

But the absolute nadir of the discussion had been reached four months earlier in the pages of *Commonweal*, where columnist Frank Getlein wrote:

> We have an American disease and have had for several decades now. Our disease is called Arms and we pass it along to the nations of the world as promiscuously and enthusiastically as any hooker with a heavy quota and no notion at all of hygiene. In fact, of course, our understanding of our arms disease is much more primitive than that poor hooker's grasp of the itchings and suppurations going on in her principal business asset.[169]

From America as Proposition, to America as Amerika, to America as Antichrist, to America as whore: all in one generation.

Getlein's grotesqueries were not representative of American Catholic attentive public opinion on the question of America in the generation after Vatican II. Yet they were published in the most distinguished lay journal in the American Church. That itself says something about the contours of the debate: that the depiction of America as Antichrist or whore was considered within the bounds of argument, even if at its outer edges. The results were clear. Such formulations dragged the discussion so far in one direction that its center of gravity ineluctably shifted, to the point where Murray's charge to the American Church—to be the instrument that further, and critically, developed the American proposition while defending its fundamental worthiness—would be seen, not merely as undesirable, but as morally obtuse. The new center of gravity in the debate was a profound moral and political skepticism about the American capacity to act for peace, security, freedom, or justice in the world.

The transition at this eighth key choice point was the most significant in the entire contextual debate on war and peace in American Catholicism. A radically altered view of the moral worthiness of the American experiment not only had an impact on thinking about the questions of intervention, the use of military force, the boundaries of political obligation, and the Soviet Union; it fundamentally altered the debate over peace itself. The peace of dynamic political community on an international scale would look less likely, and less desirable, through a prism in which the peace of political community that had been achieved in the American experiment seemed faded at best, and an expression of imperial evil at worst.

If the results of this transition—which amply justifies the description "the heritage abandoned"—seemed apparent, its causes were less clear. Themes from the New Left were at work, and there was truth in Michael Novak's observation that many "*nouveaux gauches* hated their past and had to expel immense self-hatred."[170] But psychological factors do not satisfactorily explain the susceptibility of American Catholicism, particularly in its intellectual elites, to New Left currents of thought. Part of the answer may lie in the general postconciliar mood: a radical questioning of virtually everything in the preconciliar Church was a significant part of the American Catholic experience after 1965. The optimism and enthusiasm for Vatican II could easily turn (and in many cases, did turn) into a deprecation of the Catholic past: a past that was not itself well understood, particularly in terms of Catholic social theory. In such an atmosphere, John Courtney Murray, the patrician theologian of "cool and dry" moral discourse, could (and in many cases, did) look "irrelevant." Yet this explanation is itself ultimately unsatisfactory.

The fault lines in the heritage itself may provide more fertile ground for explanatory hypothesis. The incompleteness of the Murray Project at its grandest level—his unfulfilled call for the development of a new public philosophy—created some of the conditions for the possibility of the maelstrom that followed. Absent such a defense of the American experiment, the path was open for the anti-Amerika currents of thought to define the questions at issue (even if they did not succeed completely in their more ambitious program of delegitimation). But the heritage was also vulnerable on the specific question of war and peace. It did not have an answer, or at least an answer that would be publicly persuasive under intense pressure, to the question of how the concept of peace as dynamic political community might be applied to the horrific dilemma of Vietnam.

Those who cherished the tradition tried to mount such an answer. But their effort did not result in an American Catholic horizon of moral analysis on Vietnam that combined rejection of mass violence with rejection of totalitarian designs; that could address Hanoi's murderous intention as well as the military efforts of the United States and South Vietnam; that could point the way to a genuine antiwar movement, rather than an anti-America-in-Vietnam movement. Camus's challenge had not been met. The choice was still, for many influential American Catholics, between being a victim or an executioner. Choosing, over time, not to be executioners, they helped contribute, however unwillingly or unwittingly, to the victimization of those they wished to save.

CHAPTER 7

Vietnam: Vehicle for the Abandonment of the Heritage

The debate over U.S. policy in Vietnam occasioned the virtual abandonment of the American Catholic heritage of *tranquillitas ordinis* in the decade after the Second Vatican Council. The course of American Catholic commentary on the war was not unilinear. The American role in Vietnam was hotly debated up through (and even after) the Paris accords of 1973. Yet in the opposition to U.S. policy (which came to dominate the debate), a certain trajectory with important parallels in the wider American political/intellectual culture developed.

Among American political intellectuals, opposition to U.S. policy evolved through three stages. The first stage, which ran from the early 1960s to February 1965, saw the war as a lapse in policy judgment. In the second stage, from February 1965 through December 1966, U.S. policy was held to be an exercise in immorality. In the final stage of disillusionment, from January 1967 through December 1968, U.S. policy was taken to reflect the illegitimacy of America itself.[1]

A similar path of disillusionment, running from critique of policy to abandonment of American legitimacy, can be discerned in the evolving American Catholic critique of U.S. policy in Southeast Asia. The tides of abandonment did not run as swiftly in the American Catholic attentive public as they did on the American intellectual Left; but the stages of disillusionment, and ultimately moral despair, were remarkably congruent.

Sorting through the U.S. Catholic debate over Vietnam is a complicated business. *America* and *Commonweal* alone ran over 225 essays and editorials on Vietnam and related issues between 1960 and 1983. A chronological approach brings the debate into focus, and permits an orderly review of the unfolding of American Catholic dissent from U.S. policy. Themes in the post-Vietnam debate over the war, as well as countercurrents to the voices of disillusionment during the war itself, complete the necessary portrait and set the ground for an evaluation of the impact of the Vietnam debate on the evolution of American Catholic thought on the moral problem of war and peace, security and freedom.

1963–1966: THE CONSENSUS UNDER PRESSURE

America's pages had long been open to the argument that U.S. intervention in Southeast Asia was morally and politically justifiable as an exercise in the containment of communism and an effort at building and defending democracy on an endangered front in the Cold War.[2] In November 1963 William V. Kennedy took a soberly realistic view of the American task in Vietnam. The U.S. army liaison officers with whom Kennedy had spoken were confident that they could help their South Vietnamese allies stabilize at least part of the country. But South Vietnam would remain "an armed, continually embattled, and largely American-financed outpost" so long as North Vietnam and Laos provided sanctuaries for the enemy.[3]

America's editors were highly critical of the overthrow of President Diem, arguing that Americans had too quickly forgotten his "signal contribution" to the free world. The editors abruptly dismissed the "starry-eyed liberalism" that called for instantaneous governmental virtue and democratization throughout Asia: "Before Vietnam can afford that kind of luxury, a war must be won."[4] The editors did not mean by North Vietnam and the National Liberation Front.

During the early stages of U.S. involvement in Vietnam, William V. Kennedy tried to analyze the new difficulties of counterinsurgency warfare. Sooner or later, there would have to be a direct confrontation with the North Vietnamese. Before then, South Vietnamese capabilities had to be significantly strengthened for the war against the Viet Cong. The odd circumstances of the American involvement were being felt as early as 1964; Kennedy wrote that "the most important lesson we have learned in Vietnam to date is never again to enter a counter-guerilla war, or any kind of war, in an 'advisory' capacity."[5]

As U.S. forces assumed a larger combat role in Vietnam in 1965, public dissent from Johnson administration policy intensified, which worried *America*. In May 1965 its editors acknowledged the good intentions of such critics as Senators Morse, Gruening, McGovern, and Fulbright, but criticized the senators, and political scientist Hans Morgenthau, for keeping company with the "radical left" and with those susceptible to the sentimentalities of pacifism.[6] *America*'s editors were also skeptical of Hanoi's willingness to accept a negotiated settlement in Vietnam. Although welcoming Johnson administration diplomatic initiatives in early 1966, the editors argued:

> The Communists have been adamant in their refusal to entertain even the idea of negotiation. The reply of Ho Chi Minh to the message of Pope Paul VI on December 29 [1965] underlines Hanoi's intransigence. The only acceptable terms for a Vietnam peace, in the Communist view, are the now famous four points of the so-called Democratic Republic of Vietnam, an acceptance of which would be tantamount to the abject surrender of the South Vietnamese people to the Communists. This . . . could not lead to peace with honor.[7]

America also criticized Senator Everett Dirksen's call for Hanoi's "capitulation" and Dirksen's summoning up of the ghost of General MacArthur with the phrase "there is no substitute for victory." The editors argued:

> In Vietnam, there is a substitute for victory, if by victory is understood the dictated peace that Senator Dirksen seems to be urging. It is the achievement of the limited objective we set for ourselves from the very start of our involvement in South Vietnam. Our military policy has been one of limited, measured response to an act of aggression in an effort to convince Hanoi that it cannot take over the South by force. It is called the responsible use of force in world affairs.[8]

But other voices began to be heard in 1966. Howard Zinn, writing in *Commonweal*, noted that young American blacks were drawing analogies between "civil rights in America and the war in Vietnam," from which had come the view that American national power was "streaked with cold opportunism, from the belated and weak Emancipation Proclamation until today."[9]

In April 1966 John C. Bennett, president of Union Theological Seminary in New York, and a leader in the 1930s break of liberal Protestant social ethicists with pacifism, condemned America's "anticommunist obsession," which seemed blind to the fact that "communism does not mean unchanging slavery." Hitler was a military threat and had to be resisted militarily, but communism was "only secondarily" a military threat. Bennett concluded that the United States was engaged in "acts of inhumanity that are morally intolerable" and "in all probability politically and morally self-defeating."[10] Bennett posed his essay as an exercise in "Christian realism." Paul Ramsey's rejoinder in *America* charged that Bennett's politics had gotten considerably in front of his ethical analysis. The result was not realism, but its abandonment.[11]

America's editors still agreed with Ramsey more than with Bennett. George F. Kennan's criticism of U.S intervention in Vietnam had not sat well at America House. The editors argued that the central issue in Vietnam was not escalation or deescalation, bombing of the North or a bombing halt, acceptance of the Viet Cong as a legitimate political party or not, but "whether we are right and the Communists are wrong. And if we are right, whether we can still bring ourselves, as Mr. [John] Roche puts it, 'to turn 15 million South Vietnamese over to the terror régime in Hanoi.'"[12]

Similarly, Christopher Emmet proposed Vietnam as the "supreme test" of a democracy's staying power in a protracted, defensive war fought for limited purposes. Vietnam was the "acid test" of whether containment could work against communist "wars of national liberation."[13] *America*'s editors also argued, later in 1966, that the issues were larger than the fate of South Vietnam. World order would be "dealt a most serious blow" if the United States scuttled its commitment.[14]

These arguments were no longer persuasive at *Commonweal*. In a landmark editorial on December 23, 1966, the editors stated flatly:

> The United States should get out of Vietnam: it should seek whatever safety it can for our allies; it should arrange whatever international facesaving is possible; and, even at the cost of a Communist victory, the United States should withdraw. The war in Vietnam is an unjust one. We mean that in its most profound sense: what is being done there, despite the almost certain good intentions of those doing it, is a crime and a sin.

Commonweal had not argued from pacifist premises, nor did the editors claim that nothing was at stake in Southeast Asia. The outcome in Southeast Asia would make a difference:

> But not the decisive difference needed to justify a war which may last longer than any America has ever fought, employ more U.S. troops than in Korea, cost more than all the aid we have ever given to developing nations, drop more bombs than were used against the Japanese in World War II, and kill and maim more Vietnamese than a Communist regime would have liquidated—and still not promise a definite outcome.

Commonweal thus broke ranks because of a judgment about proportionality:

> The disproportion between ends and means has grown so extreme, the consequent deformation of American foreign and domestic policy so radical, that the Christian cannot consider the Vietnam war merely a mistaken government measure to be amended eventually but tolerated meanwhile. The evil outweighs the good. This is an unjust war. The United States should get out.

The editors urged a prolongation of the holiday cease-fire, a halt to the bombing of North Vietnam, land reform in South Vietnam coupled with a more honest administration, and a deescalation of U.S. "settlement demands." There had to be a "frank and open" willingness to negotiate with the Viet Cong, and an acknowledgment of their claim to a role in South Vietnamese political life. But even should these suggestions for deescalation fail, U.S. participation in the war would remain unjustified. The war was immoral, and the United States should get out.[15] *Commonweal* thus moved the question of opposition from a debate over prudential judgments to arguments over the basic morality of U.S. policy. *America*'s editors remained unconvinced, writing that same week that the U.S. dilemma in Vietnam would be resolved only when Hanoi demonstrated interest in an honorable settlement.[16] But *Commonweal*'s perspective would become increasingly dominant as the debate sharpened.

1967-1971: FROM DISSENT TO DISILLUSIONMENT

As dissent from Lyndon Johnson's Vietnam policy intensified, so did the pressure on American Catholic bishops. In a *Commonweal* "Open Letter" in February 1967, Protestant theologian Robert McAfee Brown chided the bishops for their failure to participate in a "mobilization" intended to be "a focal point at which the entire American religious community could make a common witness that all of us deplore the slaughter and killing that Vietnam has come to represent, and that all of us together, without presuming to offer cheap and easy panaceas, could lay more directly on the conscience of the nation and on our own consciences the absolute necessity of finding a way to initiate negotiations." Brown wondered whether the bishops disapproved of a meeting for peace as thoroughly as their silence indicated, and claimed that many American Catholics had not yet thought out a "clear-cut moral stand on Vietnam." This had to be the first priority for all American religious communities.[17]

Auxiliary Bishop James Shannon of St. Paul-Minneapolis replied that Brown was being "less than fair with the American hierarchy": Brown did not have the right to determine how others' moral concern should be expressed. Brown's style of public protest was "simply not the style of Catholic bishops." Further, Bishop Shannon argued, Brown betrayed his own liberalism by denying that the American bishops were concerned about Vietnam, a charge that smacked of a play for "cheap popularity."[18]

Brown subsequently admitted that his rhetoric may have gotten ahead of his argument. But his rejoinder to Shannon's reply closed on a telling note:

> I believe Bishop Shannon when he says that the American bishops "share the heartfelt prayer of Pope Paul VI that peace will come—and soon—in Vietnam." But I

do not think Lyndon Johnson believes it, and I do not think he is going to be forced to believe it until some kind of style is found that is more effective than any of us have found yet.[19]

Themes of delegitimation came to the fore in 1967. Writing in *Commonweal*, Sidney Lens claimed that the United States was doing in Vietnam what the Soviet Union had done a decade earlier, and "on a smaller scale," in Hungary. Moreover, "Russia did have a pact with Hungary, under which it was accorded the right to station troops there. This legalistic facade was certainly as weighty as the Johnson-Rusk fiction that American troops are in Vietnam under the strictures of the SEATO pact."[20]

Further light on the evolving argument was shed by a *Commonweal* special issue, "The War," published in September 1967. James O'Gara argued that Vietnam was basically a revolutionary civil war into which America had stumbled because of its sense of "omniscience" and "omnipotence." Without a settlement, America ran the risk of World War III and casualties in the millions. Throughout his essay, O'Gara claimed that the Viet Cong were an independent force, not an instrument of North Vietnam. The American choice lay, not between good and evil, but between two evils. A less-than-perfect settlement in Vietnam was preferable to turning all of Vietnam into a "vast wasteland."[21]

John Moriarty, a former State Department analyst, disagreed with O'Gara's assessment of the Viet Cong, but argued that the American commitment in Vietnam exceeded the importance of our objectives, and the clarity with which the American public understood those objectives.[22]

Peter Steinfels was not persuaded by the "mistaken judgment" model, and charged that the United States was supporting "an anti-Communist satellite state in the South," and offsetting its weaknesses by "raining bombs on North Vietnam" and "sending a massive expeditionary force into Asia." A communist regime in Saigon would do less harm than the war was wreaking. Nor was Steinfels much impressed by the idea that following his course would mean abandoning our allies:

> There are many ways to abandon a people. For two decades the Vietnamese have been abandoned to civil strife, military destruction, national division, individual suffering. We have pledged our willingness to abandon them, if necessary, to a vast war between American armies and hordes of their ancient Chinese oppressors. One of our ex-Presidents has approved of abandoning them, if necessary, to the pleasures of atomic weapons. The Vietnamese are being abandoned to American needs and notions today, have been so abandoned for a decade past, and may be so abandoned for years to come."[23]

In a similar vein, Philip Berrigan charged that the "major reason for America's dilemma in Vietnam . . . is, of course, our wealth and the type of society necessary to produce it." Vietnam was seen in Washington, D.C., "within the context of Pax Americana, within terms that flow from the imposition of the American cultural, technological, and political miracle upon the world." Behind U.S. policy lay "ruthless economics and overwhelming military force; and behind that, a Way of Life which resolves around the highest standard of living in the world."[24]

Such arguments puzzled Vincent Kearney, who applauded the dissenters' compassion for the people of Vietnam, but could not understand the "almost masochistic tendency" to fix all blame for the savagery on the United States. The sufferings of the Vietnamese had been primarily caused by the communists. "Ought not our human compassion for the victims to be an expression of protest against the ruthless regime in Hanoi?"[25]

Commonweal columnist William V. Shannon was another voice against the delegitimators. Shannon still argued that the great issue in the world was freedom vs. tyranny; that America was the chief defender of freedom; and that the only answer in Vietnam was to stay the course of a limited war that might last up to ten years. Shannon opposed the U.S. bombing of North Vietnam not only because it was indiscriminate but because it was militarily ineffective, and was as critical of the U.S. government's seeming obsession with finding a "quick solution" as he was of similar prescriptions from the peace movement. But Shannon was worried; he saw little public ground on which his moderate policy of sustained, discriminate use of military force toward the objective of a free South Vietnam could be built.[26]

Such worries were well founded. Robert McAfee Brown had begun to raise the spectre of Nazism and the failure of the German churches to confront Hitler. Brown wanted American churches to speak a "decisive word" that would put "prophetic bite" into the Gospel, and to "take immense risks in the arena of involvement in the secular order." If the churches were going to err in 1967, then let it be through overinvolvement rather than silence. The churches ought to be specific rather than general, risking sins of commission rather than omission. The danger was not being too radical, but being too conservative.[27]

Brown's view, not Kearney's or Shannon's, won that particular argument, and in short order. For in December 1967, the editors at *U.S. Catholic* joined their *Commonweal* colleagues in arguing that the war was immoral. The editors claimed that "the overarching fallibility of the 'we must penalize aggression' argument is its dependence on the naive belief that wars today can be 'won,' that ultimately they settle anything." Without specifying means or even norms, *U.S. Catholic*, which charged the United States with a "mindless imperialism that is willing to lay waste to another country," concluded with this flat statement:

> We believe that the war in Vietnam must be ended before it does irreparable harm to our nation. We believe that it is immoral and that the American people, imploring the grace and mercy of Almighty God, must insist that it be concluded without delay.[28]

With its urban riots, the assassinations of Martin Luther King, Jr., and Robert F. Kennedy, the virtual destruction of the Democratic Party at its Chicago convention, and the Tet offensive in Vietnam, 1968 may well have been the most dreadful year in the United States since 1861. Divisions within the American Catholic attentive public on the war in Vietnam remained, but the trajectory toward rejection of U.S. policy and toward a profound moral unease with the American experiment accelerated. The trial of the Catonsville Nine was a signal event in the American Catholic debate, and marked the point at which the call for a measured approach to the discriminate use of military force in Southeast Asia essentially collapsed.

The editorial pages of *America* still held fast to the old consensus, though. *America*'s editors called on "the so-called peace movement" to stop "being an apologist for Hanoi" and to support the South Vietnamese bishops' plea for a cessation of arms infiltration into the South as well as a halt to U.S. bombing of the North.[29] *America* argued that the Tet offensive had once again revealed "the face of the enemy . . . in the agony being visited on the people of South Vietnam. What breed of animal is it that can condemn 21,000 of its own to certain death in such suicidal acts as were perpetrated throughout South Vietnam?"

Yet *America*'s firmness had begun to weaken. The editors accepted the standard journalistic portrait of Tet as a crushing defeat for the United States and its South Vietnamese allies, and argued that if the government of South Vietnam could not provide for its citizens' security, then its right to govern had to be questioned.[30] *America* defended the right of peaceful dissent against U.S. policy, but hoped Hanoi, where no public dissent was allowed, did not misread the importance of dissent in America, for that would only prolong the agony American dissenters wanted to end.[31] No longer would the phrase "so-called peace movement" issue from America House, although the editors charged that North Vietnamese negotiator Xuan Thuy sounded "like the principal speaker at a U.S. peace rally inveighing against the 'immorality' of U.S. foreign policy."[32] A "proposed solution" to the Vietnam imbroglio, published in the Jesuit weekly prior to the opening of the Paris negotiations, emphasized the role of the North Vietnamese and Vietcong for the continued killing, and urged that democratic process not be abandoned in pursuit of a way out.[33]

But the new mainstream of Catholic commentary was exemplified by a January 1968 *Commonweal* editorial claiming that American intransigence over the bombing of the North, not North Vietnamese or Viet Cong preconditions, blocked a negotiated settlement. The United States had escalated the war so often that the "Johnson-Rusk team" now believed it had to win in order to justify its policies. The recently announced candidacy of Senator Eugene McCarthy was a "slight" indication that there might be hope on the horizon.[34] Daniel Berrigan's "mission to Hanoi" to accept the release of three U.S. POW's, reported in both *America* and *Worldview*, helped drive home the message that U.S. policy, not North Vietnamese purposes, stood in the way of peace.[35]

Commonweal's March 21, 1969, editorial illustrated the shift from dissent on policy to disillusionment with the American experiment itself. The premise on which U.S. policy rested was the:

> Blind conviction that History has ordained America with a world mission, to control the pace and means of others nations' development, to "build nations," to chastise "romantic revolutionaries." . . . Our Vietnam intervention is a Gordian knot of lies, deceptions, and delusions which we can disentangle only with a clean sweep of the truth. The truth is that the people of South Vietnam are divided by a struggle to determine who shall rule their land and in what direction, a struggle prolonged and envenomed by American intervention; and that the United States has no right to arbitrate, by means of a vast war, the outcome of that struggle. The truth admitted, the diplomatic formulas for deescalation, for coalition governments, for elections, for safeguards against post-war purges could all be swiftly devised. Would such a reversal of American aims be a surrender? Yes, in exactly the same way as a Soviet withdrawal from Czechoslovakia would be a surrender. A surrender to wisdom and morality.[36]

The lessons of Vietnam were already being drawn in 1969. In a *Commonweal* symposium, "After Vietnam, What?," Harvard political scientist Stanley Hoffmann (who would become a major influence on the American bishops in the 1970s through his former student, J. Bryan Hehir) argued that the chief lesson of Vietnam for America was self-restraint. The United States must "curtail drastically the interventionist tendencies of our foreign policy" and revise its national priorities. Hoffmann believed that Hanoi would accept international guarantees for the neutralization of South Vietnam, and argued that nationalism invalidated the domino theory.[37]

The May 1970 Cambodian incursion intensified the anger in American Catholic commentary on the war. *Commonweal* described the incursion as a "ruthless, offensive act" which followed as logically from the "twisted premises of our nation's leaders" as had the invasion of Czechoslovakia from the assumptions of Moscow. The "heart of the problem" with the war was in the United States, which was the last place our leaders would look.[38]

America was less harsh. If the issue was military strategy alone, it was hard to argue with President Nixon's decision to go into Cambodia. But *America*, too, was concerned about the meaning of the Cambodian incident for the American character. Nixon's appeal to patriotism suggested that military capability was the only index of national credibility, and that restraint was a sign of weakness. Concern for avoiding humiliation had overshadowed concern for the millions of innocents at risk in Southeast Asia. American leaders had a special obligation to apply "unusual moral sensitivity" to the exercise of U.S. power and the rhetoric with which such exercises were explained to the American people.[39]

A week later, after the shootings at Kent State University *Commonweal* charged President Nixon and Vice-President Agnew with responsibility for the deaths of the four students, who were killed because the National Guard saw them as "Nixon's 'bums' and Agnew's 'rotten apples.' " Nixon was an unconverted hawk who, "like an unrepentant alcoholic," was "always tempted to just one more pull on the old escalation bottle."[40]

Commonweal closed 1970 by describing the failed U.S. commando raid on a recently abandoned POW camp in North Vietnam as "a reckless and cruel stunt, not deserving even the ritual accolades its critics were obliged to deliver. . . . Lives have been cruelly played with, not so much to rescue prisoners as to rescue Nixon, Agnew, Laird, and Associates."[41]

Delegitimations of the American experiment were paralleled in 1971 by a new argument: that American Catholicism was itself being drastically compromised by its allegedly acquiescent approach to the war. Gordon Zahn charged that American churches, and especially American Catholicism, had become accomplices to war crimes and atrocities.[42] Later that year, in an essay tracing the American bishops' response to Vietnam, Zahn wrote:

> It would be too much to say that the American bishops have duplicated the failure of their German counterparts under Hitler if only in that they deserve credit for not being as fervent in their support for the nation's war effort. On the other hand, they have been just as intent upon avoiding confrontation with the secular power.[43]

The students at Notre Dame University seemed to agree. They voted the annual Senior Fellow Award to "an individual who has the integrity, the decency, and the willingness to work for the best values and finest traditions of our society": in this case, William Kunstler, the "defender of Stokely Carmichael, H. Rap Brown, the Chicago 7 and, more recently, the Berrigans."[44]

America had had enough of Vietnam by April 1971. An editorial entitled "Easter and the American Conscience" linked Lt. William Calley with Charles Manson, and argued:

> Whatever the savagery and stubbornness of the North Vietnamese, there is no point in trying to match them on those grounds. More important even in terms of traditional "just war" morality, the loss of innocent life—so tragically dramatized in, but not confined to, the My Lai affair—has mounted beyond the point of any possible proportionate gain in the name of justice. Since responsibility for the war rests on the entire nation, each of us should accept a share in the task of making it clear that our national will is to end it now.[45]

The drumbeat of radical critique continued, and in fact came to shape the basic contours of the debate. Todd Gitlin argued that Daniel Ellsberg was an "authentic hero," whose release of the Pentagon Papers had been a rejection of his "past participation in evil." Heroes like Ellsberg had antecedents; Gitlin thought of the 1944 German generals' plot against Hitler. But the gravamen of Gitlin's celebration of Ellsberg was not Ellsberg, but the United States. Ellsberg "is a man of his word, and so in this America he becomes a criminal."[46]

1972–1975: FROM DISILLUSIONMENT TO ILLEGITIMACY

The lessons of Vietnam continued to be drawn even as the "Vietnamization" of the war ground on. Msgr. Marvin Bordelon, director of the Department of International Affairs of the United States Catholic Conference (the public policy agency of the U.S. bishops), argued within a just-war framework that the United States had engaged in "the massive indiscriminate use of air power," and claimed that the first lesson of Vietnam had to be "a sober reconsideration of the very nature of war in our times." Citing Vatican II's charge that war be evaluated with an entirely new attitude, Bordelon suggested that modern weapons technology raised severe problems under the criterion of proportionality. Bordelon also argued for a strengthening of the United Nations, but described such a process as "consulting the entire human family." The key to absorbing the lessons of Vietnam, according to Msgr. Bordelon, was to enter into the spirit of Isaiah: "Behold, now is the acceptable time for a change of heart."[47] Wars would cease when humanity underwent conversion. Wars would cease when men and women refused to fight.

The debate over America itself continued. *America* argued that the hung jury at the 1972 Berrigan conspiracy trial in Harrisburg illustrated the basic soundness of the U.S. justice system.[48] But *Commonweal* insisted that "this country is engaged in a policy of mass extermination in Vietnam," and worried whether, in an election year, the American people would "really endorse four more years of slaughter from the air." The editors hoped that polls predicting Richard Nixon's reelection were wrong.[49]

The 1972 Christmas bombing of Hanoi provoked fury. "A Christian has no choice but to condemn such actions," wrote *America*.[50] *Commonweal* described the raids as "the terror bombers over North Vietnam," and suggested that the bombing illustrated a serious constitutional weakness in America, the inability to double back on a "particularly unfortunate election result." *Commonweal* noted, without demur, that Richard Nixon's "splenetic decisions . . . put some in mind of Guernica, Lidice, Babi Yar, and Treblinka," and caused "others to wonder about his very sanity."[51]

The 1973 Paris accords, under which U.S. combat troops were withdrawn from Vietnam, did not stop the debate, nor did they soothe the anger, disillusionment, and discord that had marked it. *Commonweal* welcomed any agreement, however imperfect, that ended the slaughter, and hoped that the United States would "get out 'clean'," without trying to do covertly what it had failed to accomplish overtly. The Vietnamese should be left to settle their own problems without outside interference.[52]

Amnesty and the returned American prisoners of war were on many minds. *Commonweal* noted that "war criminals" like Dean Rusk, Robert McNamara, and McGeorge Bundy seemed to be enjoying an unofficial amnesty, and urged unconditional amnesty for those draft resisters and deserters who were "guilty only of being 'prematurely moral' on Vietnam." The editors concluded by noting that Czechoslovakia had recently offered amnesty to all those who had fled during the Soviet invasion in 1968. "If an East European nation can show this kind of mercy, cannot we do the same, and without niggling conditions?"[53]

Nor were all American Catholic commentators chastened by the return of the POWs and the tales of their abuse while in North Vietnamese captivity. Writing in the *Pittsburgh Catholic*, Msgr. Charles Owen Rice discussed the POWs in these terms:

> About those prisoners of war and their stories of mistreatment: remember what they were doing; remember the stuff they dropped on people—napalm, fragmentation bombs, etc.; remember that prisoners from the other side were being tortured by our allies as we stood by; such prisoners were sometimes dropped by our helicopters so that others would talk and that hardly made pilots lovable. . . . Killer aviators are heroes only to the peoples who possess planes. . . . Our military lost the battle of propaganda, along with some other battles, while the war was going on; it is renewing that battle now. I do hope that a bit of counterinformation will squeeze into print or onto the tubes.

Msgr. Rice's column was reprinted by John Deedy in *Commonweal*.[54]

Similar themes were struck by Bernard Sklar. Sklar approved of Jane Fonda's suggestion that the POWs had been less than candid about their situation [Fonda had called the POWs "liars and hypocrites"], by which Sklar seemed to mean that the POWs did not agree with his and Fonda's description of the U.S. role in the war: that "a modern Juggernaut"—America—had been "on the rampage in Southeast Asia, escalating its revenge in tonnage of bombs and meaninglessness of devastation in geometric ratio to its frustration and anger at itself and its failure to achieve even one of its reconstituted purposes, civilian or military, in that hapless country." Sklar would not even concede the facts of POW torture, demeaning "how much they [the POWs] suffered in the way of 'torture' and other cruelties in the hands of the enemy."[55]

Recriminatory debate continued as military pressure increased on the Thieu government in 1974 and Watergate tore America at home. *Commonweal* argued that the failures of the Paris accords were largely the responsibility of the Nixon administration in collusion with Thieu. But eager as *Commonweal*'s editors were for the

United States to cease supporting Thieu, their primary concern remained that America learn *the* lesson of Vietnam, which had to do with America itself:

> It sometimes seems that the American capacity for moral evasion and collective self-deception is so great that there is little chance, within the lifetime of these so-called best and brightest Americans most responsible for our folly, that we will face the full implications of what our pride and violence have done to this once rich and beautiful little land. That will be the task of future presidents, professors, priests, journalists, and novelists who must guide the nation through public recognition of its guilt.[56]

As support for enforcement of the Paris accords collapsed, Gabriel Kolko, a prominent revisionist historian, claimed that peace with justice was at hand in the Vietcong-controlled Quang Tri province of northern South Vietnam:

> It is this political triumph, and the depth to which it penetrates into every crevice and the very heart of a nation's values and commitments, that emerges most vividly as one moves about a province of 300,000 people under a unified PRG [Provisional Revolutionary Government] authority. Its economic progress is formidable, even monumental, in light of the extraordinary devastation of every standing physical object. Yet its very survival and accomplishments are due to its political strength and its profound interaction with a heroic people who have made the existence of the NLF [National Liberation Front], much less its present mastery of all the elements that will decide the final outcome of the Vietnam war, the most significant assertion of the capacity for humanity—one grounded in clarity and people's needs as well as heroism—to triumph over the worst adversities inflicted on any society in modern history.

Kolko believed that the Quang Tri model would help produce final Vietcong victory in a war that was shifting from military to social confrontation.[57]

But less than a year after Kolko's essay appeared, Saigon fell, not to "social confrontation," but to those North Vietnamese tanks on which the Provisional Revolutionary Government traveled south from Quang Tri province. During the reduction of South Vietnam, *Commonweal* readers were informed that the Thieu government "tortures more people more systematically than any regime of modern times."[58] It seems a remarkable assertion in the age of Hitler, Stalin, Castro, Mao. But such was the state of debate in the premier American Catholic lay journal as the tragedy of Vietnam continued—as the killing, so often assumed to be primarily the product of American intervention, continued, and as the desperate flight of refugees began.

Commonweal did agree that arranging safe passage and resettlement for the refugees were moral imperatives:

> But some of the figures mentioned—200,000, and perhaps as many as a million—seem unreal. . . . Apart from enormous problems of logistics, there is the question of propriety. An indiscriminate and wholesale expatriation could touch off new panic, with God knows what consequences; it could also sum to an indefensible depletion of the resource the country needs most in order to recover—its own people.[59]

A year later, Susan Abrams, in *Commonweal*, described the Vietnam babylift of April 1975 as a "cruel . . . manipulative . . . public relations stunt" that "more closely resembled a kidnapping."[60]

THE AFTERMATH

American Catholic debate over the moral meaning of Vietnam continues to this day. Although the tempo of the discussion decreased after the fall of Saigon and the forced reunification of Vietnam under communist leadership, the pages of American Catholic journals continued to be receptive to arguments over this or that "lesson" of the war.

Curious views of communism persisted. Father Tissa Balasuriya wondered why Marxists were "almost the sole force" who took justice issues seriously enough to "combat injustice radically."[61] But the postwar experience of reunified Vietnam made such claims harder and harder to sustain. The former superior of the Jesuits in Vietnam, Father Sesto Quercetti, wrote in 1976 and again in 1979 of the persecution of the Church under the new regime.[62] The boat people were, for some commentators, a heart-rending witness to the fact that things may have gotten worse in Vietnam. Yet Father Robert Drinan, S.J., who had sought Richard Nixon's impeachment for the invasion of Cambodia before the House Judiciary Committee in 1974, still did not agree that people were fleeing from communism. In an article urging greater Western openness to Indochinese refugees, he mourned "the economic conditions that prompted so many people to abandon their homeland."[63]

An important incident in the aftermath of the war was the petition addressed by American antiwar leaders to the Hanoi government in December 1976, urging respect for human rights throughout Vietnam, and asking the release of those imprisoned in jail or "reeducation camps" because of their political or religious views.[64] The "Appeal to the Government of Vietnam" was initiated by activist James Forest, a Catholic who had participated in the Milwaukee 14 draft board raid, and Richard John Neuhaus, a Lutheran pastor who had been a co-founder of Clergy and Laity Concerned. Prominent Catholic signatories included Thomas Cornell, Dorothy Day, Bishop John Dougherty, James Douglass, Bishop Carroll Dozier, Robert Ellsberg, James Finn, Bishop Thomas Gumbleton, and Gordon Zahn. The appeal was attacked by other activists, including Richard Barnet, Robert McAfee Brown, David Dellinger, Richard Falk, Don Luce, Paul McCleary, and Cora Weiss, who argued that "the present government of Vietnam should be hailed for its moderation and for its extraordinary effort to achieve reconciliation among all of its people."[65]

The activists thus divided, reflecting arguments during the war between those who opposed America's war, and those who opposed the war in its entirety. Even within the latter group, there were certainly signatories of the "Appeal to the Government of Vietnam" who had placed primary responsibility for the tragedy of Vietnam on the United States. Some of them may have changed their judgment on relative responsibility in the wake of the North Vietnamese victory; others certainly had not. What did seem clear was that there remained those for whom a communist victory in Vietnam was desirable. This point had not been made forthrightly during the intense, bitter debate over U.S. policy in Southeast Asia.

The "lessons of Vietnam" have been controverted ever since the fall of Saigon. One representative lesson was drawn by *America* editor Father Joseph O'Hare, S.J., who feared the impact of the "Vietnam syndrome" on U.S. policy in Central America. Writing in the early days of the Reagan administration, O'Hare was concerned about

Secretary of State Haig's "enthusiasm" for a confrontation with the U.S.S.R. over El Salvador. O'Hare worried that Haig saw in Central America the chance to end the "losing streak" that the United States had been on since the last helicopter lifted off from the roof of the U.S. Embassy in Saigon. This "Vietnam syndrome" prevented Haig from stopping right-wing violence in El Salvador, O'Hare charged.[66] Writing on the Vietnam Veterans' Memorial, David Hoekema struck a similar note. The lessons of Vietnam were that American arms could turn an internal struggle into a superpower confrontation, and that some wars could be lost in many ways but never won. The Reagan administration seemed immune to these lessons, and was thus heading down the path to Vietnam in Central America. And that policy threatened a new world war whose victims would never be memorialized.[67]

Michael Novak, who had been a strong, and at times bitter, opponent of America's war in Vietnam had learned a different lesson. Writing just before the 1980 presidential election, Novak argued that "the left was correct in its resistance to the war in Vietnam, but absorbed from that war the wrong lesson." The wrong lesson was that American military power was a threat to peace. That "tragic error of judgment" was compounded by the tendency to minimize Soviet superiority. The Left's misperception of the lessons of Vietnam made war more, not less, likely. It was time to change the balance of forces in American political culture; the Left had had the field to itself since Vietnam, with grim results. "For this reason, I stand with those who want to redress the errors of the past eight years (including the strategic retreat engineered by Nixon and Kissinger)."[68]

What one misses in this post-Vietnam commentary, and what seemed so starkly absent from the debate over the war itself, is the classic Catholic heritage of *tranquillitas ordinis*. Elements of it appeared throughout the Vietnam-era debate, and in both of the dominant post-Vietnam camps. But the heritage itself, as a comprehensive whole, seemed missing in inaction. Was it?

OTHER VOICES

The classic heritage made cameo appearances throughout the Vietnam debate. Elements of its method of moral reasoning can be found in articles by William V. O'Brien, Quentin L. Quade, and Father James V. Schall defending the measured use of military force to create the circumstances in which a peaceful resolution of the conflict in South Vietnam might be obtained.[69] Pacifist opponents of the war, such as Gordon Zahn, often appealed to the just-war theory (especially the criterion of proportionality) in debate with supporters of the war, as did the editors of *Commonweal*, *America*, and *U.S. Catholic* in their dissent from U.S. policy.[70]

But the most imaginative application of the heritage of *tranquillitas ordinis* to the dilemma of Vietnam came from a little-remembered peace effort, the National Committee for a Political Settlement in Vietnam, familiarly known as "Negotiation Now." Launched in May 1967 as a coalition of those opposed to both Hanoi's war and America's war in Vietnam, Negotiation Now proposed an eight-point program for achieving a political settlement of the war: (1) a halt to U.S. bombing of North Vietnam and other American initiatives toward a political settlement; (2) reciprocation from Hanoi and the Viet Cong, and their agreement to negotiate; (3) negotiations among all concerned parties, including the Viet Cong; (4) a standstill cease-fire by all sides; (5) an international peace-keeping force to police the cease-fire, the political settlement, and the withdrawal of all outside armed forces, and to protect minorities; (6) a political settlement based on democratic self-determination through early, free elections supervised by a representative electoral commission, with all parties agreeing to abide by the election's results; (7) a major land-tenure reform, giving title to tenant farmers and compensation to landlords; (8) humanitarian aid to all victims of the war, and development assistance channeled through multilateral agencies.[71]

Negotiation Now was an attempt to broaden the American peace effort beyond what were, by 1967, the dominant "resistance movement" categories of analysis and action. A distinguished group of citizens, including Clark Kerr, Walter Reuther, and Arthur Flemming, served as co-chairmen.[72] Catholic signatories of the Negotiation Now petition to end the war on the terms described above included Bishop John Dougherty, president of Seton Hall University, Archbishop Paul Hallinan of Atlanta, Archbishop James Peter Davis of Santa Fe, Bishop John J. Wright of Pittsburgh, Bishop James Shannon, auxiliary of St. Paul-Minneapolis, Bishop Victor Reed of Oklahoma City, and Bishop Joseph P. Donnelly, auxiliary of Hartford.[73]

Negotiation Now broke with the main ideological currents in the Vietnam-era peace movement on several key points. It sought, not American withdrawal from Vietnam, but a political solution to the issues posed by the war. It condemned the violence of Hanoi and the Viet Cong, while criticizing the Thieu government in South Vietnam for its failures in economic, social, and political reform. It was critical of U.S. policy under Presidents Johnson and Nixon without branding the war an expression of "Amerika." It was anticommunist as well as antiwar. And it sought to apply political pressure, not merely on the U.S. government, but on North Vietnam and the Viet Cong in a quest for altered circumstances in Vietnam that would create conditions for the possibility of a political settlement. Negotiation Now did not argue questions of military strategy and tactics. Although it included both pacifists and just-war theorists, its primary purpose was to apply the remedies of politics and law—in short, the heritage of *tranquillitas ordinis*—to the torture of Vietnam. The principal creator of Negotiation Now's distinctive perspective was Robert Pickus, a Jewish pacifist who had had a long career in American peace movement circles. Pickus believed that Negotiation Now was an opportunity to create a citizen peace effort that would meet Camus's standard that we be neither victims nor executioners.[74]

Negotiation Now was bitterly attacked by those for whom the central issue in Vietnam was the evil of America. To one critic, distressed by the "reflexive anti-communism"[75] of Negotiation Now, the "simple truth" that had to be faced was that the war was a "depraved act" deserving moral condemnation. America's image around the world had been stained by the "scorched earth policy" the U.S. was pursuing in Vietnam.[76] This critic, like many others, was not impressed by Negotiation Now's argument that the problem in Vietnam was not simply one of protest against the United States government. Inasmuch as every nation relied on military power for its security, a peace effort condemning only one side was making a fundamental error. Mindless support of either U.S. policy or Ho Chi Minh was not going to contribute to a solution. The war would end when both sides ceased military action and substituted democratic political processes as means to resolve the conflict.[77]

Negotiation Now, concerned not only with a political settlement in Vietnam but with the health of an American political culture scarred by the polarization of national debate around right/left, "win" or "get out" extremisms,[78] failed in both of its efforts: its effort to end the war according to the standards of *tranquillitas ordinis* and its effort to create an American peace effort that would hold itself accountable to those standards. Why, is a complicated business.

On the war itself, Negotiation Now may be reasonably charged with too optimistic a view of the prospects for bringing North Vietnam into serious negotiations aimed at a democratic political solution to the conflict in South Vietnam—and this despite Negotiation Now's more realistic understanding of the Hanoi regime than those dominant in resistance movement circles. Supporters of Negotiation Now could reply that, insofar as their program was not adopted, it can hardly be charged with failure. There is truth in this. And yet, given the Stalinist cast of mind in Hanoi so clearly evident in the aftermath of the war, it is difficult to imagine the configuration of political pressures that would have caused the government of North Vietnam to abandon its ambitions to communist domination of a united Vietnam (and Laos, and Cambodia).

These are matters of argument, almost certainly unresolvable now. Negotiation Now's adaptation of the tradition of "moderate realism" was a brave, honorable, and ultimately futile exercise, given the state of American political culture in the late 1960s. By the end of the 1968 Tet offensive, at the latest, the "resistance" currents had won the contextual debate: the issue had been posed, squarely, as "America, in or out of Vietnam?" The polarization that Negotiation Now tried to ameliorate had hardened beyond the point of significant alteration. American Catholicism's intellectual and activist elites shared a considerable responsibility for that polarization. The central issue on which the debate came to rest was not on the question of a just and peaceful settlement in Vietnam; it was on the question of the American experiment and its moral worthiness.

Negotiation Now's attempt to apply central themes in the heritage of peace as political community to the dilemma of Vietnam failed in American Catholicism because that heritage, especially on the central question of the American proposition, had been largely abandoned. From that abandonment, led by men and women who, in the main, were genuinely conscience-stricken by the face of modern war revealed in the rice paddies, villages, and cities of Southeast Asia, much else flowed and would continue to flow.

AN ASSESSMENT

Any assessment of the American Catholic contextual debate on the war in Vietnam must address at least three questions. Were the signs of the times read accurately and insightfully—that is, did the American Catholic debate reflect the empirical realities of the war? Was the heritage of peace as dynamic political community developed as a horizon for moral analysis? What did the debate teach a Church called by the fathers of Vatican II to assess the moral problem of war "with an entirely new attitude?"

The American Catholic debate on Vietnam was distorted by several notable empirical errors. *America*'s enthusiasm for Ngo Dinh Diem was almost certainly misplaced. The corruptions of Diem's administration, and the South Vietnamese government's tardiness in addressing social and economic reform, helped make a political solution (much less a successful military prosecution) of the war much more difficult. One need not approve of the coup against Diem in 1963, mounted with Kennedy administration support, to understand the weakness of this reed on which so many hopes rested. The portrait of Diem conveyed by *America* was false.

But this failure was paralleled by the persistent inability (or refusal) of many American Catholic publicists to understand the nature of the North Vietnamese regime, and the relationship between North Vietnam and the National Liberation Front. A recurring theme in the resistance movement, and in the post-1967 prestige Catholic press, was that the NLF/Viet Cong forces were independent agents conducting a civil war against the South Vietnamese government. North Vietnam's forceful entry into the war was thus interpreted as a response to the intervention of U.S. forces. History has shown this to be false. Stanley Karnow, no supporter of the war, admits that "the North Vietnamese were engaged in battle against Saigon government detachments months before the U.S. marines splashed ashore at Danang in March 1965.[79] Records captured during the 1968 Tet offensive also indicate that the Viet Cong were, and had been, under significant North Vietnamese control since the 1954 Geneva accords. The postwar testimony of Truong Nhu Tang, minister of justice in the NLF-established Provisional Revolutionary Government, who fled Vietnam in 1979, suggests the nature of the North Vietnamese regime that manipulated the NLF and the Viet Cong to its own ends: "Never has any previous regime brought such masses of people to such a desperation. Not the military dictators, not the colonialists, not even the ancient Chinese warlords."[80] The NLF was a complex phenomenon, composed of many different forces. But the picture of it, and its relationship to North Vietnam, that dominated the American Catholic debate—that of an indigenous force fighting for its rightful share in the governance of South Vietnam—was simply wrong. Whatever the complexities of its composition, the NLF was an instrument of North Vietnam for the forced reunification of Vietnam under communist control.

Those in the American Catholic debate who mocked the "domino theory"—that the fall of South Vietnam would mean the loss of all Southeast Asia to communism—were both right and wrong. As of 1986, North Vietnam was in control of all of Vietnam, Laos, and most of Cambodia, and its tanks make occasional forays against Cambodian rebel forces encamped in Thailand. "Dominoes" fell, and quickly, after the fall of Saigon. Yet the worst fears of the domino theorists have not come to pass. Thailand remains an independent, if threatened, state, and the tide of communism has not rolled through to Singapore or Jakarta. Catholic opponents of the war were also correct in their claim that ancient antagonisms between Vietnam and China were more significant than any congruence of Marxist purpose between Peking and Hanoi. In retrospect, though, what was most significant about the attack on the domino theory by American Catholics opposed to the war was not their correct perception about the ultimate limits of Hanoi's purposes, but their misperception of Hanoi's proximate interests and intentions. Anti-anticommunism played its part here.

The American Catholic debate, like much American public controversy, badly misrepresented two key military actions of the war. The Tet offensive was almost universally portrayed as a crushing defeat for the U.S. and the South Vietnamese government, a defeat supposed to have illustrated the utter bankruptcy of American policy and purpose in Vietnam. The truth, in fact, was almost precisely the opposite on the narrow military issue. The Tet offensive was a catastrophe for the Viet Cong, who suffered massive casualties to the point where it was never again the significant military force it had been in the conflict. (Some commentators believe that this was not wholly displeasing to Hanoi, whose ultimate political purposes were thereby served.) Viet Cong atrocities during Tet—massive civilian executions in the ancient Vietnamese captial of Hue, for example—provoked nothing like the moral outrage in the American Catholic debate that followed the revelation of the horrible My Lai massacre. Lt. Calley was a criminal, certainly; but a morally unworthy selectivity shaped responses to the crimes of My Lai and Hue.[81]

Similar outrage followed the Christmas bombing of Hanoi in 1972. Yet Hanoi's own figures indicate that the Christmas bombing resulted in 1,500 civilian deaths, as compared to 35,000 in Dresden, 84,000 in Tokyo, and over 100,000 at Hiroshima, World War II "parallels" regularly cited in the 1972 and 1973 commentary on the bombing.

How was it that the Christmas bombing, which was certainly morally questionable under just-war criteria, managed to elicit a rhetoric of condemnation that was astounding, even by Vietnam-era standards? American press reports that systematically misinformed the public on the nature of the attack and its probable results in casualties played their part. But why were they so quickly, even eagerly, believed in the American Catholic attentive public? This may well have had to do with the image of "Amerika" that had become so engrained in many quarters of resistance to the war. The contextual debate on the question of America, as it had worked itself out during the Vietnam era, thus served as a filter of falsification for reading the realities of the times.[82]

The entire debate over Vietnam was thus beset by false images and analyses. The relationships among the North Vietnamese, Viet Cong, Chinese, and Soviets were never clearly established, as one falsification (e.g., the Viet Cong were an independent, indigenous force) was compounded by another (e.g., a North Vietnamese victory would expand Chinese power). The domino theory was right on certain questions and wrong on others. American troops did engage in violations of the laws of war; and yet American forces (and especially American pilots) operated in some circumstances under stricter rules of engagement and targeting than had ever been applied before. Cicero's axiom, that in war, truth is the first casualty, was never more apposite than in Vietnam, and in the American domestic debate over Vietnam. Selective outrage was matched by selective outrage; prisms of analysis often distorted facts, rather than being informed by them. This is almost always true in war, but it was particularly true in Vietnam. No one was "right from the start," and few, if any, were "right" throughout the conflict. American Catholicism bears its share of responsibility for that tragedy of Vietnam.

The war in Vietnam, a paradigmatic, modern, counterinsurgency struggle involving great- and regional-power intervention in a desperate, localized conflict, put immense stress on the classic American Catholic heritage of peace as public order in political community, and on the just-war theory. Yet that time of testing did not result in a development of the heritage and a deeper sophistication of just-war analysis; it led to the virtual abandonment of the heritage as it had evolved through Vatican II. This was most evident in the American Catholic resistance movement, but it was also apparent in the American bishops' response to the war. The Vietnam debate had deleterious effects on Bishop Dozier's "peace pastoral" of Christmas 1971. And Bishop Dozier was surely not the only member of the American hierarchy to lose touch with essential elements in the Church's classic approach to the moral problem of war and peace, security and freedom.

The bishops' annual statements in 1966 and 1967 mildly endorsed U.S. policy in Vietnam. By 1968, in a major pastoral letter entitled "Human Life in Our Day," the bishops were questioning whether the *ad bellum* criterion of proportionality was being violated in Vietnam. Yet even at this date the bishops worried whether "an untimely withdrawal" would not be "equally disastrous." The bishops welcomed President Johnson's bombing halt and the opening of the Paris negotiations, but insisted that they themselves were not going to propose "technical formulas" for ending the war. That may well have been a reasonable posture, but it hardly exhausted the bishops' pastoral and civic responsibilities. By November 1968, the main contours of the American Catholic debate on Vietnam had been shaped, largely through the themes of the resistance movement; the issue posed was "America: in or out?" Were not the bishops under some obligation to review the comprehensive approach to the moral problem of war that was to be found in their own heritage, and thereby set a better framework for the American Catholic debate?

If the bishops were searching for a peace position that met many of the standards of the heritage, they could have found it in Negotiation Now (as some bishops, indeed, did). Whether the adoption of the Negotiation Now framework by the American bishops would have led to a more peaceful, free, and secure South Vietnam is certainly arguable; but it could have had a healthy effect on the debate within the American Catholic attentive public. It would have helped distinguish between a genuine peace position and a falsely labeled "antiwar" politics that had come to focus on one side's violence in Vietnam. It would have made important linkages between the possibility of peace and the advance and defense of democratic (or even predemocratic) institutions. It would have helped the Church in the United States face the weaknesses of its own government's policy and the murderous intent of the leaders of North Vietnam. It could have helped establish a moral framework for relating the discriminate use of military force toward morally worthy political ends, while urging the resolution of basic problems of economic and social injustice.

The bishops' pastoral letter of 1968 hints at some of these themes, but it does not develop them. Neither did the November 1971 National Conference of Catholic Bishops' "Resolution on Southeast Asia," which, although calling for a "speedy ending of this war," failed to discuss how America might disengage wisely, or what the relevant moral standards to guide such a policy might be. The bishops themselves, then, cannot escape a measure of responsibility for the abandonment of the heritage of *tranquillitas ordinis* during the traumatic debate over Vietnam. Opportunities to develop and apply the heritage abounded. They were not taken.[83]

And the results? The concept of peace as public order in dynamic political community was largely absent from the Vietnam debate, a situation that has not been rectified since. "Intervention," rather than a given life in an interdependent world, became a term of opprobrium. Rather than being argued in terms of appropriate ends and means, the question of intervention was answered by various neo-isolationisms, hard and soft. The pain of Vietnam, daily broadcast into American living rooms in full color, helped to truncate the contextual debate over the possibly appropriate use of military force. Since Vietnam, one looks hard to find any answer other than a simple no to this contextual question within the American Catholic attentive public. But this has not been, nor was it during the Vietnam debate, the no of a principled pacifism willing to face the moral dilemmas posed by the refusal to use military force in the defense of rights. It is, and was, an intellectually softer no, more reflective of judgments on the questions of intervention and America than of a sophisticated set of standards for resolving questions of the possible use of military force.

The Vietnam-era Catholic debate solidified the dominance of anti-anticommunism among Catholic intellectuals and publicists, and, increasingly, among bishops. The sophisticated anticommunism called for by John Courtney Murray remains to be developed. Yet one result of the Vietnam trauma is that such a development does not seem a significant issue to those most vocally concerned about the moral problem of war.

American Catholic apprehensions of America were severely distorted during Vietnam. Analogies were frequently drawn between Vietnam-era America and Germany under the Third Reich. Calls to confront the nation's collective guilt were widespread in liberal and radical Catholic circles. Such calls, which were paralleled in (and in many cases, adopted out of whole cloth from) secular currents of thought in American public life were "one of the more bizarre cultural phenomena of the Vietnam period," as one commentator has put it, inasmuch as "nothing could be more repugnant to the moral and legal tradition of liberalism than the idea that an individual can be held responsible for an act he himself did not commit." (Nothing could be more repugnant to Catholic moral theory, either.)[84]

Why this happened seems clear: the atrocity of My Lai, for example, revealed the truth about the war itself, which in turn revealed the truth about "Amerika." That spelling does not appear in any of the essays from the prestige American Catholic press reviewed above; but the message was similar. American greed, American racism, American imperialism, the insane American bent on imposing our will on the world—these were the reasons for Vietnam. The drastic alteration of America would be one principal "lesson" learned from the war. One need not attempt to defend U.S. policy during Vietnam (as I would not) to see how overdrawn this blanket condemnation was. It was not only false in its portrait of the United States. It falsified the reality of war, transforming the Viet Cong into peasant revolutionaries committed to agrarian reform and democracy. It de-Stalinized the forces of Ho Chi Minh. And it demonized American political and military leaders.

Some went further in this demonology than others. But the Vietnam debate was conducted in the American Catholic attentive public in such a way that themes of delegitimation were regularly taught in the most prestigious Catholic journals in the United States. The editors of *Commonweal* may well have disagreed with Frank Getlein's description of U.S. foreign policy as a venereal disease, and may well have wondered about Gabriel Kolko's praise of life in Quang Tri province under the Provisional Revolutionary Government. But their editorial judgment in publishing such material suggests how far, and in what direction, the American Catholic debate on the moral problem of war had been bent during the Vietnam years.

The argument over amnesty for draft resisters, exiles, and military deserters illustrated this trend. *Commonweal* sought unconditional amnesty for those who had been "prematurely moral" about Vietnam, as did numerous other Catholic commentators. But here, too, the American bishops missed an important teaching opportunity. In November 1972 the National Conference of Catholic Bishops adopted a "Resolution on Imperatives of Peace" that, among other things, addressed the question of amnesty:

> In a spirit of reconciliation, all possible consideration should be given to those young men who, because of sincere conscientious belief, refused to participate in the war. A year ago, we urged "that the civil authorities grant generous pardon of convictions incurred under the Selective Service Act, with the understanding that sincere conscientious objectors should remain open in principle to some form of service to the community" ("Resolution on Southeast Asia," National Conference of Catholic Bishops, November 1971). We again urge government officials and all Americans to respond in this spirit to the conspicuous need to find a solution to the problems of these men. Generosity represents the best of the American tradition and should characterize our response to this urgent challenge.[85]

An approach to the amnesty question that reflected the classic Catholic heritage would have been framed differently. It would have acknowledged the fact that sincere men could, in good conscience, reach different moral judgments on the war in Vietnam, and on the question of their own participation in it. It would have recalled, as the bishops did, the American tradition of amnesty. But it would have made an equally strong statement about the importance of law in a rightly ordered political community as an indispensable element of peace—and it would have set its proposal for amnesty-with-community-service in *that* framework, rather than solely in terms of the evangelical requirements of generosity and mercy.

Such a statement could have helped bind the domestic wounds of the war. It would have created ground on which the bishops could address the problems of personal conscience and civic responsibility in a democracy in the future. But most importantly, it would have demonstrated that the bishops, in a crucial test case, still held fast to the classic Catholic understanding of law and political community as alternatives to violence in the resolution of conflict. In doing so, the bishops would have set a standard for a post-Vietnam Catholic peace concern that did not focus primarily on the evil of the American experiment, and that related the requirements of public order in dynamic political community at home to the imperatives of peace in international affairs. The American bishops' failure to seize this opportunity for reclaiming their heritage was as important an omission as their failure to help establish more adequate standards for the Catholic debate on Vietnam in 1968. It is a reasonable speculation that this failure can be at least partially attributed to the bishops' susceptibility to dominant themes in the Catholic debate as defined by the resistance movement—particularly teachings on the evils of American power.

In sum, then, the Vietnam debate within American Catholicism did not lead to a genuine development of the heritage of *tranquillitas ordinis*. Rather, Vietnam was a prism through which the main themes in that heritage—the central components of a distinctively Catholic context for thinking about the moral problem of war and peace—were twisted and bent in such a way that the net effect was a virtual abandonment of the heritage as a horizon for moral analysis. Other events had their impact on this abandonment. But Vietnam was decisive. Evaluating the moral problem of war "with an entirely new attitude" had come to mean precisely that: the heritage developed from Augustine through Murray was to be dismissed, however regretfully, as irrelevant to those who would be neither victims nor executioners.

CHAPTER 8

The Moral Horizon of "An Entirely New Attitude"

Willam Butler Yeats's description of twentieth-century life as one in which "the centre cannot hold" may be read as the stark portrait of a world caught between the fire of war and the pit of totalitarianism, and as an anticipation of the moral task of the Catholic heritage of *tranquillitas ordinis* in such a world. The "center" represented by the classic Catholic heritage and its moderate realism did not "hold" in American Catholicism after the Second Vatican Council. Understanding the nature of that failure requires a sense of what sort of "center" the heritage, as it had been received and developed up through Vatican II and the work of Murray, represented.

It was not a "center" in the sense of a compromising, accommodating point between the poles of an argument. Rather, it was a "center" that, by the sophistication and comprehensiveness of its view, set a standard for Catholic debate on the moral problem of peace and freedom, and thereby disciplined the poles of the argument (i.e., those tempted to dissolve one or another side of the Janus-headed problem of war and totalitarianism). The abandonment of that "center" of moral discourse allowed one pole of the Catholic debate to predominate. The result was a new moral horizon against which to address the problem of war.

Eight important teachings that comprised the context that gave flesh to that new moral horizon were identified in the previous chapter. The most important of these teachings were those that emphasized a personal conversion and/or *shalom* definition of "peace" (and consequently deemphasized the concept of peace as *tranquillitas ordinis*); that taught a neo-isolationist view of American foreign policy; that minimized the possibility of the legitimate use of military force; that urged anti-anticommunism on the Church; and that questioned the moral worthiness of the American experiment. None of these themes became dominant in its most radical form. But in more modified terms, these teachings set the context for the postconciliar American Catholic debate on the moral problems of war and peace, security and freedom.

Other themes, more general in scope, were wedded to these contextual teachings and helped create a new horizon of moral analysis as American Catholicism's intellectual and episcopal leadership sought to meet the Vatican Council's charge that the problem of war be evaluated "with an entirely new attitude."

THE CHURCH AND THE WORLD

Classic Catholic social ethics taught an understanding of the church/world relationship that fit comfortably within H. Richard Niebuhr's model of "Christ the Transformer of Culture"[1]—a model that shaped the conciliar "Pastoral Constitution on the Church in the Modern World." Rooted in Catholicism's traditionally incarnational imagination, which stressed the continuity of creation and redemption, this approach emphasized the Church's role as a leaven in "the world," and particularly in civil society.

The Church would act as a leaven in several ways. It would maintain, develop, and teach its own social ethic, rooted in a long experience of men and affairs, but open to evolution and growth through a dialectical encounter with the signs of the times. Through that teaching, the Church would bear witness to the truth as it had come to understand it, and help form the consciences of those Catholics, chiefly lay men and women, who had primary responsibility for determining the course of public life. The universal Church, acting through the Holy See, would remain an important (if discreet) diplomatic activist. The national conferences of bishops mandated by Vatican II would work with their respective civil governments toward the peace of political community in truth, charity, freedom, and justice.

The premise of these various forms of engagement was that the Church had a legitimate mission in the world: that it should be, not merely a judge of what was inadequate or evil, but an active agent in facilitating the common good. To do so, according to Catholic theory, was not an abandonment of the Church's distinctiveness as a communion of faith, but an expression of its evangelical mission as herald of the Gospel.

This transformationalist perspective remained a significant force in American Catholicism after Vatican II. But it was also challenged by other ecclesiological models, and other understandings of the church/world relationship. Sectarian currents of thought, emphasizing the corruption of the Church through its interaction with civil authority, came to the fore. Viewed from one angle, this development was the latest example of a classic tension in Catholic Christianity between "prophetic" and "civic" understandings of the Church's mission in the world. But the new sectarian currents were shaped by theological and political judgments that gave them a distinctive cast.

The radically eschatological view of history and the apocalyptic nature of the present moment taught by Dorothy Day, Thomas Merton, Daniel and Philip Berrigan, and James Douglass had a significant impact. There were gradations of judgment on the present circumstances of "the world" among these thinkers. But the notions that "the world" was a problem in a decisive way, that days of judgment were imminent, and that the Church's very integrity was threatened by a transformationalist approach to the radical corruptions of the present order—these themes pointed toward a more sectarian ecclesiological self-understanding that emphasized the Church's role as a "witness" community over against the world and its sin, rather than as an agent of transformation and healing in a broken and divided human community.

Various theological vectors of influence contributed to American Catholic sectarianism. Liberation theology was not classically sectarian in its ecclesiology (because it envisioned a important public role for the Church as an agent of revolutionary change); but its radical critique of the organization of power in the West raised the question of whether the Church had become an unwitting accomplice to oppression by taking a transforming approach to its mission in the world. Ecumenical theological dialogue contributed to this evolution of thought. Mennonite theologian John Howard Yoder, for example, brought the classically sectarian ecclesiology of his traditional "peace church" into active conversation with Catholic theologians, publicists, and activists.[2]

New Left political themes were influential in driving the ecclesiology of the Berrigan brothers in a sectarian direction. If America was really Babylon, a principal bearer of the world's evil, then the Church had no business in making common cause—even in a critical vein—with the principalities and powers of that oppressor state. Rather, the Church's essential business was judgment. Its mission was to shout a judgment from the rooftops: to call down fire from heaven (in the form of revolutionary action) on the present order of corrupt power.

Confusions abounded here. Whereas classic, sectarian, "witness" ecclesiology stood radically over against "the world," eschewing political prescription, the Berrigans and many others who shared their view of the Church did not hesitate to make numerous public policy prescriptions. Ecclesiological sectarianism was married to political activism in a new hybrid. However one sorts out the various influences at work in the creation of that hybrid, it clearly represented a break with the transformationalist perspective of the Catholic heritage on the relationship between Church and world.

Such teachings were not exclusive to American Catholicism; similar approaches were evident in the mainline American Protestant denominations and their principal ecumenical agency, the National Council of Churches. Here, too, the influence of liberation theology with its Marxist or quasi-Marxist view of the nature of "liberation" was great. Some critics argued that the general thrust of liberation theology, while claiming to rescue the Church from an unworthy accommodation to the principalities and powers of the world, was itself a radical accommodation to a different form of worldliness.

Why should this have happened, in American Protestantism or in American Catholicism? Lutheran theologian Richard John Neuhaus offered a hard, but suggestive, answer:

> Wearied of wandering and waiting upon the promise, we long for the fleshpots of Egypt. In this case, the fleshpots offer not physical sustenance so much as a cause, a movement, a struggle that gives us the feeling of importance and usefulness in the real world. When we are no longer sure that the Gospel is true we are eager to prove it is useful. And so, in the name of Christian discipleship, we end up subscribing to one of the great heresies of the modern world, the heresy that utility is the measure of truth.[3]

It is a hard judgment, but one worth considering. Did the optimism of *Pacem in Terris* and *Gaudium et Spes* so quickly turn to ashes in the mouths of many American Catholic intellectuals and publicists (especially in light of Vietnam) that the very possibility of speaking a word of Gospel truth into the modern world, and particularly into contemporary America, became radically questionable? Did the notion that the

Church's credibility was established, not by the truth of its evangelical message but by its utility to a revolutionary cause, enter some minds? Was the widespread assumption that "secularization" defined the character of our times—an assumption decisively refuted by empirical study, but nonetheless immensely influential in American religion—a part of the problem here?[4]

Neuhaus, a Lutheran with deep Catholic sympathies, rejected both sectarianism and accommodationism, left- or right-wing in character. While he believed that most Christians rejected the identification of the Gospel with Marxist or quasi-Marxist liberation, he also argued that "their rejection is of little religious significance . . . if it is derived only from their attachment to other social and political securities." What was needed, Neuhaus claimed, was:

> To think anew about how the Church ought to be in the world. The outcome of that could be a revivified ecumenical vision of a community of transcendent faith embracing those who espouse different and even conflicting views of our mundane tasks, including our political tasks. Pointing beyond all the political options of the present age, such a reconciled and reconciling Church would signal more believably the genuinely New Politics of the Kingdom of God.[5]

That seemed a more than adequate encapsulization of the ecclesiological vision and sense of worldly mission that permeated *Gaudium et Spes*. But that vision and sense of mission, if not entirely rejected, came under severe pressure from those who argued for a more "radical" option: the Church as witness over against the principalities and powers of the American *imperium*. The sectarian character of this ecclesiology was often masked by its proponents' inclinations to propound a host of public policy prescriptions; but the basic trajectory of thought was away from "Christ the transformer of culture" and toward a more sectarian model of church-against-world. This new Catholic sectarian sensibility was not limited to intellectual and activist circles: by the late 1970s, it had made an impact on a number of American bishops, as debate over the morality of nuclear strategy began to intensify.[6]

MORALITY AND FOREIGN POLICY

One of the most significant mandates of the Second Vatican Council was its charge that the Church become more fully attuned to its biblical heritage. Postconciliar American Catholicism witnessed an explosion of biblical scholarship, much of it of high quality. Scripture study was not limited to intellectual circles; it became an important dimension of the pastoral activity of the Church and of postconciliar Catholic spirituality. Yet, paradoxically, this new biblical awareness would alter the debate over morality and foreign policy in a decidedly nonconciliar way.

John Courtney Murray's warnings against moral*ism* in foreign policy may be recalled here. Murray argued that one of the distinguishing characteristics of the moralism rejected by foreign policy realists was its biblical fundamentalism:

> In order to find the will of God for man it went directly to the Bible . . . [whose teachings] are to be taken at their immediate face value without further exegetical ado. When, for instance, the Gospel tells the Christian not to resist evil but to turn

the other cheek, the precept is clear and absolute. The true Christian abdicates the use of force even in the face of injury.[7]

Murray rejected this approach because it violated the canons of the tradition of reason. His rejection was confirmed by a later generation of Catholic biblical scholars who saw in literalism a misunderstanding of the theology of revelation and a distortion of Scripture itself. Modern Catholic biblical scholarship placed a higher value on the use of Scripture in moral reflection on the problem of war and peace than Murray allowed. But it also insisted that the Old and New Testaments not be ransacked for proof texts in support of predetermined political judgments. Moreover, modern Catholic exegesis denied that the New Testament contained clear, univocal instructions to the Catholic community on the conduct of U.S. foreign policy. Rather, Scripture suggested principles for moral analysis and images through which present reality might be understood through the mediation of reason and argument. No less than Murray would the great majority of Catholic exegetes argue that the Sermon on the Mount did not contain a foreign policy agenda.[8]

That exegetical warning was not heeded in many influential American Catholic circles.

The pastoral letters of Bishop Dozier of Memphis and Archbishop Hunthausen of Seattle illustrate the point. Both men relied on a literalist reading of the New Testament to buttress their moral claims: Dozier's case for U.S. withdrawal from Vietnam, and Hunthausen's argument for unilateral disarmament. Dozier's and Hunthausen's positions were influenced by other sources. But the sharpest points of their respective critiques of the immorality of U.S. foreign policy were drawn from biblical (and especially New Testament) teachings. Dozier argued that Jesus' injunction to walk the second mile with an adversary required immediate U.S. withdrawal from the conflict in Vietnam. Hunthausen claimed that "we need to return to the Gospel with open hearts to learn once again what it is to have faith," because "politics is itself powerless to overcome the demonic in its midst." To the archbishop of Seattle, "one obvious meaning of the cross is unilateral disarmament." The archbishop did not argue that one *possible* meaning of the cross was unilateral disarmament, but rather that one *obvious* meaning of the cross was that particular prescription.[9]

It seems paradoxical at the very least that a renaissance in Catholic biblical scholarship, which rejected literalist exegesis, should be paralleled by such literalist readings of the Gospel mandate. One possible explanation lay in the priority that *Gaudium et Spes* gave to evangelical criteria by proclaiming the Church's duty to scrutinize the signs of the times and interpret them "in the light of the gospel." The translator of *Gaudium et Spes*, in urging the hierarchy in 1966 to a more vocal opposition to the war in Vietnam, argued that the bishops were failing to their responsibilities to "represent that gospel by which all human governments and human motivations must be judged."[10]

A second possible explanation may lie in the fact that Scripture came to play an increasingly important role in postconciliar Catholic liturgy and spirituality; but that welcome evolution was not matched by much exegetical sophistication. The Catholic charismatic renewal, which emphasized the immediate presence of the Holy Spirit to individual Catholics in their encounters with the word of God, may also have been influential, as were the debates on the "prophetic" or "witness" role of the Church vis-à-vis civil society. However one sorts out the vectors of influence, however, one result was clear: American bishops, in the exercise of their teaching office, adopted an approach to the question of morality and foreign policy that, in its use of Scripture, represented precisely the style of moral reasoning rejected by John Courtney Murray and modern Catholic biblical scholarship.

The problem of Scriptural literalism in moral judgment was complicated by a further transition in American Catholic thought on morality and foreign policy. The classic Catholic heritage understood moral reason to make its most persuasive claims at the contextual level of the foreign policy debate. The heritage did not thereby decline prudential prescription; but the style of moral reasoning recommended by Murray (a recommendation fully congruent with the mainstream of Catholic moral theory) held that the empirical complexity of specific policy choices made them less susceptible to moral certainty. The heritage was most concerned with shaping the context of the foreign policy debate: by establishing the claims of moral reason to a voice in the argument, by adumbrating key moral principles, and then by offering its own best judgments on the appropriate answers to key contextual questions. Prescription was deemed less central than context-setting. Culture-formation through moral education and argument was a more appropriate exercise of the Church's teaching mission, because of the nature of that mission and the nature of political choice.

One of the Church's most prominent theologians argued this perspective during Vatican II. In a June 1965 interview with *America*, Karl Rahner distinguished Catholic moral theory from both situation ethics and political prescriptivitis. Situation ethics taught the absolute uniqueness of individual circumstances, and thus the impossibility of a close application of principled norms to them. Because the individual Christian was a member of a community of faith and reason, Rahner argued, it could never be the case that one's conscience was formed entirely by oneself. On the other hand, there were situations of such complexity that the Church could only refer to the key principles involved:

> Then the Church must say: "I cannot tell you from these principles what you should do here and now. You yourself must work it out with God and your conscience." For example, if President Kennedy had come to Pope John with the question: "Should I as a Christian cut back the production of nuclear weapons, or should I extend their production?" In such a complex and vastly important moral issue, I do not believe that the Church could give a definite answer. The Pope would have had to say to Mr. Kennedy: "We're sorry. You should take into consideration such and such principles, but what follows from them as far as this concrete decision is concerned, you have to decide for yourself."[11]

Nine years later, Rahner's view had not changed. He told *Commonweal* in 1974 that theology could not deliver practical applications to dogmatic or moral issues like "freshly baked breakfast rolls."[12]

Rahner's warning was not heeded. Prescription became the order of the day, and culture-formation came to be viewed as somehow unsavory, perhaps an abandonment of the Gospel imperative to judge the signs of the times. Innumerable prescriptions, on virtually every aspect of U.S. foreign policy, were heard from American Catholic intellectuals, publicists, and religious leaders in the decades after Vatican II. In many cases, a brief bow was made to the notion that prescriptive judgment from a religious institution carried less weight, in the nature of things, than judgment on guiding moral principles; but the trajectory of interest and intention was clear.

What is less apparent is, why? Richard Neuhaus's critique of the churches' susceptibility to a criterion of utility is worth recalling here. So, too, is the fact that the dominant historiography of American Catholicism continued to emphasize that "most American Catholics have consistently regarded their nation's foreign policy as a matter to which their faith is irrelevant."[13] Given such a self-understanding, it is little wonder that many American Catholic intellectual and religious leaders wanted to redress what must have seemed a scandalous imbalance in the scales of public responsibility. That self-understanding was based on a limited and flawed historiographic principle. But the point remains: American Catholics were taught that their's had been a voice of silence and accommodation. Vatican II's call for a deepened encounter with the modern world, received in the supercharged politics of the 1960s, led rather easily to a heightened focus on prescription and a consequent diminishment of culture-formation.

The dominant teachings in the American Catholic debate since Vatican II emphasized an ethics of intention over an ethics of consequences. This fit comfortably into the style of engagement favored by those who urged a "prophetic" stance; Robert McAfee Brown's call that the Church "be specific rather than general, making mistakes of commission rather than of omission" illustrates the point.[14] Brown clearly regarded a "sin of omission" as graver than a sin of commission, despite the consequences of the latter. Archbishop Hunthausen of Seattle agreed; a consequential concern for the possible results of unilateral nuclear disarmament did not affect his analysis or prescription. "A choice has been put before us," the archbishop argued in 1981; "anyone who wants to save one's own life by nuclear arms will lose it; but anyone who loses one's life by giving up those arms for Jesus' sake, and for the sake of the Gospel of love, will save it."[15]

The "prophetic" stance toward morality and foreign policy urged by Brown and Hunthausen was not adopted in full by the Catholic leadership of the United States. The 1983 pastoral letter, "The Challenge of Peace," taught that both intentions and consequences, as well as the nature of specific acts, must be weighed in forming moral judgments about matters of war and peace. But that official acknowledgment of the classic structure of Catholic moral thinking on foreign policy issues would not shape the discussion. A hybrid form of Catholic sectarianism, a literalist application of Scripture, and a concern for the recovery of "prophetic witness" combined to yield a greater emphasis on intentions than had been known before in Catholic social ethics.

PACIFISM

American Catholicism had been touched by pacifism before Vatican II, most notably by the Catholic Worker movement and the writings of Gordon Zahn. But in the postconciliar period, a renascent American Catholic pacifism stood, not on the margins of the debate over war and peace, but close to its very center. Several factors were responsible for this signal development.

First, one must acknowledge the cumulative impact of the Catholic Worker movement and Dorothy Day, not only in terms of their witness to the whole Church, but also because of the network of relationships that tied the Worker movement into increasingly influential American Catholic thought and action centers. One line of influence ran from the Catholic Worker to several *Commonweal* editors who, whatever their personal rejection of pacifism on theoretical grounds, were deeply influenced by Worker pacifism in their reading of the signs of the times and in their analysis of American power in the world. Another line of influence ran from the Worker to Daniel Berrigan. Dorothy Day's own steadfast loyalty to the institutional Church has its impact on the Church's official leadership over time. Finally, the *Catholic Worker* provided a journalistic vehicle of large scope for propagating the movement's pacifist teachings and political judgments in American Catholic intellectual and activist circles.

In addition to the Catholic Worker movement, one must also note the influence of Gordon Zahn, Thomas Merton, andd James Douglass. Zahn, like Dorothy Day a devout Catholic deeply commited to the institutional Church, provided a less countercultural, more intellectually developed approach to pacifism in a Catholic context. Merton, though declining the title "pacifist" on intellectual grounds, lent his immense authority to the emergence of pacifism in American Catholicism, primarily through brokering Gandhian political thought into the debate. James Douglass, a principled pacifist and another Gandhian, played an important role in the evolution of American Catholic pacifism through his first book, *The Non-Violent Cross*, which became a kind of primer for Catholic pacifists (and for all those interested in nonviolent resistance to public authority). Douglass's critique of just-war theory was especially influential.

But it seems unlikely that even these forceful personalities would have made the impact they did had it not been for the traumatizing of the American Catholic debate during the war in Vietnam. The perceived inability of classic just-war theory to appropriately analyze or set limits on a bloody, counterinsurgency guerilla war was of crucial importance. Vietnam, to a generation of American Catholics enjoined to evaluate war "with an entirely new attitude," seemed a graphic expression of everything morally repugnant about modern warfare: its immense destructiveness, its apparent disconnection from achievable political ends, its pitting of the wealthy against the poor and whites against "people of color," its capacity to warp domestic political life thousands of miles from the combat zone. Against these grim realities it increasingly struck many American Catholic leaders as morally obtuse to speak of the proportionate and discriminate use of military force for worthy political ends. To note this is not to argue the correctness of such a judgment; only that it was understandable. Vietnam, in short, created the cultural conditions for the possibility of the emergence of pacifism as a full partner in the American Catholic debate over war and peace, security and freedom. That partnership would be even more vigorously asserted—and this time, by official leaders of the church—in the debate over nuclear weapons that erupted less than half a decade after the fall of Saigon.

What did postconciliar Catholic pacifism teach the American Church, and thus contribute to the moral horizon of "an entirely new attitude" toward war?

Its connection of the pacifist conscience to the claims of the Sermon on the Mount reflected a literalist approach to Scripture. Archbishop Hunthausen was paradigmatic of this dimension of the new American Catholic pacifism, as were the writings of Catholic Worker veteran Eileen Egan.[16] But not all American Catholic pacifists fit within that ambit. Gordon Zahn reflected a form of Catholic pacifism that, in many debates, worked via an argument from negation: given contemporary political and technological circumstances, war cannot be morally tolerable under the criteria of the just-war theory, especially the *ius in bello* but also the *ius ad bellum*. Because limited war in an interdependent world could (and probably would) lead to nuclear war, and because a morally acceptable nuclear war is an oxymoron, even the *ad bellum* argument for limited war collapses, on this analysis. Since just war is an impossibility, and the crusade even more an impossibility, the only possible Catholic response to the problem of war is "no war," or pacifism. Merton's writings often reflected a similar analysis. This prudential defense of pacifism was often connected to the evangelical themes found in Dorothy Day's Christian personalism or in the theology of James Douglass, so that the *via negativa* led to a positive statement of Christian conviction. The net result in either case—the originally Gospel-centered argument, or the prudential argument with evangelical buttressing—was to posit pacifism as the only option for Catholicism.

American Catholic pacifism since Vatican II was also concerned with the theory and practice of nonviolence. Nonviolence and pacifism are often equated; but they are distinct. A Christian committed to just-war theory as a moral calculus may opt, in certain circumstances, for nonviolent forms of resistance to aggression or injustice, inasmuch as the circumstances may make the use of limited armed force impossible; Lech Wałesa and the Polish resistance movement Solidarity illustrate this path to nonviolence. Nor was the heritage of *tranquillitas ordinis* unaware of the claims of nonviolence; law and governance are, after all, the most widely applied means of nonviolent social change in history. Just-war theory's insistence that limited military force be a last resort also implies that nonviolent means of defending violated rights have been tried and found wanting; the key point is that they must be tried.

Postconciliar American Catholic pacifism turned to the theory and practice of nonviolence through several influences. Gandhian resistance as applied to the American civil rights movement by Martin Luther King, Jr., made a great impact.[17] But in conceptual terms, nonviolence is pacifism's answer to the inescapable problem of power in human affairs, and especially in politics. If power is understood as the capacity to achieve purpose, the moral problem of power cannot be avoided, because power stands at the heart of any political community. Nonviolent theory, especially in its Gandhian form, insisted that power not be identified exclusively with violence. Gandhian thought, then, did not burke the problem of power, but tried to develop a theory and practice of power that avoided, on principle, violent sanctions.[18] This proved immensely attractive to those committed to the development of either an evangelical or prudential Catholic pacifism.

Whether the emergent Catholic pacifism of the postconciliar period remained faithful to Gandhian principles is a different matter.

Thomas Merton's nervousness over the Berrigans' redefinition of nonviolence to include the possibility of violence against property illustrated an important fault line in American Catholic pacifism. But the problem with the Berrigans' nonviolence was not limited to the question of tactics ("Is burning public property a violent act?") or the acceptance of the consequences of civil disobedience ("Is 'going underground' congruent with Gandhian insistence on the importance of affirming law, even while resisting unjust laws?"). Daniel Berrigan's refusal to dismiss the possibility of violence in Third World revolutionary situations suggested a serious compromise in pacifist commitments. So, too, did the disinclination of American Catholic pacifists to critically evaluate the class struggle model of social change and the occasional blessings put on revolutionary violence by some liberation theologians. Could there be a selective pacifism?[19]

Despite the emergence of a profoundly felt pacifist conscience in American Catholicism since Vatican II, that conscience has yet to be given a sophisticated theological and political-philosophical explication. Zahn's writings come closest to this standard; but even Zahn was unwilling, perhaps unable, to distinguish between a principled pacifism and a pacifism that had become dominated by New Left currents of thought and styles of political action during Vietnam. Nor was Zahn, or any other American Catholic pacifist theorist, able to argue persuasively the moral requirement of pacifism in an non-pacifist world. The call to personal conversion at the heart of the new American Catholic pacifism makes powerful claims on individual consciences. But it remains to be shown how such an act of conversion can be expected of a government that has the *moral* responsibility of providing for its citizens' security in a persistently violent world.

These intellectual weaknesses in the pacifist case had disturbing results in American Catholicism.

There was the problem of selectivity—that is, a pacifism applied to American military actions or programs, but not to revolutionary violence in the name of "liberation." But there were other confusions involved in the new American Catholic pacifist call for a "witness" ecclesiology. A principled pacifism of religious witness, affirming immutable and absolute moral norms (e.g., the proscription of killing, even in the defense of values) and the equally absolute claims of the Sermon on the Mount and the Beatitudes, would necessarily stand over against the political order. It would decline prudential political prescription, understanding that such prescription inevitably entailed a measure of compromise or casuistry in the application of evangelical and moral norms in a world in which those norms were not universally accepted.

But it would further insist on the prior importance of its own witness, seeing it as a standard by which present public policy might be judged, and the future orientation of policy determined. A pacifism of religious witness would thus opt for a horizonal responsibility: by witnessing to the truth of how things ought to be, it would set a standard for measuring the gap between that moral and political horizon and things as they are. This would yield a benign sectarianism, an ecclesiology that posed Church over against world, but linked Church to world through the creation of a horizon of moral judgment, and through the exercise of charity in the world as it is.

This was not the form of witness adopted by postconciliar American Catholic pacifism, which opted instead for intense involvement with public policy as diagnostician, judge, and prescriptor. As diagnosticians, many American Catholic pacifists drew their primary teachings from New Left currents of thought, emphasizing the evil of American power and its roots in American political economy and culture. As judges, Catholic pacifists applied Gospel criteria to life in a pluralistic America and an even more pluralistic world without much "exegetical ado," as Murray would have put it. As prescriptors, American Catholic pacifists proposed detailed and specific policy options, ranging from immediate withdrawal of U.S. forces from Vietnam, through radical proposals for the reorganization of power in American society, and on to unilateral disarmament. In a curious role reversal, pacifists became politicians, military strategists, and tacticians.

Gordon Zahn, throughout his career, insisted that such prescription was entirely in order, because there was no necessary connection between pacifism and sectarian ecclesiology. Zahn worked to create a transformationalist pacifism which blended the ecclesiology of *Gaudium et Spes* with the rigors of the pacifist conscience. Zahn thereby opened up the pacifist option for the Church's episcopal leadership, an option taken by several dozen American bishops by the late 1970s. And yet even so persistent and able a pacifist as Zahn failed to make a finally persuasive case.

Zahn's pacifism was an attempt to have it both ways: to maintain an absolutist moral stance in social ethics while being fully engaged in politics. There are numerous problems here. As moral theory, pacifism is, if faithful to its own best instincts, a radical ethics of intention. It hears, and wishes to act on, an evangelical call to perfection that transvalues any ethical attention to a criterion of consequences. In this sense, pacifism's ethics of intention stands within a prophetic tradition that calls for justice to be done, though the heavens fall as a result. The difficulty is that the morality of political judgment must include a consequential criterion. To argue, for example, that unilateral disarmament is the sole moral option, even if its results would be to make war more likely, is not an act of prophetic witness, but of moral absurdity. Nor is this an extreme example: several American bishops argued precisely that case in the early 1980s.

The problem is not only one of consequences though: the confused application of a radical ethics of intention to the public order fundamentally denies the capacity of the tradition of reason to shape the human world in a Godly way. That the unilateralist bishops understood this was made clear by one bishop's claim that, whatever the seeming absurdity of unilateralist disarmament, "God would make up for our folly."[20] Thus the terminus of this approach was a classic *deus ex machina* appeal that fractured the continuity of creation and redemption, nature and grace—that is, the fundamental claim of the Catholic incarnational imagination. The appeal to "Gospel standards" ended up denying the possibility of a rational, morally meaningful human expression of the God-touchedness of creation in the political order. By a paradoxical route of argument and sentiment, American Catholic pacifists landed in a position similar to that of a realist like Hans Morgenthau, arguing that any political act is "inevitably evil."[21]

Postconciliar American Catholic pacifism paid little attention to law and governance as morally worthy, nonviolent means for resolving the world's inevitable conflicts. Here, one might well have argued, was the way in which the pacifist conscience might express itself in the political order, as Zahn and others wished: abandoning military analysis, and focusing its political imagination and prescriptive attention on the evolution of the peace of dynamic political community on an international scale.

A model of this kind of transformationalist pacifism was available, in the study sponsored by the American Friends Service Committee (AFSC), *Speak Truth to Power: A Quaker Search for an Alternative to Violence.*[22] Published in 1955, *Speak Truth to Power* remains an impressive attempt to apply the pacifist conscience to the dilemmas of governance without burking the realities of power or conflict. The AFSC document distinguished between "conflict" and "war" (as did the classic Catholic heritage); it was anticommunist as well as antiwar; it sketched the prospects for a world in which conflict was resolved through international legal and political mechanisms, and the means by which American foreign policy might contribute to such an evolution in world affairs. Most significantly, *Speak Truth to Power* affirmed the possibility that the United States might lead the world in such a direction, given its own experience of political community.

That these themes were not adopted and developed by postconciliar American Catholic pacifism was a tragedy of major proportions. The 1955 AFSC approach came as close to the heritage of *tranquillitas ordinis* in its fundamental moral and political intuitions as one is likely to get from a pacifist starting point. And yet it seems to have been virtually ignored in the postconciliar American Catholic debate (as it has been in the post-1965 AFSC).[23] The result was an American Catholic pacifism more attuned to New Left themes than to classic pacifist understandings in its basic politics. Postconciliar American Catholic pacifism taught the American Church an oversimplified approach to the relationship between the Scriptural witness and the dilemmas of peace and freedom; emphasized the evil of the American "system" over the possibilities of democratic political community as an alternative to mass violence in the resolution of conflict; enmeshed itself in arguments over military strategy and tactics, and their relationship to specific conflicts and to America's world role; and, concurrently, failed to shed light upon its own ecclesiological conundrum.

Postconciliar American Catholic pacifism did not, in short, contribute to the necessary development of the heritage of *tranquillitas ordinis*, but was a prime mover in the abandonment of that heritage. This was particularly evident in the pacifist-influenced discussion of post-Vietnam amnesty, where conscientious objection, a possibility affirmed by *Gaudium et Spes*, was detached from a parallel affirmation of the necessity of law in a morally sound political community. What counted was the right intention of the conscientious objector. Alternative service, rather than being seen as expressing a commitment to law as a nonviolent means for adjudicating conflict, was an obtuse imposition on those who had been "prematurely moral" on the question of Vietnam.

JUST-WAR THEORY

An impressive scholarly development of just-war theory was paralleled, in the postconciliar years, by the deliquescence of just-war forms of reasoning in the American Catholic contextual debate on war and peace, security and freedom. Despite the important work of James Childress and Ralph Potter on the nature of moral reasoning within the theory of the just war,[24] the notion that just-war reasoning represented a disastrous compromise with Gospel values became widespread. Despite the detailed historical studies of James Turner Johnson, LeRoy Walters, and F.H. Russell on the origins and development of just war reasoning,[25] American Catholics were regularly taught in their principal opinion journals, and by an increasing number of their bishops, that just-war theory had been developed in order to sooth the Catholic conscience during the time of Christendom, and had subsequently functioned as an ex post facto rationalization for whatever kind and degree of violence public authorities had deemed necessary for *raisons d' ètat*. (The bishop of Richmond, for example, described just-war theory as "an excuse to go to war, mental gymnastics, casuistry of the worst sort."[26]) Vietnam was a decisive factor in this deterioration of discourse.

Yet Vietnam was also the occasion for other voices, less contemptuous of the just-war tradition, to make themselves heard. One vehicle for this was the debate over selective conscientious objection. Unlike general conscientious objection to all wars, which is based on pacifist convictions, selective conscientious objection is a child of just-war theory. The selective conscientious objector claims, not that all wars are morally wrong and thus impermissible in terms of his participation, but that *this* war is wrong, and thus he may not take part in it. The arguments of a selective conscientious objector can be either *ad bellum* or *in bello*, involving what William V. O'Brien had called "war-decision law" or "war-conduct law."[27]

John Courtney Murray supported reforms in U.S. selective service legislation to permit selective conscientious objection. Murray believed that selective conscientious objection could be defended within the boundaries of the "tradition of reason." But he also worried that the American public debate on war and peace had been so corrupted during Vietnam that the climate of the true City, necessary for a society to make reasonable judgments about the claims of the selective conscientious objector, no longer existed.[28]

Two points are important here. First, the breakdown of civic culture necessary to sustain society's judgment on the claims of the selective conscientious objector was clearly related to the themes of American illegitimacy taught by the Vietnam-era resistance movement. Once again, through the prism of selective conscientious objection, we may see how Murray's grand project, the moral defense of democratic political community, bore heavily on specific moral questions of war and peace. The latter could not be well addressed within the heritage of *tranquillitas ordinis* if the former effort had been abandoned. Secondly, one must admit, on the evidence above, that major intellectual and religious leaders in the American Church significantly contributed to the deterioration of the forensic climate of the true City that Murray believed essential to an evolution of selective conscientious objection that would not lead to anarchy. The Catholic attack on the moral worth of the American experiment had, ironically, helped create circumstances in which selective conscientious objection was, if theoretically sustainable, perhaps culturally impossible.

The American bishops did not treat the problem of selective conscientious objection within the full horizon of Murray's concerns. The bishops' October 1971 call for legalization of selective conscientious objection was cast more narrowly, within the strict limits of just-war theory. The bishops did not discuss how the evolution of selective conscientious objection in American law might contribute to deeper moral understandings about the nature of the democratic experiment or to building the peace of political community in the world.[29]

James Finn's 1972 discussion of amnesty recovered key elements in Murray's approach. The dominant current in the debate on this question taught that draft resisters were "prematurely moral" on Vietnam, and should thus receive unconditional amnesty. The American bishops took a more moderate position in their 1971 "Resolution on Southeast Asia," urging amnesty plus community service as a means of national reconciliation. But Finn posed the relevant questions best.

Finn argued that the amnesty debate had not only revealed the divisive force of the war in Vietnam; it had also demonstrated the sad condition of moral/political discourse in modern America. The amnesty debate had obscured, not cast light upon, the complex requirements of political community. And it had further confused our understanding of the kinds of decisions individuals had to make when they found themselves in conflict with the community's law.[30] Many religious leaders shared in the responsibility for this forensic swamp. After years of teaching about the social nature of the human person and the responsibilities of individual conscience, they now, faced with draft exiles, posed the individual as the "sovereign arbiter," and then, faced with war criminals, displaced personal responsibility onto society ("if they are guilty we are more guilty").[31]

The debate had to return to the heritage of *tranquillitas ordinis*; American religious leaders had to reclaim the importance of law as a means of adjudicating conflicts over claims to justice within a society. Finn knew that some draft exiles had acted out of a sense of deep moral obligation. But he also argued that a personal moral obligation could not be immediately translated into a political right. What *political* obligations had the exiles fulfilled? Did they recognize that civic responsibility was essential in a democracy? In choosing exile, they had effectively rejected the possibility of making a contribution to American life, even radically changing it. To stay would have been very costly, to be sure. "But if one holds principles that place him at odds with the community that is his in a democratic society, it is unlikely that these principles will be maintained without some cost."[32] Thus Finn, who supported amnesty with alternative service, concluded that those resisters who "do wish to return can be expected to respond to the political obligations all citizens share. . . . Those exiles who wish to return to this less than perfect society, and those who wish them to return, must recognize that it *is* a society, not a collection of atomistic consciences, and that it is the proper function of a legitimate government to legislate on significant issues where conscientious citizens are in conflict."[33]

The phrase "legitimate government" sums up the issue. Calls for unconditional amnesty from the Catholic resistance movement, on the grounds that resisters were only "prematurely moral" on Vietnam, were often based on a sense of the illegitimacy of American governance as revealed by U.S. policy in Vietnam. Here, again, a deterioration in just-war reasoning was compounded by a confused pacifism that did not take seriously the role of law in the creation and maintenance of peace—a point that had been insisted upon by both Gandhi and King.

In addition to the issues of selective conscientious objection and amnesty, just-war theory was severely tested by the debate over nuclear weapons and strategy. That argument will be reviewed in detail in the following chapter; here, we may note that the debate included an attempt to formulate a third position in the classic just-war/pacifism argument, "nuclear pacifism." Nuclear pacifists looked to the condemnation of indiscriminate acts of war in Vatican II's "Pastoral Constitution on the Church in the Modern World" as the *locus classicus* for their argument: "Any act of war aimed indiscriminately at the destruction of entire cities or of extensive areas along with their population is a crime against God and man himself. It merits unequivocal and unhesitating condemnation."[34] From this text, which left a small margin for argument in support of the limited use of tactical nuclear weapons but which clearly condemned any possible use of strategic nuclear weapons, nuclear pacifists argued for radically rejecting all use of nuclear weapons in combat; many went even further, to a condemnation of deterrence. The notion of maintaining the nuclear "firebreak"—on the theory that even a limited use of tactical nuclear weapons would inevitably lead to a strategic exchange—bore heavily in this argument.[35]

Despite its adherents' claims, it is hard to see how nuclear pacifism constituted a distinctive third voice in the debate. The nuclear pacifists' condemnation of nuclear weapons use was based on classic just-war criteria of proportionality and discrimination. To conclude that nuclear weapons inevitably violated these principles did not require a pacifist judgment (that *all* resort to armed force was wrong); it only required a demonstration that the use of nuclear weapons violated central principles of just-war theory. "Nuclear pacifism" was less a coherent, distinctive moral position than an attempt to buttress prudential judgments about nuclear weapons, their nature, and their political/military utility, with the presumed moral high ground of pacifism. The nuclear pacifist position thus represented a continuing confusion in the postconciliar American Catholic debate: the hope that there might evolve a synthetic position that subsumed within itself both the pacifist conscience and just-war forms of moral reasoning.

At the level of moral theory, the two classic positions are not reconcilable. The attempt to merge them, as James Finn would argue in the wake of "The Challenge of Peace," tends to corrupt both traditions. Pacifists end up making judgments on military strategy and tactics, and just-war theorists are held accountable to arguments they previously declined to accept (e.g., that the Gospel injunction to "turn the other cheek" had direct implications for the conduct of statecraft). The net result is to bifurcate morality and politics, and to abandon attempts to apply the tradition of reason to the limit case of the use of armed force in the defense of values—which is precisely the result Murray most feared in his strictures against moral*ism*.

Why did this happen? The alterations in the Catholic contextual debate detailed in this chapter played their part. But the failure of just-war theorists to take up Paul Ramsey's challenge in the wake of *Pacem in Terris* and *Gaudium et Spes*—to apply their method of moral reasoning to the large question of the organization of power in the world according to the canons of the heritage of *tranquillitas ordinis*—was also crucial. Had Ramsey's call been followed, American Catholicism might have evolved a comprehensive approach to the twinned problems of war and totalitarianism that would have been proof against the moral and political confusions that followed attempts to construct a hybrid compromise between the pacifist and just-war traditions. Taking up Ramsey's challenge would not only have been congruent with the implicit trajectory of just-war reasoning itself; it would have given Catholic pacifists a more useful place in the debate, neither politically irresponsible nor ecclesiologically sectarian. The pursuit of legal and political alternatives to war could have been seen as an exercise in practical wisdom in which both pacifists and just-war theorists could work—a task that required no inevitably corrupting attempt to "blend" the two traditions at the theoretical level. Work on this common enterprise would have contributed mightily to the evolution of the transformationalist pacifism sought by Gordon Zahn and others.

But, while such a "manifold work of political imagination and will" might have been beneficial to pacifists, the general contours of the debate would have had to have been set by just-war theorists attuned to Ramsey's call to "specify the moral economy governing the use of force, which must always be employed in politics—domestically within nations; internationally in the postfeudal, preatomic era; and under the public authority of any possible world community that may be established during the nuclear age."[36] Ramsey's sense that the popes of the twentieth century had increasingly (if not yet definitively) withdrawn the right of war, by radically circumscribing the meaning of just cause, was not only correct in its intuition of what would unfold in the American Catholic contextual debate. Ramsey was equally right in asserting that a withdrawal of the *ius ad bellum* could not be made morally or politically meaningful unless it were complemented by the parallel development of legal and political structures that could perform the functions of security and peacemaking that were now the task of sovereign armed force.

The tendency of American Catholic just-war theorists to abandon discussion of Ramsey's challenge was a serious mistake; it created a vacuum in the moral imagination of American Catholicism. That vacuum was filled by a host of flawed teachings. The failure of just-war theorists to develop the heritage of *tranquillitas ordinis* along the lines proposed by Ramsey—to open a sober discussion on the nature and organization of the peace that we should, and could, seek in this world as it is—was a contributing factor in the abandonment of that heritage in the postconciliar Church.

PEACE AND JUSTICE

The fifth element in the moral horizon of "an entirely new attitude" toward the problem of war was a renovated definition of the relationship between peace and justice.

The classic Catholic heritage had a complex understanding of this relationship. The peace of *tranquillitas ordinis* emphasized *rightly ordered* political community as a means to adjudicate conflicts over justice within and among nations. That the right ordering of the political community involved questions of justice was not doubted. In *Pacem in Terris* and *Gaudium et Spes*, for example, the heritage had come to understand the importance of democratic values and institutions in the evolution of the peace of political community, even at the international level. Human freedom was thus a primary meaning of the justice that should obtain in civil society, national or international. But the Catholic heritage as developed through Vatican II did not claim that all controverted issues of justice (e.g., the question of how the canons of distributive justice should be applied to the organization of a modern political economy) had to be settled before there could be a morally worthy peace of dynamic political community. Some arguments over justice could remain open. Others (e.g., religious liberty) could not.

This more measured approach was routed in the postconciliar American Catholic debate. It was replaced by a maximalist understanding, emblematically summarized in a famous phrase of Pope Paul VI, "If you want peace, work for justice."[37] Pope Paul's perspective was shared not only by American Catholic activists, but by the American bishops who argued in a 1972 pastoral letter that "in the absence of justice, no enduring peace is possible."[38] The bishops thus mirrored the concerns of those activists in the Catholic Peace Fellowship who were told at their 1974 meeting that their common task was "to reflect upon Christ, who teaches us to be just and to proclaim the good news of justice."[39]

The notion that peace required the prior attainment of justice in the world had roots in the eschatological concept of peace as *shalom*, the peace of the Kingdom in which conflict had ended, and the lions lay with the lambs. Viewed from one angle, then, the prioritizing of justice was a further element in the disappearance of the concept of peace as dynamic political community. But other influences shaped the maximalist understanding of the peace/justice dialectic. The teachings of liberation theologies on the "structural violence" of Western political and economic systems played an important role, and were welcomed in a forensic atmosphere prepared by the vulgarized Marxism of the Vietnam-era New Left. The 1971 report of the international Synod of Bishops, "Justice in the World," was also influential. The bishops claimed that "action on behalf of justice and participation in the transformation of the world fully appear to us as a constitutive dimension of the preaching of the Gospel, or, in other words, of the Church's mission for the redemption of the human race, and its liberation from every oppressive situation."[40] Although the precise meaning of "constitutive dimension" remained theologically controverted,[41] the impact of this teaching on the American Catholic contextual debate was clear: it gave an unprecedented priority to addressing and resolving virtually all issues of justice before there could be a morally worthy peace, and thereby contributed to the absence of the concept of peace as *tranquillitas ordinis* from its accustomed role in the debate.

Numerous public statements of Paul VI reinforced this change, of which perhaps the most significant were his annual messages on the January 1 "World Day of Peace," the 1967 encyclical *Populorum Progressio*, and the Apostolic Letter *Octogesima Adveniens*. In *Populorum Progressio* ("On the Development of Peoples"), Paul VI wrote that "the new name for peace is development. . . . Peace cannot be limited to a mere absence of war, the result of an ever-precarious balance of forces. No, peace is something that is built up day after day, in the pursuit of an order intended by God, which implies a more perfect form of justice among men."[42]

Five years later, the pope urged the same maximalist understanding of the relationship between justice and peace: "This is the right way to come to the genuine discovery of peace: if we look for its true source, we find that it is rooted in a sincere feeling for man. A peace that is not the result of true respect for man is not true peace. And what do we call this sincere feeling for man? We call it justice."[43]

Paul VI emphasized parallel themes in his annual World Day of Peace messages. In 1975 he argued that interior conversion was a prerequisite to ending war.[44] In 1976 he criticized "egotistical" capitalism as an obstacle to peace. In 1977, he condemned the "absurd cold war."[45] The notion of justice as precondition to peace was the consistent thread binding these statements together. And the obverse of this maximalist understanding of the relationship between justice and peace was the argument that peace was not merely the absence of war.

This was not a new theme in Catholicism. But what Paul VI failed to acknowledge was the moral worthiness of the peace of rightly ordered political community that had been achieved in and among democratic nations, a peace in which "the absence of war" was no mean achievement. It can be argued that the pope's intention in deprecating peace as the absence of war was to call the nations beyond realpolitik, and particularly beyond the deterrence system. But whatever the pope's intention, the effect of his maximalist teaching on the relationship between justice and peace was to reinforce those currents of thought in the American Catholic debate most eager to abandon the classic Catholic heritage, not least because of their hostility to the political and economic structures of the American experiment. Pope Paul's insistence that the attainment of justice was the precondition to peace also, if unintentionally, gave support to those for whom revolutionary violence was of more moment than the peace of democratically governed politically community.

As formulated by Paul VI and the American bishops in "To Teach as Jesus Did," the idea that justice was the precondition to any morally worthy peace bordered on utopianism. It implied that this world, short of the Kingdom, could be a world without conflict. Both Paul VI and the American bishops would have rejected such an implication; but that seems to have been the precise impact of their teaching on the American Catholic contextual debate over war and peace. The difficulty is not only that a utopian view of the human prospect sets an unattainable, this-worldly goal. It also results in a deprecation of what *is* possible in this world, according to the classic Catholic heritage: the achievement of political community of sufficient moral worth so that many conflicts over justice can be adjudicated without resort to mass violence or its threat.

The problem of utopianism was exacerbated by conceptual confusion over the multiple meanings of justice itself. "Social justice," the preferred term in the Catholic peace-and-resistance movement (as well as in papal and American episcopal teaching), is a notoriously slippery concept. In Catholic theory, it has included elements of all three classic understandings of justice (i.e., commutative justice, legal justice, and distributive justice). But the failure to define precisely the contours of that "social justice" necessary for a morally worthy peace created the circumstances in which the term "social justice" acted as a kind of political Rorschach blot, into which many schemes of political economy could be read.

It does not seem accidental that this positing of justice as the precondition to peace was paralleled in the American Catholic debate by a diminution of concern for the relationship between peace and *freedom*. Virtually every American diocese now has a "Peace and Justice" or "Justice and Peace" commission or office; there is no diocese with a "Peace, Freedom, and Justice" office. While Pope Paul VI and the American bishops undoubtedly were concerned, for example, with the problem of religious liberty and its relationship to peace, and understood religious liberty to involve elementary issues of justice, the net effect of their new definition of the relationship between justice and peace was the virtual abandonment of serious and sustained thought and action on the relationship between human freedom and peace. This abandonment was influenced by other themes in the postconciliar Catholic debate: the tendency to diminish the moral worthiness of the American experiment, the drift toward anti-anticommunism. But the teaching that justice was the prerequisite to peace was especially significant, for it tended to focus primary attention on the "unjust social structures" of the democratic West in general, and the United States in particular. The radical injustice and tyranny of totalitarian states thereby faded from the dominant horizon of moral concern in American Catholicism, as did concern for the advance and defense of democratic institutions (national and international) as crucial components in the evolution of peace in the world.

"AN ENTIRELY NEW ATTITUDE"

Vatican II's injunction that the problem of war be evaluated with "an entirely new attitude" was a badly flawed formulation of what was required of Catholic theorists and activists. The phrase "entirely new" implied that the classic heritage was somehow fatally deficient and thus of little moment in present circumstances. No reasonable person would argue that those circumstances were frought with unprecedented peril, from the technology of warfare and the spectre of ultimate tyranny. The Council fathers were quite right to remind the Church and the world that the events at Petrograd in 1917 and Hiroshima in 1945 had created new and sinister problems, never before faced by humanity. But were these circumstances so radically different, so utterly divorced from previous human experience, that they required an "entirely new" appraisal of the problem of war?

It seems an excessive claim. The destructive capacities of nuclear weapons did create an unprecedented danger. But that did not necessarily mean that nuclear weapons were without political meaning in the world, or did not convey usable forms of power. Clausewitz's larger understanding of the relationship between military force and political power had not been completely falsified by the advent of nuclear weapons.[46] Nor had nuclear weapons and totalitarianism falsified the understanding of the Catholic heritage that the peace of political community was the only this-worldly answer to the moral problem of war. What Petrograd and Hiroshima had done was give that heritage a new cogency and urgency, as John XXIII understood in *Pacem in Terris* (however much he minimized the totalitarian obstacle to the evolution of an international public authority).

The heritage received and developed through Vatican II had much, not little, to offer to the world's dilemma. It had a balanced view of the human prospect, and a sophisticated understanding of the relationship between moral principles and prudential judgment. It offered a horizon of political possibility that could give guidance to decision makers in the here and now. No doubt the classic heritage needed further refinement and development. But refinement and development are one thing, and abandonment something entirely different. Abandonment was implied by the call for an "entirely new" attitude toward the moral problem of war. Even if that was not the Council fathers' intention, it was what was learned.

The notion that a change in "attitude" was required also had pernicious effects on the American Catholic debate. It suggested that there had been no previous felt concern for the moral problem of war, and that the first requisite was thus personal conversion. The suggestion and the prescription were both mistaken. Undoubtedly the concern that existed needed further expansion in the American Catholic community. But that expansion would be facilitated, not by a change in attitudes, but by a deepening of thought. The idea that the problem of war somehow came down to a question of attitudes involved a basic falsification of reality: conflict was largely epiphenomenal, a psychological rather than political reality. Wedded to the call for an "entirely new" (or "completely fresh," in some translations) approach to the problem of war, the result was a distorted and distorting definition of the problem itself, and a virtual abandonment of the idea that political community might provide the horizon of this-worldly possibility against which the dilemma of peace and freedom could be meaningfully and successfully addressed.

What, then, was the horizon of moral and political imagination that evolved among American Catholic intellectual and religious leaders in the decade after Vatican II? It was a horizon that tended to emphasize sinful social structures over individual moral responsibility. It taught a personal conversionist and/or *shalom* definition of peace. It was anti-interventionist, and in some cases neoisolationist. It found it hard to conceive of the morally appropriate use of military force. It carried a Third World–influenced view of international organization and the relationships among disarmament, international law, human rights, and economic and social development. It minimized moral obligation to existing political community while emphasizing "global citizenship." It was anti-anticommunist. And it taught the radical limits, rather than the impressive prospects, of the American experiment.

Further, the moral horizon of an "entirely new attitude" toward the problem of war contained sectarian ecclesiological intuitions; was susceptible to biblical literalism; had been shaped by a morally urgent but politically deficient pacifism; was increasingly uncomfortable with the classic calculus of just-war reasoning; and argued for the prior attainment of justice as the essential prerequisite to peace.

These were not the only themes in the Catholic debate; but these were the teachings that set its basic direction. The orientation of the debate had thus been decisively shifted, within a decade, at the very latest, after the Second Vatican Council. The results of that shift should not have been surprising.

CHAPTER 9

"The Challenge of Peace": American Catholicism and the New Nuclear Debate

"The Challenge of Peace: God's Promise and Our Response," the 1983 pastoral letter of the National Conference of Catholic Bishops, is widely and correctly regarded as a watershed event in the history of American Catholicism. According to a familiar interpretation of the pastoral, the bishops of the United States, acting bravely in the face of conservative lay opposition and stoutly resisting governmental pressure, applied the ancient wisdom of the Catholic tradition to the moral, strategic, and political conundra posed by nuclear weapons and deterrence policy. The results were a reassertion of the claims of moral reason in national security policy; a new and welcome episcopal involvement in the national security debate; a strategic vision that pointed the way beyond our present peace-and-security dilemma to a force structure more congruent with Catholic social ethics and more conducive to arms control and disarmament; and a wiser, morally healthier American political culture, better equipped to think through and act upon its responsibilities for peace in the nuclear age.[1]

The counterhermeneutic on the pastoral argues that the bishops were engaged in a monumental folly. In addition to confusing their office as religious and moral teachers with the role of military strategists or tacticians, the bishops gave aid and comfort to the renascent American peace movement of the late 1970s, with its single-minded focus on American weapons and its consequent blindness to the threat of a vastly expanded Soviet nuclear arsenal. The bishops weakened the West's will to resist the totalitarian threat, cheapened their own credibility, abandoned the just-war theory in fact if not in name, and quite possibly made war more, rather than less, likely.[2]

Were the bishops, in writing "The Challenge of Peace," engaged in a corporate act of prophetic witness, or were they acting as "useful idiots"? Neither sobriquet is accurate, for both assume that "The Challenge of Peace" emerged as a kind of intellectual virgin birth, without significant antecedents. This is an impossible image to maintain, on the evidence. "The Challenge of Peace" was profoundly influenced in substance, in tone, and in the very fact of its existence by the transformation of the basic themes in the Catholic contextual debate on war and peace, security and freedom, since Vatican II. Those who were "surprised" by the fact and content of the bishops' address to nuclear weapons issues can be said, quite simply, not to have been paying very much attention to American Catholicism since 1965. This holds true for those pleasantly surprised to find the American bishops as political partners, and for those appalled by the bishops' new position in the ideological and lobbying wars of antinuclear activism.

Assessing "The Challenge of Peace" requires locating its ideas in time and space. For contrary to the hermeneutic claims of many of its supporters and not a few of its opponents, the 1983 pastoral letter did not emerge from an intellectual, ideological, or political vacuum. The principal ideas constituting the horizon of moral imagination that shaped "The Challenge of Peace" have been reviewed in detail in the previous three chapters. A sober reading of that intellectual history suggests that the real significance of "The Challenge of Peace" was this: the nuclear weapons-and-strategy debate preceding and shaping the bishops' pastoral was the occasion for those themes that represented an abandonment of the classic Catholic heritage to move from the American Catholic attentive public to the very centers of Catholic leadership in the United States.

THE AMERICAN BISHOPS AND NUCLEAR STRATEGY: PRE-1980

Contrary to widespread media reports, "The Challenge of Peace" was not the American bishops' first statement on nuclear weapons and strategy.

The 1968 pastoral letter, "Human Life in Our Day," included a lengthy discussion of arms control issues. The bishops began by noting that nothing suggested the antilife direction of technological warfare more than the neutron bomb: it would leave property, but not people, intact.[3] The bishops recalled Vatican II's concerns about deterrence, quoting *Gaudium et Spes* that "the arms race is an utterly treacherous trap for humanity, and one which ensnares the poor to an intolerable degree."[4] The Council did not require unilateral disarmament, for Christian morality was not without realism.[5] What the Council envisioned, and the American bishops endorsed, was "reciprocal or collective disarmament 'proceeding at an equal pace according to agreement and backed up by authentic and workable safeguards.'"[6] The bishops were encouraged that the Partial Test Ban Treaty and the Nuclear Non-Proliferation Treaty were steps toward disarmament; but they also argued that the U.S. decision to build a thin antiballistic missile (ABM) shield was the latest episode in the arms race, and the prelude to a thick ABM system that might make nuclear war-fighting conceivable.[7] The ABM was a defensive weapon, but the bishops worried that "by upsetting the present strategic balance, the so-called balance of terror, there is grave danger that a United States ABM system will incite other nations to increase their offensive nuclear forces with the seeming excuse of a need to restore the balance."[8] By 1968, then, the bishops were describing arms competition as an action/reaction cycle that decreased both security and stability.[9]

The bishops' 1976 statement, "To Live in Christ Jesus: A Pastoral Reflection on the Moral Life," went even further than the 1968 pastoral. "Going significantly beyond *Gaudium et Spes*," according to J. Bryan Hehir, the bishops wrote: "With respect to nuclear weapons, at least those with massive destructive capability, the first imperative is to prevent their use. As possessors of a vast nuclear arsenal, we must also be aware that not only is it wrong to attack civilian populations, but it is also wrong to threaten to attack them as part of a strategy of deterrence."[10]

By 1976, several key themes were well established within the National Conference of Catholic Bishops (NCCB). There was an arms race, with two equally culpable partners, the United States and the Soviet Union, caught in an action/reaction cycle that threatened the fragile peace and had brought "the whole human family" to "an hour of supreme crisis."[11] The indisputably first requisite was to see that nuclear weapons were never used. The deterrence system had grave moral flaws, involving a conditional intention to do immorally indiscriminate acts of mass violence. The way out of this dilemma was through arms control leading to disarmament. Systems, such as the ABM, which jeopardized traditional arms control, were to be avoided. Mutual Assured Destruction (MAD) was a morally unworthy form of deterrence, but inasmuch as defensive systems threatened the fragile peace of the "so-called balance of terror," they could not be indulged.

Similar themes shaped the bishops' response to the Senate debate over ratification of the SALT II treaty. Testifying on behalf of the bishops' public policy agency, the United States Catholic Conference (USCC), Cardinal John J. Krol of Philadelphia began on a critical note: "The Catholic bishops of this country believe that too long have we Americans been preoccupied with preparations for war; too long have we been guided by the false criterion of equivalence or superiority of armaments; too long have we allowed other nations to virtually dictate how much we should spend on stockpiling weapons of destruction."[12] Krol argued that the SALT II treaty could not be regarded as a great arms control achievement, but urged its ratification "as a partial and imperfect step in the direction of halting the proliferation of nuclear weapons and as part of an ongoing process, begun in 1972, to negotiate actual reductions in nuclear arms."[13] Cardinal Krol then sketched the reasoning that had led the USCC to its grudging support of SALT II ratification.

Following *Gaudium et Spes*, Krol juxtaposed the arms race with the issue of world poverty. Together, these issues formed "the heart of the moral agenda of foreign policy."[14] Nuclear weapons had given "the moral sanctions against war . . . a qualitatively new character."[15] Given the inadequate political structure of international public life, some forms of war could be morally legitimate; but "nuclear war surpasses the boundaries of legitimate self-defense."[16] On these understandings, then, "the primary moral imperative of the nuclear age is to prevent any use of strategic nuclear weapons," and it was this imperative that led to the USCC endorsement of SALT II, which was meant "to illustrate our support for any reasonable effort which is designed to make nuclear war in any form less likely."[17]

The grudging quality of the USCC's support for SALT II did not rest solely on the minimal accomplishments of the treaty; it also reflected concern about the morality of deterrence. Krol cited the 1976 bishops' statement, and its "moral judgment that not only the *use* of strategic nuclear weapons, but also the *declared intent* to use them involved in our deterrence policy is wrong."[18] For that reason, "it is of the utmost importance that negotiations proceed to meaningful and continuing reductions in nuclear stockpiles, and eventually, to the phasing out altogether of nuclear deterrence and the threat of mutual assured destruction."[19]

SALT II had not led to meaningful reductions, and Krol put the U.S. government on notice that USCC patience with deterrence was wearing thin: "As long as there is hope [of meaningful and continuing reductions], Catholic moral teaching is willing, while negotiations proceed, to tolerate the possession of nuclear weapons for deterrence as the lesser of two evils. If that hope were to disappear, the moral attitude of the Catholic Church would almost certainly have to shift to one of uncompromising condemnation of both use *and* possession of such weapons."[20]

Krol's testimony also deprecated the political utility of nuclear weapons,[21] and concluded by questioning the political and strategic importance of nuclear equivalence between the United States and the U.S.S.R.: "Are we not justified in asking today if strategic equivalence is an absolute necessity? Is this doctrine not an infallible recipe for continuing the strategic arms race?" The cardinal ended with a question that was really an answer: "We, the Catholic bishops, find ourselves under the obligation of questioning fundamentally the logic of the pattern of events implied by determined pursuit of strategic equivalence."

That Krol's intention was declarative rather than interrogatory was made quite clear in his next sentence: "Our purpose in coming before this distinguished committee is to speak on moral-religious grounds in support of arms control designed to be a step toward real measures of disarmament. It would radically distort our intention and purpose if our support of SALT II were in any way coupled with plans for new military expenditures."[22] The cardinal's immediate targets were the MX and Trident II missiles; but the logic of his statement put the United States Catholic Conference on record as opposing virtually any nuclear force modernizations.

Cardinal Krol's SALT II testimony was noteworthy on five counts. First, its identification of "the heart of the moral agenda of foreign policy" as the "correlative" issues of the arms race and global poverty effectively dissolved one side of the Janus-headed problem of war and totalitarianism. Human freedom, understood as freedom from totalitarian tyranny, was not at the heart of the moral agenda of foreign policy (and this despite Krol's acknowledgment of the Soviet nuclear weapons buildup). Cardinal Krol drove this point home even more directly in an address to the "Religious Committee on SALT" three days after his Senate testimony:

> The history of certain countries under communist rule today shows that not only are human means of resistance available and effective but also that human life does not lose all meaning with replacement of one political system by another. History goes on and political systems are subject to change. As long as life exists there is hope, hope that God's grace will enable suffering and oppressed peoples to endure. A nuclear holocaust would wipe out that hope.[23]

Secondly, the heritage of *tranquillitas ordinis* was virtually absent from the Krol testimony. The USCC statement included the briefest reference to "the inadequate nature of the political structure of the international community," but failed to relate this to arms control and disarmament. The notion that disarmament could not proceed faster than the evolution of international legal and political instruments for resolving conflict was notable by its absence from the USCC approach to SALT II. On this analysis, arms control was, in fact if not in principle, an independent variable in international affairs.

The third striking element in Krol's SALT II testimony was its claim that Catholic "toleration" of deterrence was hostage to the "hope" that arms control would be successful. The cardinal did not suggest how one might know when that hope would be dashed; nor did he suggest alternatives to deterrence as a regulatory mechanism in the superpower relationship. Political judgment—that arms control had "failed"—would determine moral judgment on the ethical merits or deficiencies of deterrence.

A fourth, and related, element was the claim that the bishops spoke to the issue of SALT II "on moral-religious grounds," when in fact the bulk of the testimony was crafted in terms of strategic analysis: on the nature of deterrence, the meaning of "equivalence," the role of counterforce-capable weapons, the relationship between nuclear weapons and political power in the world, and the Soviet strategic agenda. The dialectic between moral principles and prudential political-strategic judgments was weighted heavily in favor of political-strategic judgment in Cardinal Krol's SALT II testimony.

Finally, the USCC approached SALT II as an unamendable package, a position taken during the debate by many arms control supporters. Given the cardinal's expressed interest in SALT II as a vehicle toward real reductions, it is striking that the testimony did not even mention, much less endorse, the Moynihan amendment to the treaty, which would have linked the continuation of SALT II limits and ceilings to measurable progress, within three years, on actual reductions in the superpower arsenals.[24]

The USCC position on SALT II was not supported by all the American bishops, nor by all those in the Catholic peace movement. Pax Christi–USA opposed the treaty on the grounds that it would legitimate the nuclear arms race. Bishop Thomas Gumbleton argued the Pax Christi case before the administrative board of the USCC, describing SALT II as a "cruel hoax," and claiming that the bishops could not "support this agreement and still be called Catholic peacemakers." The minimalist case for the treaty was argued before the administrative board by Fr. J. Bryan Hehir, USCC associate secretary for international justice and peace. Voices opposed to SALT II on the grounds that it opened the possibility of a destabilizing first-strike capability, given Soviet heavy missiles with multiple warheads (MIRVs), were absent from the bishops' internal debate, and may account for the failure to acknowledge Moynihan's effort to link SALT II's version of arms control to real disarmament.[25]

THE NEW NUCLEAR DEBATE

Bishop Gumbleton's arguments became increasingly prevalent in the Catholic debate over nuclear weapons and strategy after SALT II, and constituted the essential foreground to "The Challenge of Peace." Joseph J. Fahey, a member of the executive council of Pax Christi–USA, summarized these themes in early 1979. Fahey argued that Americans were now involved in "the arms race to end all arms races," an exercise that would surely "end in war much the way the arms race of the 19th century ended in World War I," because "some 35 nations will have" nuclear weapons "in 10 years."[26] Fahey drew on the condemnation of indiscriminate acts of war in *Gaudium et Spes* to buttress his claim that "if it is immoral to use nuclear weapons, it is also immoral to make them . . . because by their cost alone they kill people."[27] What prevented a widespread public recognition of this? Fahey blamed American "political reactionaries" who opposed SALT II because they wanted to pursue the arms race and achieve permanent nuclear superiority over the Soviet Union.[28]

Like his colleagues in Pax Christi–USA, Fahey, too, opposed SALT II, but on different grounds. His opposition was meant to create a "new moral basis for American foreign policy toward the Soviet Union, which allows for true détente and the possibility of reconciliation."[29] The arms race was basically a "spiritual crisis," Fahey concluded; as the arms race escalated, morality deteriorated. American Christians had to avoid the pre–World War I trap of believing that an arms race would prevent war. "We must end the arms race and replace it with a philosophy to conquer our true enemy: the lack of development."[30]

That ending the arms race was primarily a matter of American will was driven home to *Commonweal* readers in a June 1979 essay reporting on a conference of Soviet and American church leaders who "with one heart and one voice, sounded an alarm over the escalating arms race and called on the peoples of the world to exorcise the demon of nuclear holocaust."[31] Among the key themes of the conference, which involved leaders of the National Council of Churches and their counterparts from the Soviet Union, were the U.S.S.R.'s support for arms control, exemplified by "seventy Soviet initiatives on disarmament in the postwar years" (counted by Metropolitan Juvenaly of Krutitsky and Kolomna);[32] the "madness" of the arms race (analyzed by Dr. Bruce Rigdon of McCormick Theological Seminary);[33] the openness of the Soviet regime to "a rough *modus vivendi* with the churches" of the U.S.S.R.;[34] and the history of Catholic acquiescence to the national security policies of the U.S. government ("Held in suspicion by a dominant Protestant society, it is no wonder that Catholic leaders became stout champions of the American way of life, and that the catechism Jimmy Breslin learned in Brooklyn was as much an exercise in Americanism as in Christianity").[35] This was taken as a precise parallel to the pressures under which the Church operated in the U.S.S.R. after the Bolshevik revolution: "Under the Communist government, Russian Christians were placed in a dilemma similar to that of many Catholic immigrants to America: the need to preserve one's cultural heritage while proving one's loyalty and patriotism toward the established order."[36]

Similar themes rang throughout the American Catholic debate on nuclear weapons and strategy in the late 1970s and early 1980s.[37] The most significant development of the period, though, was the degree to which these teachings were promoted by many American bishops. Archbishop Hunthausen's role has been noted at several points above. Although the archbishop's description of the Trident submarine as "the Auschwitz of Puget Sound," his prescription for unilateral disarmament, and his call for tax resistance as a form of protest against preparations for "the global crucifixion of Jesus" constituted a rhetorical apogee in the American hierarchy, Hunthausen was by no means alone.

Archbishop John R. Quinn of San Francisco, preaching in his cathedral on "Instruments of Peace, Weapons of War," claimed that U.S. nuclear force modernization was a "madness" that "takes lives just as surely as if the weapons produced had actually been put to use," for "the extreme poverty that is endured by one-third of the human race is in large part a direct by-product of an arms race out of control." Quinn praised the work of groups like Physicians for Social Responsibility, which he identified as a campaign for public awareness that nuclear war had no cure; urged his congregants to support the Nuclear Weapons Freeze Campaign; called on Catholic hospital administrators and staff to resist civil defense planning; proclaimed his support for "developing creative proposals for converting military weapons technology to civilian production uses"; and offered St. Francis of Assisi as a model by which we might "replace violence and mistrust and hate with confidence and caring."[38]

The question of the bishops' role in the nuclear debate was sharpened by Bishop Michael Kenny of Juneau in an address to the 1981 NCCB meeting that captured the new sectarianism alive in the American hierarchy, a sectarianism that minimized the threat of totalitarianism and deprecated the moral significance of the American experiment:

> In our teaching on war and peace, I believe it is critically important that we distance ourselves, at least to a degree, from time and place; that is, from this particular age and this particular nation. As Bishops of the Catholic Church it is critical for us to take a global and long-range point of view. I say this for two reasons. First: with the advent of nuclear weapons, what is at stake is not just this or that form of government, this or that philosophy, but the continued existence of all life, especially human. Second: whenever we as Church leaders ally ourselves too closely with a particular period or nation we tend to weaken our prophetic role. In some notable instances, such alliance has gotten us into serious trouble. As followers of the Lord whose naked image hangs before us on the cross, our kingdom is not of this world. We are a faith community, and whatever favor we receive from one government will not do us all that much good; whatever opposition and oppression we get from another government will not destroy us. We are his Church and have his promise that the very gates of hell will not prevail against us. We as Church should never be overly concerned even about our own safety and security. Our primary concern should be proclaiming the Gospel with courage and conviction. Our concern should be the well-being of all our brothers and sisters whoever and wherever they may be.[39]

The pastoral implications of this trajectory of thought were expounded forcefully by Bishop Leroy P. Matthiesen of Amarillo, in whose diocese nuclear warheads were assembled. Bishop Matthiessen was particularly angry at the Reagan administration's decision to deploy neutron warheads on tactical nuclear weapons. This was "the latest in a series of tragic anti-life decisions taken by our government." The "madness" had to stop. "We beg our administration to stop accelerating the arms race. We beg our military to use common sense and moderation in our defense posture." And then the proposal that would win Matthiessen national attention: "We urge individuals involved in the production and stockpiling of nuclear bombs to consider what they are doing, to resign from such activities, and to seek employment in peaceful pursuits."[40]

Matthiessen's call to the members of his diocese to leave the defense industry caused a storm of protest in Amarillo. But three weeks after Matthiessen spoke, his position was endorsed by the Catholic bishops of Texas, whose conference secretary, Bishop Joseph Fiorenza of San Antonio, told reporters that "the bishops wanted to support Bishop Matthiessen in order to dispel any impression that his stand did not reflect the opinion of the other bishops of Texas or the teachings of the church on nuclear weapons."[41]

By late 1981, episcopal criticism of U.S. nuclear weapons strategy had become a fire storm. According to Bishop Raymond Lucker of New Ulm, Minnesota, "It is immoral to possess nuclear weapons. . . . Nuclear weapons may not be used for attack or for first strike. . . . They may not be used in defense. They may not be threatened to be used. Therefore it seems to me that even to possess them is wrong."[42] Bishop Frank Rodimer of Patterson, New Jersey, called the U.S. decision to build the neutron warhead "a sin against humanity."[43] Bishop Walter Sullivan of Richmond argued that "it is immoral to be associated with the production or use of nuclear weapons."[44] Bishop Phillip Straling of San Bernardino issued a statement supporting Archbishop Quinn's proposals, and claimed that "as our nation puts greater emphasis on mightier and more destructive instruments of war, especially the neutron warhead . . . it is a contradiction to the message and spirit of the word of God."[45]

Even bishops critical of the unilateralism of Pax Christi–USA were caught up in the general mood of condemnation. Bishop Edward O'Rourke of Peoria, though warning that unilateral disarmament was "dangerous, unrealistic, and not in accord with papal teachings on the right of nations to self-defense," also stated in his diocesan paper that he had "reached the conclusion that present diplomatic policies and negotiating procedures are inadequate to meet the seriousness and urgency of this crisis. . . . We must boldly suggest a radically different manner in which disarmament and peace can be discussed and pursued. Failure to do so may occasion a nuclear war which would destroy civilization as we now know it."[46] The bishop suggested the development of strategic defense systems, a concept that would be vigorously resisted by other episcopal critics of U.S. nuclear strategy.

Several important themes were prominent in the bishops' debate two years before the completion of "The Challenge of Peace." An apocalyptic sense permeated their rhetoric. The bishops seem to have believed that they stood in an utterly unique historical moment, one in which, as Fr. Theodore Hesburgh would put it, "there were no precedents to invoke, no history to depend upon for a wise lesson, no real body of theology except for that which dates back to pacifism or a just-war doctrine that was first applied in a day of spears, swords, bows and arrows, not ICBMs."[47] This in itself was an extraordinary claim for bishops who carried a fifteen-hundred-year-old heritage of thought on the moral problem of war and peace. The bishops also seemed to believe that their prophetic task was to alert the nation and the world to a peril of which men and women were unaware. The counter-proposition, in fact a worse situation, that men and women were indeed painfully aware of the dilemma of peace and security, and saw little way out save for muddling through the next crisis, does not seem to have occurred to the bishops. Given this reading of the level of public awareness, the bishops were thus in a position to apply to forty years of strategic debate the critique of just-war theory once argued by Bishop Sullivan: "mental gymnastics," or "casuistry of the worst sort."

A third element in the bishops' prepastoral letter critique was a tendency to assign primary (in some cases, exclusive) blame for the world's perilous state to U.S. policy. This judgment may well have had to do with many bishops' disaffection for the newly elected president, Ronald Reagan. Jim Castelli, a reporter given access to the USCC records involved in the drafting of "The Challenge of Peace," claimed, not unpersuasively on the evidence above, that "Reagan's election—with the rhetoric and policies he brought to office—was the single greatest factor influencing the bishops' discussion."[48]

Other bishops took a more measured approach to responsibility for the arms race. Speaking to the November 1981 NCCB meeting, Archbishop Joseph L. Bernardin, who had been appointed the year before as chairman of an ad hoc committee on war and peace to draft a pastoral letter on the subject, stated that his committee was "fully aware that current tensions are by no means attributable to U.S. policy alone.

Clearly the enormous build-up of nuclear and conventional arms pursued by the Soviet Union in recent years has done more than its share to heighten the peril of the present moment. The duty of responsible moral action falls equally on both superpowers." Bernardin's parallelism of responsibility was immediately followed, though, by a unilateralist sense of where the bishops' duty lay: "But if we direct our attention particularly to the United States, it is for the simple reason that we are American citizens and have a right and duty to address our government."[49] The missing point was that the content and tone of the bishops' address to the United States government had an important impact on Soviet perceptions and policy.

A fourth key element in the bishops' debate was its inclination toward a mechanistic, even deterministic, view of the relationship between arms competition and war. Joseph Fahey's citation of the arms race that preceded World War I was repeated by many bishops. It is true as far as it goes. But the parallel lesson of the 1930s—that *not* arming in the face of an aggressive power could lead to war—was rarely drawn. Nor was the appropriate conclusion drawn from the "lessons" of both World Wars I and II: that inasmuch as arming and not arming could both lead to war, a primary focus on weapons as the chief issue to be debated was a false entry point for moral and political reasoning (as Murray had insisted a generation before).

The lines from the "themes in the abandonment of the heritage" to the prepastoral letter nuclear debate in American Catholicism were clear. The new nuclear debate was influenced by both *shalom* and personalist/conversionist understandings of peace, but rarely by the concept of peace as dynamic, rightly ordered political community. Anti-interventionism shaped perceptions of the American responsibility for the arms race, as did the new Catholic sense of the limitations of an American experiment too beholden to the "military-industrial complex." Concepts of conflict drawn from personal psychology tended to replace political understandings of a genuine conflict of values and interests in world affairs, and intersected with American Catholic anti-anticommunism to yield a minimalist sense of the Soviet Union's responsibility for nuclear arms competition. The hardest statement was that the United States and the Soviet Union were "equally" responsible. Pacifist readings of Gospel imperatives figured more largely in the new nuclear debate than just-war understandings, and a sectarian ecclesiology emerged as a significant factor in the National Conference of Catholic Bishops. The ground of receptivity for these themes had been well prepared in American Catholicism since Vatican II.

But the bishops were also influenced by currents of thought in the wider American political culture, in a manner remarkably parallel to the vulnerability of American Catholic commentators during the Vietnam War. Apocalyptic themes were regularly taught by renascent antinuclear weapons organizations (actually, anti American weapons program organizations) like Physicians for Social Responsibility, and dominated the basic text of the new antinuclear movement, Jonathan Schell's *The Fate of the Earth*.[50] These teachings surely made an impact on the bishops.

The nuclear freeze campaign was another instrument for teaching the parallelism of U.S./Soviet responsibility for the arms race, and offered a simple answer to the dilemma of peace and security: stop now. It was an answer reminiscent of the claim that immediate U.S. withdrawal from Vietnam would end the killing; but that lesson of Vietnam was not drawn by freeze supporters. The freeze movement's emphasis on the budgetary implications of the arms race was an agreeable leitmotif to bishops taught by *Gaudium et Spes* that the arms race constituted a theft from the poor.[51] Freeze support was strong in mainline Protestant churches, and this, too, played its part in influencing the new nuclear debate among the bishops.[52] The theme of public unawareness about the nuclear weapons dilemma was also prominent among freeze activists, and buttressed the bishops' inclinations toward a similar analysis.

What seems particularly striking in this review of the pre-pastoral debate among the American bishops, the foreground to "The Challenge of Peace," is the tendency of the debate to uncouple the problem of nuclear weapons from larger questions of international order and world politics. The new nuclear debate among American bishops did not take full advantage of the complex moral reasoning embedded in the heritage of *tranquillitas ordinis.* Occasional references were made to the connection between disarmament and effective international institutions of law and governance; but far more prominent was the oft-asserted connection between the economic costs of the arms race and the problem of gross poverty in the Third World. No serious analyst claims that there is no connection between these situations; but, in the bishops' heritage, they were linked through a complex set of understandings about the connections among disarmament, international institutions, human rights, and development. On the evidence cited here, which is a mere sampling of emblematic statements from American bishops, the Catholic debate portrayed nuclear weapons as an independent variable in world affairs: an issue that could be forcefully addressed without much reference to the problem of totalitarianism, and without a vision of international politics in which war (or, in the case of nuclear weapons, the threat of war) could be replaced as a primary instrument of conflict-resolution. The new Catholic nuclear debate was thus marked by a disconcerting tendency toward ahistoricity, and bore many of the marks of that moral*ism* that had so exercised John Courtney Murray.

This abandonment of the politics of *tranquillitas ordinis* as a horizon for judgment on issues of nuclear weapons and strategy was reinforced by the traditional disinclination of the arms control community to see the problem of nuclear weapons as a function of the larger problem of war. Classic arms control theory had long since abandoned the goal of disarmament, and taught that nuclear weapons issues could be largely detached from the wider superpower political context and dealt with rather independently. "Linkage," a word with positive (in fact, inescapable) connotations in the bishops' heritage, was a term of opprobrium to classic arms control theorists. Thus the strategists to whom the bishops ultimately turned in formulating their policy prescriptions in the prepastoral letter debate and in "The Challenge of Peace" would be precisely those most opposed, on theoretical grounds, to the larger contextual approach to the moral problem of war of which the bishops were the assumed bearers.[53]

Finally, the profound sense of urgency in the new Catholic nuclear debate may well have had to do with some bishops' judgment that the NCCB had been too little and too late engaged in the public discussion of Vietnam. Gordon Zahn's critique of the bishops' role during that crisis had been taken to heart, and many bishops did not wish to be late again. Several influential bishops had absorbed more than Zahn's critique of their sense of timing during Vietnam, though, and had adopted views of American perversity and incapacity in the world that drew heavily on Vietnam-era Catholic New Left teachings. This residual sense of Catholic "failure" during the Vietnam War, intersecting with the election of a president who dared call U.S.

intentions in Southeast Asia "noble," helped shape the fact *and the content* of the bishops' determination to be fully engaged with the public debate on war and peace in the future. That determination, substantively influenced by currents of thought avowedly concerned with the global threat of nuclear weapons, but primarily critical of U.S. nuclear weapons strategy and policy, led to "The Challenge of Peace."[54]

THE EVOLUTION AND PRINCIPAL TEACHINGS OF "THE CHALLENGE OF PEACE"

An adequate history of the bureaucratic and forensic evolution of "The Challenge of Peace," from the establishment of the NCCB ad hoc committee on war and peace in January 1981 to the adoption of the pastoral letter at a special NCCB meeting in May 1983, remains to be written.[55] Because the primary interest of this chapter is the teaching of "The Challenge of Peace" and its relationship to the evolution of ideas in the American Catholic debate on war and peace since Vatican II, a brief sketch of the process that produced the document will suffice.

While "The Challenge of Peace" resulted from a lengthy, complex intellectual transformation in American Catholicism, its immediate impetus was a *varium* or "proposal for new business" submitted to Bishop Thomas C. Kelly, general secretary of the NCCB/USCC, in the summer of 1980 by Auxiliary Bishop P. Francis Murphy of Baltimore. Murphy proposed that the bishops consider a statement reiterating Catholic teaching in the morality of war and peace, and urging heightened educational efforts to make that teaching more widely known.[56] Kelly accepted Murphy's proposal as new business for the November 1980 NCCB meeting where, after discussion focused specifically on nuclear weapons questions, the bishops formally agreed to address the issue.[57] Two months later, Archbishop John Roach of St. Paul-Minneapolis, president of the NCCB, appointed an ad hoc committee on war and peace to draft the bishops' statement. The chairman of the committee was Archbishop Joseph L. Bernardin of Cincinnati, a former general secretary and president of the NCCB/USCC who would, during the course of the committee work, be appointed archbishop of Chicago and subsequently raised to the rank of cardinal. The four other committee members were Auxiliary Bishop Thomas Gumbleton of Detroit (an active leader of Pax Christi–USA), Auxiliary Bishop John J. O'Connor (of the Military Ordinariate, the "diocese" of American military personnel), Bishop Daniel Reilly of Norwich, Connecticut, and Auxiliary Bishop George Fulcher of Columbus, Ohio. Neither Reilly nor Fulcher had been previously identified with the new nuclear debate in American Catholicism, in which both O'Connor and Gumbleton had vigorously participated. Bernardin was chosen by Roach because of his immense personal prestige among his fellow bishops, a prestige based in part on his ability to secure consensus decisions amid ecclesiastical controversies.

The chief USCC staff for the committee were Fr. J. Bryan Hehir and Edward Doherty, a retired foreign service officer. After failing to secure former *Commonweal* columnist William V. Shannon as principal outside consultant and drafter, Bernardin accepted Hehir's recommendation that this role be given to Dr. Bruce Russett of Yale University, editor of the *Journal of Conflict Resolution.* The committee staff was completed by the appointments of Fr. Richard Warner, Indiana provincial of the

Congregation of the Holy Cross, as the representative of the Conference of Major Superiors of Men, and Sister Juliana Casey of the Order of the Immaculate Heart of Mary as representative of the Leadership Conference of Women Religious.

The committee heard testimony from thirty-six formal witnesses.[58] The witnesses' testimony and the committee's internal debates resulted in three drafts of "The Challenge of Peace," each of which produced further volumes of commentary from members of the NCCB, the formal witnesses, and dozens of interested parties inside and outside the Church. Bernardin, Hehir, and Roach were also invited to a Vatican-sponsored consultation on the pastoral, held in early 1983 between the second and third drafts of the pastoral letter. The two key figures throughout this complex process were Cardinal Bernardin and Father Hehir. Bernardin, who announced at the first committee meeting that the group's one ground rule would be that "it would not, under any circumstances, support unilateral nuclear disarmament,"[59] continually urged the committee, particularly as it wrote the crucial third draft, to move "back toward center."[60] This concern reflected Bernardin's eagerness for consensus within the NCCB. But it may also have been insufficient in its analysis of where the "center" lay, given the themes that had restructured the spectrum of American Catholic debate on war and peace in general and on nuclear weapons issues in particular. One might also note that consensus, an admirable goal, is no guarantee of truth, only of the agreeability of propositions within a complex organization characterized by different views.

Fr. Hehir's role was different, but no less important, for Hehir, according to Gumbleton, defined what had to be decided: "He was always saying, 'This is the issue, this is what you have to decide—but you have to decide.'" Edward Doherty confirmed that "that must have happened thirty times."[61] The power to define the issues to be decided was an immense power indeed. Hehir's commitment to nuclear pacifism as a distinctive third voice in the strategic debate would be among his significant contributions to defining just what it was that the bishops on the ad hoc committee had to decide.

What, then, did "The Challenge of Peace" teach?

The bishops' entry point was itself significant. After citing *Gaudium et Spes* on the "moment of supreme crisis" faced by "the whole human race . . . in its advance toward maturity," the pastoral began with a remarkable confession: that the bishops, as pastors in a nuclear power, had met "terror in the minds and hearts of our people," and shared that "terror" themselves.[62] The terror came from the fact that nuclear war threatened the very existence of the earth, and was the gravest danger the world had ever faced. It was "neither tolerable nor necessary that human beings live under this threat," although removing it would require "a major effort of intelligence, courage, and faith."[63]

The Introduction to the pastoral echoed Cardinal Bernardin's claim that the bishops' first responsibility was to speak to the U.S. government. As citizens of the first country to produce, and the only country to use, nuclear weapons, and as religious leaders in a nation that could decisively influence the course of the nuclear age, "we have grave human, moral, and political responsibilities to see that a 'conscious choice' is made to save humanity." The pastoral would thus be an "invitation and a challenge" to American Catholics who had to take increased responsibility for public policy in this "moment of supreme crisis."[64]

The body of "The Challenge of Peace" had four parts; Part One was entitled "Peace in the Modern World: Religious Perspectives and Principles." At the outset of Part One, the bishops distinguished the multiple levels of teaching authority in their pastoral letter. Noting that they would speak to many specific issues of policy, the bishops warned that they did "not intend that our treatment of each of these issues carry the same moral authority as our statement of universal moral principles and formal Church teaching." Not every statement in the pastoral had the same moral authority. Different prudential judgments on specific questions could be reached by people of good will. The bishops cited their endorsement of a NATO "no first use" of nuclear weapons pledge as an example of a legitimately controverted prudential judgment. But they also admonished American Catholics that the bishops' prudential judgments, although not binding, were to be taken seriously as Catholics formed their own consciences in the light of the Gospel.[65]

Part One continued by charting the signs of the times that had particularly influenced the bishops: the universal longing for peace; the judgment of Vatican II that the "arms race is one of the greatest curses on the human race and the harm it inflicts on the poor is more than can be endured"; and the "qualitatively new problems" raised by nuclear weapons, which required new applications of classic moral principles.[66]

Catholic teaching on peace, as all Catholic social theory, began with the transcendence of God and the dignity of the human person.[67] The purpose of Catholic social teaching on war and peace was twofold: it was meant to help Catholics form their own consciences, and it was intended to influence the public policy debate in society at large. The bishops thus spoke to two different, but overlapping, audiences: their own congregants, and the "wider civil community," which, though not sharing the Church's faith, was still bound by universal moral norms. All men and women could find the natural law in the depths of their consciences. The moral norms that the bishops applied in their letter were reason's effort to create a normative ethical context for the public debate over war and peace.[68]

These preliminary distinctions drawn, the bishops began their formal presentation by asserting that the Scriptures provided believers with a foundation for confronting war and peace in the modern world.[69] There were multiple understandings of peace in the Old and New Testaments; but the bishops emphasized that the eschatological peace of the *shalom* Kingdom and the peace of personal right relationship with God gave direction to the concepts of peace as individual physical well-being or the cessation of hostilities among nations.[70] Peace in the Old Testament was believed to be a gift from God that came from fidelity to the covenant. In the new covenant, Jesus Christ was himself our peace, and his death and resurrection brought God's peace to the world.[71] The Scriptures did not provide detailed answers to contemporary questions; but they did propose an "urgent direction." The fullness of the peace of the Kingdom remained before us in hope, even while we enjoyed the peace of reconciliation with God through Christ.[72]

We thus lived in an interim period of human history, between the definitive revelation of God's purpose in the Resurrection and the completion of that purpose in the establishment of the Kingdom. In this "in-between" time, "peace must always be built on justice in a world where the personal and social consequences of sin are evident."[73] War had many causes in this kind of world. But Vatican II's call for a "completely fresh appraisal of war" (or, "an entirely new attitude") required some description of the peace that was possible.[74]

Catholic social theory had always construed peace in positive terms. Peace was both God's gift and the work of human hands. Peace had to be built on a foundation of truth, justice, freedom, and love. This positive conception of peace meant that the moral burden of proof had to be met by those who chose for war. Catholic theory had developed a calculus for determining when the presumption for peace might be overridden, "precisely in the name of preserving the kind of peace which protects human dignity and human rights."[75]

The bishops acknowledged Vatican II's call for a "universally recognized public authority with effective power 'to safeguard, on the behalf of all, security, regard for justice, and respect for rights.'" "*But what of the present?*" the bishops immediately asked, in italics. Short of the achievement of institutions capable of advancing the universal common good, the Christian was obliged to defend peace against aggression. "This is an inalienable obligation."[76] The obligation could be met through two moral options: for pacifism, acting through nonviolent means of resistance and social change, or through just-war reasoning and the proportionate and discriminate use of armed force in the defense of rights.

According to the bishops, Catholic teaching saw pacifism and the just war as "complementary," for both sought the common good. The pacifist option, however, was open only to individuals. Councils and popes had taught that governments threatened by aggression had a moral obligation to defend their people, by armed force if necessary as a last resort.[77] Armed force was not the only defense against aggressors; nonviolent means of defense and conflict-resolution "best reflect the call of Jesus both to love and to justice."[78] Christians lived in a paradox, this side of the Kingdom: we are obligated to proclaim that "love is possible and the only real hope for all human relations." And yet we had to accept that even deadly force is occasionally justified in legitimate national defense. "An even greater commitment to Christ and his message" was the Christian call in the face of this conundrum.[79]

Part One ended with a review of just-war theory, a defense of "the value of nonviolence" (which the bishops identified with pacifism), and a further claim that these two positions represented "distinct but interdependent methods of evaluating warfare."[80]

Part Two was entitled "War and Peace in the Modern World: Problems and Principles." It began with a familiar assertion of the "unique challenge" posed by nuclear weapons. Moral reflection had to begin with the fact that nuclear weapons posed new moral questions.[81] The destructive power of nuclear weapons, the bishops continued, "threatens the human person, the civilization we have slowly constructed, and even the created order itself."[82] Echoing themes raised by some physicians and scientists, the bishops argued that this "new moment" required prevention, rather than cure; and prevention of the holocaust required the bishops to help "build a barrier" against the idea that a nuclear war could be fought and won. There ought to be "clear and public resistance" to any proposals along these lines.[83]

In their conversations and study, the bishops had become convinced of the "overwhelming probability" that a major nuclear exchange would be unlimited, and agreed with the judgment of the Pontifical Academy of Sciences that even a counterforce nuclear exchange would involve such immediate and residual civilian casualties as to make "prevention . . . our only recourse."[84]

The bishops then stated flatly that "under no circumstances may nuclear weapons or other instruments of mass slaughter be used for the purpose of destroying population centers or other predominantly civilian targets." This condemnation applied to retaliatory second strikes as well as preemptive first strikes. No Christian could morally carry out orders or policies deliberately aimed at killing noncombatants.[85] The bishops also declared that they did "not perceive any situation in which the deliberate initiation of nuclear warfare, on however restricted a scale, can be morally justified." Conventional attacks had to be resisted by conventional forces.

The bishops repeated their "extreme skepticism" that a nuclear exchange could be controlled, no matter how limited first use might have been. Thus tactical nuclear responses to conventional attack were "morally unjustifiable." The moral necessity of a "barrier against any use of nuclear weapons" set the context for the bishops' support of a "no-first-use" policy.[86] NATO should move in that direction, "in tandem with the development of an adequate alternative defense posture."[87] To cross the barrier between conventional and nuclear war was to enter an unknown world where there was no experience of control, little prospect of its possibility, and therefore no moral justification for taking the risk. Therefore, the "first imperative" was to prevent any use of nuclear weapons.[88]

In sum, although the bishops did not flatly reject the possible retaliatory use of small-yield nuclear weapons against clearly definable military targets, they believed that the burden of proof rested on those who asserted that limiting a nuclear war was possible. On this "centimeter of ambiguity," as Father Hehir subsequently called it, deterrence would have to rest.[89]

The bishops acknowledged that deterrence strategies antedated the nuclear age. But deterrence had taken on a new meaning and importance since 1945.[90] "Stable nuclear deterrence" was a situation in which both superpowers deployed their retaliatory forces in ways that were invulnerable to preemptive attack. Deterrent stability thus required that both sides avoid deployment of weapons "which appear to have a first strike capability."[91] The bishops noted Pope John Paul II's affirmation of deterrence, "certainly not as an end in itself but as a step toward a progressive disarmament;"[92] reviewed targeting options and the relationship between deterrence strategy and the prevention of war;[93] and then made their summary judgment: deterrence could be afforded "a strictly conditioned moral acceptance," but could not be considered "adequate as a long-term basis for peace."[94] This judgment led to several "specific evaluations": that proposals to "go beyond" deterrence and plan for prolonged nuclear war-fighting were not acceptable; that deterrent sufficiency was adequate; the nuclear superiority must not be pursued; and that every nuclear force modernization had to be evaluated in terms of whether it would make "progressive disarmament" more or less likely.[95]

In light of these general principles, the bishops opposed a number of current or potential U.S. policies. They opposed "the addition of weapons which are likely to be vulnerable to attack, yet also possess a 'prompt hard-target kill' capability that threatens to make the other side's forces vulnerable." (The bishops urged the Soviet Union not to deploy such weapons, but the footnote to this text specifically named the

MX and Pershing II missiles as fitting into the proscribed category according to "several experts in strategic theory.")[96] The bishops also opposed strategic planning for a war-fighting capability, and systems that blurred the distinction between conventional and nuclear weapons.[97] In the context of their support for deterrent "sufficiency," the bishops supported "immediate, bilateral, verifiable agreements to halt the testing, production, and deployment of new nuclear weapons systems" (i.e., the nuclear freeze); negotiated deep cuts in both superpower arsenals, with particular focus on destabilizing weapons; early and successful negotiation of a comprehensive nuclear test ban; removal of short-range nuclear weapons from front-line positions in Europe; and strengthened command and control systems to prevent unauthorized nuclear weapons use.[98]

Rethinking deterrence, reducing the possibility of nuclear war, and moving toward stable security demanded "a substantial intellectual, political, and moral effort." But it also required a willingness to open ourselves to God's power in the world, "which calls us to recognize our common humanity and the bonds of mutual responsibility which exist in the international community in spite of political differences and nuclear arsenals." The bishops acknowledged the unilateralist voices in the NCCB and throughout American Catholicism, and called them "a prophetic challenge to the community of faith." The unilateralist case had its own intellectual integrity and religious power.[99] The pastoral agreed with the unilateralists that the nuclear danger and the difficulties of international instability required "steps beyond our present conceptions of security and defense policy."[100] Outlining those steps was the business of Part Three, entitled "The Promotion of Peace: Proposals and Policies."

The bishops adopted numerous proposals for accelerated arms control, arms reduction, and disarmament. Although rejecting unilateral disarmament, they reiterated their support for the nuclear freeze (without labeling it as such), and noted the importance of verifiability in arms control agreements. The bishops also encouraged "independent initiatives aimed at tension-reduction and confidence-building between the superpowers, defining initiatives as "carefully chosen limited steps which the United States could take for a defined period of time, seeking to elicit a comparable step from the Soviet Union. If an appropriate response is not forthcoming, the United States would no longer be bound by steps taken."[101]

The bishops believed in summitry and détente, arguing again that arms control had to be complemented by efforts to reduce political tension. Toward this end, the bishops recommended periodic meetings between senior U.S. and Soviet officials, and regular summits.[102] The bishops also endorsed efforts to limit nuclear proliferation; urged the observance of prohibitions against chemical and biological weapons; called for a "reversal of . . . course" in U.S. conventional arms sales abroad; and linked nuclear weapons reductions to balanced conventional force reductions.[103] The bishops were skeptical of Reagan administration civil defense proposals, and called for an independent commission of scientists, engineers, and weapons experts to examine whether such efforts held out any prospect of saving the nation's population or its values.[104]

The bishops then discussed "non-violent means of conflict resolution," including nonviolent civilian-based defense. Such methods did not insure against loss of life, the bishops admitted; but because contemporary weapons and strategies involved "a very real threat to the future existence of humankind itself," both reason and faith demanded that civilian-based nonviolent defense be seriously explored.[105] This would be facilitated by the establishment of a National Peace Academy.[106] The bishops then repeated their endorsement of both pacifist conscientious objection and just-war-based selective conscientious objection.[107]

At this point, two-thirds through the pastoral letter, the bishops turned their attention to "shaping a peaceful world." The bishops reaffirmed the teaching of *Pacem in Terris* on the universal common good and the fact of interdependence, and argued that "mutual security and survival require a new vision of the world as one interdependent planet."[108] Humanity would learn to resolve the problems of interdependence in common, or "we shall destroy one another."[109]

This brought the bishops to the question of the Soviet Union:

> The fact of a Soviet threat, as well as the existence of a Soviet imperial drive for hegemony, cannot be denied. . . . Many peoples are forcibly kept under communist domination despite their manifest wishes to be free. Soviet power is very great. Whether the Soviet Union's pursuit of military might is motivated primarily by defensive or aggressive aims might be debated, but the effect is nevertheless to leave profoundly insecure those who must live in the shadow of that might.

Americans should have no delusions about human rights repression in the U.S.S.R., or about Soviet active measures throughout the world. The bishops did not wish to pin all the blame for the world's ills on one superpower: "our government has sometimes supported repressive governments in the name of freedom, has carried out repugnant covert operations of its own, and remains imperfect in its domestic record of insuring equal rights for all." Still, there was a difference. NATO was a free alliance of democracies; the Warsaw Pact was not. U.S. foreign policy was imperfect, but the facts did not support "the invidious comparisons made at times, even in our own society, between our way of life, in which most basic human rights are at least recognized if they are not always adequately supported, and those totalitarian and tyrannical regimes in which such rights are either denied or systematically repressed."

The bishops knew that the Soviet system was itself an obstacle to successful arms negotiations, which had to take account of the "radically different" understanding of morality that Soviet diplomats brought to the bargaining table. "This is no reason for not negotiating. It is a very good reason for not negotiating blindly or naively." On the other hand, successful diplomacy required that the American people and their government not fall prey to a "form of anti-Sovietism" that missed the centrality of the threat of nuclear weapons and the superpowers' "common interest" in avoiding nuclear war.

The bishops concluded on a note of hope: "Soviet behavior in some cases merits the adjective reprehensible, but the Soviet people and their leaders are human beings created in the image and likeness of God." We were not condemned to a static pattern of hostility. Creative diplomacy and God's power could open doors we now imagined to be bolted shut. The path to peace and stability would not be easy. But that was no excuse for a "hardness of heart" that could close our minds and imaginations to the changes we had a responsibility to initiate.[110]

Part Three ended with reflections on economic interdependence and the problem of development. Americans had to reconceive the meaning of national interest in an interdependent world. American generosity toward the world's poor had to be completed by "a more systematic response" to the major issues of underdevelopment. U.S.

policy should promote those "profound structural reforms called for by recent papal teaching."[111] The relationship between disarmament and development was not simple; but "the fact of a massive distortion of resources in the face of crying human need creates a moral question." The threats to peace and stability did not all arise from modern weapons.[112] The United States ought to take a stronger leadership role in the United Nations, which would require realism about its limitations and imagination about its possible reform.[113]

Part Four was addressed specifically to the Church and entitled "The Pastoral Challenge and Response." The bishops called for educational programs that would "explain clearly those principles or teachings about which there is little question. Those teachings, which seek to make explicit the gospel call to peace and the tradition of the Church, should then be applied to concrete situations. They must indicate what the possible legitimate options are and what the consequences of those options may be." Worried about selectivity, the bishops noted that "while this approach should be self-evident, it needs to be emphasized."[114]

The bishops urged "reverence for life" throughout American society. Abortion blunted our sense of the sacredness of human life. In a country where the unborn were wantonly killed, "How can we expect people to feel righteous revulsion at the act or threat of killing non-combatants in war?"[115] There were differences between taking life in abortion and taking life in war. Still, the bishops pleaded "with all who would work to end the scourge of war to begin by defending life at its most defenseless, the life of the unborn."[116]

The bishops called American Catholics to increased prayer and penance for peace, and committed themselves to fast and abstain from meat on Fridays, a practice in which they urged others to join.[117]

The pastoral ended with exhortations to various Catholic groups: priests, deacons, religious, and pastoral ministers; educators (and theologians, urged to help create a "theology of peace"); parents; young people; men and women in the armed forces; men and women in defense industries;[118] scientists; the media, public officials; and finally all Catholics as citizens. "While some other countries also possess nuclear weapons, we may not forget that the United States was the first to build and use them. Like the Soviet Union, this country now possesses so many weapons as to imperil the continuation of civilization. Americans share responsibility for the current situation, and cannot evade responsibility for trying to resolve it."[119]

The formal Conclusion of the pastoral tried to answer two questions: Why did the bishops speak, and what did they say? The bishops insisted that they spoke as pastors and teachers, not politicians or technicians. They addressed issues of war and peace because they were trying to live up to the call of the Beatitudes in their own time and circumstances.[120]

What had the bishops said in the previous thirty-eight thousand words?

> Fundamentally, we are saying that the decisions about nuclear weapons are among the most pressing moral questions of our age. . . . In simple terms, we are saying that good ends (defending one's country, protecting freedom, etc.) cannot justify immoral means (the use of weapons which kill indiscriminately and threaten whole societies). . . . In our quest for security, we fear we are actually becoming less and less secure.[121]

The bishops wanted to close on a visionary note. In the 334th section of their letter, they argued that John XXIII's and Paul VI's hopes for a "truly effective international authority" were not unrealistic. There was a substitute for war: "negotiation under the supervision of a global body realistically fashioned to do its job." The task of creating such an instrument was immense. But could not human genius and God's grace combine to do what had to be done?

> We turn to our own government and we beg it to propose to the United Nations that it begin this work immediately; that it create an international task force for peace; that this task force, with membership open to every nation, meet daily through the years ahead with one sole agenda: the creation of a world that will one day be safe from war. Freed from the bondage of war that holds it captive in its threat, the world will at least be able to address its problems and to make genuine human progress, so that every day there can be more freedom, more food, and more opportunity for every human being who walks the face of the earth.[122]

God willed, not a perfect world, but a better one. That better world was a task for human hands and hearts.[123] Faith in the risen Christ sustained Christians in their responsibilities. For before us was the vision of that time when God was all in all, when there would be "a new heaven and a new earth" in which "he who sat upon the throne said, 'Behold, I make all things new' (Rev. 21:1–5)."[124]

THE VATICAN CONSULTATION

"The Challenge of Peace," like the documents of Vatican II, was the result of extensive debate and compromise. The public debate over the letter continues to this day; the prepastoral debate was the most vigorous in American Catholicism since Pope Paul VI's 1968 encyclical on contraception, *Humanae Vitae*.[125] But perhaps the most decisive intervention in shaping the final text was a consultation held at the Vatican on January 18–19, 1983. The meeting involved bishops and specialists from the United States, France, West Germany, Great Britain, Belgium, Italy, and the Netherlands. Vatican participants included senior representatives and consultors from the Sacred Congregation for the Doctrine of the Faith (including its proprefect, Cardinal Joseph Ratzinger, who chaired the meetings), the Council for the Public Affairs of the Church (including the Vatican Secretary of State, Cardinal Agostino Casaroli), and the Pontifical Justice and Peace Commission. American participants in the consultation were Cardinal-elect Bernardin, Archbishop Roach, Father Hehir, and Msgr. Daniel Hoye, General Secretary of the NCCB/USCC.[126]

The proceedings of the consultation were confidential, but the participants agreed that a synthesis prepared by Rev. Jan Schotte, secretary of the Pontifical Justice and Peace Commission, would be published as "a point of reference and a guide to the U.S. bishops in preparing the next draft of their pastoral letter."[127] The text of the Schotte synthesis was mailed to all U.S. bishops, and subsequently released to the NC News documentary service, *Origins*.

The synthesis was a masterpiece of ecclesiastical discretion and indirection, which characterized the consultation as "centered on five main themes: the precise teaching role of a bishops' conference; the application of moral principles to the nuclear weapons debate; the use of scripture; the relationship between just-war theory and pacifism in Catholic tradition; the morality of deterrence."[128] On each of these points, the consultation preceded crucial alterations in the third draft of "The Challenge of Peace."

On the question of the teaching authority of the pastoral letter, a close reading of the Schotte synthesis suggests that there was considerable discussion about the various levels of authority in the draft document, and concern that the NCCB had not made these distinct levels of authority and their relationship to Catholic consciences sufficiently clear. The American bishops, according to the synthesis, argued that "the nuclear threat cannot be adequately addressed solely on the basis of proposing principles. In order to be effective and to be heard the bishops must be specific."[129] The Americans conceded that their text spoke at different levels of episcopal authority, and that these different levels were "a. proposing the teachings of Vatican Council II and of the popes (e.g., on the kingdom, on deterrence); b. referring to theological traditions (e.g., nonviolent tradition, the just-war theory) and applying them; c. using a moral principle (e.g., proportionality) and applying it to concrete situations by combining it with a judgment of fact (e.g., first use, limited nuclear war)."[130]

The participants in the consultation were evidently worried that the second draft mixed these different levels of authority to the point where it would be difficult for readers to make the necessary distinctions. The summary statement on this discussion contained, in its interstices, a serious critique of the NCCB's second draft:

> It was the consensus of the meeting—also seconded by the U.S. bishops—that the pastoral letter in its present draft makes it nearly impossible for the reader to make the necessary distinctions between the different levels of authority that are intertwined. The desire of the meeting—offered as a contribution in the spirit of collegiality—was that the pastoral letter at least be rewritten to state clearly the different levels of authority, and this for several reasons. First, in respect for the freedom of the Christian so that he or she be clearly informed about what is binding in conscience. Second, in respect for the integrity of the Catholic faith so that nothing be proposed as doctrine of the church that pertains to prudential judgment or alternative choices. Third, for reason of ecclesiology, that the teaching authority which belongs to each bishop not be wrongly applied and therefore obscure its credibility.[131]

These concerns resulted in a major alteration of the third draft of the pastoral letter, which began with a distinction between the binding character of moral principles and the nonbinding nature of the bishops' prudential judgments.

On the application of moral principles to nuclear weapons issues, consultation participants also worried about confusions in the teaching authority of the bishops, and "feared appearing to substitute themselves as bishops in the place of the conscience of the individual persons (e.g., government officials, politicians, military, etc.)." As "teachers of the faith," the bishops ought not take sides when various prudential applications were possible.[132]

On the question of first use of nuclear weapons, for example, the American participants explained that their position was meant to illustrate the qualitative difference between conventional and nuclear war; the Americans also wanted to take a stand against "tendencies in their country to present nuclear war as more likely and more acceptable." Yet other participants, who "recognized that the bishops have a duty to admonish those in authority and to warn about possible consequences," still argued that the second draft contained an "apodictic moral judgment" on no first use, even though the text also acknowledged that there was serious public controversy on this point. The flat moral condemnation of first use was, according to some participants, a "highly contingent" judgment. "Others saw the possibility of first use as still necessary at this stage within the context of deterrence."[133]

The upshot of this particular discussion was an agreement that the third draft had to make "clearer distinctions" in regard to the first use issue, which indeed the third draft did. But beneath this agreement other points seemed to lurk. It is not clear from the Schotte synthesis, for example, that the non-American participants in the consultation held so firmly to a radical and qualitative distinction between conventional and nuclear arms as did the Americans, at least at the level of moral analysis. The American insistence that nuclear weapons constituted a drastically "new moment" in both empirical and moral terms was evidently under scrutiny, and was the source of some disagreement.

Further, the Schotte synthesis seems to imply that some participants in the consultation did not take the draconian view of the Reagan administration's disinclination to make a "no first use" pledge that shaped the American debate. The third draft of "The Challenge of Peace" did connect a no-first-use pledge to conventional force balances in Europe, a point that concerned the European and Vatican participants in the consultation.

The discussion on the use of Scripture in the pastoral letter had been preceded by an American opening statement that pledged "a more comprehensive review of relevant texts" and a "better integration of the biblical data and theological reflection on it in the pastoral."[134] This remark presumably anticipated sharp criticism of the use of Scripture in the previous drafts. The chief point of concern, as taken from the Schotte synthesis, was that "the text should clearly avoid mixing up two distinct levels and differing realities: our faith that the Kingdom of God will come and the realization that it is not certain if and when true peace will effectively exist in the world that is ours."[135]

The consultation included a pointed discussion on the relationship between pacifism and just-war theory in the Catholic tradition. The first and second drafts of the pastoral had posed these as essentially equivalent streams of Catholic reflection. Some participants in the consultation, though, argued that the Church had never taught that nonviolence was a substitute for just-war theory:

> The affirmation in the draft that "the witnesses to nonviolence and to Christian pacifism run from some church fathers through Francis of Assisi to Dorothy Day and Martin Luther King" is factually incorrect and does not support the affirmation that there is a pacifist tradition which holds "that any use of military force is incompatible with the Christian vocation," nor that it is clear from the writings of leading theologians in the first four centuries that "there was a certain level of opposition to military service based upon particular gospel passages."

The Vatican consultors also charged that the positioning of pacifism in modern Catholic teaching in the second draft went beyond what *Gaudium et Spes* actually said about conscientious objection as a right of personal conscience.[136] After what must have been an interestingly brisk discussion on these controverted points, the participants agreed that "there is only one Catholic tradition: the just-war theory, but that this tradition was subject to inner tensions coming from an ever-present desire for peace." The U.S. bishops further "denied affirming in the pastoral that a nation as such could adopt a pacifist stance."[137] The fact of the denial implies that some involved in the consultation had read the second draft as suggesting precisely that. In any event, the final text explicitly rejects pacifism as an option for governments, while continuing to affirm it an option for individual consciences.

The cryptic quality of the Schotte synthesis makes it difficult to determine precisely the points at issue here. The non-American participants seemed eager to challenge the Americans' historical judgments about "two traditions." But at a deeper level, the key critique may have involved the American draft's suggestion that pacifism and just-war theory were somehow reconcilable, perhaps minimally in the "nuclear pacifist" position. The final text of the pastoral does suggest that both options flow from a concern for the common good, and are thereby related. And no doubt the participants in the consultation appreciated the American bishops' pastoral concern for American Catholics dedicated to pacifism. But the gist of the consultation was an affirmation that there were not "parallel traditions," either historically or theoretically. The integrity of both positions required that their distinctiveness be maintained.

The consultation's work on deterrence decisively shaped the final text of "The Challenge of Peace." The fulcrum of the discussion was Pope John Paul II's teaching which, according to the second draft of the pastoral letter, "was designed to limit the acceptable function of deterrence precisely to the one positive value it is said to have—preventing the use of nuclear weapons in any form."[138] The Schotte synthesis implied that several participants in the consultation considered this too limited an exegesis of the pope's meaning.

According to the synthesis, the Vatican consultation wanted a more nuanced position on deterrence in the Americans' final text. Among the points raised during the consultation were that deterrence "is one of the more difficult questions placed before the moral judgment of pastors and faithful"; that deterrence "must be seen in the wider context of the geopolitical situations and considerations"; that the evaluation of deterrence "must take into account the existence of actual and probable threats of aggression," and "cannot be separated from moral considerations and from prudential judgment on military and political facts and strategies"; and that deterrence has "a psychological component since it requires credibility to be effective."[139]

Each of these themes could be found in the second draft, but evidently not to the full satisfaction of the participants in the Vatican consultation. Cardinal Casaroli, offering a "personal commentary" on the pope's message to the Second UN Special Session on Disarmament, suggested, obliquely, that the American second draft did not sufficiently link the value of deterrence to the threat of aggression (by which one can infer, at least primarily, the threat of Soviet aggression). Casaroli noted that the question of deterrence involved *both* "the danger of nuclear conflict" *and* "the endangering of the independence and freedom of entire peoples" and "in the West, the fear . . . of an imposition of an ideology and of a 'socialist' regime."[140] Casaroli went on to suggest that "in both cases the concern is about vital values for peoples and for humanity. A question *worthy of careful examination* is whether there exists a relationship of equality or of superiority between the two above-mentioned points of consideration."[141]

This exceptionally discreet formulation is susceptible to several interpretations. But it seems reasonable to suggest that Casaroli, in his exegesis of John Paul II's views, was hinting that the American second draft was insufficient in its understanding of the relationship between the threat of nuclear war and the threat of totalitarianism, and that the Americans had not devoted adequate attention to the problem of Soviet power and purpose. This latter point was driven home later in Casaroli's remark that "one must not give the impression that moral principles are to be affirmed (even with explicit condemnations) for one side, while forgetting as it were that these same principles have universal value and applicability (i.e., for all sides)."[142] The formulation is general and discreet, but the message was clear.

Casaroli was also worried about the tone of the second draft toward U.S. policy. According to Casaroli:

> One must not give the impression that the church does not take sufficiently into account the magnitude of the problems and the seriousness of the responsibilities of government authorities who have to make decisions in these matters. This does not mean that the church cannot and must not clearly enunciate the certain and seriously obligatory moral principles that the authorities themselves must keep in mind and follow. This must be done, however, in such a manner that it helps those authorities to get a correct orientation according to the basic principles of human and Christian morals and not to create even greater difficulties for them in an area so enormously difficult and so full of responsibility. The same observations apply also to public opinion.[143]

Again, the language was oblique, but the message seems clear. Why would Cardinal Casaroli have raised such concerns, unless he (and, presumably, the pope whose views he was "personally" interpreting) believed that the second draft was deficient on precisely these points?

The impact of the Vatican consultation on the final text of "The Challenge of Peace" can only be inferred. In their report to the NCCB, Cardinal Bernardin and Archbishop Roach emphasized the "positive tone of the meeting" and the way in which members of other episcopal conferences had commended "the NCCB for both the courage to undertake the pastoral and the seriousness with which it has been proposed."[144] Yet it does not seem to have been an accident that the Vatican consultation was held before the third draft of "The Challenge of Peace," which contained significant changes from predecessor drafts. In most cases, these differences reflected themes in the consultation. The third draft cited statements of Pope John Paul II much more frequently than drafts one and two, and the pope's statement to the UN Special Session on Disarmament was exegeted in line with Cardinal Casaroli's "personal commentary" at the consultation. One result was the formulation on deterrence, which was accorded a "strictly conditioned moral acceptance," given its importance in creating the circumstances in which disarmament might proceed. This represented a subtle, but important, shift from the second draft. The clear distinction in the third draft between levels of teaching authority, its considerably tougher view of the Soviet

Union, its linkage of a no-first-use pledge to conventional force balances in Europe, and its argument that pacifism was not an option for governments all reflected themes in the Vatican consultation.

While a direct and unambiguous line between the Vatican consultation and these alterations in the final text of the American letter cannot be drawn, it seems unlikely that these results were coincidental, or simply reflected arguments already advanced within the ad hoc committee on war and peace. The latter were certainly a factor; but they were given more weight by the Vatican consultation.

In sum, then, the Vatican consultation was a moderating influence on "The Challenge of Peace." It was not the only such influence. But the Vatican consultation graphically illustrated the Holy See's concerns for the American bishops' work. By setting the boundaries for the final debate over the pastoral letter, the Vatican consultation was a crucial part of the overall deliberative process.[145]

"THE CHALLENGE OF PEACE" AND THE ABANDONMENT OF THE HERITAGE

Did "The Challenge of Peace" represent a genuine evolution of the Catholic heritage as developed through Vatican II and Murray, or a decisive moment in its abandonment? There are many positive things about the 1983 NCCB pastoral letter. The enormous public visibility it received helped establish the claims of moral reason to a central place in the national debate over nuclear weapons and strategy. The argument in the American Church preceding and following the letter was vigorous (and ongoing), and became the occasion for important new scholarship in both the just-war and pacifist traditions. The consultative process established by the NCCB ad hoc committee on war and peace was a welcome expression of Pope John XXIII's call for an "open church," and brought bishops, theologians, and lay persons together in an unprecedented common reflection on an issue of the gravest importance for both Church and society. The pastoral letter itself contained important truths, and helped teach the American Church much of what it did, and did not, know about the moral problem of war and peace.

All this can be admitted; and yet the summary judgment on "The Challenge of Peace" must be that it represented a continuation of the abandonment of the classic Catholic heritage. One can even go further, and argue that "The Challenge of Peace" was a decisive moment in that process, because it involved the adoption, by the National Conference of Catholic Bishops, of key themes of abandonment that had become pervasive in American Catholicism in the years following the Second Vatican Council.

These are hard judgments, and require justification.

The principal deficiency of the bishops' pastoral letter was its virtual detachment of the problem of nuclear weapons from the political context in which they are best analyzed, morally and strategically. The bishops' insistence on the "unique" dangers of nuclear weapons was taken far beyond the level of empirical judgment (on which there could be little disagreement; nuclear weapons are qualitatively different in their destructive capacity), and made into a kind of moral and strategic absolute. The result was a curiously ahistorical document in both tone and substance, despite its authors' efforts to the contrary.

But the most damaging dimension of the pastoral's claim that the threat of nuclear weapons was unique and absolute was its impact on the structure of the bishops' moral argument. Unlike the heritage it hoped to develop (which began with the interpenetration of morality and politics and located the question of the legitimate use or threat of armed force within that larger context), "The Challenge of Peace" began with the weapons themselves. This entry point distorted the entire analysis.

It made the bishops vulnerable to intellectual currents and emotional passions that were not only external to their own tradition, but fundamentally opposed to its central claims. The bishops' choice of the word "terror" to describe both their congregants' and their own response to the signs of the times was emblematic of this problem. Why was this the place to begin? Did it reflect a renascent, New Testament-based apocalypticism within the NCCB? Was it an attempt to bring the eschatological teachings of the Judeo-Christian tradition to bear on contemporary issues (no small task, given the complexity of that tradition)? Or, much more likely, was it a reflection of those survivalist currents in American political culture that had played a notable role in shaping the nuclear debate of the late 1970s and early 1980s? No one can be sure in any final sense why some words (such as "terror") were chosen rather than others. But it seems more than coincidental that the bishops' choice of "terror" as the starting point for their analysis mirrored the language and imagery used so successfully by antinuclear publicists such as Jonathan Schell, Helen Caldicott, and Carl Sagan.

The final text of "The Challenge of Peace" tempered the bishops' tacit adoption of survivalist imagery. But that survivalism had made its impact on the bishops' deliberations was clear from the crucial second draft, which claimed that "the destructive potential of the nuclear powers threatens the sovereignty of God over the world he has brought into being. We could destroy his work."[146] Indeed, we could, should the worst happen. But how would this "threaten the sovereignty of God," which is presumably not a function of our works? This extraordinary claim was deleted from the final text of "The Challenge of Peace"; but that it appeared at all suggests the influences at work.

The notion that sheer physical survival, in either personal or species terms, is the highest good to which all other goods must be subordinated is not a theme compatible with Catholic ethics. As the pastoral rightly reminds us, we are called, in this interim time, to help build the Kingdom. But from its earliest days, Christianity has taught that we have here no earthly home; that we are a pilgrim people, living in the tension between what has been revealed in the Resurrection and what will be completed in glory at the final establishment of the Kingdom by God. Why should Catholic bishops, leaders of a Church that traces its origins from the blood of Christ on the cross and regularly celebrates its martyrs in the rhythm of its liturgical life, have adopted language and imagery that is at cross-purposes with basic Christian insights into the human condition and the moral norms that can be read from that condition? Why should they have failed to see, and state, that survivalism was incompatible with peace: that a politics of fear can lead only to a politics of hate and the disintegration of the political community that provides us the instrument with which to address the tangled web of peace, security, and freedom? Why should bishops charged with the task of helping form Catholic consciences adopt language and imagery from a survivalist antiethic that, by absolutizing the value of sheer physical survival, so transvalues all other values that they become relative, and ultimately irrelevant?

Whatever emotional, psychological, or cultural factors may have shaped this remarkable turn of events, the bishops' single-minded insistence on the uniqueness of the nuclear threat was the key intellectual move opening the door to a survivalism they would finally reject, but which was crucial in shaping their reflections.

The nuclear weapons entry point had three other, equally distorting, effects on "The Challenge of Peace."

First, it was the instrument by which the question of "just cause," normally the first order of business in just-war reasoning, was deferred to the end of the pastoral. The bishops' description of the threat of Soviet power and purpose was unexceptionable, as it was finally thrashed out in the third and final draft of the pastoral letter (after the Vatican consultation). But why was this essential analysis left so far back in the letter? Why did not the signs of the times include, first, the brute fact of totalitarianism as an inescapable corollary to the contemporary problem of war, and secondly, the fact that nuclear consternation in the West followed close on the heels of Soviet parity in some classes of nuclear weapons, and superiority in others? Why did the bishops fail to see that such a relativizing of the totalitarian dimension of the problem could (as it did in the 1930s) help make war more likely, were it to reinforce a psychology of appeasement already well advanced in some elite circles in American religious and political culture? Did not the hard and inescapable facts of Soviet power help form the "new moment" that called for the clear voice of moral reason the bishops wished to raise?

The positioning of the bishops' discussion of the U.S.S.R. reflected at least three prominent features in the contemporary Catholic debate on war and peace: its disaffection with the possibilities of the American experiment (What would the country that produced "Vietnam" do with nuclear weapons?); its anti-anticommunism; and its visceral distaste for the Reagan administration. These three themes, focused through the prism of a nuclear weapons entry point for moral discourse, resulted in the bishops' painting a flawed portrait of contemporary reality, which acknowledged one facet of the dilemma of moral imagination in the modern world (the threat of global holocaust) but not the other (the threat of totalitarianism). The groundwork for this development had been laid in the bishops' 1968 and 1976 pastoral letters, and particularly in Cardinal Krol's SALT II testimony and subsequent address to the Religious Committee on SALT. But there was no reason why the bishops, in "The Challenge of Peace," could not have described the relationship between the nuclear and totalitarian threats more accurately, as did their West German colleagues in a pastoral letter issued three weeks prior to the American bishops' May 1983 meeting.[147]

Secondly, the insistence on a nuclear weapons entry point led to a decontextualized discussion of the morality of deterrence. No one who followed the debates preceding and during the work of the NCCB ad hoc committee on war and peace can doubt that deterrence was the central issue the bishops (and the USCC staff) wished to address. But why could it not have been located within a wider context: one that accurately portrayed the Janus-headed problem of war and tyranny; affirmed the possibility of peace through dynamic political community and law at the international level; related the necessity of arms control and disarmament to this larger problem of the formation of morally and politically effective international political community; and only then dealt with deterrence and its capacity to act as a regulatory mechanism in international affairs between what presently obtained and what had to be sought? Had the bishops pursued this structure of argument, they would have been in a stronger position to address the morality of deterrence and the specific prudential options that would make stable deterrence contribute to arms control and disarmament. The decontextualization of deterrence in "The Challenge of Peace" thus not only distorted the structure of the argument; it helped create the circumstances in which the bishops "prudential judgments" (e.g., on the nuclear freeze, on no first use) received primary public attention, despite the bishops' statement that these were the least morally weighty recommendations in the entire pastoral letter.[148]

The priority focus on nuclear weapons was also responsible, in the third place, for the letter's confusions on the compatibility of the just-war and pacifist traditions. Here, the phantom of "nuclear pacifism" was decisive; it was posited, tacitly, as a bridge between the poles of the moral argument. But it is not, and insupportable claims followed. The pastoral letter's argument that just-war theory and pacifism were "interdependent methods of evaluating warfare" is, to put it plainly, false. They are, as James Finn pointed out in a commentary on the pastoral, quite opposite methods of evaluating warfare. The principled pacifist opposes all resort to armed force; the just-war theorist allows the proportionate and discriminate use of armed force in carefully defined circumstances.[149]

The confusion of these two positions leads to the corruption of both, Finn persuasively argued, and to a disjunction between the quest for peace and the use of armed force. It also led to a certain incongruity in the debate and voting on "The Challenge of Peace," with pacifist bishops such as Gumbleton and Hunthausen arguing questions of military strategy and tactics, and then voting to accept a document that rejected their prescriptions of unilateral disarmament.[150]

The pastoral's muddled thinking on the relationship between pacifism and just-war theory also had an unhappy practical effect. Pacifist and just-war claims are not ultimately reconcilable at the level of moral theory, but there is no reason why pacifists and just-war theorists cannot work together in the practical order on building international political community sufficient to sustain legal and political means of resolving conflict. The pacifist would do so on the grounds that law and governance are nonviolent means of adjudicating conflict; the just-war theorist would join in such work because, as Paul Ramsey noted in his commentary on *Pacem in Terris*, the claims of law and governance as means of conflict-resolution are embedded in the interstices and the logical trajectory of the just-war theory. But "The Challenge of Peace" never explicitly identified democratic governance as the world's most successful and widespread form of nonviolent conflict-resolution. And the pastoral's quixotic search for a higher viewpoint that would unite pacifism and just-war theory at the level of moral theory drove the bishops off onto a path of reasoning in which the potentially crucial cooperation of pacifist and just-war theorists at the level of practical politics never rose on the horizon—save in calls for a nuclear freeze.

The minimal attention paid to the concept of peace as governed community in "The Challenge of Peace" (which led to an entirely too soft view of the present U.N.)

was paralleled by serious confusions on the relationships among the various meanings of peace in Catholic tradition. The bishops admitted that biblical understandings of peace—personal right relationship with God, and the eschatological vision of *shalom*—do not yield either moral norms or political prescriptions for the contemporary world. And yet the bishops insisted that these concepts defined the "paradox" in which we live: "We must continue to articulate our belief that love is possible and the only real hope for all human relations, and yet accept that force, even deadly force, is sometimes justified and that nations must provide for their defense."[151] Why is "the paradox" defined in these terms?

The implicit identification of "peace" with "love" (in either personalist or eschatological terms) marked the key symbolic point at which "The Challenge of Peace" definitively broke with ther heritage of *tranquillitas òrdinis*, which taught the possibility of a morally worth peace of political community *prior* to the world's final conversion. The "paradox" of our times may be defined in many ways: in Christian terms, as the tension between the life of faith available in the Spirit given to the Church by the risen Christ, and the incompletion of God's plan for salvation history; or in political terms as the tension between a world that is now an interdependent political arena but not yet a political community; or, most narrowly, as the tension between the threat of global holocaust and the threat of global tyranny. But none of these accurate descriptions of the paradoxical character of life between the fire and the pit has very much to do with the relationship between "love" and an achievable, this-worldly peace.

Why, then, did the bishops choose such a formulation? It had been anticipated in John XXIII's claim that "mutual trust alone" was the fundamental principle on which to build the "true and solid peace of nations."[152] It may well have had to do with the contemporary Church's continuing struggle to integrate biblical themes into moral reflection. It may have involved the cumulative influence of those teachings that just-war theory was "mental gymnastics, the worst form of casuistry." But whatever the taxonomy of influences at work, the result in "The Challenge of Peace" was an important conceptual confusion whose outcome amounted to the virtual abandonment of the heritage as Murray would have understood it.

"The Challenge of Peace" was not a "peace pastoral" so much as it was a "weapons pastoral." Considerably more attention was paid, in the text and in the debates that preceded it, to questions of deterrence strategy and nuclear force modernizations than to the peace that was possible in this world—and that would thereby set the horizon of moral and political judgment against which weapons issues could be analyzed.

On specific weapons-related issues, the bishops' prudential judgments fell into a predictable pattern that reflected the primary agenda of the renascent antinuclear movement of the early 1980s. It could conceivably be argued that this agenda and the Catholic tradition were happily and coincidentally congruent. But it seems much more likely that the bishops, eager to "provide a moral assessment of existing policy which would both set limits to political action and provide direction for a policy designed to lead us out of the dilemma of deterrence,"[153] were considerably more influenced by the extra-ecclesiastical political culture than they influenced it. Both the entry point the bishops chose and their specific policy recommendations illustrated this. No doubt a dialectic of mutual influence was at work. But the dialectic seems to have been weighted considerably more *from* the antinuclear movement *to* the bishops than vice versa.[154]

The results of these manifold flaws in "The Challenge of Peace" continue to work themselves out in American Catholicism. The bishops' position against the MX missile and the resistance within both the hierarchy and the USCC staff to the Strategic Defense Initiative were predictable reflections of the analytic structure of "The Challenge of Peace." On these issues, as on the question of the nuclear freeze and no first use, the bishops, rather than being the creators of a new spectrum of public debate, had become another cluster of protagonists at one specific pole of the existing argument.[155]

The full pastoral impact of "The Challenge of Peace" can be measured only over decades. But certain trajectories of influence were clear within a year of its adoption in May 1983. The pastoral was followed by a flood of "study guides" whose avowed intention was to facilitate education on the pastoral's teachings. Yet an analysis of representative guides suggests that three principal themes were at the forefront of the postpastoral discussion. First, many study guides taught that a literal reading of the New Testament teaching of Jesus could be directly applied to problems of international life today. One series, published by Paulist Press, suggested that Jesus' action toward the woman caught in adultery might be applied in one-to-one correspondence to contemporary U.S./Soviet relations. Others looked toward the eschatologically expectant nonresistance of the primitive Church in Acts as the only model of authentic Christianity in its relationship to the wider society.

The second theme regularly driven home was the equivalence of pacifism, historically and theologically, with the just-war theory in the Catholic social-ethical tradition: a theme surely influenced by the pastoral's own weaknesses in discussing this issue, and its refusal to criticize the defects in pacifist analysis and prescription today. One looks long and hard in these study guides to find reference to the pastoral's teaching that pacifism is a *morally* inappropriate posture for a government.

Finally, the study guides often suggest that international conflict can be understood and responded to through examples drawn from personalist psychology—a suggestion dependent on the pastoral's affirmation of "trust" and "love" as the primary meanings of "peace."[156] The United States Catholic Conference has not developed or applied standards for such study guides. And it has remained silent in the face of the ubiquity of distorted presentations of the teaching of "The Challenge of Peace."

"The Challenge of Peace" was a tragically lost opportunity. Much that is commendable in it was swallowed up by the document's central deficiencies. The bishops were quite right in their sense that the times required a powerful statement of the Catholic heritage of moral reason and political vision on the central problem of war and peace. What they delivered, though, was a document that, for all its soundness of intention, summed up the abandonment of the heritage that had been underway in the American Church for fifteen years. In doing so, the bishops missed an extraordinary opportunity to reshape the contours of the American public debate on war and peace, security and freedom—a debate badly in need of the kind of moral horizon the bishops could have provided, had they remained the bearers and developers of the heritage of *tranquillitas ordinis* rather than officiants at what became a crucial moment in its internment.

CHAPTER 10

American Catholicism and the Dilemma of Peace and Freedom in Central America

The same themes of abandonment that, in a tempered form, shaped "The Challenge of Peace" dominated American Catholic debate over the dilemma of peace and freedom in Central America in the 1970s and 1980s—and usually in an untempered form. As on the issue of nuclear weapons and strategy, these themes had a decisive effect on positions taken by the National Conference of Catholic Bishops and its public policy arm, the United States Catholic Conference. Once again, the result would not be the development of the classic Catholic heritage as a horizon for moral and political judgment, but its abandonment.

THE AMERICAN CATHOLIC DEBATE ON CENTRAL AMERICA

The Rise of the Theologies of Liberation

A crucial determinant of the course of the U.S. Catholic debate on Central America was the rise, in Latin American Catholicism, of the theologies of liberation. Imported into the United States, these became the prism through which events south of the border were read, analyzed, and interpreted.

The theology of liberation made its first major appearance on the horizon of U.S. Catholicism through the 1968 meeting of the Latin American episcopal conference (CELAM) in Medellín, Colombia. CELAM met in Medellín to apply the teachings of Vatican II to the Church in Latin America. But the Latin American bishops ventured far beyond the Council in their reading of the signs of the times, and in their prescriptions for the role of the church in society. The 1968 date was itself important, according to one sympathetic commentator; although Pope Paul VI's *Populorum Progressio* anticipated the teachings of Medellín, other influences were also at work, among which "one should cite the general atmosphere of 1968, the Paris May, the proliferation of political and revolutionary theologies, the radicalization of Latin American social scientists."[1]

The Medellín documents followed the standard methodology in theologies of liberation. "Social reality" was described and analyzed; theological principles were identified; pastoral options were proposed.[2] It was at the first stage, "social reality," that Medellín broke with the liberalism of Vatican II and *Gaudium et Spes*. Developmental categories of analysis and prescription were explicitly rejected; in their place were substituted that the bishops believed to be "revolutionary" perspectives. The bishops at Medellín emphasized the disjunction between the "dominant" and "oppressed" classes of Latin American society, and taught that this "social reality" reflected a larger international disorder: Latin American countries were victims of an "international system of dependence,"[3] which, due to "international monopolies and the international imperialism of money,"[4] condemned their nations to the periphery of international economic life.

At Medellín, the Latin American bishops thus chose a class-struggle model for understanding "social reality." One of the principal Medellín theologians, Fr. Gustavo Gutiérrez of Peru, was quite explicit about this, even as he acknowledged the tensions it posed for Catholicism: "It is undeniable that the class struggle plants problems for the universality of Christian love and church unity. But every consideration on this matter ought to begin with two elemental attestations: class struggle is a fact and neutrality in this matter is impossible."[5] The class-struggle model led the bishops to altered understandings of violence, which Phillip Berryman frankly described as a "reversal of meaning":[6] "If the Christian believes in the fecundity of peace in order to arrive at justice, he also believes that justice is an unavoidable condition for peace. He cannot but see that Latin America finds itself in many places in a situation that can be called *institutionalized violence*."[7] The "fundamental violence" was the violence of those who maintained their privilege at the expense of the majority.[8] Thus the "liberal capitalist system" was an ineluctably violent and reprehensible distortion of "social reality."

The concept of capitalistic "structural violence" was married to the dependency theory of Latin American underdevelopment by (among others) Gustavo Gutiérrez, who taught that the underdevelopment of the poor was a "historical by-product of the development of other countries." Capitalism led to a "center" and a "periphery," which meant progress and wealth for the few and deprivation for the many.[9] The wretched of the earth in Latin America owed their condition to the self-aggrandizing greed of the developed "North."

Such a situation required, not the remedial efforts of "development," but a genuinely revolutionary restructuring of society, according to the documents of Medellín. The Church would participate in this restructuring by exercising a "preferential option for the poor," who would be organized and evangelized (through consciousness-raising) in small "base communities," which would be "the initial cell of ecclesial structuring, focal point of evangelization, and [thus] . . . a primordial fact of human promotion and development."[10] In these base communities, the poor would be enabled to comprehend their victimization and, motivated by the biblical image of the Exodus and a dedication to Jesus Christ as liberator, would take up the task of re-creating society.

The Medellín conference was a watershed in the history of Latin American Catholicism. Enthusiastic observers like Phillip Berryman described it as "undoubtedly . . . more meaningful than Vatican II" for at least "the 'liberationist' sectors of the Church."[11] Energized by the experience of the Medellín conference, Latin American theologians brought the theology of liberation to full flower over the next decade.

The starting point for many liberation theologians was a world in conflict, or "the world as conflictivity," as Hugo Assman put it.[12] To understand the conflicted nature of current "social reality" was to comprehend the "institutionalized violence . . . practiced routinely by the power structure, [which] takes place in the haciendas and factories, banks and government ministries, the White House or Pentagon."[13] Counterposed to this "first violence" was revolutionary "second violence," practiced in order to seize power and establish justice.[14] Phillip Berryman believed that "second violence" should be judged ethically according to just-war concepts of self-defense.[15]

James Finn, among others, was not so sure. Many statements in liberation theology rang of a "crusade mentality: moral certitude in the justice of one's cause; radical divisions between good and evil forces; the rejection of mediating positions; the willingness to take great risks, even to the final sacrifice; a disinclination to compromise. In brief, the transformation of political action into a religious cause, a holy war."[16] Some theologians of liberation worked to incorporate the theory and practice of nonviolence into their work; but the very distinction between "first violence" and "second violence," with its weakening of the strictures against violence, lent itself to the problems Finn described.

Liberation theologies were self-consciously political theologies, and as such could be presented as a necessary corrective to the privatized understandings of Christianity that had long dominated Latin American Catholicism. But most theologians of liberation meant something far beyond deprivatization in their concept of political theology. The Medellín documents demonstrated that a particular politics was being welded to Christian themes and symbols. Development was rejected in favor of revolutionary social change. Consciousness-raising was the essential precondition to such change. Dependency theory explained the cruel realities of Latin American poverty and revealed the equally cruel face of bourgeois liberal capitalism. The essential division in the world was not East/West but North/South, or center-and-periphery. Violence was to be located first in social and economic structures. Moderation was a sign of accommodation and false consciousness. Class struggle was built into the very fabric of social life, and should not be ameliorated, but intensified in the minds of the oppressed who would then rise up and throw off the shackles of dependence.

The proximate origin of these themes in Marxism was not denied by liberation theologians, but celebrated. One prominent hierarchical exponent of liberation theology, Archbishop Hélder Câmara, argued that theologians should "do with Karl Marx what St. Thomas, in his day, did with Aristotle."[17] All of this led to a fascination with and commitment to socialism among theologians of liberation.[18]

The Church's role, on this social/political analysis, was "prophetic denunciation."[19] The internal renewal of the Church was secondary; for theologians of liberation, human liberation had priority over *aggiornamento*, as Pope John called the Church's self-renewal.[20] The task of liberation required a reinterpretation of basic Christian symbols. Like many conciliar and postconciliar Catholic theologians, liberation theologians stressed the unity of creation and redemption. But liberation theologians concluded that the lordship of Christ "demands the socialization of the means of production in the service of all."[21]

Liberation theologians also emphasized the "collective nature" of sin.[22] On the analogy of "first violence," theologies of liberation gave priority, not to the sin that infects the human heart and renders all societies subject to judgment, but to those sinful social structures that led to dependency and impoverishment. As Gutiérrez put it, "Sin takes place in oppressive structures, in the exploitation of man by man, in the domination and slavery of peoples, races, and social classes. Sin emerges, then, as the fundamental alienation, as the root of a situation of injustice and exploitation."[23] Freedom from sin was not a matter of sacramental grace, but had to be "mediated through political and historical liberation."[24] Personal conversion was of less theological moment than the socialization of the means of production in the service of all.

Liberation Christologies flowed naturally from these basic understandings, and included efforts to use "the revolutionary elements in the life of Jesus, his conflicts with established authorities, his poverty, preaching of brotherhood, of conversion, his self-image as Isaian liberator (Luke 4)."[25] Jesus, according to Gutiérrez, was not so much preaching personal conversion as a "permanent principle of revolution." Personal conversion, even the messianic salvation of the Jewish people, were "but an aspect of a universal and permanent revolution."[26] In liberation theology, then, Jesus was read, not as the herald and embodiment of the Kingdom of God that transcends all time and every place, but as the liberator who calls us first to social conversion through revolutionary praxis,[27] and who calls his Church to "prophetic denunciation."

The Church was not a gathering point where mutual understanding was sought, ideas exchanged, and communion celebrated among those on different sides of politics. "The church must decide, it must make an unambiguously partisan commitment."[28] Or as Gutiérrez put it with admirable frankness, "The Gospel announces the love of God for all people and calls us to love as he loves. But to accept class struggle means to decide for some people and against others."[29] The Peruvian theologian described this as a challenge to a more mature charity, but such sentiments seemed secondary to his primary concerns.[30] Liberation ecclesiology was self-consciously at the service of a partisan Church. The Church must be a partisan in the creation of a this-worldly utopia, the Kingdom of justice that would result in peace.[31]

The issue of the "partisan Church" was one point at which theologies of liberation were seriously criticized by those who shared the liberation theologians' commitment to a Church engaged in history and in the quest for freedom with justice. Peter Berger, for example, worried that liberation theology resulted in a Church that was not so much freed from the domination of a particular culture as bound to a different cultural (and social/political) absolute.[32] The history of altar-and-throne arrangements in Latin Catholicism may have been at play here, according to some critics, with the throne of monarchy or oligarchy being exchanged for the bureaucratic "throne" that would necessarily accompany the socialization of the means of production in the service of all.[33]

Richard Neuhaus was concerned, in a critique of Gutiérrez's concept of the "social appropriation of freedom," that the Peruvian had paid singularly little attention to the question of the "freedom that nourishes the dialectic between individual and community, between the existent and the possible, between present and future."[34] Neuhaus also criticized Gutiérrez's failure to provide a calculus for weighing the means and ends of a revolutionary struggle. "As it is, almost any struggle that fashions itself a liberation struggle is reinforced by Gutiérrez with all the moral warrants appropriate to Christ's work in the world."[35]

The concerns of these two prominent and ecumenical Lutherans were echoed by Pope John Paul II when he attended the 1979 Puebla, Mexico, meeting of CELAM, called to reflect on the work that had flowed from Medellín. The pope was critical of those "rereadings" of the Gospel that "cause confusion by diverging from central criteria of the faith of the Church." He opposed teachings in which "the Kingdom of God is emptied of its full content and is understood in a rather secularist sense," as if the Kingdom might be reached "by mere changing of structures and social and political involvement." And he challenged those who "claim to show Jesus as politically committed, as one who fought against Roman oppression and the authorities, and also as one involved in the class struggle. This idea of Christ as a political figure, a revolutionary, as the subversive man from Nazareth, does not tally with the Church's catechesis."[36]

Contrary to the media commentary at the time, which claimed that the pope wanted to disengage the Church from social and political questions, John Paul clearly wished his Church to stand fast on the side of the oppressed. But he also argued that the Church did not need to look to "ideological systems" in order to collaborate in the work of human liberation and to resist the forces of domination, slavery, and violence. The Church had a profound commitment to the poor; it was because of this commitment that the Church rejected a partisan ideology or ecclesiology.[37]

The pope then cut to the central definitional issue: "Liberation . . . in the framework of the Church's proper mission is not reduced to the simple and narrow economic, political, social, or cultural dimension, and is not sacrificed to the demands of any strategy, practice, or short-term solution." Christian liberation had its own integrity and originality, which gave the Church a fundamental meaning it dared not lose. Without it, the Church would be vulnerable to the manipulations of "ideological systems and political parties."[38]

Michael Novak argued that liberation theology, in its fondness for Marxist analysis, was ignoring important empirical evidence, despite the claims of Gutiérrez and others that Marxism was the "science" to be wedded to Christian faith. Novak noted Fr. Juan Luis Segundo's defense of Marxist solutions before an audience of American Jesuits, in which Segundo had argued that, because there was no perfect solution, one had to choose between two oppressions. "And the history of Marxism, even oppressive, offers right now more hope than the history of existing capitalism. . . . Marx did not create the class struggle, international capitalism did."[39] But it was precisely on his historical judgment that Novak wished to challenge Segundo. Marxist praxis was no longer a theoretical matter; the world had had considerable experience of it. But one would not know this from reading theologies of liberation. Liberation theologians freely criticized capitalist practice, but were strangely silent on the performance of Marxist regimes:

> Since almost three-quarters of the world's nations are, officially, Marxist in design, and since most have had upwards of thirty years to prove themselves, it should not have been beyond the capacity of theologians to work out an assessment, even a theological assessment, of their actual daily "praxis," and judge these in the light of the gospels. But this the liberation theologians have conspicuously not done.[40]

Novak also marked the silence of liberation theology on Third World economic success stories such as Taiwan, Hong Kong, South Korea, and Singapore, and argued that liberation theology had failed to study the comparative inequalities between rich and poor in Marxist and capitalist countries. Nor did liberation theologians take much interest in the human rights differences between, for example, North and South Korea.[41] Novak also challenged, on empirical grounds, liberation theology's insistence that the poor were getting poorer, and suggested that liberation theology had paid too little attention to the cultural requisites for creating wealth. Resources, Novak insisted, were not simply inanimate objects; resources were things for which human creativity had found a purpose. Why so little attention to this dimension of creativity in the theologies of liberation?

In what sense were theologians of liberation "Marxist"? They were not atheists, and claimed not to be materialists or totalitarians. Liberation theologians were Marxists, according to Novak, "by faith." As Leszek Kolakowski had shown,[42] Marxism was increasingly sterile, dogmatic, and out of touch with the cutting edges of modern life. But Marxism could still inspire "fantasies of utopian fulfillment," and it could still give definition to the enemy who blocked the path to liberation.[43] Liberation theologians had confused the utopian visions of Marxism's world-historical outlook with the promises of the Kingdom of God; and this was their great tragedy. The Church ought to be a voice for the oppressed; the Latin American Church had suffered from an excessively privatized view of the Gospel. But liberation theologies were terribly susceptible to a false infinity. That susceptibility, as the history of Marxist regimes had shown, could lead only to more suffering. The Grand Inquisitor's classic temptation—the exchange of freedom for bread—had raised its head again in the Church.

In 1984 an official Vatican "Instruction on Certain Aspects of the Theology of Liberation" was issued. Its main points paralleled John Paul II's critique at Puebla. The Instruction agreed that liberation was an important Christian theme, recognized the conditions of desperate poverty in much of Latin America, and argued that the Church did indeed have a special concern for the poor.[44] The Instruction also acknowledged that there were many theologies of liberation and multiple forms of contemporary Marxism.[45]

But certain key themes in some theologies of liberation had to be rejected. Among them the Instruction cited the reduction of the Exodus to a narrowly political meaning;[46] the primary location of sin in social, economic, and political structures;[47] the "partisan" understanding of truth;[48] the class struggle model of social analysis and related understandings of "structural violence";[49] the subordination of the individual to the collectivity;[50] the loss of a transcendent concept of the distinction between good and evil, by its transformation into strictly political categories;[51] the confusion of the Gospel poor with the Marxist proletariat;[52] the identification of God with history and a resultant understanding of faith as "fidelity to history";[53] the concept of a partisan Church of the people, meaning the Church of a particular class;[54] Christologies that offered "an exclusively political interpretation" of the death of Christ, in which Jesus becomes "a kind of symbol who sums up in himself the requirements of the struggle of the oppressed";[55] and Eucharistic theologies that denied the unity of the Church by portraying the Eucharist as "a celebration of the people in their struggle," rather than as "the real sacramental presence of the reconciling sacrifice and as the gift of the body and blood of Christ."[56]

The Instruction concluded with a summary statement of the essential Christian truths that theologies of liberation tended to misinterpret or dismiss:

> The transcendence and gratuity of liberation in Jesus Christ, true God and true man; the sovereignty of grace; and the true nature of the means of salvation, especially of the church and the sacraments. One should also keep in mind the true meaning of ethics, in which the distinction between good and evil is not relativized, the real meaning of sin, the necessity for conversion and the universality of the law of fraternal love. One needs to be on guard against the politicization of existence, which, misunderstanding the entire meaning of the kingdom of God and the transcendence of the person, begins to sacralize politics and betray the religion of the people in favor of the projects of the revolution.[57]

The 1984 Vatican Instruction on liberation theology was often portrayed as a reactionary document which misunderstood the primary themes of the theology of liberation and thus attacked straw men in the name of orthodoxy.[58] But a survey of the American Catholic debate over U.S. policy in Central America suggests that, whatever theological nuances the Instruction may have missed, its sense of the principal themes that had been *learned* from the theologies of liberation—and that had formed the prism through which American Catholic opinion elites read the reality of Central America—was far from inaccurate.

The Transformation of the American Catholic Debate

During the Kennedy administration and Vatican II, there was great enthusiasm among American Catholic commentators and publicists for the classically liberal politics and economics of the Alliance for Progress. American Catholic commentators saw the Alliance as a welcome change in their country's traditional policy toward the problems of Latin America and a useful alternative to the siren songs of Fidel Castro. Castro had demonstrated that Latin America's long festering problems of poverty and tyranny could no longer be ignored. The Alliance and the Peace Corps were meant to "irretrievably" change the nature of hemispheric relations.[59]

Appreciation for the Alliance continued after Kennedy's death. In 1964, for example, Gerard Mangone argued that Americans had to admit their traditional disinterest in Latin America, and their reluctance to help facilitate the "fundamental reforms" that were needed. U.S. policy had too frequently been aligned with repressive dictators and military regimes. The Alliance had changed all that. Some might still look back fondly to the days of Calvin Coolidge, but "they do not represent the mainstream of American foreign policy."[60] Mangone's critique of the past did not lead him away from the classic, liberal, "development" agenda. Victory in the Alliance for Progress, he argued, "will depend not only upon continued investment, but upon a kind of community virtue in Latin America that will recognize the value of mass education, the justice of an equitable distribution of income, the duty of supporting through taxation the total needs of a nation, and the worth of honest, efficient, and responsible government."[61]

The Church had an important role in developing community virtue; Church support for "reformist, left-of-center, pro-Western" Christian Democratic parties would help create a genuine option to the policies of oligarchs and Marxists. Special attention should be focused on developing reformist leadership cadres worthy of their peoples' trust and capable of nurturing democracy.[62] Thus, the classic, liberal Catholic approach: economic development coupled with political reform aimed at the creation of stable, democratic, peaceful societies capable of providing freedom, bread, and nonviolent means to settle claims of injustice.

The break with this approach can be symbolically dated with some precision: the publication of Msgr. Ivan Illich's famous essay "The Seamy Side of Charity," in *America* on January 21, 1967.[63] Illich's immediate target was the call for large numbers of North American clergy and religious to undertake missionary work in Latin America; but his villain was the development/reformist economic and political agenda. North American missioners carried "a foreign Christian image, a foreign pastoral approach, and a foreign political message" into Latin America, in a project that "relied on an impulse supported by uncritical imagination and sentimental judgment."[64] But this merely reflected the deficiencies of the general U.S. approach to Latin America, by which Illich meant "the perversity of our power politics and the destructive direction of our warped efforts to impose unilaterally 'our way of life' on all."[65]

Illich warned that missionaries could become pawns in a "world ideological struggle," and claimed that it was "blasphemous to use the gospel to prop up any social or political system."[66] Which system he had in mind was illustrated by Illich's charge that the Alliance for Progress was a "deception designed to maintain the status quo," which Illich described as "U.S. socio-political aggression in Latin America."[67] American missionaries were thus placed in the "traditional role of a colonial power's lackey chaplain."[68] A year later, Illich sharpened his critique by arguing that both the missioner and the development-concerned U.S. official were "seen by 90 percent of mankind as the exploiting outsider who shores up his privilege by promoting a delusive belief in the ideals of democracy, equal opportunity, and free enterprise among people who haven't a remote possibility of profiting from these."[69]

Illich's rejection of the Christian Democratic/reformist/developmental model rapidly became common wisdom among American Catholic intellectual and opinion elites. Christian Democracy in Brazil, for example, was accused of "moralism," of being the "middle-class party par excellence," of foolishly rejecting the notion that "centralized planning and initiatives were indispensable for the continuity of the developmental process."[70] Christian Democracy was too preoccupied with the alleged communist danger in Latin America, and was out of phase with a Catholic Left which believed the United States a greater threat to human dignity and development than communism. For the Catholic Left, the development models deserving sympathy were China, Cuba, and Yugoslavia.[71]

The revolutionary option in Latin America included a deep commitment to consciousness-raising, which was being forcefully propounded to American Catholicism by 1968. "Gradualist" approaches to consciousness-raising existed, but they were deficient; they only instilled "a self-help, reformist mentality."[72] If the Church wanted to justify its presence in Latin America, argued Brazilian Jesuit Henrique de Lima Vaz, it had to approach conscientization as a "political problem," in which questions of ideology could not be avoided. "What is the difficulty in giving the masses an ideology?"[73]

The content of that ideology was sketched by the prominent liberation theologian Juan Luis Segundo. Segundo believed that the suffering masses of Latin America were becoming increasingly desperate as their champions, the guerilla revolutionaries, were "crushed by the defenders of the status quo."[74] The local crushing may have been done by Latin American governments, but the real villain lay to the north: Latin America constituted "the principal external proletariat of the American Empire."[75]

Gary MacEoin raised a crucial and related point in 1977, when he argued that "communism thrives in Latin America, not through externally inspired agitation, but through the reality of masses of hungry, unemployed, and frustrated people."[76] Nor was that poverty an accident of history; it was the result of American policies that had created the modern order of injustice in Latin America.[77]

Understandings of freedom changed drastically during the post-conciliar American Catholic debate on Latin America. One essayist argued that the more fundamental meaning of freedom was "the right to act collectively as an organized and politicized people"; this is what really constituted self-determination. Jeffrey Klaiber was not impressed by claims that freedom had to do with free speech, religious liberty, or property rights; freedom "can never be considered as a mere individual 'right,' but rather as a collective goal to be attained for all."[78] These truths were being grasped by American blacks and Latin American *campesinos* who were both "exploited colonial subjects within societies that proclaim themselves to be free and democratic."[79]

Dependency theory was the dominant mode of economic analysis in the debate. Gary MacEoin reported in 1971 that developmental theories had been rejected by the "best thinkers" in Latin America. The poor were poor because of someone else's development: "poor because the rich made them poor, and poor because the rich keep them poor by regulating world trade and finances to give themselves an inordinate share of the benefit."[80] Dependency theory had specific ideological origins; it rejected partnership models, and preferred "the class war concept which Marx developed to explain exploitation of the poor by the rich within capitalist societies. Underdevelopment under this light emerges not as an accident of history but as the result of a deliberate process. . . . The poor are poor because they are kept poor."[81] MacEoin's conclusion was not surprising. The revolution had to begin in North America: "The problem is not down there or over there. It is not even next door. It is right here at home."[82]

The first of many heroes of this point of view was the Colombian priest Camilo Torres, killed in 1966 while under arms with Marxist guerilla forces in his country. To Phillip Berryman, Torres's thought represented a theological watershed, but in a new form: fragments of thought connected by the drama of a life. Berryman admitted that Torres had left no theological research, but the Colombian had grasped the "central theological concerns" of the modern Latin American Church. Revolution was a "Christian imperative." Empowering the masses through revolutionary praxis had to take precedence over the evangelical reform of the Church. If theology is to be measured by insightfulness, rather than by scholarly product, then Torres produced theology.[83] Torres's life represented the seminal insight that "a liberating theology can be discovered as a participant in the struggle."[84] His death was a martyr's gauntlet thrown down before North American Christianity.

These themes were occasionally brought to bear in the American Catholic debate over the Panama Canal treaties. The charge that the Canal Zone was a U.S. colony was frequently made; the canal existed so that the United States could expropriate the wealth of the Panamanian people.[85] Yet the general tone of the discussion was reasonably moderate. *America* editorially celebrated the treaties as an expression of "our better concept of ourselves." The treaties expressed the American belief that even those nations important to our interests should retain their own integrity.[86] Paul Fitzgerald saw the treaties as a rare event in human history: a powerful nation voluntarily recognizing the claims of a weaker neighbor, and thereby creating a partnership rather than a relationship of dependency. Ratifying the treaties would "give the United States the moral leadership in the world that it has often sought, but which it has not always earned."[87]

Still, the Catholic debate over the Panama Canal treaties did not give serious attention to the question of the canal as an international resource that might best be protected by international guarantees (or, more radically, by internationalizing its operating authority, with sovereignty over the Canal Zone reverting to Panama).[88] The key issues were defined as the prerogatives of Panamanian sovereignty and the need for a symbolic demonstration that the U.S. claim to be an anti-imperial power was true.

Moderation did not mark the American Catholic debate over Nicaragua in the days immediately before, and continuing after, the overthrow of the Somoza dictatorship. In December 1978 *America* editorially argued that the United States must side with the Nicaraguan people, who had evidently decided that life was not so precious as to be purchased at the cost of slavery.[89] *America* understood "the Nicaraguan people" to mean the Sandinista National Liberation Front [FSLN], who represented "the only opponents of . . . [Somoza] who have wide popular support and who are armed."[90] Nonviolent solutions to the conflict in Nicaragua seemed unlikely, and perhaps even undesirable: mediation was irrelevant to the revolutionary situation, James Brockman claimed in January 1979.[91]

Seven months later, Anastasio Somoza had been ousted by a broad opposition whose principal military elements were the Sandinistas. Reporting on the aftermath of the insurrection, Chris Gjording, S.J., admitted that the board front employed to overthrow Somoza could turn out to be fragile,[92] but the main thrust of his essay was a paean to the Sandinista regime. Sandinista defense committees in the barrios were not instruments for the imposition of one-party rule, but means to meet reconstruction needs. The committees (which admittedly had spies and which controlled food-ration cards) were the basis on which to build labor and peasant unions, and a strong political party.[93] The FSLN was flexible, and had demonstrated much common sense. Marxist analysis was used "with considerable precision" in FSLN discussions of national and international politics, but this should not cause undue alarm: a "humanistic ethic" was the FSLN's principal characteristic.[94] Moreover, the Sandinistas were not anti-religious; "a Mass celebrated in the open-air plaza of a marginal barrio of Managua, shortly after the end of the war, illustrates the ease with which certain Sandinist symbols lend themselves to an explicitly Christian reading."[95] *Sandinismo*, Catholicism, and the popular faith of the barrio made for an "easy mix."[96]

This interpretation of the Sandinista regime was driven home by prominent clerics in the government of Nicaragua, most explicitly by Minister of Culture Fr. Ernesto Cardenal. Five months after Somoza's ouster, Tennent Wright, S.J., reported on his conversations with Cardenal: "Father Cardenal's mind was clear: There need be no separation between Christianity and communism. He spoke of himself as a Christian communist in the way the earliest Christians spoke of themselves as such."[97] Cardenal later argued that "capitalism isn't compatible with the ideals of Jesus," and that "Marxism is a scientific method for studying and changing society." Christianity set the goal of social change: "a perfect humanity."[98]

The celebration of *Sandinismo* continued unabated in American Catholic opinion journals in the years immediately following the revolution. Could the FSLN government achieve a stable democracy? Father Robert Drinan, S.J., was "optimistic and hopeful" in early 1980.[99] Drinan suggested that the Nicaraguan revolution was a unique combination of genuine populist revolt and Catholic social activism. Some in Managua, Drinan reported, saw the Sandinista government as the "first fruit" of the 1968 CELAM meeting in Medellín.[100] Drinan's worries about the Nicaraguan future were external: reactionary elements in El Salvador and Guatemala might mount an invasion. The Sandinistas' recent abstention from the U.N. vote condemning the Soviet invasion of Afghanistan was troublesome because it would hurt the FSLN regime with the U.S. Congress.[101] Still, Drinan was encouraged by his sense that the Sandinistas had avoided the violence and polarization that so often marred Latin American politics.[102] Nicaragua under the Sandinistas might well represent "the future of Christian democracy and social justice in Latin America."[103]

Another Jesuit, Arthur McGovern, reported a year later that any transformation of Nicaragua into another Cuba would be largely the fault of the United States.[104] The FSLN did not want to repeat the Cuban mistake, and exchange dependency on America with dependency on the Soviet Union.[105] Moreover, the Sandinistas had worked to create democratic institutions open to all sectors of Nicaraguan society. "They may even emerge with a far more democratic system than we have in the United States where lack of campaign funds and lack of prestige generally prevent factory workers, truckers, secretaries, and other mainstream Americans from being elected."[106] The Sandinista army was growing exponentially; but what struck Fr. McGovern was the "absence of traditional Marxist rhetoric" among the FSLN officials with whom he had spoken.[107] Nicaragua also had "a strong record on human rights—especially when compared to any other country in Latin America."[108] There were strains within Nicaraguan society; but what McGovern and his companions had seen was "fundamentally just and good."[109]

These commentaries on the Nicaraguan revolution were paralleled by a fire storm of protest against U.S. policy in neighboring El Salvador. To Jesuit Philip Land, it was Archbishop Oscar Romero who spoke for the country's masses, not the Christian Democrats in the Salvadoran government who had overthrown the traditional oligarchy.[110] Land had no truck with the argument that the Christian Democrats represented a middle ground between violent extremists on the oligarchic right and the guerilla left; the Christian Democrats were a "nonsolution." They had nothing to offer the people.[111] Land joined Romero in condemning Marxist guerilla violence. But Americans had to understand that this was violence provoked by "desperation" over "the violence of the system."[112] There were parallels between El Salvador and Nicaragua. Land's contacts in Nicaragua assured him that the government would turn out to be on the democratic left; if El Salvador had to "go Nicaragua's route of revolution," it, too, could establish a "people's state, at once revolutionary and democratic."[113]

Archbishop Romero was assassinated at the altar a month after Land's article appeared. Romero's death confirmed to *Commonweal*'s editors that there was no middle in El Salvador. The Christian Democratic option represented the Grand

Inquisitor's temptation: bread, but no freedom; land reform, but elite rule. This was not wrong only on its own terms; it had been cobbled together by the threat of revolution, and only that threat gave the government sanction over the oligarchy. What would happen when the revolutionary option was bloodily suppressed?[114] Eight months later, when four American churchwomen had been murdered in El Salvador, *America* charged that José Napoleón Duarte's appointment as president of El Salvador did not represent a real change in the structure of power in El Salvador.[115] A month later, in early 1981, after the inauguration of Ronald Reagan, *America*'s view hardened even further: U.S. support for the Duarte government was unjustifiable. The Duarte junta was not a center: depredations by government troops continued. U.S. military assistance to those who were murdering civilians with impunity violated both national values and national interests.[116]

Commonweal agreed that the Duarte government was neither centrist nor reformist, and asked, in what would become a familiar analogy, whether "the U.S., like the Soviet Union in Poland, [is] really determined to do permanent sentry duty for oppression in Central America?"[117] What of the Salvadoran left, which was doing its own share of the bloodletting? *Commonweal* knew that the guerillas were Marxist-Leninists; but those principles were balanced by "about equal amounts of Christian liberation theology."[118] In October 1981 *Commonweal* charged that Duarte was "more than ever the generals' window dressing." The failure of the Reagan administration to understand this was attributable to "ideological fanaticism" against U.S. liberals and their alternative approach to U.S. policy in Central America.[119]

Commonweal did acknowledge some of the cruel complexities of El Salvador in early 1982. An editorial entitled "America's Poland" admitted that the parallels were not complete, although some dissimilarities were in the Soviets' favor: repression in Poland had been "less arbitrary and less brutal (although probably more thorough) than in El Salvador." Still, there was not much to be said for the argument that the Salvadoran guerillas' Marxism was homegrown; so, after all, was the "fascism" of the death squads. *Commonweal* was particularly critical of propaganda films such as "El Salvador: Another Vietnam?," which portrayed the guerillas as the only representatives of the people of that tortured country; "the guerillas . . . are not Solidarity." Yet the primary focus of *Commonweal*'s criticism remained the Reagan administration:

> [Its] sanctimonious protests against "outside meddling" are shamefully similar to Moscow's displays of brutal pressure *cum* protests of innocence. When the administration certifies the Salvadoran government, with its thousands upon thousands of civilian victims, as making progress on human rights, and simultaneously characterizes the Nicaraguan regime, which for all its ominous aspects remains remarkably free of bloodshed, as "totalitarian," then Doublespeak reigns.[120]

El Salvador's Constituent Assembly election in March 1982 drew decidedly mixed reviews. John Garvey's argument, that the causes of reform and justice had been set back by the election, was not atypical. Garvey thought the massive turnout was "interesting, but hardly encouraging," and warned that "the electoral process should be seen in the light of Plato's *Gorgias*, which describes the limits of democracy."[121] *Commonweal*'s editors were less grudging; they believed that the election demonstrated massive popular support for a political resolution of the Salvadoran conflict. The election had also shown the weakness of the guerillas. *Commonweal* still worried about the Reagan administration, though; Washington did not really want negotiations, and would step back from land reform if it meant clashing with hardliners like Roberto D'Aubuisson.[122]

A year later, a *Commonweal* essay, "The El Salvador labyrinth," summed up the main currents of Catholic elite debate on the issue and the bitterness that the issue had engendered. "Latin America already *is* Eastern Europe, *our* Eastern Europe," Jim Chapin and Jack Clark argued. The United States was every bit as ruthless as the U.S.S.R. in imposing its will within its sphere of influence; countries could indeed leave the Soviet bloc (Albania, Yugoslavia): the fate of Salvador Allende revealed what happened when nations tried to get distance from America. There was a distinction between authoritarian and totalitarian governments. "Poland is an authoritarian country, El Salvador totalitarian."

The Reagan administration policy in Central America was partially attributable to "blind and nearly paranoid anticommunism," but was also influenced by a quixotic search for a "third force" in world affairs. The search for a third force between right- and left-wing extremists explained the administration's "endless search for a political center on the far right of El Salvador's political spectrum" and the "persistent myth-making" that had surrounded the 1982 elections in El Salvador. The election had only settled things within the ruling coalition (which had, in fact, come in second in the balloting).

Chapin and Clark were eager to burnish the reputation of those on the left who had allied themselves with the guerillas: "The moderate left has formed a coalition with the far left because it has nowhere else to go." What, then, did the authors propose? Nothing else but the "Finlandization" of Central America: an agreement in which Central American countries would determine their own futures but not be bound to join an anti-U.S. alliance.[123]

Thus the state of the American Catholic debate on the dilemma of peace and freedom in Central America by 1983. Some, such as the editors of *Commonweal*, had come to recognize the Leninist character of the Sandinista regime in Nicaragua. But these same voices also argued that the American "secret war" against the FSLN was feeding "the militarists and totalitarians" in the Nicaraguan regime.[124] The prospects of democratization in El Salvador were regularly deprecated. Anti-anticommunism was an important influence in the debate. *America* argued that the Reagan administration often seemed more interested in the form than the substance of democracy. Communist governments did not necessarily mean Soviet military bases in the hemisphere. Economic development had to be the first priority of any revolutionary government, and influence could easily outstrip Soviet influence on this front. Would Soviet influence in Nicaragua, or Cuba, have grown to its present extent had America not economically isolated these countries? "The answer to such a question may be problematic, but surely the possibility must be tested before the answer can be given."[125]

The notion that Nicaraguan Marxists had become Marxists because of U.S. intransigence dominated many influential minds and circles. The democratic prospect was not viewed very optimistically, nor were linkages drawn between democratization and peace. *America*'s editorial response to the Kissinger Commission report did not mention the commission's clear endorsement of the twin processes of democratization and economic development in the region, and focused instead on the commission's recommendation for increased military assistance to the government of El Salvador. Readers of the editorial would not have learned there that the ratio of economic to military aid recommended by the commission was 16:1.[126] The idea that the Reagan administration preferred a military solution in Central America reached the status of an unquestionable first principle.[127]

It was against this backdrop of impassioned, partisan debate that the American bishops addressed the question of U.S. policy in Central America from the mid-1970s on. Perhaps not surprisingly, their commentary reflected, in more tempered language, the themes just explored.

THE AMERICAN BISHOPS AND U.S. POLICY IN CENTRAL AMERICA

The bishops of the United States had been involved in the life of the Latin American Church long before the Central American crises of the 1970s and 1980s. Missionary activity had been the primary focus of this involvement. The coming of the Second World War, for example, had reoriented the work of the Catholic Foreign Mission Society of America (popularly known as "Maryknoll") from Asia to Latin America. In the late 1950s and early 1960s, the concept of a particular U.S. Catholic responsibility to mission activity in Latin America came to full flower. Agencies such as the Missionary Society of St. James the Apostle, the Papal Volunteers for Latin America (PAVLA), and the Association for International Development created an extensive U.S. Catholic network of evangelization and pastoral activism in Central and South America. That activism was built on several assumptions, prominent among them that the Latin American Church suffered from an ecclesiastical underdevelopment that could be ameliorated by large-scale infusions of volunteer missionary clergy and laity from North America. North American Catholicism had many things to offer Latin American Catholicism: volunteer clergy and religious missionaries, lay evangelists, and lay technical experts could help Latin American Catholics in their immense task of ministry to the world's largest Catholic continent.[128]

The belief that North American Catholicism had something to offer Latin American Catholicism was not only challenged, but inverted, in the postconciliar years. The Latin American Church, particularly the post-Medellín Church, would teach North Americans about authentic Catholicism in the Americas. Liberation theology, emphasizing the "structural violence" visited on the Latin American poor by North American liberal capitalism, played an obvious and important role in this reversal. But so did important themes in the abandonment of the classic Catholic heritage of *tranquillitas ordinis*. The moral questionableness of American society was the key teaching shaping the reversal.

A 1972 meeting of Jesuits at LeMoyne College in Syracuse illustrates this point. The LeMoyne meeting was called to assess the U.S. impact on world politics and economics. Its participants were eager to abandon classic liberal notions of development, because the ambiguous character of all U.S. aid programs, even the Peace Corps, had been made clear over the past decade. The LeMoyne participants agreed that "in some serious ways the United States was a severely underdeveloped country."

This conviction led to a "reverse mission": the best thing North Americans could do for the Church in Latin America was to remain at home and work to change the values and policy-direction of the United States, conforming them in the future to the ethos of the post-Medellín Church. PAVLA, the Alliance for Progress, and the St. James Society had only increased Latin American dependence, and had resulted in an "imperialism of mass culture" that taught the "false values" of North American society. Several participants at the LeMoyne conference argued that "the communitarian values of such socialist countries as Cuba and China were in many ways closer to the gospel ethic than the individualistic, consumer values of North America." As Joseph O'Hare put it, a significant number of the conferees believed that "there did not seem too much left to what John Courtney Murray liked to call 'the American proposition.'" Dependency theory also played a major role at LeMoyne, with particular focus on North American consumerism as the cause of Latin American poverty.[129]

The effects of the "reverse mission" concept on the leadership of U.S. Catholicism were profound. As reported by Msgr. Joseph Gremillion (former secretary of the Vatican Justice and Peace Commission) in 1976:

> The North American bishops and religious orders were exposed to the third world and awakened to its realities. It *gave them an opportunity to realize what poverty was like*, to see the way so many people live in misery, to understand the frustrations of political instability. How many bishops have visited Latin America? I'd guess over a hundred. *It's opened their eyes to the reality of human misery.* It's helped them to understand phenomena like the war on poverty, black power, the Chavez movement, the Campaign for Human Development. They've seen that they can take stands on these things. They've said to themselves, "Look how the Latin American bishops stick *their* necks out!"
>
> And another thing: until Vatican II, our bishops attempted no role in relation to the foreign policy of the U.S. Government. They supported the country without question; it was part of their assimilation into the U.S. mainstream. . . . But then came the Latin American experience. . . . We were awakened to the relationships between U.S. multinational corporations, U.S. embassies, the U.S. Information Service and military power in Latin America. We became aware of the fact that we had to exercise some influence on U.S. policy in these matters.[130]

Gremillion captured the essence of the experience and themes of the reverse mission. American bishops and religious leaders, sheltered from the harsh realities of poverty and human misery, had suddenly and forcefully had their noses rubbed in the degradation that marked much of daily life in Latin America. What had been learned from this experience was as important, in fact *more* important, than the experience itself: the poor were poor because of the affluence of the rich. American consumerism, anticommunism, and multinational corporations caused the misery of the Latin American underclass. The bishops shared a measure of responsibility for this because of their traditional acquiescence to U.S. foreign policy. Therefore, the U.S. bishops, to right a historic wrong and to challenge the structure of American imperialism in Latin America, had to take an adversary stance vis-à-vis U.S. government policy in Latin America. None of these assertions stands up under empirical or historical scrutiny: the causes of Latin American poverty, for example, are immensely complex, and have no little to do with values brought to Latin America by Spanish and Portuguese missionaries long before the arrival of American corporations. But Msgr. Gremillion's portrait remains an accurate (and devastating) summary of what many American bishops and leaders of religious orders had *learned* about Latin America in the decade after

Vatican II. There did not, indeed, "seem too much left to what John Courtney Murray liked to call 'the American proposition.'"

What had been learned was first applied in the public policy arena as debate over U.S. policy in Central America came to the forefront of public and ecclesiastical attention in the late 1970s.

USCC commentary on Central America between 1978 and 1985 was characterized by six dominant motifs. The USCC stressed the unhappy history of previous U.S. intervention in the region, and feared that further intervention would lead only to war. The struggles in Central America were essentially indigenous, and, although reflecting East/West tensions, were not to be understood primarily as expressions of U.S./Soviet geopolitical competition. Military aid and intervention from all outside powers was illegitimate (the USCC thus teaching a kind of parallelism between U.S. and Soviet intentions in the region). Poverty and human rights abuses took absolute priority over security concerns in the region. A democratic center was an unlikely vehicle for the evolution of peace, security, freedom, and justice. Finally, the bishops claimed that, acting primarily as pastors, they were reflecting, in the U.S. domestic debate, the concerns of the Central American bishops.

These motifs were displayed in the bishops' benchmark "Statement of the United States Catholic Conference on Central America," of November 19, 1981.[131] The USCC statement claimed that the Central American Church had taken its lead from the teachings of Vatican II, Popes Paul VI and John Paul II, and the CELAM conferences at Medellín and Puebla. From these teachings, the church had evolved a pastoral witness that stressed its own need for conversion, and identified with the poor in their struggle for justice. This had produced "a new and challenging style of ministry."[132] That ministry had been costly; the USCC mentioned the murders of Archbishop Romero, four U.S. churchwomen, and Fr. Stanley Rother, recently killed in Guatemala.[133]

According to the USCC, the causes of strife in Central America were internal:

> The Latin American church has repeatedly stated in the last decade that external subversion is not the primary threat or principal cause of conflict in these countries. The dominant challenge is the internal conditions of poverty and the denial of basic human rights which characterize many of these societies. These conditions, if unattended, become an invitation for intervention.[134]

Any analysis of Central America primarily cast in terms of US/Soviet security issues was "profoundly mistaken."[135] Elections might provide a solution in El Salvador, but only after "appropriate preconditions" were met.[136]

Two years later, the same themes dominated USCC testimony before the National Bipartisan Commission on U.S. Policy in Central America (the Kissinger Commission). The USCC charged that the United States still sought a military solution to the problems of Central America.[137] This had led to the "imminent possibility" of a regional war.[138] The basic misconception remained; until the U.S. government understood that questions of social justice were at the root of the Central American conflict, there could be no wise policy.[139] U.S. policy abused human rights criteria by using them for "tactical advantage or propaganda points"; and there was concern whether in fact human rights criteria were being selectively applied.[140] On specific cases, the USCC charged that the American role in El Salvador "continues primarily in a military direction," and that the administration gave "the appearance of encouraging war in Nicaragua." U.S. policy "contributed" to the internal problems of Nicaragua, and provided "precisely the pretext" for Sandinista repression.[141]

The USCC acknowledged, before the Kissinger Commission, that there were geopolitical dimensions to the crises of Central America, but argued, in a curious parallelism, that "direct Soviet intervention in Central America is no more welcome, *legitimate*, or tolerable than direct U.S. intervention in Eastern Europe."[142] The complex situation required a regional approach such as the Contadora process, because the United States could not play a mediator's role in the region.[143] Thus the first requisite for U.S. policy was to stop "the drift toward a regional war" and seek political solutions to the region's conflicts, for such solutions "must precede large-scale and lasting economic programs."[144] America should "strive to be seen as a mature, democratic, stabilizing force in the region, not a destabilizing bully." This would involve, "the acceptance, and more than that, the welcoming of dramatic change to achieve social justice and human rights in the region."[145] The contents of "social justice" and "human rights" were not defined.

These themes continued to shape USCC Congressional testimony on Central America in 1984 and 1985. In 1984 Fr. J. Bryan Hehir argued that actual U.S. policy (as distinguished from declaratory policy) seemed "fixated on military pressures, coercive moves, and the role of threat and intimidation." Such a policy could not make for peace, and heightened the danger of a regional war.[146] U.S. policy had to be redirected toward a political-diplomatic role in Central America. The Kissinger Commission had failed to set this direction. Congress ought to act "to prevent further militarization of U.S. policy."[147] Hehir concluded that "the last few years have witnessed a new kind of bold assertiveness in the conduct of U.S. foreign policy in this hemisphere. Some would describe it as aggressive, truculent, even belligerent."[148] Ecclesiastical indirection notwithstanding, among those "some" were the framers of policy at the United States Catholic Conference.

The impact of these themes on the USCC's analysis of and prescription for particular situations was illustrated by USCC testimony on El Salvador and Nicaragua.

USCC commentary on El Salvador between 1980 and 1984 portrayed the Salvadoran civil war as a struggle between the traditional oligarchy and the guerillas of the Faribundo Marti National Liberation Front (FMLN), with the people of El Salvador caught in the middle. The idea that the democratic center in El Salvador ought to be developed and nurtured was not prominent in the USCC view of the conflict.

In March 1980, for example, USCC/NCCB general secretary Bishop Thomas Kelly, O.P., urged Congress to cut off military assistance to the Salvadoran government, which, two years before, had ousted the oligarchy's selected president. Kelly claimed that the U.S. bishops were basing their position on information coming from the Church in El Salvador. There was no democratic center caught between death squads and guerillas; the "military arm of the government in El Salvador is itself an instrument of terror and repression, quite unable to win the political support needed to govern in peace."[149] Archbishop Romero of San Salvador was murdered the day Kelly wrote his letter. NCCB president Archbishop John Quinn subsequently described the late archbishop's request that President Carter terminate U.S. military assistance as "prophetic."[150] Nine months later, after the brutal murders of four U.S.

churchwomen in El Salvador, NCCB president Archbishop John Roach endorsed suspension of all U.S. assistance, economic and military, until the murders had been investigated.[151]

The USCC held fast against military assistance for three more years, until March 1983 when, evidently under pressure from the Salvadoran bishops' conference (and certainly without the opposition of Romero's successor, Archbishop Arturo Rivera y Damas), the U.S. bishops grudgingly endorsed military assistance "conditioned on stringent requirements linking it to a pursuit of dialogue and cease-fire."[152] Yet the bishops' shifting stance on military assistance is perhaps the least interesting aspect of their commentary on El Salvador. For their initial rejection of military aid derived from prior political and ideological judgments, and the 1983 shift marked no real change in perceptions at the USCC, only a changed (and unignorable) message from the Salvadoran hierarchy.

Fears of American military intervention in El Salvador pervaded USCC commentary from 1980 on. From a relatively mild statement in January 1981 ("we fear that the kind of military aid now proposed will increase the possibility of further direct U.S. involvement in the conflict in El Salvador in the months ahead"),[153] the bishops argued in March of that year that the dispatch of more U.S. military advisors to El Salvador was "a particularly ominous development." Archbishop James Hickey of Washington, D.C., testifying for the USCC, described the advisor policy as "risky to the point of being reckless." It confirmed the worst fears of Latin Americans that the Reagan administration was slowly preparing to invade El Salvador. The U.S. government still misunderstood the nature of the conflict, which was "not principally a matter of guns but of justice."[154] Archbishop Hickey testified that he came before a subcommittee of the House Foreign Affairs Committee "as a pastor." But the bulk of his statement was a political-strategic analysis of the Salvadoran civil war and an analysis of the policy assumptions of the U.S. government.[155]

Testifying in Congress a year later, Fr. J. Bryan Hehir laid out four principles the USCC thought should guide U.S. policy toward El Salvador. The "principle of nonintervention" was first. Hehir acknowledged international dimensions to the conflict, and called Soviet/Cuban intervention "illegitimate." But the USCC did not believe that the "driving force" of the Salvadoran civil war was in Moscow, Havana, or Managua. "We fear the U.S. threats to go 'to the source' may mistake the source of assistance for the roots of the war."[156] Fr. Hehir did not suggest how the arms flow from Cuba and Nicaragua might actually be stopped, or how America might act to make clear its convictions about the illegitimacy of Soviet or Cuban intervention in the region. The previous year, Hehir had testified that stopping the arms flow to the Salvadoran guerilla forces "would best be accomplished by some regionally or internationally agreed upon strategy."[157]

The second principle was "the primacy of the internal situation," at the heart of which, in El Salvador, was the question of human rights. There had been a "slight" improvement in the human rights situation in El Salvador, but Hehir was eager to dissociate himself from the view that U.S. military assistance had contributed to that improvement.[158]

The third principle was "the necessity of a political solution." Hehir cited the November 1981 USCC statement that "appropriate preconditions" had to be met if elections were to lead to peace. These conditions presumably included a place for the

FMLN in the pre-electoral government of the country. The USCC welcomed the elections as a "sign of hope," but Hehir also lamented "the absence of . . . political dialogue among the contending parties." The FMLN refusal to participate in the 1982 Constituent Assembly elections went unremarked. A political solution, according to the USCC, had less to do with the emergence of a democratic electoral process in El Salvador than with meeting the preconditions set by the FMLN.[159]

The fourth principle, "to oppose military assistance from all sources to any party in El Salvador," was crafted to appear directed to both U.S. and Soviet-bloc intervention, but the remaining testimony on this point attacked U.S. military assistance programs alone. "Political measures, preferably of a multilateral nature" would have to stem the flow of communist arms into El Salvador; how, was not detailed. Once again, Fr. Hehir argued that U.S. military assistance, "especially when liberally granted, reduces our leverage for human rights reforms."[160]

Fr. Hehir's December 1982 Congressional testimony on human rights and U.S. foreign policy made it even clearer that the USCC saw the contestants in El Salvador as moral equals. It was now eight months after the Salvadoran Constituent Assembly elections, in which the guerilla opposition had refused to participate (and had in fact tried to disrupt). But Hehir argued that "the only solution to the human rights problem in El Salvador is to end the civil war and reincorporate into the political life of the country *those whose commitment to reform and social justice has led them into armed opposition.*"[161]

USCC testimony before the Kissinger Commission continued, if obliquely, to deprecate the electoral route to political community in El Salvador. Archbishop Hickey testified that "the tactics of the leftist opposition become more and more destructive as the war drags on," as if others were to blame for FMLN decisions to attack the economic infrastructure of El Salvador. But the USCC's primary concern remained to drive home the messages that "the U.S. role in El Salvador continues primarily in a military direction," and that "the political option, a negotiated settlement, is the humane and wise way to end this brutal conflict."[162] A year later, after the commission had recommended economic assistance to Central America in a 16:1 ratio over military assistance, the USCC still argued that Congress should prevent "the further militarization of U.S. policy."[163] The same theme shaped the USCC response to the beginning of negotiations between the FMLN and its political affiliate, and the government of President José Napoleón Duarte; although the USCC "strongly supports" the dialogue begun between Duarte and the opposition, it also continued to "oppose the militarization of U.S. policy toward El Salvador."[164]

If the USCC approach to the travail of El Salvador was notable for its original mis-statement of the nature of the conflict (i.e., as between right-wing death squads and left-wing guerillas "whose commitment to reform and social justice [had] led them into armed opposition"), and a concomitant disinclination to perceive, let alone vigorously support, the nascent democratic center in El Salvador represented by Duarte, the bishops' commentary on postrevolution events in Nicaragua gave every possible benefit of the doubt to the Sandinista regime.

In its November 1981 "Statement on Central America," the USCC claimed to share "the concerns expressed recently by our brother bishops in Nicaragua about increasing restrictions on human rights." But, while arguing that "the rights of free association, speech, press, and freedom of education [must] be protected," the bishops accepted the Sandinista claim that the social and economic needs of the people were being met.[165]

Sandinista repression of the Church increased in 1982, but a September 1982 statement by NCCB president Archbishop John Roach assured both Americans and Nicaraguans that "we [i.e., the U.S. bishops] have some awareness of the complexity of these unhappy events." Roach forthrightly condemned "the attempted defamation and acts of physical abuse directed at prominent clerics, the inappropriate exercise of state control over the communications media, including those of the church, the apparent threats to the church's role in education, and, most ominous of all, the increasing tendency of public demonstrations to result in bloody conflict."

But the archbishop prefaced his protest by asserting that "the sources of these attacks [i.e., on the church and on prominent church officials] or the motivation of those perpetrating them is not always clear." He further stated that the U.S. bishops were "not unmindful of the serious problems created both by external aggression and internal opposition" in bringing Nicaragua to "a most difficult moment."[166] Roach condemned specific acts of repression; but the tendency to try to find ameliorating circumstances that would explain Sandinista attacks on the Nicaraguan Church remained pronounced.

Rodrigo Reyes, secretary of the ruling junta in Nicaragua, responded to Archbishop Roach, acknowledging the "mistakes" that "have in some cases" been committed. But Reyes's primary concern was "the way in which such incidents have been taken advantage of in a distorted manner in order to defame and calumniate the Popular Sandinist Revolution as part of the destabilization campaign by the present U.S. administration."[167]

Archbishop Roach quickly thanked Reyes for the "serious manner" in which he had responded to Roach's September statement; repeated his argument that "the source and motivation for the conflicts in Nicaragua were not always clear"; and assured the FSLN leader that "we here do appreciate the complex and conflicted nature of the present situation, and that the attitude and actions of other governments including our own have contributed to the tension." Then, after a mild reminder that the government of Nicaragua had a "unique responsibility for public order" and should "prevent abuses that inflame public sentiment," Roach praised the Nicaraguan for his "commendable and forthright" acknowledgment of "certain errors which are not reflective of government policy." Archbishop Roach thus accepted at face value the FSLN claim that attacks on the Church and its leaders were not "reflective of government policy." In his response to Reyes, the archbishop also discussed briefly "the far more serious conflicts which sadly characterize much of the Central American region today."[168] A systematic attack on the leadership of the Nicaraguan Church did not appear to the president of the NCCB, in December 1982, as among the "more serious conflicts" in Central America.

Father Hehir struck a similar note in his December 1982 Congressional testimony: "an initial decision [by the Sandinistas] not to allow publication of a papal letter in early August, the travel restrictions imposed on a bishop whose vicariate includes a militarized zone where fighting has taken place, and the free rein given to the media in covering a particularly distasteful story involving a priest *represent bad judgment and possibly bad faith on the part of the government authorities.*"[169]

Accommodationist language and analysis continued to mark USCC testimony on

Nicaragua in 1983. After noting the U.S. bishops' disagreement with Sandinista claims to an educational monopoly, and their concern over the "maltreatment of prisoners and persons suspected of actions hostile to the new regime," Archbishop Hickey still argued that "the situation is complex and capable of moving in a positive or negative direction," and that U.S. policy had "served as a continuous provocation which has given a pretext for ever-increasing governmental attempts to control important elements of Nicaraguan life."[170] Four years after the Nicaraguan Revolution, the USCC still insisted that part of the blame for Sandinista attacks on the Church should be laid at the feet of the U.S. government.

But what was most striking about the 1983 USCC testimony was what it did *not* describe: the continuing vilification of Archbishop Obando y Bravo in the Sandinista press. Even more astonishingly, Archbishop Hickey did not mention the defamation of Pope John Paul II at an outdoor Mass in Managua three days before the USCC testimony.[171] This extraordinary omission paralleled the policy priorities sketched a month earlier by three U.S. archbishops (Hickey of Washington, D.C., Patrick Flores of San Antonio, and Peter Gerety of Newark) after a nine-day trip to Central America. The archbishops' statement on Nicaragua stressed their "conviction that the cause of peace would not be best served by isolating Nicaragua from access to critically needed resources," and their call to the U.S. government "to avoid actions or statements that would tend to further such isolation." Only after this did the archbishops cite "the need to support the church in Nicaragua in its efforts to maintain those basic human freedoms essential to its Christian heritage."[172]

The USCC opposed aid to the Nicaraguan democratic opposition, the *contras*. Although the United States had a legitimate interest in Nicaragua's regional role and internal human rights problems, U.S. policy toward the country was still "misguided and fundamentally flawed," because "instead of ameliorating tensions and seeking to influence through effective diplomacy, present U.S. policy is moving in a contrary and increasingly dangerous direction," Fr. Hehir testified in 1984. Hehir opposed "all covert aid to forces seeking by violence to overthrow the present government," because such aid would "corrupt" our own standards and provide "convenient justification" for further repression within Nicaragua.[173] Again, there was no mention of the Sandinista vilification of the pope the year before. The argument that Sandinista repression was partially the fault of the United States rather than the expected behavior of Marxist-Leninists held firm.

THE CENTRAL AMERICA DEBATE AND THE HERITAGE ABANDONED: AN EVALUATION

An aphorism once attributed to Reinhold Niebuhr captures the tragedy of the American Catholic debate on Central America over the past decade: "The stones which begin a Tower of Babel are usually laid with more justice than those that complete it." It seems an apt image, for no one can deny the humanly painful, morally unacceptable, and politically explosive facts of Central American life in the mid-1970s: the harsh realities of poverty, disease, and ignorance, each compounded by the depredations of dictatorial governments. These were perennial in the region; but they had been brought to a boiling point by the phenomena of mass communications and (paradoxically) rapid economic development throughout Latin America. The poor had come to realize that their condition was not fixed in the order of things; life could change, progress was possible, a minimally decent standard of living was not beyond one's grasp. North American missionaries had been one important instrument of this new realization. There was no way in which their fellow Catholics in the United States could, or should, cast a blind or paternalistic eye southward. The "cry of the people" was real. It could not be ignored.

The argument here, then, is not meant to deny the realities ot Central American poverty or tyranny; the question is the response of American Catholicism, in its intellectual and leadership elites, to those realities. The issue is not *whether* the bishops should have tried to influence the public policy process, but *how* they undertook that necessary task. Their response was a graphic illustration of the abandonment of the heritage of *tranquillitas ordinis* as a moral and political horizon for the Church in the United States.

The American Catholic debate on Central America, and the American bishops' commentary on U.S. policy in the region, were marred by crucial empirical distortions. The constant theme in both the debate and the commentary, that the sources of the conflict in Central America were indigenous, was only a partial truth. Poverty was indeed a "seed of conflict" in the region; but poverty had been the chronic condition of Central America for centuries (and in fact the economic circumstances of the region had been significantly improving since the early 1960s). Poverty was a necessary but not sufficient explanation of the social, political, and military conflict in Central America. The argument from "indigenous sources of conflict" confused the *cause* of the region's turmoil with the *forms* in which that turmoil unfolded after 1978.

Traditional authoritarian and oligarchic resistance to social, economic, and political change was surely a factor in Central America's crisis of crises; but so was the rise of Marxist-Leninist political and miltary forces. Marxist-Leninist guerillas were not a *necessary* result of poverty and degradation. Their political power came from their capacity to expropriate the language of resistance to social injustice; their military power came from Cuba and, by extension, from Moscow (locales not notable for their commitment to social justice). If some in the American policy community erred by seeing the conflicts of Central America exclusively through the prism of East/West geopolitics (and it was considerably harder to find such crude analyses than could have been surmised from reading the prestige Catholic press during the 1980s), it is at least as true that many American Catholic intellectual and religious leaders erred in denying the impact of these external factors on the shape and course of the region's strife. U.S. "external intervention" was perceived, and roundly condemned. Soviet/Cuban intervention in and militarization of Central America were decidedly downplayed.

The elite debate and the bishops' commentary were also distorted by a false view of the dynamics and purposes of the Central American Left. This was particularly true in regard to the Sandinista regime in Nicaragua. The persistent claim that the FSLN became increasingly repressive because of U.S. pressure ignored the history of the FSLN, the depth of its leadership's ideological commitment, and the nature of the regime's relationship to Cuba and the U.S.S.R.[174] U.S. policy under Presidents Carter and Reagan no more made the Sandinista leadership Marxist-Leninists than U.S. policy under Presidents Eisenhower and Kennedy had made Fidel Castro a Marxist-

Leninist. One need not defend the details of U.S. policy in Nicaragua to understand that the FSLN leadership was quite capable of defining its ideological position without pressure from the United States.

The idea that the FSLN had expropriated the broad-based, popular revolution against the dictator Somoza was largely absent from the American Catholic debate and from the bishops' commentary on U.S. policy (and this despite the clear witness of the Nicaraguan hierarchy from at least 1982 on). Similar blindness marred U.S. Catholic understandings of the ideological stance of the Salvadoran FMLN guerilla forces. Ample evidence of the Marxist-Leninist core of both the FSLN and FMLN was available to American Catholic intellectuals, publicists, and religious leaders. That evidence was deprecated (on the grounds that these were Marxists influenced by theologies of liberation), minimized, or ignored.[175]

This disinclination to accurately describe the communist forces at work in Central America was paralleled by a failure to perceive and support the democratic center in the region. This was most apparent in the debate over El Salvador. The Salvadoran forces led by Christian Democrat José Napoleón Duarte found precious little support among American Catholic intellectuals and journalists. Rather, Duarte and his colleagues were regularly portrayed as tools of the traditional Salvadoran oligarchy (and this despite the fact that Duarte had been tortured and exiled by precisely those forces). Duarte's and others' calls to develop and sustain a democratic center in El Salvador went, in the main, unheeded in the leadership elites of American Catholicism, which preferred to see the Salvadoran civil war as between oligarch-funded death squads and "those whose commitment to reform and social justice [had] led them into armed opposition." On that analysis, El Salvador needed a political solution, which meant the incorporation of the FMLN into a provisional government. The Salvadoran people's evident preference for evolutionary democracy, made movingly clear in their participation in three elections, did not figure prominently in the calculations of many American Catholic commentators. It was only in 1985, for example, after a democratically elected President Duarte had courageously opened negotiations with the FMLN and its political partners that the USCC commended Duarte for his efforts in behalf of peace in his tortured country.

Similar predispositions skewed Catholic commentators' perceptions of the anti-Sandinista rebels in Nicaragua. No one familiar with the postrevolution history of Nicaragua doubts that there were substantial numbers of *Somocista* forces in the original composition of the *contras.* Yet these were a clear minority by 1983 and 1984, and the political leadership of the *contra* forces was in the hands of anti-Somoza democrats such as Arturo Cruz, Alfonso Robelo, and Adolfo Calero. Robelo and Cruz were former officials of a Sandinista regime that, on their testimony, had betrayed the Nicaraguan people's revolution. But as late as 1985, even after Cruz had made a powerful plea for peace and democracy in Nicaragua in the pages of *The New Republic* and elsewhere, the USCC failed to even acknowledge the emergence of a legitimate, democratic, Nicaraguan opposition—a powerful moral claimant to the mantle of the Nicaraguan revolution—in its testimony before Congress.[176]

The U.S. Catholic debate on Central America, and the bishops' commentary on U.S. policy in the region, were also wrong in their antiphonally repeated warnings of imminent, direct, U.S. military involvement in the region. But the worst example of empirical misperception was on the issue of religious persecution in Nicaragua. As late as April 1985, USCC Congressional testimony still insisted that the root problem in Nicaragua, "the dominant fact of Nicaraguan life today," was "the war being waged against the government by the insurgent forces."[177] That this war had been caused by Sandinista repression, of which the attack on the Church was surely a crucial factor in the minds of the Nicaraguan people (as it was in the mind of Archbishop Miguel Obando y Bravo, who was raised to the cardinalate by Pope John Paul II within three weeks of the 1985 USCC Congressional testimony), was not a prominent, or even very evident, theme in either the American Catholic debate or USCC testimony.[178] USCC commentary on Nicaragua stiffened from 1983 on; but virtually every benefit of the doubt was offered to the Sandinista regime in the course of the testimony, so that its net effect was to weaken perceptions of Sandinista repression of the church (as well as of the independent press and trade union movements in the country).

On November 12, 1985, the independent Nicaraguan Permanent Commission on Human Rights wrote the leadership of the NCCB, informing the U.S. bishops that the Sandinista newspaper *Barricada* had been highlighting "the supposed support of the North American bishops for the policies of the Sandinista Government and their condemnation of the policies of . . . their country." These statements were then used to condemn the policy of Cardinal Obando y Bravo and the Nicaraguan bishops. USCC testimony was, in other words, being actively used as an instrument of repression in Nicaragua. The Sandinista talent for mendacity notwithstanding, there was, as is unhappily clear from the survey above, considerable material for *Barricada* to work with.[179]

Empirical errors were compounded by analytic deficiencies. Both the elite debate and the bishops' commentary put heavy emphasis on the past history of U.S. intervention in Central America, but paid little attention to the shifts in policy that had marked the Carter administration's approach to the Nicaraguan revolution (withholding support from Somoza during his downfall, and then providing large-scale aid to Nicaragua in the eighteen months after the revolution), or the Carter administration's support for the 1978 coup that dislodged the traditional oligarchic leadership in El Salvador. Nor did the Catholic debate and commentary reflect a sense that the Panama Canal treaties, which the American bishops had vigorously supported, reflected a post-imperial approach to the region.

If one were to believe the dominant themes in the American Catholic elite debate and the USCC Congressional testimony, contemporary American policy-makers were still in thrall to the mistakes of the Coolidge administration; the wrongheadedness of U.S. policy in the 1920s precluded a positive U.S. role in Central America sixty years later. The "lessons of Vietnam," particularly the identification of intervention with military action, loomed large in this analysis, and reinforced the alleged lessons of previous U.S. policy errors in Central America.

A soft neo-isolationism shaped both the American Catholic elite debate and the USCC's policy prescriptions. The bishops insisted, from 1978 through 1985, that they entered the debate over U.S. policy in Central America as pastors; yet their contributions to that debate were primarily cast as policy analysis and prescription. The six dominant motifs in that analysis and prescription were bound together by a radically weakened sense of the U.S. capacity to act for good ends in Central America. When this pattern was broken (as in the bishops' reluctant 1983 endorsement of military assistance to El Salvador), it was only under pressure from the official Church leadership of the region. But this was the exception, not the rule. The predominant themes in the bishops' testimony to the Kissinger Commission, for example, were America's historical and contemporary incapacities as a force for anything resembling peace with freedom and a measure of justice in Central America.

One should not deny the difficulty of the analytic and prescriptive task that faced American Catholicism in Central America. The Salvadoran civil war had a horrific character in the early 1980s, and the task of moral and political imagination required to chart a way to peace with freedom through the bloody divisions of Salvadoran society was large indeed. So, too, with the task of national reconstruction in post-Somoza Nicaragua: under any circumstances, it loomed as a colossal trial of imagination and will.

Yet the Catholic heritage of peace as dynamic political community had resources with which to address even these enormous difficulties, had that heritage been taken seriously. The heritage could have helped prevent empirical and analytic misperceptions. It would have accurately portrayed the various political forces at work in the region: José Napoleón Duarte would not have been described as a front for the Salvadoran oligarchy and death squads; Archbishop Obando y Bravo and Arturo Cruz would not have been vilified as counter-revolutionaries; the ideological convictions of the Sandinistas and the Salvadoran FMLN would have been made clear, as would their connections to Cuba and their place in Soviet geopolitical maneuverings. The history of American involvement in the region would have been forthrightly acknowledged, warts and all; but the errors of the 1920s would not have been thought to incapacitate positive American involvement in the 1980s. The possibility of sustaining (where it existed) or helping to create (where it was very weak) a democratic center in Nicaragua, El Salvador, Honduras, and Guatemala would have been of prime concern, rather than being seen as a peripheral issue at best. The relationship between gross poverty and the forms of civil strife in Central America (and particularly the relationship between poverty and the rise of Marxist-Leninist guerilla forces) would have been analyzed and explained with sophistication rather than sophistry.

More accurate empirical perceptions and less distorted analytic frameworks could have led to an official Catholic commentary on the crises of Central America that clearly identified the goals that should inform public policy: America should help facilitate the emergence of economically viable societies in Central America whose leadership respected basic human rights, helped create conditions in which basic human needs were being met, conducted the business of governance according to the peoples' will, and were not active agents of the geopolitical aspirations of the Soviet Union. Such goals would have set a moral and political horizon, describing a Central America that had evolved over at least a generation. But had such a set of goals been articulated by American Catholic intellectuals, publicists, activists, and bishops, they would have created a standard by which U.S. policy could be charted so that peace, freedom, security, and justice were being served simultaneously.

These goals would also have led to a more accurate identification of the obstacles that blocked progress toward their achievement. Traditional patterns of political economy, social life, and governance in the region presented such obstacles; but so did the forces of Marxism-Leninism in the region. Would a Central America shorn of the threat posed by the FMLN in El Salvador and the Sandinista regime in Nicaragua have been a Central America returned to the days of Somoza and the Salvadoran oligarchy? This was an impossibly difficult case to argue, at least by 1983. It was much likelier that such a state of affairs would have been more conducive, not less, to the evolution of peace with freedom, security, and opportunity in Central America. Yet that was not the teaching of the major thought centers in American Catholicism, which persisted in seeing the Sandinistas, in particular, as forces for social justice who had made unfortunate mistakes during their consolidation of power. Recognizing the falsity of this teaching would *not* make the case for U.S. military intervention in the region. It would, however, have been conducive to understanding the sources of armed conflict in Central America with more accuracy than was evident in either the Catholic elites' debate or the U.S. bishops' commentary. Proposals for a political-diplomatic "regional solution" could then have been built upon a realistic foundation, rather than on mistaken notions of the Salvadoran FMLN as "those whose commitment to reform and social justice [had] led them into armed opposition," or of the Sandinista regime as increasingly totalitarian by reason of U.S. pressure rather than ideological conviction.

The classic Catholic heritage, had it been developed, would have suggested an approach to Central American politics that stressed the paramount importance of the region's nascent democratic center. It would have urged U.S. support, through a variety of private and public instrumentalities, for those intermediate institutions on which popular democracy could be built: trade unions, peasant cooperatives, political parties, a free press, and, perhaps above all, a free Church.

Central America was blessed during the 1980s with two outstanding Catholic leaders in Archbishop Rivera y Damas of San Salvador and Cardinal Obando y Bravo of Managua. Yet Rivera y Damas was never accorded the status of his predecessor, Archbishop Romero, by the American Catholic intellectual and religious elite, and Obando y Bravo was often dismissed in the American Catholic debate as a counterrevolutionary, while his brother bishops in the United States spoke out in his defense only when the evidence of Sandinista repression of the Nicaraguan Church had become overwhelming.[180]

Above all else, American Catholicism should have stood fast for religious freedom as an essential building block of a morally worthy peace in Central America. But as late as 1982, the president of the NCCB was assuring the Sandinista leadership that the bishops of the United States understood the "complexities" of church/state relations in Nicaragua, and official USCC Congressional testimony four months later failed to mention the FSLN attack on the pope's Mass in Managua. It is little wonder that the Nicaraguan Permanent Commission on Human Rights felt compelled to inform the NCCB and USCC that their statements were being used against, rather than in defense of, religious liberty.

José Napoleón Duarte and the forces he represented did constitute a democratic center in El Salvador; and this was missed by many American Catholic leaders. Archbishop Obando y Bravo and the forces he represented were legitimate claimants to the mantle of the Nicaraguan people's revolution; but this, too, was missed by many American Catholic leaders. Those voices dismissive of the options symbolically represented by Duarte and Obando y Bravo were not the only voices in the American Catholic debate, nor in the official leadership circles of the Church.[181] But they were the dominant voices.

Why?

Msgr. Gremillion's portrait of American Catholic leaders shocked by the brutal realities of Latin American poverty, although meant as a celebration of consciousness-raising, could in fact be better interpreted as a portrait of profound naivete. The radicalization of the American Catholic elite through contact with the social and political realities of Central American life says more about the problems of ignorance compounded by a compassion divorced from political and ideological sophistication than it does about raised consciousnesses.

Naivete surely played its part in the abandonment of the heritage of *tranquillitas ordinis* during the Central America debate. But this naivete led in a specific direction, toward the kind of commentary on Central America propounded by the American bishops. Why? Two themes in the postconciliar abandonment of the heritage were crucial. The deterioration of the idea of America was a dominant force in shaping the American Catholic debate on Central America and the bishops' commentary on U.S. policy in the region. That remarkable phrase from the 1972 Jesuit conference at LeMoyne College—that "there did not seem to be too much left to what John Courtney Murray liked to call 'the American proposition' "—captures this starkly and perfectly. Other voices in American Catholicism may have believed (and, in fact, did) that this was a badly mistaken judgment. But it was this theme of abandonment that set the analytic starting point for the intellectuals, journalists, publicists, and, eventually, bishops, who came to dominate American Catholic involvement with the crises of Central America. The United States was not, even on balance and considering the alternatives, much of a force for good in Central America. From this conviction, much would flow.

The Catholic debate on Central America was also profoundly influenced by the anti-anticommunism of the postconciliar years. But in this instance, the negativity of anti-anticommunism was blended with positive teachings drawn from various currents in the theologies of liberation. The absolute priority of the achievement of "justice" (rarely, if ever, defined with precision) before there could be a morally worthy peace; the idea that Marxism provided the scientific understanding that, in Third World situations, could complement Christian revelation; the tacit (and sometimes overt) deprecation of democratic values and institutions as means to peace—these teachings were coupled with anti-anticommunism and a soured view of American possibilities to yield the kind of commentary on Central America just reviewed. Romantic in its view of the Central American Left, and softly neo-isolationist in its approach to American policy in the region: whatever else may or may not be said for the substance and form of American Catholic commentary on Central America, it can hardly be argued that it met the intellectual (and moral) standards of its heritage.[182]

The results were similar to the results of the new nuclear debate. The abandonment of its heritage crippled the American Church's ability to do what its official leadership claimed time and again it most wanted to do: create a new spectrum of morally sensitive public debate capable of leading to a more humane future in Central America. As on the issue of nuclear weapons and strategy, the American bishops, rather than being creators of a new debate, had simply (and tragically) become partisans at one pole of the debate already underway in American political culture. They had not reshaped public understanding, nor had they acted primarily as pastors. Rather, they had come to mirror one narrow band of opinion and had become partisan actors rather than moral teachers. They had not helped to link the goals of peace, security, freedom, and justice in the public debate. They had not even accurately reflected the reality of the multiple forces at work in Central America. Nor had they developed the heritage of which they were the bearers. In a word, they had abandoned their heritage. By doing so, they had missed an exceptional opportunity to shape the course of American public understanding and U.S. foreign policy.

INTERLUDE: PORTRAIT

J. Bryan Hehir

The single most influential figure in the transformation of the American bishops' public commentary on issues of war and peace has been the Rev. J. Bryan Hehir. A priest of the archdiocese of Boston, Hehir was associate secretary in charge of the USCC Office of International Justice and Peace from 1973 until 1984, when he was promoted to become secretary of the more comprehensive USCC Department of Social Development and World Peace. Fr. Hehir's pivotal role in the bishops' commentary on nuclear weapons and Central American issues has not gone unnoticed; Hehir has received honorary doctorates from Catholic colleges and universities, the Institute for Policy Studies' Letelier/Moffit Award, and a major grant from the MacArthur Foundation.[1]

Fr. Hehir's immense influence with the American bishops derives from several sources, not the least important of which are his bureaucratic skills and his dedication to priestly ministry outside the confines of the USCC. Unlike other USCC priest-staffers, Hehir lives in a parish where he handles a full load of liturgical and counseling responsibilities; and until 1984, he commuted to Boston every week in order to teach at the archdiocesan seminary. Hehir is extremely reticent about publicity and has dutifully avoided the media spotlight, preferring to remain behind the stage on which his bishops have played an increasingly vocal role. Bryan Hehir's influence also reflects his orderly, synthetic mind. Hehir typically dissects issues into easily assimilable outline form, a Godsend to overloaded bishops grateful for a staff person who can carve complex materials into digestible portions. As Cardinal Krol of Philadelphia once put it, Hehir is a "valuable resource person" who "does the digging and gives us the research."[2]

Yet Fr. Hehir has done much more than give the bishops the research. Hehir has skillfully created a framework for foreign policy analysis at the USCC. "The research," like any research, has to be organized, and he who sets the interpretive framework through which the research is read holds great power over the policy prescriptions that result from the bishops' deliberations. What is most important about J. Bryan Hehir and his influence on American Catholic thought on war and peace is not, then, his priestly dedication or his bureaucratic abilities; what is important are Hehir's ideas. For, in large measure and despite the public disclaimers, Hehir's ideas have become the bishops' ideas.

Bryan Hehir acknowledges intellectual debts to John Courtney Murray, Yves Congar, and Karl Rahner, but the foreign policy framework he has created at the USCC has less to do with these theological giants than with Hehir's frankly admitted admiration for Harvard political scientist Stanley Hoffmann, his former teacher.[3] The influence of Hoffmann's approach was apparent in a 1975 lecture Hehir delivered to the Catholic Theological Society of America (CTSA). Following Hoffmann, Hehir argued that the "structure of power" in the world had changed considerably since World War II. The postwar, bipolar world had first become pentagonal, and economic power vied with traditional forms of military power for predominance. The power disparities among the First, Second, and Third Worlds, and the gap in material prosperity between the Third World and the rest, had led to new problems.[4] This pentagonal world of complex relationships among the United States, the Soviet Union, Western Europe, Japan, and the Peoples Republic of China had ultimately given way to an even more complex world of global interdependence, in which three elements were of primary importance.

First, following Hoffmann, Hehir argued that questions of military security had been "displaced" from their accustomed position at the center of foreign policy concerns. Secondly, international economics had become high politics; interdependence meant that economics could no longer be conceived as independent of foreign policy. Finally, the modern world was characterized by the increasing number of transnational problems, on the one hand, and transnational "actors" (multinational corporations, religious institutions, regional and international organizations and agencies), on the other.[5]

Hehir's conclusion, drawn explicitly from Hoffmann, was that "the manipulation of interdependence has become the core of interstate politics."[6] Bipolar competition between the United States and the Soviet Union or, more broadly, between the democratic West and the Soviet-dominated Eastern bloc, no longer defined the center of world affairs. The United States retained significant influence and maneuvering room in the world of interdependence. But American policy-makers had to learn, as a new first principle, the limits on U.S. power. Traditional usages of power could now rebound back on the nation that exercised them. Nuclear weapons use was the obvious example. But economic coercion with allies was also dangerous, because it threatened the whole structure of international economics and finance. Hehir admitted that interdependence was ambiguous. It could lead to greater cooperation, even community, in international public life; but it could also lead to conflict and chaos. "Our mutual need means mutual vulnerability; the manipulation of interdependence can mean seeking to maximize the dependence of others on us; interrelated issues can be used as weapons rather than as bonds of cooperation." The transition from material to moral interdependence would be neither short nor simple.[7]

Nuclear weapons, economic interdependence, and environmental issues had broken down the notion of state sovereignty as it had evolved since the Peace of Westphalia in 1648. The basic problem of international public life today was the increasing gap between what the independent, sovereign state was expected to do, and what it could actually achieve, politically, economically, and strategically. The world's key problems were transnational in nature, but there was little indication of an evolution toward transnational institutions that could "supplant the state." How to live in this structural gap, "how to act effectively with inadequate means of action," was the statesman's fundamental dilemma in the late twentieth century.[8] This was not simply a technical or political question; it was also a moral issue that ought to draw the particular interest of Catholic theology.[9]

Indeed it should have, given classic Catholic understandings of the peace of dynamic political community. But with Bryan Hehir, the central focus of American Catholic moral thought and political action decisively shifted from the horizontal task of developing the theory and practice of *tranquillitas ordinis* to determining "how to act effectively with inadequate means of action." Hehir, like so many others, declined to take up Paul Ramsey's charge to Catholic theology in the wake of *Pacem in Terris*. Because there was little evidence pointing toward the evolution of the kind of transnational political authority urged by John XXIII, the task of Catholic theology, Hehir argued, was to help determine the rules by which the dilemma of interdependence could be managed. That the classic Catholic heritage, disciplined and developed by Ramsey's insistence that just-war norms chart the path from chaotic interdependence to morally sound international political community, might provide an analytic and moral horizon for the determination of these rules has not been a prominent theme in Hehir's work.

Fr. Hehir's 1975 CTSA lecture, which stressed the "justice claims" of Third and Fourth World countries, was notable for its omissions as well as for its positive definition of the new international agenda. Hehir did not discuss the dangers to an interdependent world posed by a massively armed Leninist state, the Soviet Union, or by Soviet-armed revolutionary forces in the Third World. Ideology is strikingly absent from Hehir's analytic model. Just as Hehir, with Hoffmann, viewed the United States as a Gulliver, a clumsy giant loose in a fragile and confusing new world, so too would Hehir, following Hoffmann, minimize the role of the Second World, save when nuclear weapons issues were engaged. Ideology, in a typically Gaullist view of the Soviet Union, reinforced by model analysis, was epiphenomenal. The U.S.S.R. was, at bottom, not a revolutionary state with an ideologically validated world-historical agenda, but Russia: a great power, to be sure, but not a distinctive presence or threat.

Hehir's CTSA lecture was also notable for its insistence that moral claims ran from the Third World to the West, but not vice versa. The Third World definition of the core of the world *problématique*—that is, the gap between rich and poor—was accepted, if nuanced by the residual claims of strategic security issues. The analysis of the causes of poverty in the U.N.'s New International Economic Order manifesto might also have to be nuanced (a favorite Hehir idiom), but they did set the terms of the debate correctly. The moral questions involved were ones of "social and distributive justice," not "relief and charity."[10]

Finally, Bryan Hehir's CTSA lecture did not analyze the moral claims of democracy, or the possibility of democratization as a means of meeting the legitimate justice claims of the developing world—an omission that recurs throughout Hehir's essays on the general contours of the foreign policy debate.[11]

Bryan Hehir's analytic framework is thus built around several key themes. Hehir's "Gaullist" or Hoffmannesque view of the United States and the Soviet Union has already been noted. Both superpowers are clumsy giants abroad in an increasingly complex world whose interdependence has not adequately entered American and Soviet policy calculations. Hehir is no apologist for internal Soviet repression, nor does he defend such Soviet aggressions as the invasion of Afghanistan or the intense Soviet pressure on Poland. Yet his principal concern seems to be that "too often the United States adopts an obstructionist posture on all sorts of issues—arms control, law of the sea, economic questions."[12] And while Fr. Hehir has never indulged in the kind of crude U.S./Soviet parallelism that marred the American Catholic debate in the years after Vatican II, his defense of a legitimate Soviet role in a Mideast settlement, his arguments on the weight of U.S. responsibility for nuclear arms competition, his discomfort with linkage between Soviet human rights performance and the arms control process, his discussion of the relative Soviet and American roles in Central America, and his tendency to illustrate human rights concerns by primary reference to countries like South Korea, Chile, and the Philippines—all suggest that Hehir-the-strategist views the United States and the Soviet Union analytically as equally culpable and equally blundering mastodons in a world that neither of them understands very well. Détente should thus guide the conduct of U.S./Soviet relations.[13]

Interdependence is another primary motif in Hehir's analytic framework, as his 1975 CTSA lecture illustrated. By 1980, Hehir admitted that "it is too much to say that interdependence has transformed international relations so that the role of the state and resort to force are outmoded. . . . A series of events has illustrated the staying power of political-strategic issues in world politics." Yet Hehir also argued that the world of the 1980s could not be adequately explained in a classical, Westphalian model, because transnational problems were not susceptible of resolution by "a resurgence of strategic concerns."[14]

In 1980, Hehir conceded that U.S./Soviet relations would rise to the top of the foreign policy agenda in the new decade, an expectation that sharply diverged from many analysts' (including Hehir's) expectations in the 1970s. Yet Hehir also implied that this new focus on the superpower relationship was somehow a failure of will or imagination:

> [President] Carter's Notre Dame speech proposed to reorient the priorities of policy to achieve greater balance between the superpower rivalry and other issues. Nongovernmental voices, some within the churches, argued that the North-South question had replaced the East-West question in world politics. Each of these statements reflected a truth to be heeded but together *they were not sufficient to sink the superpower rivalry*.[15]

In Hehir's 1975 CTSA framework, such things were not supposed to happen. By 1980, the centrality (not, surely, exclusivity) of the U.S./Soviet competition in shaping world affairs could not be denied as a general proposition. Where it could, and would, be denied was in the specific case of Central America.

J. Bryan Hehir has not been merely another voice in the general debate over grand strategy in foreign policy; his has been a voice from a particular locale, the public policy agency of the Catholic bishops of the United States. Thus, in addition to the general strategic framework Hehir has developed for the USCC, his work for the bishops has been shaped by his understanding of the Church's role in America and in the world. Fr. Hehir has emphasized the role of the Catholic Church as a venerable transnational actor with particularly valuable sources of information on such issues as human rights.[16] But Hehir has also been concerned about the Church's specific role in the U.S. foreign policy debate. With virtually all Catholic moral theologians, Hehir argues that the Church's primary task in this arena is moral education: establishing the fact of a moral dimension to foreign policy debate, and propounding the relevant moral principles that both suggest the goals foreign policy should seek and the limits on the means appropriate to the achievement of those goals. Yet Hehir, with many others, has also insisted that this educational task be completed by the explication of prudential judgments that show how principles can be applied in particular cases. Hehir defined one approach to this latter role in 1976:

> What religious organizations are uniquely equipped to do . . . is to form a constituency of conscience on key foreign policy issues. By a constituency of conscience I mean a body of citizenry in the midst of the larger society which has a specific angle of vision on foreign policy questions involving significant moral content. Such a constituency of conscience has to be cultivated; it cannot be coerced. The cultivation of such a constituency, however, grows out of the very terms of meaning and motivation in life which are the main themes of our religious traditions. The challenge is to cast the meaning and motivational themes in structural terms of policy and politics.
>
> The power of a constituency of conscience approach lies not only in the unique assets religious bodies have to build such a consensus but also in the legitimate role we can play with our own people on these policy issues. As soon as we enter the social field, questions arise about the legitimacy of our role. In the model I propose we are not trying to tell people how to vote but seeking to help people to think and decide. *The dynamics of the model, as I see it, involve the religious institutions taking specific policy positions and then going to their constituencies with the position to see if they can garner support.*
>
> *In this approach, we should make no public claims to speak for 50 million Catholics when we take a position. We speak for institutional Catholicism, and we are going to the community of the Church to form a consensus on a given issue.*[17]

Hehir, in 1976, thus affirmed a dialogical process within the Church—but one in which those who had already formed, not merely their conscience but their policy prescriptions, went "to their constituencies to see if they can garner support." Moreover, the very terms of the model suggest that, with due regard for the primacy of the Church's teaching task, what really engages Hehir's attention is the policy process at its narrowest end—that is, specific judgments on specific questions. Hehir doubtless agrees in theory with the statement in "The Challenge of Peace" that religious institutions' claims to moral wisdom are weakest at precisely this level of prescription. Yet the impression created by Hehir's USCC activism and his writings is that the policy arena most engages his imagination and energies. Thus, Hehir-crafted USCC Congressional testimony, which typically begins with an assertion that the bishops come before the Congress as pastors and teachers, is then primarily devoted to strategic analysis and policy prescription (in some instances down to the fine points of funding the "soft-loan window" of international lending institutions).[18]

Father Hehir's approach to specific issues is thus shaped by a distinctive analytic framework and a particular view of the Church's role in the foreign policy debate (in which policy prescription is largely, although not exclusively, predominant). Hehir's prescriptive thought can, in many cases, be inferred from a careful reading of USCC policy statements and Congressional testimony. But it is more directly accessible in his personal commentary on nuclear weapons issues and questions of U.S. policy in Central America.

On nuclear weapons questions Hehir has been a prominent defender of the claim that nuclear pacifism constitutes a distinctive third voice between traditional pacifist and just-war thought. Hehir would not describe himself, technically, as a nuclear pacifist, inasmuch as, with the bishops, he would hold to a "centimeter of ambiguity" on the question of whether any use of small-yield nuclear weapons might be morally tolerable. Yet his attraction to nuclear pacifism led him, throughout the late 1970s and early 1980s, to argue that prevention of nuclear weapons *use* was the primary moral imperative.[19] Nuclear weapons use was the first priority because virtually any conceivable use of any nuclear weapon would lead, ineluctably, to a general nuclear war.[20]

This "no use" entry point led Hehir to a host of prescriptions on strategic policy; tempered support for SALT II;[21] opposition to counterforce targeting;[22] opposition to the neutron bomb;[23] profound skepticism about cruise missiles;[24] opposition to the MX missile;[25] opposition to the development of ballistic missile defense;[26] endorsement of the Comprehensive Test Ban Treaty;[27] and support for a no-first-use of nuclear weapons pledge by the United States.[28] Hehir's prescriptions fit comfortably within the traditional arms control community,[29] but Hehir has added to this framework the concept of a "bluff deterrent" (i.e., a deterrent force that can never be used, but which, by its existence, threatens an adversary and thus provides for deterrent stability).[30] Fr. Hehir has also shown a marked skepticism about the strategic triad[31] and an appreciation for the new antinuclear movement's "democratization" of the nuclear debate.[32]

Fr. Hehir vigorously defended his bishops' vocal entry into the nuclear weapons argument in the early 1980s. He attacked Navy Secretary John Lehman's critique of some bishops' rhetoric, claiming that Lehman's rebuttal had "the literary quality of a petulant attack rather than a reasoned argument."[33] Hehir's attack on Lehman was a function of his broader appreciation for the renascent peace movement, which had "posed the nuclear issue for public debate in a fashion which cannot be ignored," by "engaging large numbers of people in a discussion previously confined to government circles and a few research institutes." By 1982, Hehir argued, antinuclear activism constituted a "fragile but effective alliance of diverse groups focused on reversing the nuclear arms race."[34]

It is important to note what Hehir emphasized and what he minimized or ignored in discussing "the popular movement . . . exemplified in the Nuclear Freeze initiatives, the Ground Zero Week, the Peace Pentecost sponsored on Memorial Weekend and the spate of activities surrounding the U.N. Special Session on Disarmament."[35] The survivalist current so prominent in the antinuclear activism of the early 1980s went unremarked in Hehir's analysis—a notable omission for a Catholic theologian whose fundamental understandings of the meaning and end of human life were being undercut by such antinuclear survivalists as Helen Caldicott of Physicians for Social Responsibility and Carl Sagan of the Union of Concerned Scientists. Nor did Fr. Hehir challenge the activist theme of parallel (i.e., equivalent) U.S./Soviet responsibility for the arms race.

Hehir's understanding of the Reagan administration's nuclear weapons and strategy policy resembled the concerns of those who seemed to believe that the administration was almost eager for a nuclear confrontation with the Soviet Union. Hehir argued in 1983 that the administration "zero option" proposal for eliminating intermediate-range nuclear forces in Europe, the administration May 1982 START proposals, and the establishment of the Scowcroft Commission had "all been generated partially in response to strong public calls for a serious U.S. arms control policy."[36] Typically, Hehir nuanced his claim by the modifier "partially." Yet the drift of his message was clear: the Reagan administration had not been interested in arms control or reductions, and had been forced to put these issues on the agenda by the weight of public opinion. Throughout this commentary, Hehir did not offer any sustained critique of the arms control policies of the 1970s, which, whatever their other accomplishments, had seen an expansion of nuclear weapons capability, especially on the part of the U.S.S.R.

Hehir was also in step with the new antinuclear activism in rejecting linkage between arms control and other issues of Soviet international conduct. Reflecting back on "The Challenge of Peace," Hehir argued in 1984 that the "political context" of U.S./Soviet relations was as important as technological and military issues:

> The [pastoral] Letter argued that despite numerous divisions, the Soviet Union and the United States still share a strong common interest in the nuclear question. Thus the logic of the Letter . . . argues against linkage, against tying the nuclear issue to other issues such as Soviet behavior in the Third World. Establishing this delinkage argument in the American political system, in the minds of the citizenry, would have enormous political significance for how we handle arms control and the nuclear question.[37]

The word "détente" was not used here, but seems implied by Hehir's formulation. In discussing linkage, Hehir did not mention alleged Soviet violations of existing arms control agreements. Nor did he suggest how one could resolve central security issues with an adversary violating its Helsinki Accord commitments domestically and its obligations under the U.N. Charter in the invasion and occupation of Afghanistan. The "absolute moral imperative that nuclear weapons never be used" rendered such considerations (which could not, of course, be related to arms control and reduction in a simplistic one-to-one correspondence) less than urgent.[38]

Fr. Hehir's approach to Central America was illustrated in USCC Congressional testimonies. In his own writing, Hehir, who once described the travail of Central America as a "class conflict,"[39] consistently maintained throughout the early 1980s that "analyzing Central America through the superpower lens will inevitably align the United States with repressive forces in the region."[40] Substantial change was imperative in El Salvador, Honduras, and Guatemala. Prioritizing the geopolitical dimension of the conflict necessarily meant minimizing the requirements of change. "The Central American case requires a policy which gives priority to interdependence issues and which interprets security policy in light of these."[41]

This capsule summary of Hehir's analytic framework for Central America raises several questions. Why, for example, would acknowledging the facts of Soviet-Cuban involvement in the region *necessarily* mean subordinating the requirements of substantial change? The classic Catholic heritage would have suggested that Marxist-Leninist regimes might present formidable obstacles to the substantial change Hehir correctly perceived as essential in Central American societies. Why, in other words, did one have to think of "interdependence issues" and "security issues" in prioritized terms (in either direction)? Why not see them as inextricably bound together in the contemporary historical reality of Central America? Why, also, did Hehir's commentary on Central America consistently fail to cite democratization as a means toward nonviolent change?

In USCC Central America testimony before Congress, and in his own writing on the subject, Bryan Hehir regularly argued that "a decisive military solution is not possible in political or moral terms."[42] By this, Hehir meant that negotiated settlements, perhaps through the Contadora process, were the only morally sound and political viable options available. Negotiated settlements that resulted in Central American societies able to meet basic human needs, capable of governing themselves through democratic institutions, and practicing noninterference in their neighbors' affairs were, of course, morally and politically desirable. Yet Hehir was surely mistaken in his claim that a "military solution" was "not possible in political . . . terms." Such a "solution" would indeed be possible were the forces of the Nicaraguan FSLN and the Salvadoran FMLN allowed free rein in the region. This threat of "military solution" did not appear in Fr. Hehir's commentary, although, in his nuanced way, he would usually acknowledge it as a factor in the region's turmoil. What did appear, regularly, was Hehir's fear of U.S. military intervention. Time and again the bishops (using Hehir-crafted testimony) warned Congress against U.S. "militarization" of the region's conflicts. Soviet-Cuban intervention was occasionally noted; but any objective reading of the weight of the testimony suggests that Hehir's principal fear was the use of American military force in the region.

Thus, in the aftermath of the liberation of Grenada, Hehir warned:

> Grenada and Central America [are] . . . the heart of the matter. If either the public reaction to or the policy conclusions drawn from the Grenada exercise are used to move the Reagan Administration even further toward direct military action against Nicaragua or El Salvador, then Grenada will have long-term significance. Those of us with doubts about or decisive opposition to the Grenada action should understand where the focus of attention should be: establishing a barrier against future interventions, not winning the debate about the last one.[43]

Hehir genuinely seems to have believed (as late as the 1984 USCC Congressional testimony) that U.S. military intervention in Central America was a live policy option, if not an outright probability. Yet all the publicly available evidence suggests that an active U.S. military presence in the region was not being seriously considered by the administration. Still, the focus of Hehir's attention remained on U.S. "militarization" of the region's conflicts: first, to prevent any overt U.S. action against Nicaragua or the Salvadoran FMLN; secondly, to minimize U.S. military assistance to the Salvadoran government and, later, to the Nicaraguan rebel forces.

Grenada was also an example of superpower parallelism. Hehir noted that "Mr. Reagan resents and resists the analogy of Grenada and Afghanistan, but this may not be the key case. The United States has clearly reinforced the convention that big powers are supreme in their own region—above the logic of the law. The Poles are hardly Grenadians, but the precedent for Soviet intervention in Eastern Europe—especially if invited in by key actors—has been strengthened by U.S. actions."[44] In the world of geopolitical model analysis, the difference between a superpower action aimed at restoring democracy and a superpower action to reinforce totalitarian rule was thin indeed. Thus the focus of Hehir's concern over the militarization of Central America was Washington, D.C., not Havana, Managua, or Moscow.

Bryan Hehir often expresses a concern to "build barriers" against the use of military force in world affairs, an entirely admirable intention. Yet the thrust of his commentary on nuclear weapons issues and on Central America reinforces the view that Hehir is primarily concerned about U.S. military force: barriers must be built against U.S. use of nuclear weapons, against U.S. military intervention in Central America, against U.S. "militarization" of Central America through military assistance to the Salvadoran government or the Nicaraguan rebels. In his careful, nuanced way, Hehir always mentions the dangers of Soviet or Soviet-supported military force. Yet the net effect of his thought and his work at the USCC has been to set barriers against U.S. military programs alone. Hehir justifies this on the grounds that "the role of the U.S. Church is precise and limited: its impact on American policy"[45]—a variation on the anti-Vietnam peace movement theme that "our responsibility is to speak to our government."

What Hehir fails to address, though, is how such a unilinear approach reinforces the problem of others' militarism. A peace effort that addressed only U.S. military involvement in Vietnam reinforced the military efforts of Hanoi. So, too, did a single-minded focus on U.S. military assistance in Central America reinforce the efforts of the Sandinistas, the Salvadoran FMLN, and their Cuban and Soviet sponsors. Hehir never forthrightly addresses this side of the equation: how his and the USCC's approach to nuclear weapons and to Central America has neither "built barriers" to the military resolution of conflict, nor brought effective pressure for change to bear on the Soviet Union (in the case of nuclear weapons) or Cuba, Nicaragua, and the Salvadoran FMLN (in the case of Central America).

There are other important dimensions to Fr. Hehir's thought. On several occasions, Hehir has consciously set himself within the John Courtney Murray stream of American Catholic social and political thought.[46] Hehir, against other post-Murray commentators, has defended the use of the "tradition of reason" in relating religiously based values to the world of public policy.[47] Like Murray, Hehir has cautioned against biblical literalism in social ethics, and early in his career warned against the distortion of eschatological images and categories in the pacifist thought of James Douglass.[48] But the differences between Hehir and Murray are more apparent than the methodological similarities. Hehir can hardly be said, on the evidence of his own writing as well as the USCC testimony he crafted, to share Murray's views on the problem of communism in general or the Soviet Union in particular. Hehir's insistence on the "absolute moral imperative" of no use of nuclear weapons flatly contradicts Murray's call for the evolution of nuclear weapons that would meet the just-war standards of proportionality and discrimination. Hehir's thought on nuclear weapons also involves a kind of determinism that Murray would have rejected: Murray did not believe that there were technological or political imperatives in the strategic field, which always remained an arena of human intentionality and freedom. Hehir would not deny this; but his insistence on forms of the slippery slope analogy as applied to either evolving weapons technologies or the possible use of nuclear weapons runs across the grain of Murray's firm insistence that determinisms of all sorts be resisted. Hehir would argue, not unreasonably, that the technological and strategic circumstances surrounding the nuclear weapons debate have changed considerably since Murray's day. Yet it is difficult to imagine that Murray would have endorsed the strategic views of the traditional arms control community to the degree that Hehir has done. This seems especially true on the question of linkage. Would Murray have espoused the view that nuclear weapons issues can be effectively uncoupled over the long run from other issues in U.S./Soviet relations? It seems an extremely unlikely hypothesis.

Hehir breaks most decisively with the Murray legacy on the question of America. One searches in vain through Hehir's published work (and through the USCC testimony he has influenced) for a glimmer of Murray's confidence in the moral importance of the American experiment in democratic pluralism. This break goes further than Hehir's views of the American Gulliver loose in the world; it appears in the singular absence of democracy as an alternative to war in Hehir's writing. Bryan Hehir has written eloquently about the undoubted moral claims of the poor and the hungry within today's interdependent international system. Yet one learns little from him about the moral claims of the world's democrats, particularly those struggling for peace with freedom in countries like El Salvador and Nicaragua. Fostering the evolution of a democratic center in Third World countries as an alternative to traditional authoritarians or Leninist revolutionaries is not something that, on the public record, seriously engages Fr. Hehir's imagination. No doubt he would concede the theoretical possibility, even desirability, of such alternatives in, for example, the Philippines. But what is so striking is his lack of support for such alternatives in other circumstances where they existed, even if in fragile, embryonic form. The USCC's failure to concede, until 1985, the existence of a Salvadoran democratic center in the forces gathered around José Napoleón Duarte is but the most glaring example of this deficiency. Here, Fr. Hehir stands, not within the Murray tradition, but as a counterpoint to it. The foreign policy framework Hehir has developed in and for the USCC bears the *imprimatur* of Stanley Hoffmann more than it does that of John Courtney Murray.

There is an undercurrent in Hehir's thought which suggests that he views himself as a mediating force in the American Catholic debate on war and peace: a figure standing between the rhetoric and analyses of post-Vietnam Catholic activists and intellectuals, and the stringencies of the Catholic social ethical tradition. Yet Fr. Hehir's disinclination to criticize the dominant themes in Catholic activism on the questions of nuclear weapons and Central America has been an important element in validating those themes within the Catholic debate. The themes of the abandonment of the heritage now set the terms of debate throughout the Church's intellectual, journalistic, and episcopal elites. The center is located rather differently on the general spectrum of foreign policy argument from where it once was. Fr. Hehir's centrism is, in actuality, located at a point far along the path to one pole of the foreign policy debate.

Hehir's appropriation of the themes of the abandonment of the heritage has been, to use his term, "nuanced." Yet the result has been the same, in the cases of nuclear weapons and Central America. The leadership of American Catholicism, on which Hehir exerted such a powerful influence, has not created a new and more morally attuned spectrum of debate so much as it has become a protagonist in a definable camp in the preexisting arguments.

J. Bryan Hehir's accomplishments should be fully acknowledged. He has made the United States Catholic Conference an important voice in the foreign policy debate. He has brought a form a just-war reasoning back to life in secular environments. His thought has decisively influenced an entire generation of American Catholic political activists. And in those circles, Hehir has argued, correctly, that the American Catholic social ethical tradition on the moral question of war and peace requires development, not the repetition of old formulas.

Yet the core of Fr. Hehir's thought is an expression, in however nuanced or modulated a form, of the principal themes by which that heritage has been abandoned in American Catholicism over the past generation. Given Hehir's influence on the American hierarchy, his thought and work have been the crucial vessel through which the abandonment of the heritage was completed, not by activists or intellectuals or journalists, but by the Catholic bishops of the United States and their public policy agency, the United States Catholic Conference.

PART THREE

The Heritage Reclaimed and Developed

It is futile to discuss reform without reference to form.

Gilbert Keith Chesterton

You grant me the clear confidence that You exist, and that You will ensure that not all the ways of goodness are blocked.

A prayer of Aleksandr Solzhenitsyn

CHAPTER 11

Beyond Abandonment

Half a century ago, in a time beset by the present danger of totalitarianism and the imminent threat of a new world war, Pope Pius XI urged this prayer on his Church: "Let us thank God that He makes us live among the present problems. It is no longer permitted to anyone to be mediocre." Pius XI's call to Christian excellence, a favorite of Dorothy Day[1], has not lost its immediacy as we await the turn of the second millennium, still beset by tyranny and mass violence, still seemingly caught between the fire of war and the pit of totalitarianism.

Pius XI's charge was given a new edge by Pope John Paul II in 1981 during two addresses at Hiroshima. Karol Wojtyla is a living embodiment of the will to resist totalitarianism with the sword of an excellent faith; the man who built the church at Nowa Huta is not, literally *can*not, be an appeaser of the masters of Gulag. But John Paul II also understands the peril of modern war. And it was of this that he spoke at a site which, with the fields of Flanders, epitomizes the horror of modern mass violence and warns of one possible outcome to the human story. "It is only through a conscious choice and through a deliberate policy," said John Paul II at Hiroshima, "that humanity can survive."[2]

There are two ways to read this papal judgment. A survivalist interpretation would argue that our conscious choice must be single-mindedly fixed on the threat of nuclear weapons, even if other issues are diminished by comparison. Faced with the possibility of human extinction, there is only one real issue: nuclear weapons. The linkage between peace and freedom is dissolved under the pressure of the threat of nuclear holocaust. Hiroshima poses the primary question for humanity today; Vorkuta, however appalling, is a secondary matter. It is impossible to imagine that this survivalism is what John Paul II meant at Hiroshima.

The real meaning of the pope's claim comes clear from John Paul's address at Hiroshima's Peace Memorial Park. There is an antiphonal refrain voiced four times in this remarkable statement: "to remember the past is to commit oneself to the future."[3] The pope understood the obverse of the survivalist temptation, which is to dismiss the threat of modern war: "Some people, even among those who were alive at the time of the events we commemorate today, might prefer not to think about the horror of nuclear war and its dire consequences. . . . But there is no justification for not raising the question of the responsibility of each nation and each individual in the face of possible wars and of the nuclear threat."[4] So the past, and present, danger of war must be remembered. But to what end?

> To remember the past is *to commit oneself to the future*. . . . To remember what the people of this city suffered is *to renew our faith in man, in his capacity to do what is good, in his freedom to choose what is right, in his determination to turn disaster into a new beginning*. In the face of the man-made calamity that every war is, one must affirm and reaffirm, again and again, that the waging of war is not inevitable or unchangeable. *Humanity is not destined for self-destruction.*[5]

The counterpoint to the threat of self-destruction, the pope argued, was the heritage of peace as public order in dynamic political community:

> Clashes of ideologies, aspirations, and needs can and must be settled and resolved by means other than war and violence. Humanity owes it to itself to settle differences and conflicts by peaceful means. The great spectrum of problems facing the many peoples in varying stages of cultural, social, economic, and political development gives rise to international tension and conflict. It is vital for humanity that these problems should be solved in accordance with ethical principles of equity and justice enshrined in meaningful agreements and institutions. The international community should thus give itself a system of law that will regulate international relations and maintain peace, just as the rule of law protects national order.[6]

Humanity's most pressing moral need is for an ethics and politics of responsibility for history.

That providence often speaks in paradox is a common understanding in the Judeo-Christian tradition. No responsible person can deny the dire threats under which we live and labor. Yet the very threats that make the human future questionable force us to confront the possibility of a mature humanity in which "it is no longer permitted . . . to be mediocre," in which we must *choose* the human future. The claim that ours is the first human generation which must choose the future can be read as an apocalyptic warning. It can also open up the prospect of a New Advent of human moral and political responsibility.

In 1965, at the close of the Second Vatican Council, it seemed that Catholicism was well positioned to contribute to this necessary moral and political evolution. The Catholic heritage complemented its personal and eschatological understandings of "peace" with a complex, richly textured concept of peace as rightly ordered, dynamic political community, an achievable, this-worldly possibility. The heritage was neither quietist (counseling a retreat into interiority, what Robert Nisbet has called the modern fascination with "self-spelunking,"[7]) nor utopian (urging the possibility of human perfectibility).

With significant help from the American Church, Catholicism had come to understand the bond between peace and freedom, and had developed an appreciation of democracy as an important, this-worldly answer to issues of pluralism, conflict, and community.

The Catholic heritage had thought its way through to a coherent political agenda for the gradual creation of the peace of political community on an international scale: an agenda that linked the goals of disarmament, international legal and political institutions, human rights (as the basis of transnational political community), and development (economic, social, cultural, and political) into a comprehensive whole.

The Church, by the end of the Council, also had a clear sense of its own role in the task of building peace with freedom: the Church was the guardian of a vision of human possibility, a teacher of the moral principles that had to inform any worthy (and workable) peace, a prophetic voice reminding humanity of its potential for good as well as its propensity for evil. The concrete tasks of peacemaking were not the immediate business of the Church as Church; the future to be sought and built was not another *Pax Christiana*, and the Church claimed no special competence with respect to strategy, development, or institution-building. What it did offer, as Pope Paul VI told the United Nations in 1965, was its experience as an "expert on mankind." That was the most basic, and thus most important, level of insight required to build a peace of political community worthy of those made in the image and likeness of God.[8]

Finally, the Catholic heritage understood the instrumental character of war. War existed for a definable purpose: war or its threat settled the argument over power, over who shall rule. But that very understanding suggested (as experience had borne out) the possibility of other means for resolving conflict. The heritage thus distinguished between "conflict" and "war." Conflict was the political meaning of the doctrine of original sin; war was a present, but not inevitable, means for prosecuting conflict. The tradition of just-war reasoning had sketched the moral economy of armed force in a way that linked its limited use to the defense of legitimate political community and human rights, rather than deeming it simply another expression of human fallenness.

Thus it seemed, in 1965, that Catholic Christianity, and American Catholicism in particular, were well positioned to meet Camus's challenge: to make a major contribution to the evolution of a human condition in which, even amid conflict, human beings would not have to choose between being victims or executioners. The task was surely one for many generations. But it was a task worthy of the Church, worthy of its heritage, worthy of those made in God's image and likeness.

Twenty years later, the picture was quite different.

In the years since Vatican II, many key voices in the American Catholic elite (academic, journalistic, and even episcopal) have virtually abandoned the heritage of *tranquillitas ordinis*: the concept of peace as rightly ordered political community was notably absent from the post-1965 American Catholic debate. American Catholicism's elites came to understand the gravamen of world conflict in terms drawn more from the ideology of the Third World caucus at the United Nations (reinforced by key teachings in the theologies of liberation) than from the heritage received up through Vatican II. Freedom was disengaged from peace, and an ill-defined notion of "social justice" was conceived as the essential precondition for peace. The political agenda of the classic heritage was fractured, and linkage across issues of human rights and disarmament, for example, became a term of opprobrium. The Church's role was often understood as that of a lobbyist for specific policies, despite the admission that the claims of moral judgment were weakest at this level. American Catholicism grew uncomfortable with just-war reasoning, and increasingly tolerant of a species of pacifism in which religious witness against war was persistently confused with judgments on military procurement, nuclear force modernization, and the strategy and tactics of counterinsurgency warfare. The results of this transformation of thought, which amply warrants the description of a "heritage abandoned," were clear in two key cases: the questions of nuclear weapons, and U.S. policy in Central America. The official Church in the United States, and its most vocal elites, did not help shape a more intelligent, less demagogic, morally sensitive public debate capable of advancing peace and freedom. Rather, the official Church, following many in its intellectual and journalistic elites, became a protagonist at one pole of the public argument already underway.

Why?

The classic Catholic heritage, as received and developed in American Catholicism through the Second Vatican Council, was not without serious vulnerabilities. The incompleteness of John Courtney Murray's grand project, the failure to define the moral underpinnings of an experiment in democratic pluralism, had left the heritage vulnerable to the passions of self-hatred that swept through America's political-cultural elites during the late 1960s. Paul Ramsey's counsel—to take up the challenge of *Pacem in Terris* and *Gaudium et Spes* by harnessing those visions of human possibility to the discipline of just-war reasoning—had not been heeded. Thus the vision of peace as political community on an international scale was either ignored, or used as a vehicle for minimizing the totalitarian dimension of the Janus-headed problem of peace and freedom. Just-war thinking was focused on particular questions of the *ius in bello*, rather than on the larger question of that *ius ad pacem* suggested in the interstices and basic moral trajectory of the just-war tradition.

Because there was no developed Catholic theology or politics of pacifism, it proved impossible to critically incorporate pacifist moral claims or pacifist political insights into the evolving thought of the American church. Attempts to blend pacifism and just-war reasoning multiplied, usually corrupting both traditions. Postconciliar pacifist thought virtually ignored the concept of rightly ordered political community as an alternative to war. Just law through democratic governance was not recognized as history's most successful experiment in nonviolent conflict-resolution. The anti-communism of the American Catholic heritage had not been developed into a persuasive body of thought on how the actions of democratic nations might lead to needed change in the internal and foreign policies of the Soviet Union and its clients. This failure of political imagination was in part responsible for the anti-anticommunism that helped define the abandonment of the heritage, and that persistently denied the ideological nature of the world's basic division.

The heritage had other vulnerabilities at the time of Vatican II. Would the spirited optimism of the Council (and especially of Pope John XXIII) devolve into a sentimentality about the human prospect that failed to take account of the darker side of human nature? Would the welcome reclamation of the Church's Scriptural patrimony result in precisely the kind of moral*ism* against which Murray warned? Would the claims of individual conscience, evident in the conciliar affirmation of principled conscientious objection, be detached from a sense of obligation to existing political community, thus eroding one instrument for the creation of transnational political community? The answers were yes, it might happen and yes, it did.

The vulnerabilities and weaknesses of American Catholic thought on war and peace in 1965 opened up the possibility of the abandonment of the heritage in the generation after the Council. But the heritage cannot be held entirely responsible for its own demise. In thinking through the why of the abandonment, the extraordinarily heightened circumstances of the times have to be acknowledged.

Even the casual observer of American religion knows that the postconciliar years were times of enormous tension in American Catholicism. The task of assimilating the teachings of Vatican II was monumental (and in fact remained unfinished twenty years after the Council's close). Liturgical changes put stress on the Catholic community, but "the changes in the Mass" hardly account for the uproar in the American Church following Vatican II, for survey research clearly demonstrates that the overwhelming majority of American Catholics approve the Council-mandated alteration of their liturgical, penitential, and devotional lives.[9]

No, the furies that swept through American Catholicism after the Council revolved around a different set of issues, primarily having to do with identity and authority in the Church. Quarrels over the contemporary meaning of priestly and vowed religious life led to an unprecedented exodus of men and women from rectories, convents, and monasteries. Nor was the laity spared postconciliar trauma. Pope Paul VI's 1968 encyclical *Humanae Vitae*, which reaffirmed the Church's traditional ban on artificial means of contraception (against both the experience of the American laity and the majority report of the papal commission on the subject), caused further shock waves throughout the American Church, precisely when debate over the implementation of Vatican II was running at full tide.

One result of all this turmoil, in a Church which once prided itself that "we think in centuries here," was a parade of new enthusiasms that followed (or collided with) one another with frantic frequency. The old certainties were gone, and the quest for new certainties ran at a furious pace. The passage from inspiration to practical program was measured in weeks, not years (much less centuries):

> Salvation history, kerygmatic theology, the third world, peace and justice, Renew, Genesis II, sensitivity training, encounter groups, marriage encounter, the charismatic renewal, feminism, nuclear disarmament, self-study, the rights of gays and lesbians, evangelization, divorced-to-be-married Catholics, etc., etc., etc.—all of these banners have been vigorously raised without much thought or reflection on what they might mean or serious consideration of how they might be promulgated, or responsible attempts to see if the promulgation worked. It is not in the nature of any of these causes that they necessarily be either superficial or aprioristic. In fact, however, superficial pragmatism, romantic amateurism, and self-righteous moralism have tended to replace study, reflection, careful planning and programming, detailed evaluation and dedicated perseverance after the first few failures.[10]

Andrew Greeley's catalogue of post-conciliar causes strikes all the more home because Greeley was rightly impressed with the Council's reforms.

To deplore this apparently endless parade of new enthusiasms is not to argue that a return to the aprioristic certainties of immigrant American Catholicism is either possible or desirable; it is neither. Demographic changes, as Catholics were successfully assimilated into the American mainstream, would have brought a time of turmoil in the American Church, with or without the impetus of Vatican II and the storm over *Humanae Vitae*.[11] But what specifically concerns us here is the abandonment of the heritage of *tranquillitas ordinis* in the midst of the general uproar in the Church. Look again at Greeley's list of postconciliar enthusiasms: Which have stayed the course for more than a few years? Without exception, it is the political enthusiasms that have remained. Given the temper of the American 1960s, it should not be surprising that political causes were the most seductive for American Catholic religious and lay leaders determined to forge a new, postconciliar identity.

And thus a second point: the abandonment of the classic heritage on war and peace took place, not only in a heightened time of commotion within the Church, but during an equally agitated period in American political culture. Within a generation,

Americans lived through a civil rights revolution; the assassination of a president (followed by the assassination of the assassin); the war in Vietnam; the driving from office of Lyndon Johnson; Watergate, and the first resignation of a president; the hostage crisis in Iran and the collapse of yet another presidency; the Soviet invasion of Afghanistan and the bruising public debate over "détente"; the emergence of the counterculture, the New Left, and militant feminism; the triumph of the mass media, particularly television, as the principal vehicles of popular culture; and a time of radical questioning about the purposes and means of education. Throughout these turbulent years, the American liberal elite, deemed the pioneers of a new frontier as recently as 1963, suffered a profound loss of confidence. And, as Norman Podhoretz put it, "into this vacuum of demoralization the ideas not of the right but of the New Left came pouring in. The result was the conversion of liberalism from containment to neo-isolationism, from anti-discrimination to preferential treatment, and from economic growth to redistributionism."[12]

Given the perceived symmetry between post-New Deal American liberalism and the Conciliar optimism of Vatican II, it should not have been surprising (no matter how disconcerting) that American Catholicism's elites followed the ideological path of their counterparts in the wider political culture. Turmoil in the Church was matched by turmoil in the country at large; the zeitgeist had no patience for the careful development of a complex tradition of moral reasoning. Moreover, the vulnerabilities of the classic Catholic heritage matched, in almost perfect one-to-one correspondence, many of the primary teachings of Vietnam-era American political culture. The failure to develop a sophisticated anticommunism aimed at changing the structure of power in the U.S.S.R. and thereby reducing the international threat of Soviet power intersected with the Vietnam-era rise of anti-anticommunism. The tendency of just-war theorists to reduce the question of intervention to its military dimensions intersected with a deterioration of confidence in America's ability to act for good ends in the world. The lack of a developed theology of pacifism was met by the emergence of a highly politicized pacifism during Vietnam. Absent the completion of the Murray Project in its largest dimensions, the American Church's elites were vulnerable to the "Amerika" currents taught by both antiwar activists and theologians of liberation. And the failure to develop further the concept of peace as dynamic political community left American Catholicism open to utopian connotations of *shalom* and to overpsychologized understandings of both conflict and peace.

There is, then, no single answer to the question of why the classic Catholic heritage collapsed less than a generation after the presumptively reinvigorating experience of the Second Vatican Council. Its own weaknesses constitute a partial explanation. So, too, do the conditions under which those weaknesses were exacerbated—the time of troubles that beset both Church and society in America during the 1960s and 1970s. However one weighs the various influences, though, the result seems painfully clear: the intricate and abundant heritage of *tranquillitas ordinis*, as it had been developed in American Catholicism up through the time of Vatican II and the work of John Courtney Murray, was not further developed, but abandoned.

And yet the moral and political necessity of addressing the dilemma of peace and freedom is now more, rather than less, urgent than at the most heightened moments of the times of abandonment. This necessity derives from both the threats and the possibilities confronting humanity at the brink of the third millennium. The threats can be quickly summarized. Nuclear weaponry continues to proliferate throughout the world. During the times of abandonment, the Soviet Union achieved strategic equality (and, on some indices, superiority) with the West. The U.S.S.R. also gained important footholds in Southeast Asia, the Horn of Africa, and southern Africa, while seeking a similar entry into Central America—enterprises made possible by a massive increase in conventional weapons, airlift, and naval capabilities.

But there is more to be accounted for than the threat of Soviet power and purpose. Chaos, in the guise of uncontrolled international terrorism, has become a hallmark of the late twentieth century. At times, as in the attempt to assassinate Pope John Paul II, the connection between the Soviet threat and the predatory destructiveness of the terror network comes into clearer focus. At other moments it seems as if "mere anarchy" has indeed been loosed upon the world. Some might argue, in the face of these harsh realities, that ours are not the times in which to raise up again the vision of the peace of political community as a human possibility. Yet it is precisely in times like these that this vision is most necessary: for, without a horizon to set the direction for present policy and the standards to guide it, we, too, may unwittingly contribute to the reign of the kingdom of chaos.

But there are other, brighter, realities requiring the orienting vision of a reclaimed and renewed heritage of *tranquillitas ordinis.* There is the opportunity, suggested by John Paul II at Hiroshima, for a mature humanity to make a conscious choice for the future: What kind of future shall we choose, guided by what values, incarnated in what institutions? There is the social revolution of human expectation throughout the world: Toward what ends will the idea that things need not be the way they always have been be channeled? Can the knowledge that hunger, disease, poverty, racial and sexual discrimination, illiteracy, and high rates of infant mortality are not fixed in the order of things strengthen the bonds of human responsibility toward a world of peace, freedom, and justice? Or will these new understandings, and the social volatility they create, erode the fragile peace-of-a-sort that now exists?

Faced with these questions, the classic Catholic heritage, reclaimed *and developed*, is a necessity, not a luxury. That the dynamics of international public life, under the pressure of widespread demands for change, will continue to threaten an imminent explosion cannot be doubted. But there are explosions that lead to chaos, and explosions that can be channeled to ends more worthy of the images of God in history. Those charged with the stewardship of creation have no moral option but to strive toward modes of dynamic public order worthy of men and women, and away from a chaos fit only for Hobbesian brutes.

American Catholicism bears a particular responsibility toward these issues today, because of its distinctive development of the classic heritage and because of the position of the United States in world affairs. The contests between freedom and tyranny, and between right order and chaos, will be heavily influenced by American actions in the world. Whether humanity makes progress toward the peace of public order in political community will have much (if not everything) to do with the policies pursued by the world's leading democratic power. American Catholics are a formidable part of the national political culture that will shape the course of those policies. What tack, then, will American Catholicism take?

To continue down the path taken since the Council, whose landmarks have been identified above, would mean disaster, both for the Church and the United States. The ecclesiastical disaster may be defined as succumbing to the temptation of Esau: an American Catholicism whose intellectual and leadership elites continue to teach the abandonment of their heritage will have traded its birthright for a mess of pottage. The political-cultural disaster will be an American Catholic contribution to the further polarization of national debate over America's role in world affairs, at precisely the time when partial insights held at both poles of the present argument need to be gathered into a fuller, more comprehensive truth.

The electoral results of 1980 and 1984 notwithstanding, American political culture remains deeply divided on the large questions of national purpose in the world. Isolationisms old and new lurk just beneath the surface of public controversy over particular issues. The dual nature of the contemporary dilemma of world affairs has yet to be faced and addressed in ways that could advance the prospects of both peace and freedom. The temptation to dissolve the dilemma—toward a survivalist fixation on the problem of nuclear weapons or toward a single-minded focus on the totalitarian threat—remains strong, and defines much of contemporary political discourse. A voice for both peace and freedom, a voice echoing the themes of *tranquillitas ordinis*, is a pressing public need. American Catholicism could raise such a voice. This will require frankly confronting the abandonment of the heritage that has largely defined the American Catholic debate on war and peace, security and freedom, over the past twenty years.

But raising a voice for peace and freedom cannot be an exercise in antiquarianism; more than a repetition of old formulas is needed. To cite proof texts from John Courtney Murray in support of predetermined political judgments is no answer to those who would do the same with the words of Ernesto Cardenal (or Stanley Hoffmann). The American Catholic task in the present moment is to creatively reclaim its heritage of reflection on issues of war and peace in ways that lead to a genuine development of that heritage. The weaknesses of the heritage must be acknowledged and confronted as part of its reclamation. If the heritage is to be further developed, the principles and analytic patterns of the reclaimed heritage must be brought into play on the full range of present dangers and possibilities in international life, many of which could barely be imagined in Murray's day.

In their 1983 pastoral letter, the bishops of the United States urged that American Catholicism be "a Church at the service of peace."[13] Reclaiming and developing the heritage that has been so largely abandoned will require a new understanding of what it means to be a *Church* at the service of peace, and of what it means to be a Church at the service of *peace*. The first order of business is theology.

CHAPTER 12

To Be a *Church* at the Service of Peace

American Catholicism's reclamation and development of the heritage of *tranquillitas ordinis* must begin with theology. What does it mean to be "a *Church* at the service of peace"?

The Church is a communion of saints across the boundaries of time. The task of reclaiming an abandoned heritage requires an understanding that those who have gone before may be able to teach us things. Cicero's warning about the dangers of historical amnesia—"He who is ignorant of what happened before his birth is always a child"[1]—is especially pertinent for American Catholicism. The Church in the United States has not been distinguished by a commitment to the study of its own history. Thus, contemporary American Catholicism needs to pay particular heed to G.K. Chesterton's famous definition of "tradition":

> Tradition is only democracy extended through time. . . . Tradition may be defined as an extension of the franchise. Tradition means giving votes to the most obscure of all classes, our ancestors. It is the democracy of the dead. Tradition refuses to submit to the small and arrogant oligarchy of those who merely happen to be walking about. All democrats object to men being disqualified by the accident of birth; tradition objects to their being disqualified by the accident of death. Democracy tells us not to neglect a good man's opinion, even if he is our groom; tradition asks us not to neglect a good man's opinion, even if he is our father.[2]

Or our grandfather. Or our grandfather's father's archbishop. Reclaiming the American Catholic heritage requires a deepened, more sophisticated appropriation into contemporary thought of the universal Church's rich patrimony of reflection on the moral problems of war and peace, security and freedom. It also requires a special effort to reclaim the distinctive American Catholic contribution to this patrimony, with particular attention to the relationship between democratic values and institutions and the peace of dynamic political community.

But the Church is not just a transtemporal communion of saints; the Church is incarnated in the present. To reclaim the heritage is necessary, but insufficient; reclamation must lead to development. The Church's self-understanding as a religious community "at the service of peace" will thus take on a specific cast in each particular moment of human history. Certain themes will be given special prominence, in light of the conditions of time and place in which the church stands as witness to the eternal truths of God's self-revelation in Jesus Christ. Throughout fifteen hundred years of systematic reflection on the moral problem of war, the Church has persistently taught that the peace of rightly ordered political community is a human possibility. The possibility of peace sets the ground on which the Church can insist on the necessity of its ministry of peace. But to affirm that peace is a human possibility is not a simple statement of political faith; it is, most fundamentally, a statement of theological anthropology. To be a *Church* "at the service of peace" means to be a Church with a distinctive view of the human prospect.

THE INCARNATE CHRIST AND THE HUMAN PROSPECT

The theological anthropology of peace has been a leitmotif in the preaching and ministry of Pope John Paul II. In his 1981 address to the United Nations University, the pope argued that "it is only through a conscious choice and through a deliberate policy that humanity can survive." Yet the pope distinguished his remarks from those of many secular survivalists by posing, amid the dangers of modern war, the possibility of a more mature humanity: not in spite of, but precisely because of the many present dangers that surround us, "humanity is being called to take a major step forward in civilization and wisdom."[3]

John Paul II's call for the moral evolution of humanity reflects, in tempered form, the optimism about the human prospect so characteristic of *Pacem in Terris* and *Gaudium et Spes.* The pope's hearkening back to the spirit of Vatican II is itself interesting, for the postconciliar period has seen an abandonment of conciliar optimism in key intellectual sectors of the Church. Catholic theological conservatives were never comfortable with a hopeful reading of the human prospect under the conditions of modernity, and their current disenchantment with the confidence of *Pacem in Terris* and *Gaudium et Spes* is not surprising. What is surprising is the degree to which conciliar optimism has dried up among those of a more liberal theological bent, who were, after all, the main crafters of the "Pastoral Constitution on the Church in the Modern World." Liberal pessimism was one current in the American bishops' 1983 pastoral letter, where the bishops wrote that the world was becoming "increasingly estranged" from Christian values. "In order to remain a Christian, one must take a resolute stand against many commonly accepted axioms of the world."[4] There are important truths here; but the tone quality reflects hopes turned to ashes, rather than the more buoyant sense of human possibility that infused the conciliar period. John XXIII's admonitions to "prophets of doom" would have to be aimed in more than one direction today, it seems.

John Paul II has set himself and his Church the task of reclaiming and renewing the council's optimism, albeit in a more tempered form. The maturation of humanity for which the pope calls, and of which he wishes the Church to be an instrument, would combine Johannine optimism with a more Augustinian sense of the demons still abroad in human hearts and institutions. What is required, in other words, is a reclamation and development of the Catholic heritage of moderate realism, or "temperate optimism." A postconciliar idealism without illusions would face squarely, and without inhibition, the complex of threats to the human future represented in the indivisible problem of modern war and modern totalitarianism. But it would not stop there.

It would complement Augustinian sobriety with an equally frank recognition of those contemporary realities that suggest evolution, not deterioration. It would celebrate the modern social revolution in its insistence on the dignity of men and women irrespective of race, caste, sect, or sex. It would gratefully acknowledge that tolerance, personal liberty, and equality of persons before the law are among the distinctive values of modernity. It would welcome the transportation, communication, and economic revolutions that have, for the first time in human history, made it possible to speak of "a world" in a meaningful sense. Each of these facets of modern life has its shadow: a revolution of rising expectations can lead to chaos, tolerance can dissolve into license, interdependence means that no nation any longer controls its environment. But, viewed through the lens of moderate realism or temperate optimism, modernity has not simply created the tinder for holocaust (in its nuclear or Gulag forms); modernity has also created conditions for the possibility of work on the agenda of *tranquillitas ordinis* suggested by *Pacem in Terris* and reaffirmed by *Gaudium et Spes.* John Paul II's call for "a major step forward in civilization and wisdom" is more than a pious wish. Without disregarding the myriad obstacles now blocking the way, the pope's call is now, for perhaps the first time in human history, capable of modest realization on an international scale. At the very least, the circumstances are such that a realistic, nonchimerical effort can be made.

The Church most powerfully addresses the possibilities of the human condition when its anthropology is intimately related to its Christology. This, too, has been a prominent theme in the catechesis of John Paul II. As his intellectual biographer George Huntston Williams puts it, Karol Wojtyla has been arguing "that Christ is not only a revelation *of God* and his salvific will for all mankind through the Church, but also a revelation *of man*, of what man was intended to be at creation and is by reason of the Incarnation of the Son of God and by reason of the Crucifixion, Resurrection, and Ascension of the God-Man Jesus Christ."[5] John Paul II made the point himself in his inaugural encyclical, *Redemptor Hominis:*

> The Redeemer of the world! In him has been revealed in a new and more wonderful way the fundamental truth concerning creation to which the Book of Genesis gives witness when it repeats several times: "God saw that it was good" (Genesis 1, passim). The good has its sources in Wisdom and Love. In Jesus Christ the visible world which God created for man—the world that, when sin entered, "was subjected to futility" (Romans 8:20)—recovers again its original link with the divine source of Wisdom and Love. Indeed, "God so loved the world that he gave his only Son" (John 3:16). As this link was broken in the man Adam, so in the Man Christ was it reforged. . . . Rightly, therefore, does the Second Vatican Council teach: "The truth is that only in the mystery of the Incarnate Word does the mystery of man take on light. . . . Christ the new Adam, in the very revelation of the mystery of the Father and his love, *fully reveals man to himself* and brings to light his most high calling."[6]

John Paul II's Christology rests against the background of Poland's historical suffering and the Nazi genocide of the people of Abraham, Moses, and Jesus.[7] There is nothing of the Pollyanna in Karol Wojtyla, the man or the theologian. Yet the pope's Christology is so infused with the mystic experience of God's ever-present care for the world that Wojtyla can locate even the Fall within the ambit of divine providential love:

> Through the Incarnation God gave human life the dimension that he intended man to have from his first beginning; he has granted that dimension definitively—in the way that is peculiar to him alone, in keeping with his eternal love and mercy, with the full freedom of God—and he has granted it also with the bounty that enables us, in considering the original sin and the whole history of the sins of humanity, and in considering the errors of the human intellect, will, and heart, to repeat with amazement the words of the Sacred Liturgy: "O happy fault . . . which gained us so great a Redeemer!"[8]

The Christological proclamation of John Paul II is one of human possibility under divine grace and favor. Such a Christology is particularly well suited to an age in which human depravity has plumbed new depths of evil. In the face of modern war and modern totalitarianism, human beings should need little reminder of their individual and corporate capacities for the most monstrous crimes. As the pope himself put it at Auschwitz, "Can it . . . be a surprise . . . that the Pope who came to the See of St. Peter from the diocese in whose territory is situated the camp of Oswiecim, should have begun his first Encyclical with the words *Redemptor hominis* and should have dedicated it as a whole to the cause of man?"[9]

What may well be needed, though, is an affirmation of that creative and redemptive love that still bounds the human universe, calling it to a more complete wisdom, a more searching truth. For John Paul II, the sacrifice of Jesus Christ is not so much a satisfaction of the Father's righteous anger as it is of "the Father's *eternal love*, that fatherhood that from the beginning found expression in creating the world, giving man all the riches of creation, and making him 'little less than God,' in that he was created 'in the image and likeness of God.'"[10] The mystery of salvation is a mystery of love in which creation is renewed by the Son who "satisfied the fatherhood of God."[11]

The mystery of the Incarnation not only reveals God to Man, but unveils the full meaning of Man himself. We can know, through the grace and mercy of God, that we are not created for Hobbesian brutishness. Humanity, in the mystery of God's providential will, is the material through which God chooses to speak and act when God wills to express his infinite Love in a finite world. As Paul Johnson put it, John Paul II believes in the depth of his person "that man's existence and God's are inextricably entangled. All human experience, properly understood, is experience of God, and therefore to confess belief in God is to embark on the quest for truth and meaning, and the freedom that men find in God and God alone. The essence of Christian humanism is that man is driven by some interior force to realize his full personality; that the force is from God, in the form of grace; and that the realization is toward God, through the incarnation of his Son."[12] Thus *Redemptor Hominis* spoke of the present moment in human history as a "new Advent, a season of expectation" not only for the Church, but for all humanity.[13]

These central themes in the Christology of John Paul II mirror currents of thought in the theological anthropology of one of the great conciliar theologians, Karl Rahner. Like Rahner, John Paul offers a view of the human condition that emphasizes our innate, ever-openness to a divine word in history. That word is freely spoken by God, who also freely respects our freedom to be a "hearer of the word."[14] Yet, again like Rahner, John Paul argues that this process of divine invitation and human response is not an intellectual abstraction, but took definitive form in the life, death, and resurrection of Jesus Christ. The moral and spiritual evolution of humanity is not circular, but linear. The Word, decisively received and incarnated in the humanity of Jesus of Nazareth, points the trajectory of human evolution into the future.[15] We come from God, and we are meant for God; through the Incarnation, this is no longer a theoretical affirmation, but a reality that has been lived to the full in human history. The Church's evangelical task is to proclaim this good news: that faith in God and entry into God's love and wisdom does not drive us out of our creatureliness, but more deeply into it. Those most attuned to the depth of the mystery that lies at the heart of every human life are those most receptive to the divine invitation to communion offered in creation and history.[16]

John Paul II thus offers a world in mortal danger the vision of an incarnational humanism—one that faces the reality of sin and takes the full measure of the evil of our times but yet understands (through the mystery of grace) that the broken vessel of humanity was the material chosen by God for his complete self-revelation in redemptive love to the world. The Incarnation is a revelation of divine mercy *and* human possibility. This is, as the disciples of Jesus would say in another context, "a hard saying" (John 7:60). But its difficulty lies, not in its acknowledgment of human depravity (which is hardly difficult to accept), but in its affirmation of human possibility. In the twentieth century, belief in the Fall comes easily; it is belief in men and women as the images of God in history that is difficult.

Given the profound nature of the threats to human life and freedom that mar our age, should the themes of John Paul II's incarnational humanism be prominent in reclaiming and developing the Catholic heritage of *tranquillitas ordinis*? Although it is certainly true that less sober views of the human prospect (at times, bordering on the utopian notion of human perfectibility) lurked beneath the surface of the abandonment of that heritage, it is more important to recognize and challenge the pessimism that leads to survivalist distortions of the meaning of peace.

Survivalism, with its reduction of all other values to the value of sheer physical escape from the present danger of nuclear war, is rooted in historical and moral claustrophobia. Survivalism is an expression of what Chesterton called a "narrow universality . . . a small and cramped eternity." The survivalist mind "moves in a perfect but narrow circle." And while "a small circle is quite as infinite as a large circle . . . it is not so large."[17] Survivalism can only be of moment in a moral culture where the future seems closed. A Christologically ordered incarnational humanism meets our pressing moral (and psychological) need for a larger infinity, a less cramped eternity, a more open horizon of human possibility against which to shape the future.

The answer to survivalism and its draining of the moral powers of human intentionality is not to reconvict Man of original sin. Survivalists know all about human finitude, even though "sin" is not their favorite metaphor for it. No, the answer to survivalism is precisely an incarnational humanism that, fully cognizant of the human propensity for evil, still insists that we are made for larger purposes. Recreating a conviction that the future is open—defining, to reverse Chesterton, a spacious eternity as a moral horizon for action in this world—must be a central theme in the reclamation and development of the heritage of peace as dynamic political community.

The proclamation of human possibilities under grace should be a prominent motif in the Church's prophetic address to the moral problem of war and peace. Too often in the postconciliar period the prophetic quality of American Catholicism has been measured by how stridently one opposes government policy; the more strident the formulation, the more prophetic it is often judged to be. This badly distorts the meaning of prophecy, which has to do with truth rather than rhetorical fervor. No doubt there have been, are, and will be moments when the Church's prophetic mission requires a sharp and unambiguous no to the principalities and powers of this world, including those resident in Washington, D.C. Yet there is a larger prophetic task today: prophetic affirmation.

The contemporary focus of the prophetic mission of the Church should be understood as that persistent call to "a major step forward in civilization and wisdom" of which John Paul II spoke at Hiroshima. Such a prophetic call would be rooted in incarnational humanism. It would say no to those doomsayers and survivalists who proclaim narrow universalities, and yes to human capacities, in truth and under grace, to shape a future reflective of the divinely authored and incarnationally affirmed dignity of human beings. The true measure of prophetic utterance will not be its capacity to distance the Church from the American political community, but its ability to call the United States to a fuller expression of those claims about the dignity and end of the human person that are the moral substance of the American proposition and experiment, and that ought to define the moral horizon of America's inescapable involvement with the dilemma of peace and freedom in the modern world.

ESCHATOLOGY AND HISTORICAL RESPONSIBILITY

The Church's Christological proclamation about the nature and destiny of Man flows directly into eschatology. Because the historical event of the Incarnation reveals the fullness of God's creative intention, we now live, in an anticipatory way, in a new age, or as the pope prefers, a "New Advent." The mystery of the Incarnation is not a withdrawal from history but a call to historical responsibility. Human beings actualize their God-given instinct for the truth through action in history. John Paul II's Christology leads naturally into an eschatology of responsibility for history.

To view the present historical moment as a New Advent is to understand today's Christian community as the early Church: not in a sectarian re-creation of that small band of the elect hiding in the upper room in anticipation of the imminent return of the Lord, but in a longer historical perspective. To see in the present moment a New Advent opens up the prospect of a long and, hopefully, bountiful future. As we approach the third millennium of the Christian era, the temptation to see ours as the times of the end will grow. But the trajectory of Catholic incarnational humanism suggests a different angle of vision, which sees ours as times of beginning. Thus the pope's call for a "major step forward in civilization and wisdom" should not be proclaimed as the epilogue of a rapidly ending human story (putting the house in order before the new deluge) but as a challenge to the continuing maturation of humanity, whose final end has been revealed in the Incarnation and Resurrection. Understanding the Christian community of today as the early Church would not, in this perspective, call us back to the past but into the future. *We* are the early Church; the future remains to be forged according to our vocation as the images of God in history. The story, rather than ending, may be only beginning. This is the eschatological message of a New Advent most in need of vigorous proclamation in an age of narrow universalities.

Such a proclamation would allow a reconsideration of the peace of *shalom* that is proof against those utopianisms of human perfectibility that lie beneath the abandonment of the concept of peace as political community, and beneath the totalitarian temptation. The Church as witness to a New Advent would proclaim *shalom* as the horizon of human possibility, not as an achievable, this-worldly end. But it would affirm the horizonal vision. Set in the context of Catholic incarnational humanism, *shalom* points a direction for human moral agency in the present: we can know that toward which we should strive, even as we understand that its completion is a matter of God's time, not our own.

Shalom as orienting horizon would also set standards by which this-worldly efforts at the peace of dynamic political community would be judged. *Shalom* as moral horizon reminds us that a more perfect charity, a more complete justice, and a truer freedom all remain to be achieved. But *shalom*, within the ambit of incarnational humanism, would also caution against the teaching that justice is the precondition to peace. Complete justice, in the visions of Isaiah, is an eschatological reality, a hallmark of the Kingdom in its fullness. *Tranquillitas ordinis*, the achievable expression of a maturing humanity, would not, on this view, be disparaged, but understood as the condition for the possibility of work toward the justice of the Kingdom that will always stand before us as judgment on the present and direction for the future.

Eschatologies of impending disaster are often on display today; they will multiply as the talismanic year 2000 approaches. But the incarnational humanism of John Paul II and other contemporary Catholic Christologists offers an alternative vision: not an eschatology of catastrophe but of completion. Modern eschatologies of catastrophe are rarely based on a biblical sense of God's action on history; rather, they seem rooted in a profound loss of heart over the human prospect. Morris West's two novels, *The Shoes of the Fisherman* and *The Clowns of God*, suggest the shift in some religious imaginations on this point. *The Shoes of the Fisherman* was written at a moment of Teilhardian optimism in the early 1960s. Twenty years later, having failed to re-create the world in its preferred image and likeness, that optimism has lost, not only heart, but faith—faith in the possibility of a renovated, "Christified" humanity. *The Clowns of God* is a novel of disillusionment; West prefers his second fictional pope, who proclaims the imminent end of the world, over his former creation, the pope from the Gulag who proclaims human possibility, because West seems to have lost faith in his earlier vision of a transfigured human condition.

One may welcome the Augustinian sobriety that such a disillusionment entails; but only to a point. Augustine's gravity is not our only teacher. The basic Christian affirmation is that the worst in history has already happened, on Good Friday, and that God gave his answer to that radical evil on Easter. Easter faith, the faith of a moderately realistic hope for the future, is the most appropriate contemporary answer to the narrow universality of an eschatology of catastrophe, because it more adequately responds to the charge in Genesis that we be the stewards, not the undertakers, of creation. Catastrophe, in the world's terms, is a possibility. But to assume its inevitability involves, not only a loss of faith, but a loss of hope and charity as well. Nothing drains the moral power required to work intelligently on the peace of political community faster or more surely than the notion that a catastrophic end to history is inevitable, and in fact close at hand. An eschatology of completion does not deny the possibility of disaster. It denies, on grounds of faith and reason, its inevitability. And it affirms that even that which appears as catastrophe to human eyes may yet be brought to completion through the mysterious providence of a God whose ways are not our ways.

To be "a Church at the service of peace" means to be a herald of good news: that God's action in the Incarnation, and the response of the man Jesus to God's invitation to communion, have revealed the full meaning of human life; that human beings can, under the promptings of grace, order their lives and their societies in ways worthy of those who are the images of God in history. A Church at the service of peace has a powerful message to convey, not only about God, but about Man. The future is not closed, but open. The human story today need not be conceived as epilogue but as prologue, for, in Jesus Christ, Man has been offered "full awareness of his dignity, of the heights to which he has been raised, of the surpassing worth of his own humanity."[18]

TOWARD A THEOLOGY OF PEACE: FIVE ISSUES

The American bishops' call for the development of a "theology of peace"[19] will best be met by a process of reclaiming and developing the classic heritage of *tranquillitas ordinis.* Catholic incarnational humanism, flowing into an eschatology of responsibility for history, should form the base on which that theology of peace is constructed, as the household of faith brings the riches of its tradition to bear on the linked problems of war and peace, security and freedom. But a theology of peace will involve more than affirming and proclaiming central themes in Catholic doctrine; it will necessarily involve several other derivative theological questions. Five issues, now widely disputed in American Catholicism, suggest some of the most crucial problems to be sorted out in this process of theological development.

The great question of Christianity and culture, or the issue of *the church/world relationship*, needs considerable attention in the development of an American Catholic theology of peace. Which of the five models identified by H. Richard Niebuhr in *Christ and Culture* will shape an American Catholic theology of peace?

Niebuhr's model of "Christ the Transformer of Culture"[20] seems most congruent with the main trajectory of Catholic social thought, and most appropriate to the demands of our age. A transformation model of the church/world relationship sets the theological ground on which to mount a principled challenge to the voices of sectarianism that played a key role in the abandonment of the classic Catholic heritage. The Church, on this model, is not a small elite of the saved set over against the world, but a leaven in the midst of the world and its affairs.

The transformation model challenges sectarianism by affirming the continuity, in God's providential plan and favor, of the orders of nature and grace, creation and redemption. It draws its deepest inspiration from that understanding of the Incarnation as the revelation of man's God-given end that permeates the catechesis of Pope John Paul II (and a host of other conciliar Catholic theologians). Moreover, by stressing the Christian community's responsibility for history, the transformation model establishes the foundation on which a "peace church" in the service of dynamic, rightly ordered political community can be built. This understanding of the Church's ministry in the created order would draw on an eschatology of completion just as sectarianism is often related to an eschatology of imminent apocalyptic catastrophe.

The transformation model proposes the world as an arena of moral responsibility, rather than as "the enemy." It calls the People of God into history, rather than demanding their withdrawal from it as the price of Christian fidelity. There is ample room for religious witness against the evils of the present moment within this conception of the Church's ministry to the world. But the transformation model, emphasizing human capacities to order the world in a manner worthy of the images of God in history, locates witness within a theology of conversion, rather than in a narrower theology of judgment and condemnation. Judgment and condemnation are required of the pilgrim Church in certain times and places. The transformation model, however, suggests that even judgment and condemnation be related to the task of worldly stewardship given us at the creation. We judge and condemn when necessary, not merely to convict the world of its sin ("For God sent the Son into the world, not to condemn the world, but that the world might be saved through him" [John 3:17]), but to call it to that fullness of life which the Lord promised to those who were faithful disciples ("I came that they may have life, and have it abundantly" [John 10:10]). A Church seeking to reclaim and develop the heritage of *tranquillitas ordinis* would take, as its central prophetic task, the proclamation of the human possibility that entered history through the Incarnation. And from that proclamation it would derive its transforming mission in the world during this "interim time" between the Easter revelation of God's full intention and the completion of God's creative purpose in the Kingdom that is to come.

American Catholic understandings of *morality and foreign policy* shifted rapidly from 1965 to 1985. The new theology of peace would challenge many of the concepts that have been central to this shift. It would reject, flatly, the biblical literalism that marred the American Catholic debate over nuclear weapons and strategy—not for the sake of placing biblical themes and images on the margins of the Catholic heritage, but precisely for the sake of the integrity of Scripture. The theology of peace will reject that moral*ism* in which policy prescriptions are derived in a literalist fashion from the Sermon on the Mount, and will challenge American Catholics to a more sophisticated entry into the inescapable hermeneutic circle in which biblical themes and images shape, but do not univocally determine, the moral horizon against which we measure our responsibilities for peace, security, and freedom.

Most fundamentally, a developed American Catholic theology of peace will reject what Alasdair MacIntyre has called "emotivism" and will insist that morality is a function of intelligence.[21] The theology of peace must, in other words, be related to what Murray called "the tradition of reason." Doing so requires more than a reclamation of classic Thomistic moral theory; development is also required. The terms in which contemporary American Catholicism enters the tradition of reason and explores contemporary issues in its light will not be identical to Murray's. And yet, unless the Church can develop a moral language that speaks across a pluralistic society, its theology of peace will more closely resemble the Baptist's "voice crying in the wilderness" than it will a Church in active dialogue with the civic community.

How this might be accomplished is beyond our scope here. The relationship, for example, between intentions, acts, and consequences in Catholic social ethics is, and will doubtless remain, hotly controverted within the theological community. Amid that controversy, however, a central teaching must be clear: that there is no substitute for the exercise of reason and intelligence in the conduct of the moral life, in either personal or social terms.

Whatever its specific judgments on methodology in moral theology, the new theology of peace will forthrightly (and charitably) challenge those voices of abandonment which claimed that the degree of one's passion on issues of war and peace, and the purity of one's intentions, established the depth of one's moral concern and the weight of one's claims on the Church's attention. Conversely, an emerging theology of peace would challenge the utilitarian temptation to which some currents in the theologies of liberation are so vulnerable. The Church's ministry of moral intelligence on questions of war and peace, security and freedom, is, in the final analysis, measured by standards different from those of the public policy process at its narrowest end. The Church's primary responsibility is to seek the truth, and to do so in ways that illuminate the difficult choices faced by those charged with responsibility for the common good.

The theology of peace must think harder about *pacifism* than has been the case in American Catholicism since Vatican II. The legitimacy of the pacifist conscience for individual Catholics has been made plain in the teachings of the Council, of recent popes, of the American bishops, and of many Catholic theologians. That "Catholic pacifist" is not an oxymoron is no longer in dispute. But recognizing the legitimacy of a personal pacifist commitment opens, rather than closes, the needed debate. How shall the individual pacifist conscience function within the Church? How shall the pacifist conscience address the wider political community? Gordon Zahn's call for a nonsectarian Catholic pacifism taking responsibility for political action in history is welcome. But toward what ends, measured against what standards, shall that distinctive Catholic pacifism work?

A pacifism contributing to the reclamation and development of the classic Catholic heritage would have several distinguishing characteristics. It would challenge the selectivity by which some pacifists mount radical critiques of American military strategies and programs, but turn a blind eye to revolutionary violence in the Third World. The American bishops rejected such a challenge during their final debate over the 1983 pastoral letter on war and peace; one commentator called Bishop John O'Connor's call for consistency "snide."[22] It is not, of course, snide at all. A pacifism true to its own standards should have mounted the challenge of its own accord.

American Catholic pacifism would also, on principle, eschew military prescriptions. The pacifist conscience cannot reject the resort to armed force in the defense of rights and then offer counsel on the force structure of the American military. Pacifism only corrupts its distinctive message and mars the quality of its witness when it enters the arena of prudential judgment where, as "The Challenge of Peace" teaches, other criteria drawn from just-war analysis and its moral economy of the use of armed force must apply. Pacifists who wish to enter the policy arena must consciously adopt standards of judgment distinct from those they choose to apply in their own lives.

American Catholic pacifism could still play an important role in the policy debate by focusing on the development of international legal and political institutions for prosecuting conflict without the use or threat of mass violence. A pacifism that wishes to enter the arena of responsibility for history would see law and democratic governance as the most widespread, effective means for resolving conflict in ways congruent with pacifist insights and values. American Catholic pacifism has not been distinguished by its thoughtfulness on these questions. Primary attention has been focused on resistance to American military programs. And when pacifists discuss alternatives to war, themes drawn from the U.N.'s Third World caucus tend to dominate.

Yet why should not an American Catholic pacifism serious about democratic political community as a means for resolving conflict be at the forefront of work to reform international agencies like UNESCO? Why should not American Catholic pacifism take the lead in challenging the gross politicization of the U.N. General Assembly, and the obscenity of acts such as the identification of Zionism with racism? There is no reason, in principle, why pacifists should not develop the most trenchant critique of present international organizations, and then point the way to their reform (or replacement). The obstacles to such a pacifist contribution to the reclamation and development of the Catholic heritage lie, not in principled pacifism per se, but in the deterioration of principled pacifism through political understandings derived from sources external (and, in many cases, hostile) to the pacifist conscience.

Principled American Catholic pacifism would reject fundamentalist or literalist approaches to Scripture in the moral debate over war and peace, security and freedom. Principled pacifism would also make a distinctive contribution to the Church's pastoral ministry on questions of personal conscience, war, and civic responsibility. To urge draft resistance as a means for stripping the gears of the American "military machine" is not an act of pacifist witness; it is anarchism (at best; sedition at worst). Yet the question of the draft does bring issues of war and peace down to the sharpest level of personal moral choice. The Church thus has an important pastoral ministry to those faced with involuntary (or voluntary) military service. Principled pacifists (with nonpacifists, of course) could take up the ministry of draft counseling. But in doing so, they would stress not only the prerogatives of personal conscience, but the obligations of citizenship in a society whose laws protect the right of conscientious objection. This has not been a consistent theme in American Catholic pacifism. Yet the reasons why are, again, not indigenous to the principled pacifist conscience. The themes dominating the rhetoric of contemporary draft registration "resistance" are largely drawn from Vietnam-era New Left teachings about the corruption of American society and its necessarily corrupt intervention in the world.

Pacifists serious about alternatives to violence in the resolution of conflict—that is, pacifists serious about democratic law and governance—should be prominent among those challenging the idea that personal conscientious objection to military service can be uncoupled from the moral demands of citizenship. Conscientious objection without a parallel commitment to civic responsibility—without a concrete expression of one's commitment to work for the peace of political community—is not pacifism, but anarchism (however confused and muddled). Principled pacifism would bring this teaching into the life of Church and society and, in doing so, would contribute to the reclamation and development of the heritage of *tranquillitas ordinis* in a sound theology and politics of peace.

Just-war theory must also be rejuvenated as American Catholicism builds a theology of peace that reclaims and develops its abandoned heritage. Recalling Paul Ramsey's injunctions in the wake of *Pacem in Terris* and *Gaudium et Spes*, which have been frequently cited above, is of primary importance here. Ramsey's challenge to address John XXIII's and the Council's call for the evolution of international political community *through*, rather than *around*, the canons of just-war reasoning has not been well met in postconciliar American Catholicism. Welcome attention has been focused on the historical roots of just-war reasoning, and several scholars have made significant contributions to the development of the *ius in bello* under the conditions of contemporary nuclear and conventional warfare. Yet, as William V. O'Brien has persistently argued, little attention has been paid to the larger questions of the *ius ad bellum*. Creating a new spectrum of *ad bellum* argumentation would involve analyzing the specific claims that the threat of totalitarianism makes on democratic powers in the world today. But it would go further.

Catholic teachings about the possibility of peace in rightly ordered political community are closely related to just-war reasoning. The classic *ad bellum* criteria of just cause, competent authority, last resort, formal declaration, and proportionality contain, in their interstices, concepts of rightly ordered political community as a preferred means for resolving those conflicts that lead to war. Moreover, as Ramsey argued after the Council, just-war reasoning establishes the "moral economy of power" that would have to guide the evolution of *tranquillitas ordinis* on an international scale, as it once guided the evolution of peace within individual nation-states. Contemporary American Catholic just-war theory should take up, within its own canons of reasoning, those *ius ad pacem* questions that create the moral horizon against which *ius ad bellum* and *ius in bello* issues must be addressed in a world marked by the threat of nuclear or totalitarian holocaust.

Defining the meaning of peace in the modern world must be the responsibility of those who consider that morality and politics are dimensions of one human enterprise, and who wish to bring the discrete use of armed force (or its threat) within the ambit of moral reasoning. That is, the task must be taken up by theologians and political theorists grounded in the tenets and methodology of just-war reasoning. Work for the evolution of *tranquillitas ordinis* on an international scale can be a common endeavor linking those committed to the ethic of the just war with those committed to the ethic of pacifism. But the greater burden of moral reasoning on building the peace of political community will rest with thinkers grounded in just-war principles. Theirs are the intellectual tools capable of guiding work toward the gradual realization of the vision of *Pacem in Terris* and *Gaudium et Spes*. Those tools have not been placed at the disposal of that project by many American Catholic just-war theorists over the last generation. Unless that vacuum can be filled, as a central component in the development of a theology of peace, other currents of thought, less well equipped for the necessary job, will continue to shape (and misshape) the argument over the peace that can be sought, and the means appropriate to our seeking of it.

Finally, an American Catholic theology of peace requires a considerably more sophisticated understanding of *the relationship between peace and justice* than has been evident since Vatican II. During that time a maximalist conception, emblematically captured in Pope Paul VI's phrase, "If you want peace, work for justice," has dominated the American Catholic debate. That trend must be reversed. The classic Catholic heritage understood that questions of truth, charity, and justice were inextricably bound up in the creation of a rightly ordered peace of political community.

Since John XXIII, the Church's magisterium has also defined the centrality of freedom in the scheme of *tranquillitas ordinis.*

The heritage to be reclaimed and developed thus had a complex understanding of the relationship between peace and justice. It did not claim that all possibly controverted issues of justice had to be settled before there could be a morally worthy peace; on the contrary, it expected rightly ordered, dynamic political communities to resolve many questions of justice without resort to mass violence. At the same time, the heritage did not view all forms of public order as commensurate with a morally worthy peace. Tyrannical governments, which coercively compelled the orderliness of a cowed population, could not simultaneously promote truth, charity, freedom, or justice. Some questions of justice had to be settled before there could be a peace of *rightly* ordered political community; others could remain open. Peace required religious liberty; it did not require agreement on all controverted issues of political economy. One could, according to the heritage, achieve a morally worthy peace of political community without prior resolution of arguments such as those over a two-tiered minimum wage. It *was* essential that the political community provide means for conducting such arguments that allowed for the expression of popular consent—that is, means for resolving contested justice arguments short of the resort to mass violence.

Catholic social theory traditionally distinguished three understandings of justice. One modern commentator has described these as "*commutative* [justice], relating to contractual obligations between individuals involving a *ius strictum* and the obligation of restitution; *distributive* [justice], the relationship between a government and its people regulating the burdens and benefits of social life; *legal* [justice], regulating the relationship of the individual toward society."[23] Within this triadic model, the *primary* questions that must be settled in order that there be a morally worthy peace are questions of legal justice, or the constitutional foundations and framework of political community. Where these have been well built, other issues can be resolved through rightly ordered legal and political institutions and processes.

Viewed from another angle, this suggests that questions of *freedom* and its concretization in institutions of governance are considerably more central to building peace than has been accounted for in the recent American Catholic debate over the relationship between peace and justice. That debate has been dominated by the tacit (and sometimes explicit) notion that virtually all issues of distributive justice (generally understood in frames of reference drawn from the various theologies of liberation, and appealing to such warrants as the economic analysis contained in the U.N.'s New International Economic Order) must be resolved before there can be any peace worthy of the name. The next chapter will address these claims more directly. For now, it can be said that there is little foundation in the main trajectory of the classic Catholic heritage for the maximalist distortion of the relationship between justice and peace that has characterized the American Catholic debate on war and peace since Vatican II. Nor, on the evidence of conciliar and papal teaching, is there any warrant for minimalizing the claims of freedom as an integral part of the problem of war and peace. "Justice and peace" is a recurring antiphon in postconciliar American Catholicism; "peace and freedom" is a chord rarely struck.

An American Catholic theology of peace should teach that conflict over differing understandings of what is just in a given situation are a constant of the human condition. Conflict, in these terms, is the political meaning of the doctrine of original sin. This side of the reign of the Kingdom of God, ours will not be a world rid of conflict; ours will remain a world of imperfect justice. Yet this does not, according to the heritage, condemn us to the Hobbesian jungle. Political communities can be created so that conflicts over differing claims to justice can be waged through courts and elections. The result is not the peace of *shalom*; but *shalom* is a reality in God's time, not our own. The result can be, as the experience of democratic nations has shown, a peace of rightly ordered political community in which the governed have sufficient faith in the legal justice of the institutions of governance that they do not take up arms, but politics (and, increasingly, law suits). Conflict remains; but war has been ended. This is no mean accomplishment. And this is considerably more than "peace of a sort": it is a dynamic peace that can lead to fuller justice and truer freedom. It is the kind of peace, in freedom, that speaks to both the vision of *Pacem in Terris* and the call of John Paul II for "a major step forward in civilization and wisdom."

Establishing the linkage between peace and freedom is essential for any evolving American Catholic theology of peace that seeks to reclaim and develop the heritage of *tranquillitas ordinis*. This linkage does not preclude brisk argument over which questions of justice can be left open, and which must be settled, before a rightly ordered peace of political community can be realized. Resolving some claims of distributive justice can be an important part of creating institutions of legal or constitutional justice. The Salvadoran land and credit reform program illustrates the relationship between distributive justice, democracy-building, and the possible peace of political community.

But the theology of peace must acknowledge that there *are* questions to be left open. Put the other way around, it would decline the maximalist conception of the relationship between justice and peace. There can be no peace of rightly ordered political community without legal justice; but neither can there be such a peace if its prerequisite is universal agreement on what the canons of distributive justice require in all circumstances. There is, of course, a considerable ground between these two claims, and it can be best traversed through the mediation of institutions of freedom. This is not, fundamentally, a claim of political theory; it is a theological affirmation. Institutions of freedom give concrete expression to that human dignity proclaimed in creation and confirmed in the Incarnation. To understand that peace and freedom are inextricably bound together is to enter, in concrete, historical terms, into the mystery of human possibility in a way that points toward the human future intended by the Creator.

AN ECCLESIOLOGY FOR PASTORAL EDUCATION AND CIVIC ENGAGEMENT

A Church proclaiming this radical message of human possibility would have a distinctive ecclesiology, pastoral agenda, and civic strategy. The ecclesiology of pastoral education and political engagement most adequate to a reclamation and development of the classic Catholic heritage would differ from present patterns of thought and action in several ways.

American Catholicism must rid itself of the sense of being a breathless latecomer to the moral debate over war and peace, security and freedom. The American Church brings a rich, complex heritage of reflection on these issues to the contemporary argument. Although the heritage had its weaknesses and vulnerabilities, the fact of it cannot be denied. Catholicism did *not* discover the moral problem of war during Vietnam, and still less during the nuclear weapons debates of the late 1970s and early 1980s. To accept the charge that American Catholicism had been a silent bystander at the great public debates over war and peace prior to the bishops' 1983 pastoral letter is not only an indulgence in historical falsification; it is an act of self-lobotomization. The first ecclesiological requisite for American Catholicism is thus a sense of tradition, of being a community with an abundant intellectual resource. American Catholicism need not establish its credentials in the *agora* of contemporary public debate by vain appeals to being relevant. Its credentials are well established. They date back to Augustine and Aquinas. The task is to bring the intellectual resources that establish those credentials to bear on the exigencies of the present.

An American Catholic ecclesiology adequate to the task of reclaiming and developing the heritage of peace as political community would also assert forthrightly, and without fear of being thought marginal to the real issues, that its first mission in the civic arena is as a *teacher*. Perhaps the gravest temptation in contemporary American Catholicism is to reduce the Church's role to that of lobbyist for particular policies. This temptation is intensified by political and ecclesiological misunderstandings and by a misconception about moral reasoning.

The political misunderstanding is the perception of the policy process at its narrowest end—legislation and executive action—as most important. This is a mistake on at least two grounds. First, it ignores the fact that compromise is the essence of policy-making. The policy process is constructed to achieve maximum accommodation within agreed limits; it is not an arena in which apodictic judgments are received kindly.

Secondly, and more importantly, the increasing official attention of American Catholicism to the Washington, D.C., policy process fails to understand that the nation's capital is *not* the place in which the most basic arguments over America's role in world affairs are conducted—a surprising misconception for a Church which has stressed the principle of subsidiarity and the importance of mediating institutions. The capital, as those who have worked in it for any length of time come quickly (and often painfully) to know, is something like a cash register: it generally rings up transactions already made in the wider society. The nuclear freeze movement, whatever its manifold errors of perception, analysis, and prescription, was one proof of this. The intense public debate on nuclear weapons and strategy in the early 1980s did not originate in Washington, D.C., but in church basements, town hall meetings, university teach-ins, doctors' conventions, and voluntary organization arenas throughout the country. No doubt the policy process, narrowly defined as the institutions of government, played its part in shaping this debate. But the debate began, and its basic themes were framed, elsewhere—as were those of the environmental movement and the civil rights revolution. That society comes before polity is not an abstract affirmation of political theory; it is a fundamental fact of life in American political culture.

A Church committed to bringing the developed heritage of *tranquillitas ordinis* to bear on central questions of national purpose in the world will not, therefore, focus its attention so exclusively on Congressional hearings and executive-branch lobbying as to drastically subordinate its other civic missions to the goal of legislative victory. To do so is to try to reshape arguments created elsewhere.

There are also ecclesiological and moral-theological misunderstandings at work in the Church-as-lobbyist model. The claims of moral reason are weakest at the level of "prudential judgment": not because the choices to be faced here have, somehow, less moral content, but because empirical ambiguities bear most heavily on decision-making when binary choices must be made. The American bishops recognized this in principle in "The Challenge of Peace" and in their Congressional testimony. And yet the pastoral letter and the bishops' testimonies on nuclear weapons issues and U.S. policy in Central America consistently give the most intense attention to particular, discrete policy choices. This disjunction between principle and practice must be corrected.

Ecclesiological misunderstanding is also at work. To minimize the role of teacher in order to assert the priority of policy prescription is a distortion of the Church's essential mission *as Church.* The Church has no special claim to wisdom at the level of prudential prescription. It makes its most persuasive claims at the level of normative principles. But the Church's competence is not limited to the articulation of general moral norms. The Church can, and should, also claim a distinctive competence at a mediating level of the civic debate, the process of context formation. This is the point where formal principles of the moral order are crafted into middle axioms, themes that mediate between the truth of principle and the exigencies of historical circumstance. This is where the virtue of prudence begins to make its most convincing claims, as the orienting questions of foreign policy debate are clarified. An American Catholicism reclaiming and developing its classic heritage will be a Church giving primary attention to the contextual level of the debate over peace and freedom, to culture formation. This would meet a civic, as well as ecclesiological, need.

James Rosenau has aptly described the present condition of the American foreign policy debate as "fragmegration."[24] The pre-Vietnam consensus that had sustained U.S. foreign policy has not only fractured, but the shards of the fractured consensus have reintegrated into polar positions on the spectrum of argument. A "fragmegrated" foreign policy debate is shaped like a barbell. The civic result is a public discourse that more closely resembles a dialogue of the deaf than a search for truth; the policy result is oscillation and inconsistency of purpose and method. The Carter administration suffered from the effects of this "fragmegrated" public arena,[25] and the Reagan administration, despite two overwhelming electoral college mandates, has not been immune to the problem either (as witness its reliance on such ad hoc mechanisms for consensus as the Scowcroft Commission on the future of U.S. strategic forces and the Kissinger Commission on U.S. policy in Central America). The post-Vietnam period has confirmed the wisdom of John Courtney Murray's warning that "it would not be well for the American giant to go lumbering about the world . . . lost and mad."[26]

Fragmegration is not only a function of attentive public disagreement on American purpose in the world; it reflects the increasing complexities of the international environment. But the nation cannot adequately respond to these complexities unless some basic agreement on the contextual questions that shape the foreign policy debate can be rebuilt—agreement about the legitimacy of intervention and the discriminate use of armed force; about the nature and geopolitical meaning of the Soviet threat; about the definition of the peace that can and should be sought; about the viability of the present international system; finally and perhaps most importantly, about the moral worthiness of the American experiment.

An American Catholicism reclaiming and developing its classic heritage in what Murray and others have understood as a distinctive "Catholic moment" in the evolution of the American experiment would actively work to redress the problem of fragmegration in the American foreign policy debate. Such a Church would be both facilitator and proponent. Understanding that important insights exist at both poles of today's fragmegrated debate (which can be broadly described as the "peace and disarmament" pole and the "security and liberty" pole), the Church as facilitator would help create opportunities for a civic debate on contextual issues that met Murray's standards of reason and reasonability—a debate that would not be "hot and humid, like the climate of the animal kingdom," but "cool and dry, with the coolness and dryness that characterize good argument among informed and responsible men."[27]

The immediate goal of this enterprise would not be policy consensus at the least common denominator, as if agreement involved splitting the difference between the poles of the forensic barbell. Rather, the first goal would be to achieve minimal agreement on answers to the fundamental contextual questions, such that disagreement (which Murray always thought a great accomplishment) on specific policies would not lead to further fragmegration and cacophony, but to prudential insight.

To take two examples. Agreement on the nature and meaning of the Soviet threat could unleash a policy debate that, while marked by disagreement on specific choices (e.g., the use of economic sanctions), might still be commonly focused on the goal of U.S. initiatives aimed at *changing* the present course of Soviet activity in the world. Similarly, agreement on the contextual question of intervention could, by marginalizing the claims of isolationisms old and new, result in a policy debate more usefully focused on questions of the ends and means of America's inescapable intervention in world politics and economics. The Church, an institution whose primary concern is the cause of truth, is distinctively positioned to broker the creation of a new contextual debate on U.S. foreign policy. It cannot play that civic role, though, if its primary interest is (or is perceived to be) advancing the policy prescriptions of one or another pole in the current argument.

Facilitating contextual agreement in the American attentive public does not require the Church to be a neutral (or neutered) institution, stripped of its own insights and content to provide the arena for a new public debate. The Church can, and should, be developing its own distinctive set of answers to the key contextual questions out of which policy choices are shaped. But developing a context reflective of the classic Catholic heritage is of a different order of magnitude from fashioning a kit bag of foreign policy prescriptions. The former task is, if anything, more difficult. But it is more difficult because it is more fundamental. And because it is more fundamental, it is more urgent, in both ecclesiological and civic terms.

Such a task of reconstruction will require a major effort at self-education in the American Church. Catechesis of the Catholic social-ethical tradition has not been a traditional strong suit in American Catholicism. That situation must be remedied, at all levels of Catholic education—and by genuine education, not propaganda for themes reflecting one or another pole of the fragmegrated civic debate. Many of the study guides developed in the aftermath of "The Challenge of Peace" failed to meet this standard.

Finally, American Catholicism must distinguish between its roles as forensic facilitator and context proponent in a publicly visible and understandable way. The differences between these two functions have been blurred over the past generation. Moreover, Church advocacy in behalf of its own distinctive answers to the contextual questions of the foreign policy debate must be distinguished from various Church agencies' prudential judgments on what the policy implications of that context may be in particular circumstances. No doubt all of this requires considerable intellectual sophistication and programatic dexterity. But the effort must be made, if the Church is to be true to its primary mission as teacher, and if the Church is to seize the opportunity to facilitate the kind of contextual reconstruction so evidently necessary in American political culture.[28]

Two related points may be raised briefly. American Catholicism must be a Church in which the roles of laity and clergy in the Church's address to the moral problem of war and peace are distinctive and complementary. Neither distinctiveness nor complementarity has been very evident since they were mandated by Vatican II, despite efforts (as in drafting "The Challenge of Peace") to broaden the dialogue. The clear teaching of the Second Vatican Council is that the laity, by reason of baptism, has a special ministry "in the world." This conciliar teaching does not preclude an active role for clergy and religious in scholarship, public debate, pastoral programing, or policy prescription on questions of war and peace, security and freedom. It does, however, suggest a Church in which clergy, religious, and laity share responsibility for the Church's missions as teacher and civic facilitator, but primary responsibility for the development and argumentation of prudential prescriptions is a lay prerogative. This is not to suggest that anyone in holy orders or under formal religious vows is ex officio precluded from a role at the level of prudential prescription. But the *primary* obligation and responsibility at this level of the Church's engagement with issues of peace and freedom rests with lay Catholics, by reason of their distinctive charism and their mundane expertise.

A reclericalized Church, in which priestly and episcopal prescriptors are the primary fashioners and chief public exponents of the official Church's prudential judgments on questions of war and peace, would not be a Church in accord with the vision of Vatican II. This is often argued on grounds of the integrity of the special charisms of the priesthood and the vowed religious life; it is even more a question of respect for the distinctive baptismal charism of the lay Catholic.

Finally, an ecclesiology of pastoral education and civic responsibility would be thoroughly ecumenical in its search for truth. A reclaimed and developed American Catholic heritage of *tranquillitas ordinis* would have much to offer the wider Judeo-Christian community of faith in the United States. But the very process of developing the heritage would be enhanced by an active conversation with the ecumenical Church and with the American Jewish community. Postconciliar patterns of ecumenism on issues of war and peace have usually focused on those mainline Protestant and Lutheran churches with whom theological consultations on the great questions of baptism, Eucharist, ministry, and ecclesiastical authority have long been underway. The partnerships involved in that dialogue should be expanded. American Catholicism's reclamation and development of its heritage would be strengthened by active dialogue with the Orthodox churches, whose spirituality of *theosis* or the divinization of man is of great importance in the Church's proclamation of that for which humanity is intended by God. American Catholicism should also pay considerably more attention to the experience of those Uniate Churches, Orthodox in expression but in union with the bishop of Rome, that have suffered so terribly from both war and totalitarianism in the twentieth century.

But the most suggestive ecumenical axis in the next decade may well be between American Catholics and nonfundamentalist Protestant evangelicals. The public terms of the abortion debate in the early 1980s were redefined through an ad hoc coalition of Catholics and evangelical Protestants—two communities which had, traditionally, little experience of and much wariness about each other. Yet as the churches of the Protestant mainline have continued to lay down the burden of moral-cultural formation in American society, which had been theirs since the days of John Winthrop, it was Catholics and evangelical Protestants who were forcing the question on the relationship between religiously based values and public policy in America. It is now time to move beyond ad hoc political coalitions on specific issues to serious theological and political-philosophical dialogue on the full range of questions involved in the moral testing of the American experiment in the world. A wide-ranging Catholic-Evangelical consultation on the great questions of war and peace, security and freedom—questions whose profound moral content makes them central in the debate over religious values and public policy—would benefit both parties *and* American political culture as it seeks a way beyond fragmegration.[29]

SACRAMENT OF HUMAN UNITY, COMMUNITY OF CHARITY AND PRAYER

The most fundamental reality of the Church is not its presence as an institution in the world. In the strict theological sense, the Church is essentially a mystery. In Catholic theology, a mystery is not an intellectual riddle to be solved, but a multivalent, God-touched reality, never susceptible of complete human expression because of its measureless richness. The mystery of the Church, like the mystery of God, is a reality that we enter most fully, not in speculative analysis, but in love.

Contemporary Catholic thought has employed various images to express different dimensions of the mystery, never to be completely captured, of the Church: the Church as mystical communion; the Church as herald of the good news; the Church as servant of the world; the Church as the People of God on pilgrimage from and to the source and end of human life, God the Creator, Redeemer, and Holy Spirit.[30]

A number of prominent Catholic theologians, before and since the Second Vatican Council, have described the Church as a sacrament. Henri de Lubac, for example, wrote that "If Christ is the sacrament of God, the Church is for us the sacrament of Christ; she represents him, in the full and ancient meaning of the term, she really makes him present. She not only carries on his work, but she is his very continuation, in a sense far more real than that in which it can be said that any human institution is its founder's continuation."[31] The fathers of Vatican II adopted this image at the outset of *Lumen Gentium*, describing the Church as "a kind of sacrament of intimate union with God and of the unity of all mankind."[32]

The very nature of the Church, then, makes it "a Church at the service of peace." Because the Church is a sacrament of Christ's presence to the world, the Church is a witness to the divine intention for humanity revealed in the Incarnation and the Resurrection. One tangible expression of the Church's sacramentality is its universality. In the Body of Christ are gathered the contemporary equivalent of those "Parthians and Medes and Elamites and residents of Mesopotamia, Judea and Cappadocia, Pontus and Asia, Phrygia and Pamphylia, Egypt and the parts of Libya belonging to Cyrene, and visitors from Rome, both Jews and proselytes, Cretans and Arabians" who heard the apostles "telling . . . the mighty works of God" on the day of Pentecost (Acts 2:9–11). By its very nature and constitution, the Church is an expression of the capacity of human beings to form, under grace, a community in which the boundaries of race, sex, language, caste, and nationality are transcended in a unity amid plurality.

The unity of a Church that resembles the patriarch Joseph's coat of many colors is a unity of a particular kind. It is built by the Holy Spirit, calling men and women to faith, worship, and the exercise of mutual charity and service to the world. The unity of the sacramental community is of a different order than the unity that sustains community within rightly ordered polities. But Vatican II's description of the Church as a "kind of sacrament . . . of the unity of all mankind" should not be gainsaid as an important component in reclaiming and developing the heritage of *tranquillitas ordinis* in American Catholicism. The Church "represents," in de Lubac's term, the unity of the human family, not in the sense that a member of Congress represents a constituency, but in a strict sense of re-presentation: the Church makes real and visible the fundamental human unity that derives from our common status as creatures of the same Creator. In this theological sense, we can speak accurately of a "human family" and "a world."[33]

The Church's basic mission "at the service of peace" is to be this "sacrament of the unity of all mankind": the Church's mission is to be the Church. Prior to its active engagement with the affairs of the world, the Church serves the cause of peace by its very existence. A Church "at the service of peace" will therefore take care that it makes itself as clear a sacramental symbol as possible. The quality and character of the ecclesiastical debate on war and peace, security and freedom, should be of a different temper from what it is in other institutions concerned with these issues. The Church's debate should be distinguished by a mutual search for truth, and not merely for intellectual, programmatic, or prescriptive victory. Differences in analysis and prudential judgment will not become the occasion for anathematizing opponents. Hard judgments on the inadequacies, even stupidities, of others' perceptions and prescriptions will be made, but not raised to the level of *status confessionis* issues. Partial truths, from whatever point on the political spectrum they appear, will be gathered into fuller, more complete truths.

It cannot be said that the American Catholic debate on war and peace has met these standards since the Second Vatican Council. But past failures need not define future options. Were the Church in the United States to take seriously the task of reclaiming and developing the heritage of *tranquillitas ordinis* in all its richness and complexity, it would find that the heritage itself provides the framework in which a debate worthy of a Church sacramentally representing Christ in the world can unfold. The conduct of such a debate within American Catholicism would be a powerful sign in an American polity still bitterly divided on the contextual issues that shape the foreign policy choices of the American government.

To affirm that the Church's first task is to be the Church is not to opt for a quasi-sectarian ecclesiology; but it is to understand that the quality of the Church's civic activity will reflect its interior life as a community of faith, charity, and prayer. To be "a Church at the service of peace" is to be a herald of good news, and a sacramental representation of that unity for which God intended humanity at the creation.

It is also to be a Church of prayer. "Praying for peace" is often a pious afterthought in the contemporary American Catholic debate. Occasionally, the call to prayer becomes an ex post facto benediction on particular political judgments. And yet prayer for peace must be a central element in the reclamation and development of the heritage of *tranquillitas ordinis*. John Macquarrie correctly positioned prayer in the Church's ministry of peace in these terms:

> To pray for peace, Christians believe, is more than just to meditate on the meaning of peace with a view to becoming better servants of peace. It is to bring into the human situation the very power of the God of peace, or, better expressed, to open up our human situation to that power. No doubt at any given time only a tiny minority of mankind is actively praying for peace in this way. But no one can say what is being accomplished through these openings into the human situation which they provide. Those who pray for peace can take encouragement from some words in *The Cloud of Unknowing*: "The whole of mankind is wonderfully helped by what you are doing, in ways you do not understand."[34]

To be a community of faith, charity, and prayer does not exhaust the mission of "a Church at the service of peace" whose call is to be a sacrament of Christ's abiding presence in the world. But it is the beginning of that mission. Recapturing a clear sense of the distinctiveness with which the Church enters the civic debate is the indispensable foundation from which a reclaimed and developed Catholic heritage can be brought to bear on the contemporary dilemma of peace and freedom.

CHAPTER 13

To Be a Church at the Service of *Peace*

Reclaiming and developing the American Catholic heritage of *tranquillitas ordinis* requires acts of political imagination, disciplined by a sober assessment of the many obstacles that block the world's path from anarchy to political community. The Church's ministry "at the service of peace" is not confined to the quality of its internal life as a believing community of faith and charity; nor is it limited to the Church's important role in setting an arena for civic debate in which truth is the guiding criterion. The Church not only has a right, but a duty, to help develop answers to those basic contextual questions that shape the civic debate over war and peace, security and freedom. This duty finds its roots in Catholicism's classic insistence that politics is a moral enterprise, a claim that reflects the Church's incarnational humanism. The Church is called to be more than a witness to the belief that *shalom* will be a blessing in that Kingdom of which the Church is both sign and herald. The Church is obligated to contribute to the public discussion of how the peace of rightly ordered political community might be pursued in the world as it is.

This does not mean that the Church has its own foreign policy. It means that the Church assumes responsibility for vigorously entering the civic debate at the point where normative moral principles cross the exigencies of historical circumstance. This mediating level of the debate is the level of "context formation." What is the nature of the peace we are to seek? Does the present international system make such peace possible, or must it be altered? Do the boundaries of political obligation coincide with (and stop at) national borders? Should the United States intervene in the world, actively shaping world politics and economics? Is the resort to armed force ever justifiable, and if so, under what conditions and constraints? What is the nature of the threat posed by Soviet power and purpose in the world today? Finally, and perhaps most importantly, what is the moral and political worth of the American experiment? Answers to these orienting questions form the context out of which specific issues are addressed, give basic shape to policy prescriptions, and create the possibility of coherent strategy over time. It is at this level of the debate that agreement in American political culture has broken down and, in James Rosenau's neologism, "fragmegrated." But it is precisely around these issues that a politics of peace and freedom will be built, if it is to be built at all.

The Church's responsibilities at this level of discourse are twofold. The Church must take responsibility for the moral and intellectual health of the contextual debate in American society, by providing an arena in which these basic orienting questions are debated under the standard of truth, rather than of partisan or ideological victory. But the Church's responsibilities do not end there. The Church must also develop and clarify its own answers to these fundamental questions. Here, the classic heritage can be reclaimed and developed in ways that are faithful to Catholic understandings and shed important light on America's responsibilities for the pursuit of peace, security, and freedom in the world.

What would a distinctive American Catholic perspective on the problem of peace and freedom teach, at this level of the contextual debate? The suggestions that follow are illustrative rather than definitive. They do not propose to be *the* American Catholic "answer" to the moral problem of war; in the nature of things, no such "answer" exists. But the proposals here are more congruent with a needed development of the classic Catholic heritage, and more likely to lead to advances in the linked causes of peace and freedom, than the themes and prescriptions that have dominated American Catholic elite opinion over the past twenty years.

THE PEACE TO BE SOUGHT

"Peace" has multiple meanings in the Catholic tradition. Peace as interior serenity before God is surely the business of a Church called to proclaim the good news of salvation in a broken world. So, too, must the Church preach the eschatological vision of *shalom*, that peace of the Kingdom in which the creation is completed according to the Creator's design. But if the question to be addressed is neither personal spirituality nor eschatological fulfillment, but the condition of a world caught between the fire of modern war and the pit of totalitarianism, then a third meaning of "peace" must be offered: the peace of *tranquillitas ordinis*, of rightly ordered, dynamic political community.

There are ancient warrants for this concept of peace in Catholic social thought. Augustine's concept of peace as order was transformed, in the work of Thomas Aquinas and the scholastic commentators, into a positive construct: peace as "the tranquillity of order" was not a punitive reflection of human propensities for evil, but an expression of innate human sociability. Political community was a natural reality, not a function of original sin. Politics, rightly ordered in truth, charity, freedom, and justice, and oriented to the pursuit of the common good, was an expression of human possibility, a reflection of God's charge that we be the stewards of creation. Politics, this side of the Kingdom of God, inevitably involved conflict; but rightly ordered political communities provided means by which conflict could be prosecuted and resolved without resort to the organized mass violence of war. Conflict was not simply a synonym for the Fall; mediated through politics, conflict could be a source of creativity and human progress. These notions were imbedded in the interstices and basic trajectory of just-war reasoning and constituted a *ius ad pacem*, a horizon of achievable human possibility against which questions of the *ius ad bellum* and the *ius in bello* could be argued.

The peace of political community was not, and should not be, construed as the fullness of peace in Catholic spirituality or eschatology. The interior peace of personal serenity before God is, at one level, a deeper meaning of peace in the Catholic imagination. The peace of *shalom* is a more complete, perfected form of peace than peace as political community. As an eschatological reality, *shalom* stands in judgment on the achievements of peace as political community in this world, reminding us that the task of stewarding the creation is never finished. Peace as dynamic, rightly ordered political community is an interim ethic: it offers a goal that can be realistically pursued in this world. But this does not minimize the claims of *tranquillitas ordinis*, for it, far more than the peace of spiritual interiority or the peace of *shalom*, is directly relevant to the politics of a world in which human fallenness and human possibility are inextricably woven together. Moreover, viewed through the lens of our contemporary situation, scarred as it is by war and totalitarianism, the peace of dynamic political community offers the opportunity to meet Camus's challenge that we be neither victims nor executioners. The slow, hard, painful evolution of political community in the world is, for all its limitations when viewed *sub specie aeternitatis*, the only available option for a Catholic moral imagination taking seriously the threats of modern war and totalitarian tyranny.

Peace as political community has been a consistent theme in papal teaching since Benedict XV and the First World War, and reached its most dramatic expression in John XXIII's *Pacem in Terris*. Papal teaching has emphasized that the peace of political community is not static; order is not the only value in Catholic social teaching on the problem of war. The Catholic tradition, dating back to Aquinas, has insisted that *rightly* ordered political community is the goal to be sought, for politics is a matter of virtue as well as of institutional mechanics. In the last twenty-five years, the Church's magisterium has taught that *tranquillitas ordinis* must be shaped by freedom, as well as by truth, justice, and charity. The distinctive nature of totalitarianism has played a role in this evolution of thought on the composition of rightly ordered political community, as has the demonstrable ability of democratic governance to build societies in which charity, justice, and prosperity can be maximized.

Peace as political community cannot, then, be dismissed as merely "peace of a sort." It is not the peace of *shalom*; but *shalom* is for the Kingdom. The peace of rightly ordered political community creates conditions for the pursuit of a more complete justice, a nobler freedom, a society marked by deeper truth and more perfect charity. Peace as dynamic political community sets the ground on which *shalom* can become a fruitful, rather than destructive, eschatological vision.

Reclaiming and developing the concept of peace as rightly ordered political community is the foundation on which American Catholicism can build a more adequate set of answers to the key contextual questions that shape the war/peace debate today. That concept has been largely absent from the American Catholic discussion for almost a generation. Catholic political realists, understandably cautious about posing such a scheme of human possibility given the nature and capacities of the postwar Soviet threat, have concentrated on immediate questions of national security and the defense of the democracies. The wise, humane management of the balance of power has seemed the best that can be posited at present. Yet for all its commendable sobriety, this diminution of the concept of peace as political community—particularly as a framework in which to understand the linked character of the problems of war and totalitarianism—has left a gap in American Catholic moral imagination as it grapples with the responsibilities of the United States in the world. That gap has been filled by fuzzy notions of an achievable *shalom* this side of the Kingdom of God, by a grossly overpsychologized understanding of international conflict, and by a "peace politics" whose primary target is American military programs, not the problem of war. Failure to address John XXIII's call for the creation of international political community in Paul Ramsey's terms has not resulted in a more sober analysis of the threats that block the achievement of that vision; it has resulted in precisely the opposite.[1]

It is time to reclaim, in a developed form that takes full account of the many present dangers of the late twentieth century, the concept of peace as dynamic political community, as *tranquillitas ordinis* at the international level. Absent such a reclamation and development, the voices of the abandonment of the Catholic heritage will continue to shape the contextual debate and the Church's address to immediate policy options. Given such a reclamation and development, the door will be open to an American Catholicism more thoughtfully engaged with the problems of war and totalitarianism, more adequately addressing the need for both peace and freedom.

THE QUESTION OF THE PRESENT INTERNATIONAL SYSTEM

That the world of the twentieth century has become a political arena in a historically unprecedented way is a commonplace understanding; that the world must make the passage from political arena to political community is the clear teaching of the magisterium of Roman Catholicism. How shall this gap—between things as they are and things as they ought to be—be closed, and in ways that provide for peace, security, *and* freedom? What agenda for change in the present international system can be responsibly pursued, without making matters worse than they already are?

John XXIII's call for the evolution of international political community, whatever its weaknesses in analyzing the obstacles blocking progress toward that goal, contained such an agenda. It was a fourfold construct. It sought (1) mutual, balanced, and verifiable arms reduction, in both conventional and nuclear weapons, leading to genuine disarmament; (2) international legal and political institutions capable of resolving conflicts between states without either the resort to mass violence or the destruction of national identities and communities; (3) a transnational sense of political community and obligation, rooted in respect for and protection of basic human rights; and (4) economic, social, and political development in the formerly colonized world, where relief from deprivation was the highest priority. These goals were linked. There could be no real disarmament without alternative means of resolving conflict. Institutions capable of that task could not be built without a foundation of human freedom and dignity, and without a successful address to the desperate needs of the world's impoverished and tyrant-beset underclass. Progress on any one of these goals required progress on the entire agenda. "Linkage," in this sense, was a given.

This fourfold structure of goals for constructing international life according to the standards of *tranquillitas ordinis* was an agenda for the long haul. Its realization lay in the distant, but not unimaginable, future. But its four goals, taken together, created a framework for understanding the exigencies of the present, and set standards for policy and practice that allowed for a comprehensive approach to the modern dilemma of peace, security, and freedom. The fracturing of this agenda in American Catholicism over the past twenty years has been one of the principal hallmarks of the abandonment of the classic Catholic heritage. Reclaiming and developing that heritage requires fresh thought on each of the issues comprising this complex agenda of goals that chart the path from political arena to political community.

A new approach to the fourfold political agenda of *tranquillitas ordinis* would begin by frankly acknowledging our post–World War II failures to create international political institutions that met Atlantic Charter standards. To work for a rightly ordered peace of political community at the international level does not require a romantic view of the United Nations system, but the opposite. The U.N. system has failed to meet the challenge of its charter. It has not rid humanity of the scourge of war. And in some instances it has made war more, rather than less, likely. Has this been a failure of will? It seems unlikely. The primary sources of failure lie in politics, not psychology, and have to do with the impossibility of building political institutions without a foundation of political community—a problem that has been intensified by the ideological claims and world role of Marxist-Leninist states. A realistic critique of the failures of past efforts to organize the peace of political community in the world is thus an essential starting point for reclaiming and developing American Catholic thought on war and peace, security and freedom.

But that critique is a starting point, not a terminus. For, side by side with the fact of failure is the fact of necessity. No nation-state in the modern world can guarantee its citizens that minimal security which is a fundamental *raison d'être* of organized polities. Those who argue that a world armed with nuclear, chemical, and biological weapons and intercontinental delivery systems can indefinitely sustain itself under conditions of international anarchy are the real utopians, especially when the harsh facts of totalitarian purpose and capability are factored into the equation. John Paul II was right: the future of humanity requires "a major step forward in civilization and wisdom"—which is to say, progress toward rightly ordered political community at the international level. Catholic theory today cannot be forced to choose between the acknowledgment of past failures and the demands of present necessity. It must hold these two together, in a creative tension from which fresh thought and work on the fourfold structure of the politics of *tranquillitas ordinis* can flow.

Nowhere in the American Catholic debate is intellectual initiative more essential than in the area of *arms reduction and disarmament.*

The American Catholic debate on these issues was shaped by two schools of thought in the years following the Second Vatican Council. The more recent school, that of the unilateral disarmers, has been consistently rejected by both the Vatican and the National Conference of Catholic Bishops (where unilateralist currents of thought remain strong). Unilateral disarmament does not meet the standards of moral and political reasoning set by the Catholic heritage, for it is, at bottom, a denial of politics and a choice for religious and political sectarianism. The older school of thought, however, the school of arms control that dominated the analysis and prescriptions of the United States Catholic Conference, has not been subjected to such rigorous critique.

Arms control theory has shown a remarkable capacity to seize and hold the moral high ground in American political discourse despite the manifold failures of arms control practice. While it is too much to say that the "arms control process" as conducted since the late 1960s caused the fantastic acceleration of nuclear weapons capability in the world (and particularly in the Soviet Union) over the past generation, the acceleration paralleled the "process," and the "process" has proven incapable of reversing it. Commitments to a process with such unimpressive results begin, after a while, to appear more a matter of faith than reason, perhaps more a matter of psychological necessity than political wisdom.

Vehement fealty to arms control theory is especially strange in American Catholicism, for the theory evolved, not as a means to disarmament, but as an accommodation to the view that disarmament was impossible. Arms control theorists taught that the best that could be hoped for was the evolution of ground rules for conducting nuclear weapons competition. The centerpiece of that regulatory system was deterrence through Mutual Assured Destruction, a notion at utter variance with Catholic social ethics. Arms controllers thus argued that weapons aimed at weapons were "destabilizing," whereas weapons aimed at people were stabilizing. Arms control theory insisted that nuclear and conventional weaponry be uncoupled in negotiations, as if nuclear weapons were an independent variable in world affairs. Arms control advocates believed that they could teach the Soviet Union how great powers behave in a nuclear-armed world, and insisted that the arms race was a species of action/reaction cycle.[2]

Themes from classic arms control theory were prominent in "The Challenge of Peace." Nuclear weapons were qualitatively different. The idea that nuclear weapons, whatever their difference-in-kind in destructive capability, could still confer political power in world affairs was noticeably absent from the bishops' pastoral. The pastoral tried to combine classic arms control doctrine with a rejection of deterrence through Mutual Assured Destruction. But this maneuver, made even more difficult by the bishops' insistence that the overriding issue was that nuclear weapons never be used (because doing so would almost inevitably involve vertical escalation to general nuclear war), resulted in a form of bluff deterrent: a deterrent that seemed unlikely to deter for long. How the "arms control process" would lead, under such circumstances, to the disarmament the bishops sought was never made clear. Arms control, in "The Challenge of Peace," appeared more as a radical restraint on American force modernizations than as a coherent method for gaining Soviet agreement on verifiable weapons reductions. Subsequent USCC Congressional testimony on nuclear weapons issues has reinforced this impression.[3]

American Catholicism should help break the vicelike grip of arms control theory on the American strategic imagination—precisely for the sake of arms reduction. The Catholic debate should acknowledge that good and faithful men and women, deeply committed to averting nuclear war, have worked long and hard in the heat of the day for the sake of the arms control process—and have, by and large, failed. The failures of arms control would not be attributed to ill will, the military-industrial complex, or the psychological dysfunctions of phallically obsessed American male political leaders (*pace* Dr. Helen Caldicott),[4] but to basic conceptual flaws in classic arms control theory. Arms control theory was morally mistaken in its (sometimes overt, often tacit) commitment to deterrence through Mutual Assured Destruction; politically mistaken in its view of the dynamics of the U.S./Soviet nuclear weapons competition; strategically mistaken in its uncoupling of nuclear and conventional forces; and socially mistaken in its belief that American political culture could remain indefinitely committed to a deterrence system based on a realpolitik approach to world affairs that cut straight across the grain of American self-understanding (and thereby raised the spectre of isolationism).[5]

Why, given these conceptual flaws and the self-evident inadequacies of arms control practice, did arms control retain the moral high ground in American public discourse? The answer may lie in a basic public confusion, that "arms control" intended serious arms reduction. This simple equation of "arms control" with "arms reduction" must be challenged. Arms control theory evolved as a specific set of ideas, not so much for arms reduction, but for the wise management of arms *competition*. That theory has not led, on the presently available evidence, to arms reduction or to a measured superpower arms competition conducted according to agreed ground rules. When arms "control" is disengaged from the medium-term goal of arms *reduction* and the long-term goal of *disarmament*—that is, when these goals fail to inform short-term arms "control" policy—the result is precisely what has been seen over the past fifteen years: a vast expansion of nuclear weapons, with marginal restrictions that are difficult to verify and have little impact on the main trajectory of nuclear weapons expansion. Arms control, as it has been conceived and conducted since the late 1960s, does not deserve the moral high ground in American public discourse. Those who have worked so hard for arms control can be honored for their intentions, and for their modest accomplishments. But the basic flaws in their theory simply must be faced and acknowledged. An American Catholicism reclaiming and developing its heritage could help lead such an essential (and painful) process of critique and reconceptualization—not as a partisan witch-hunt, but as a search for practical wisdom.

It would do so by reestablishing the goal of arms reduction, leading to disarmament, as the criterion by which short-terms proposals for the "control" of arms are judged. It would challenge the uncoupling of nuclear and conventional weapons in arms reduction and disarmament theory and practice, understanding that this seamless web must be addressed as a whole, lest further imbalances increase the danger of war. Deterrence, on the new model, would be conceived as an interim regulatory mechanism providing the stable circumstances under which arms reduction leading to mutual, verifiable disarmament can proceed.[6] Deterrent stability, meaning the inability of either the Soviet Union or the United States to theoretically mount a crippling first strike against the other, would be one criterion by which to judge short-term decisions on nuclear weapons modernization and control, and stability would be conceived as a function of each side's force structure as a whole. American Catholicism would thus help lead a reconceptualization of arms reduction and disarmament theory that abandoned weapon-by-weapon approaches and sought reductions that solidified deterrent stability across the whole range of nuclear weapons systems.

But it would do more. American Catholicism's reclamation and development of the heritage of *tranquillitas ordinis* would actively seek ways to alter the fundamental dynamic of weapons competition so that the strategic regime would gradually evolve into a system of stability through mutual security, rather than through mutual vulnerability. Given this goal, strategic defense systems could play a crucial role in the process of arms reduction leading to genuine disarmament.

Under the most favorable arms reduction circumstances imaginable—agreement on the de-MIRVing of both superpowers' arsenals, leading to deterrent stability at lower and lower levels of nuclear capability—defensive systems would remain a necessity. At a certain point in the arms reduction process, when a preemptive counterforce attack could be mounted without an aggressor nation's running the risk of losing its entire society through a retaliatory blow, the spectre of first strike will inevitably rear its head. Reductions in offensive nuclear weapons are insufficient; shields will be necessary as swords are reduced. This is true not only over the long haul of superpower offensive weapons reductions; it is immediately true given the threat posed by the possibility of terrorists or outlaw-nations possessing nuclear weapons, and the ever-present danger of nuclear war begun by mechanical accident. Given the technological progress that now seems likely, strategic defensive weaponry will be a fact of the future—whether that future is defined in terms of a generation, or many generations. The Soviet Union understands this (hence, its aggressively pursued research and development programs in strategic defense systems); so do an increasing number of American theorists and policy-makers.[7] The crucial question, therefore, is the mode in which strategic defensive weaponry is developed. Are such systems developed unilaterally, as means for strengthening deterrence through offensive threat? Or could they be developed in a mutual or common security framework, so that the security of no major nation rests on its ability to incinerate its adversary? Such a radical alteration in the strategic regime has been repeatedly suggested by American officials, but has rarely been raised to the forefront of the attentive public debate.[8]

American Catholicism need not become embroiled in the technological fine points of strategic defense systems (although its leaders should be informed by the full range of technological expertise publicly available). The Church's role is more fundamental: helping shape the contours of the public argument over arms reduction and disarmament strategy so that the ends of both peace and freedom are served. That argument is now in considerable disarray. The heritage of *tranquillitas ordinis*, under contemporary conditions of weapons technology, would seem to suggest that mutual or common security should be the goal of arms reduction and disarmament policy. Is this feasible, given the present dynamics of the U.S./Soviet conflict? Could those dynamics be changed through the persistent pursuit of an altered strategic regime in which effective defense replaces the threat of mutual destruction as a regulatory mechanism in superpower relations? How could the evolution of defensive systems be coordinated with offensive weapons reductions, leading to verifiable disarmament? Can these goals be pursued in ways that challenge, rather than reinforce, the Leninist control of the Soviet state over Soviet society? How might alterations in the strategic regime be related to conventional weapons reductions? These are large questions. But they are precisely the questions that need exploration if we are to escape the present morass of arms control debate and policy.

American Catholicism's intellectual and leadership elites could play an important public role if these were the questions at the top of the Church's contextual agenda in the arms reduction and disarmament field. For perhaps the first time in a generation, technological possibilities have raised the prospect of a fundamental reconceptualization of arms reduction and disarmament theory. An American Catholicism reclaiming and developing its classic heritage could help ensure that such an important opportunity for altering the stagnant pattern of public debate on these issues is not missed. Such a role would be more faithful to the Church's primary mission as teacher than today's approach, in which the Church's official voice becomes another combatant, with highly patterned and predictable positions, in the existing debates.

American Catholicism could similarly help recast public debate over *interna-*

tional legal and political institutions. The failures of the United Nations system have been amply demonstrated;[9] but these failures are the beginning, not the end, of the needed discussion. An American Catholicism reclaiming and developing its classic heritage would begin with these facts of failure. But, true to its tradition, the Church would help initiate fresh thought on how the peace of political community might be effectively pursued through the creation of morally and politically sound transnational and international institutions of law and governance.

This exploration would concede, at the outset, that the notion of international law has been drastically weakened since the heady days of 1945 when the U.N. Charter was being drafted in San Francisco. The assumptions that underlie the failures of the U.N. system—particularly the assumption that legal and political *institutions* can be well constructed without a foundation in political *community*—would be vigorously examined and debunked as necessary. An exploration of the future of international institutions conducted according to the canons of the Catholic heritage would not operate on the premise that the first requisite is the maintenance of the present U.N. system. It would carefully examine the prospects for reform within the U.N.; but it would also consider the possibility that the present organizational situation at the U.N. is beyond reform.

Rather than excusing the present gross politicization of international institutions, American Catholicism should be in the forefront of those seeking reform where possible, or decent burial if inevitable. But, should the latter become necessary (for individual agencies, like UNESCO, or for the U.N. system as a whole), American Catholicism would not leave the matter there. Refusing to concede that anarchy is all that can be expected in international life, it would help open a new public debate over alternative international legal and political institutions more capable of meeting the challenges of the Preamble to the U.N. Charter. There is no necessary incompatibility, in other words, between the most stringent, far-reaching critique of the existing international legal and political system, and thoughtful work on the long-term evolution of institutions of law and governance more likely to fulfill John XXIII's call for international political authority.[10]

Such work would have to confront the claims of state sovereignty that have been the foundation of international public life since the Peace of Westphalia ended the European wars of religion in 1648. Sovereignty, according to the classic Catholic heritage, is a real, but limited, good. The limits of sovereignty are set by the principle of the common good, which, according to *Pacem in Terris*, must be construed as having an international dimension. Yet attention to the limits of sovereignty should not be at the expense of a realistic appreciation of the good accomplished by orderly national communities. Those who argue that the path to international political community lies in abrogating existing national communities need only look to the chaos of Lebanon, Northern Ireland, and many parts of Africa to understand that the collapse of sovereignty can lead to massive bloodshed, rather than to more comprehensive forms of political community. The issue most in need of exploration, therefore, is how an international political community capable of securing peace and freedom can be built *through*, rather than at the expense of, those stable national sovereignties that now exist.

Once again, a new spectrum of debate is required: the choices for the future cannot be *between* national and international political community. Rather, the primary question is how national political communities can contribute to the gradual evolution of an international political community and authority that meets the tests of truth, charity, legal justice, and human freedom. Present international legal and political institutions fail, in the main, to meet those tests. American Catholicism ought to recognize that, but then go on to explore ways in which worthier institutions of law and governance, transcending (without abrogating) national boundaries, might be constructed.[11]

American Catholicism could help, for example, lead an examination of the possibilities latent in regionally- or ideologically-based intermediate organizations that could become sturdy building blocks for reformed (or new) international institutions that meet democratic standards. Regional human rights courts, now in place in the Americas and in Western Europe, are one vehicle worthy of consideration. So, too, is the possibility of an association of democratic nations, which might work together as a political party or caucus on those issues of security, human rights, antiterrorism, and Third World economic development that are best addressed in a coordinated fashion.

No experienced person will gainsay the difficulties involved in creating international (or even regional) political and legal institutions capable of advancing the causes of peace and freedom. Many men and women once committed to the prospect of an international community under the rule of law have abandoned that goal as not just illusory, but damaging to the interests of the world's democracies. But to leave the cause of international legal and political institutions to those forces that have done the most to damage them since the founding of the United Nations is morally unworthy and politically unwise. The demands of the universal common good must be met through the reform of existing international institutions, or by their replacement with more adequate alternatives. Nor will the facts of necessity disappear if and when the world's democracies pronounce a curse on the U.N. system. The problems that system was created to address will remain, and in fact intensify.

It was not a liberal optimist, but Eugene Rostow, one of the founders of the Committee on the Present Danger, who made this case most cogently in the early 1970s:

> Unless agreement is reached soon on minimal rules of order for the conduct of rivalry among nations, the degree of fragile order achieved during the last generation will crumble at an accelerating rate. It follows that general war, or total war, would then be nearly inevitable, for reasons of panic or fear that have been familiar at least since Thucydides—reasons whose dominance, when anarchy threatens, has been demonstrated three times at least in this century. Such war, should it come, would surely destroy first the nations of greatest power in world politics, although it is hard to imagine exempted enclaves in such a holocaust. For these reasons I believe it is realistic to posit the existence of a universal world society, and the necessity for national action addressed to consolidating and confirming it as a single polity, based on a limited but effective code of public law.[12]

It is not the Church's business to design instrumentalities of international public order. It is the Church's business to affirm, in and out of season, that political community is a human possibility, even at the international level. That affirmation is strengthened, not weakened, by a tempered and critical view of contemporary international institutions. American Catholicism's reclamation and development of the concept of peace as dynamic political community must combine criticism and affirmation, sobriety and political imagination, on the question of institutions of law and governance in world affairs. In doing so, the Church could help revitalize a public debate that now seems caught between the voices of romantic optimism and chastened rejection. A new voice, grounded in the moderate realism of the classic Catholic heritage, is needed. American Catholicism's intellectual and leadership elites could help raise, and shape, that voice.

The cause of peace and the cause of *human rights*—the indispensable foundation for an international society capable of sustaining institutions of law and governance—were linked in the emphasis of the Catholic heritage on *rightly* ordered political community as the peace to be sought in a broken world. This linkage was made explicit in *Pacem in Terris*, and has been a constant theme in the teaching of Pope John's successors—none more so than Pope John Paul II, who made human rights the central motif of his address to the U.N. General Assembly in 1979.[13] Papal teaching, paralleling the human rights debate within American Catholicism's intellectual elites, has tended toward a comprehensive view of "human rights," a development with other parallels in the 1948 Universal Declaration of Human Rights. Catholic teaching has included social, economic, and cultural desiderata in its taxonomy of "human rights," as well as those civil liberties and political freedoms that have formed the core of human rights in the liberal democratic tradition.[14] Pope John Paul II's enumeration of basic human rights at the United Nations in 1979 illustrates this trend:

> The right to life, liberty, and security of person; the right to food, clothing, housing, sufficient health care, rest, and leisure; the right to freedom of expression, education, and culture; the right to freedom of thought, conscience, and religion, and the right to manifest one's religion either individually or in community, in public or in private; the right to choose a state of life, to found a family, and to enjoy all conditions necessary for family life; the right to property and work, to adequate working conditions and a just wage; the right of assembly and association; the right to freedom of movement, to internal and external migration; the right to nationality and residence; the right to political participation, and the right to participate in the free choice of the political system of the people to which one belongs.

All these rights, the pope taught, reflected the dignity of the human person "understood in his entirety, not as reduced to one dimension only."[15]

Such comprehensive lists of human rights serve a truthful, and useful, purpose as a moral horizon. They establish a framework of human goods that will be sought by any society that wishes to measure its moral and political worthiness by the standards of incarnational humanism. But comprehensive lists of human rights are also problematic. They tend to distract attention from the question of priorities. Do some rights create conditions for the possibility of institutional bases for the pursuit of other human goods? Comprehensive lists of human rights can also flatten moral understandings. Is it really true that religious liberty and secondary education are "equivalent rights," such that countries with compulsory and state-governed high schools are the moral equivalents of countries that protect religious liberty but do not provide free, state-supported secondary education?

Catalogues of human goods understood as "human rights" may also divert needed moral attention from the question of the institutionalization of rights under different forms of governance. Cuba, for example, seems a reasonably humane society if measured by economic and social "rights," yet all experience shows that, in totalitarian societies such as Castro's, there are no "rights" understood as protections from the claims of the state; there are only benefices the state distributes at its pleasure or withholds at its whim. To call such benefices "rights" may be acceptable at some level of abstraction, but does little good in helping establish models of human freedom, or in understanding the relationship between basic protections of the human person and conscience, and the pursuit of rightly ordered political community. In the absence of such models and understandings, the rhetoric of "human rights" becomes another tool in the hands of authoritarian or totalitarian tyrannies.

American Catholicism could help advance the debate over human rights and peace by giving particular and close attention to human rights priorities and institutionalization. It would affirm the horizonal truth of comprehensive visions of the human good in society; but it would think long and hard about the designation of all social and economic goods as "rights." As on the question of democracy and Catholic social theory, American Catholicism may have much to teach the universal Church here. The liberal democratic preference for limiting the language of "human rights" to civil liberties and political freedoms may be deficient when set against the horizon of social goods implied by *Pacem in Terris* and the Universal Declaration of Human Rights. But the fact that liberal democracies, on all the available empirical evidence, consistently perform better in providing economic and social goods, as well as in protecting and nurturing human freedom, than societies that give ideological or rhetorical priority to economic and social "rights" ought to impel a reconsideration of the necessity of painting "human rights" with so broad a brush.

An American Catholicism at the service of peace would thus lead an exploration of human rights priorities. It would give special attention to religious liberty as the most basic of human rights. The priority of religious liberty can be argued both theoretically and empirically. Religious liberty is *the* basic human right because it most clearly establishes that within each human person rests an inviolable sanctuary of conscience, a *sanctum sanctorum*, into which the state may not tread. Religious liberty defines the essential distinction between the person and the state that is the foundation of any scheme of human rights whose aim is the protection and nurturance of freedom.

The empirical priority of religious liberty is demonstrated more grimly: totalitarian regimes make the subjugation of religious institutions a first order of business, through outright persecution or co-optation. Totalitarians understand that the exercise of religious liberty is a fundamental (and, in principle, unanswerable) challenge to their claims. In this sense, the fact that religious liberty is often honored more in the breach than in the observance illustrates its priority in any meaningful human rights design. The institutionalization of religious liberty is not without its problems;[16] but establishing the irreducible claims of religious liberty ought to be of prime importance to a Church at the service of peace.[17]

Beyond religious liberty, are some human rights more basic than others, more necessary to protecting human freedom and advancing the common good? Peter Berger, among others, has argued that there are fundamental rights of the person so universally established in world religious and ethical systems that they constitute a central core of "human rights" and establish the possibility of meaningfully arguing for "universal human rights" in a pluralistic world. Berger's list of the "grossest cases" of human rights violations today includes, in addition to "the deliberate desecration of religious symbols and the persecution of those adhering to them," such depredations as:

> Genocide; the massacre of large numbers of innocent people by their own government or by alien conquerors; the deliberate abandonment of a population to starvation; the systematic use of terror (including torture) as government policy; the expulsion of large numbers of people from their homes; enslavement through various forms of forced labor; the forced separation of families (including the taking away of children from their parents by actions of government) . . . [and] the destruction of institutions that embody ethnic identity.

Berger acknowledged that each of these practices was routine in many modern countries. But he also argued that, in condemning these atrocities as human rights violations, we can cite warrants beyond those developed in Western civilization. The consensus on these issues "emerges from all the major world cultures, especially in their religious foundations—and it is a consensus all the more impressive in view of the vast (and partly irreconcilable) differences among the world religions in their understanding of reality and of human destiny."[18] The effective protection of innocents against these fundamental and gross depredations of the human person is essential to the peace of rightly ordered political community. However one construes the breadth (or narrowness) of the appropriate use of "human rights" language and imagery, it is hard to argue that there are rights more basic (or even equivalent) to the protections on Berger's short list.

Reclaiming and developing the Catholic heritage will thus include a reconceptualization of the relationship between various understandings of "human rights," including a careful critique of the moral and political implications of comprehensive human rights schemes; an open acknowledgment that there are, both theoretically and practically, priorities to be established in the advance and defense of human rights; a recommitment to religious liberty as *the* fundamental human right; and a thoroughgoing analysis of the ways in which various forms of governance meet the tests of human freedom and the common good at the economic, social, and cultural levels. On this last point, American Catholicism must pay more attention to the social and economic achievements of liberal democracy, founded on a narrow construction of "human rights" as civil liberties and political freedoms, than has been the case over the past generation (during which the liberal democratic disinclination to identify social and economic goods as "rights" was often considered a failure of moral imagination that cast the entire liberal democratic experiment into question).

Tranquillitas ordinis, even at the international level, is a matter of both peace and freedom. The condition for the possibility of our being neither victims nor executioners is, Camus argued, the creation of a *civilisation du dialogue*, a "sociable culture," in which the God-given integrity of the human person is protected and advanced through *both* moral affirmation *and* institutions of law and governance dedicated to fostering individual liberty and the common good. Peace and freedom will be advanced together; otherwise, it seems unlikely that either will be advanced at all.

Progress toward peace in international political community also requires, according to the Catholic heritage, the *economic, social, and political development* of the world's underclass. Here as well, fresh thought is in order. Three avenues of exploration may be mentioned briefly.

In the first instance, American Catholicism's intellectual and leadership elites must subject dependency theory to the same rigorous critique that has been expended on other analytic models. The central claim of dependency theory—that the world's rich have achieved and maintain their privileged status at the expense of the world's poor—is often received as common wisdom in American Catholic leadership, intellectual, and activist circles. But does that claim stand up to empirical scrutiny? Why has the American Catholic debate focused so much attention on the causes of poverty, and so little attention on the causes of wealth? What lessons might be learned from successful examples of Third World economic development (e.g., the ASEAN countries of Southeast Asia)? Are there ideological (and psychological) predispositions toward accepting the central images and claims of dependency theory that require serious self-examination? Such questions are now being forcefully raised by a host of development theorists and practitioners.[19] They must also be addressed by those whose primary concern is the relationship between development and the peace of dynamic political community.

The concept of "development" itself needs considerable expansion. Lack of natural resources, one simple (and simplistic) measure of a nation's capacity for development, does not provide a very satisfactory explanation of the impoverishment of Latin America, which is resource-rich. Development, in other words, involves more than raw materials. Development, understood as the creation and expansion of wealth commensurate with the advance of the common good, is a matter of human imagination, values, creativity, and will. Oil did not create wealth by the fact of its existence; the material resource became a source of wealth when it was wedded to the spiritual resources of creativity and imagination that led to the internal combustion engine and the technology of extraction and production. Human capital may be as important, and quite possibly more important, than the more easily measurable evidence of material resources. But if that is the case, then development must involve, not simply the transfer of material resources from rich to poor, but the creation and sustenance of moral cultures with specific values. Deferred gratification is a classic Christian spiritual good; but it is also an indispensable element in creating wealth. Similar links between Christian values and economic advancement may be found in a culture's support of cooperative activity toward common ends; thrift; a sense of responsibility to one's family and its future; and political stability based on a minimum of popular consent. The claim of many contemporary development theorists, that there are moral-cultural prerequisites to economic development narrowly construed, coheres at important points with Catholic incarnational humanism. American Catholicism's address to Third World economic development must, therefore, take increasing account of the importance of human capital—the moral-cultural foundations of a social community—for the creation and distribution of wealth and the consequent alleviation of poverty.[20]

The adjective "consequent" is used deliberately here, for there is a link, not only between creating wealth and alleviating poverty, but between wealth-creation, the reduction of poverty, and the possibilities of progress toward the peace of political community in the world. As Peter Berger, whose involvement in Third World issues is rooted in the demands of Christian social ethics, has written:

> Successful development presupposes sustained and self-generating economic growth. . . . We have a pretty clear idea of what a zero-growth world would look like. It would either freeze the existing inequities between rich and poor, or it would see a violent struggle to divide up a pie that is no longer growing. Neither scenario holds out the slightest promise for such values as human rights and democracy. The existing inequities would have to be brutally defended or brutally altered.[21]

On this analysis, economic growth leading to the alleviation of poverty is a *sine qua non* of the evolution of *tranquillitas ordinis.* The peace of rightly ordered political community cannot be built in a zero-growth world.[22]

Finally, American Catholicism must think more carefully about the relationship between economic development and democracy. Just as the moral-cultural elements of development cannot be divorced from economic growth and the alleviation of poverty, so, too, must the relationship between economic development and various forms of governance be rigorously explored. Empirical study is important here. That there has not been a single successful case of socialist development in the Third World, that socialist states tend toward tyranny, and that socialist states fail to deliver on their egalitarian promises are important data to be considered. "Socialist equality is shared poverty by serfs, coupled with the monopolization of both privilege and power by a small (increasingly hereditary) aristocracy," Peter Berger argues.[23] Those who argued the priority of bread before freedom have done poorly by both bread and freedom.

Such evidence does not prove an opposite case, that fully functioning parliamentary democracy is a precondition to successful development. Empirical evidence contradicts this claim as well, as in the ASEAN countries. There is, therefore, much to be explored and investigated here. In the long run, Berger claims, "democracy and development are necessarily linked realities."[24] But what about the short- and medium-run? Successful economic development seems to require limitations on the state's role in social and economic life (one indispensable foundation of democratic governance). And the miserable economic failures of totalitarian states make it clear that economic development and gross, persistent violations of basic human rights are incompatible. The moral-cultural values necessary for successful development are of little moment in totalitarian states (as the social decay of the Soviet Union makes plain);[25] yet these values are also under pressure in authoritarian societies.

Can one, then, argue for indispensably minimal "predemocratic" values and institutional arrangements as prerequisites to successful economic development? At the most rudimentary level, the institutionally protected distinction between state and society appears in all Third World development success stories; but what beyond this is necessary? Is a publicly visible political process (even if participation in it is largely restricted to a single dominant party) essential? If one important meaning of development is "the extension of human choice," what case can be made for the necessity of democratic building-block institutions like free trade unions and peasant cooperatives in economically successful development? Fully functioning, modernized economies that provide both growth and equity may indeed require democratic forms of governance in order to sustain themselves; but what means are available for helping underdeveloped countries move toward bread and freedom simultaneously?[26] The rule of law, for example, would seem crucial to economic development precisely for the poor, to whom it gives an equality they radically lack when *caudillos* or generals are in charge of a country.

These are questions in need of the most careful analysis, for the sake of the world's underclass, as well as for progress toward the peace of public order within, as well as among, nations. They are not, unhappily, the questions currently engaging the imaginations of the majority of American Catholic development theorists and commentators. A reoriented discussion is, once again, in order: on the nature and processes of development itself, and on the relationship between economic, social, and political development and the evolution of the dynamic peace of political community at the international level.

THE QUESTION OF THE SOVIET UNION

There are many obstacles now impeding progress on the fourfold agenda of the politics of *tranquillitas ordinis*, and thus blocking the path from anarchy to political community in world affairs. The morning's headlines often make it seem as if chaos is increasing its destructive sway throughout the world. International terrorism, for example, not only erodes the stability of existing national communities, but time and again has proven a bloody solvent eating away at the possibility of political community across national borders. The West's sense of impotence in the face of international terrorism is compounded by the difficulties of developing a coordinated approach to the problem among law-abiding nations, much less by the failures of international institutions to satisfactorily address, let alone resolve, the issue. If our analysis is derived from today's crisis, the list of obstacles to progress toward the peace of political community in international life can appear so endless as to render coherent, long-term policy impossible.

Viewed through the prism of the past forty years, however, a taxonomy of priority difficulties begins to emerge. In such a historical perspective, the chief obstacle to progress toward the peace of dynamic political community has been the policy and practice of the Soviet Union. An American Catholicism reclaiming and developing its heritage must craft a persuasive set of answers to the unavoidable question, "What about the Soviets?"

The starting point must be a rejection of the anti-anticommunism that shaped much of the American Catholic debate on peace, security, and freedom since 1965. A Church "at the service of peace" must be rigorously, thoughtfully, and visibly anticommunist. Such a stance is fully congruent with the National Conference of Catholic Bishops' 1980 "Pastoral Letter on Marxist Communism."[27] This clear, public posture ought not create or exacerbate a climate of fear and suspicion in American Catholicism, or in American political culture. Still less should it align the American Church with any particular political party. Although it is not widely understood among American Catholic intellectual and leadership elites today, it is quite possible to be a liberal, even social democratic, anticommunist.[28] American Catholic anticommunism needs to draw on the insights of a James Burnham and a Whittaker Chambers; but it also needs to learn from left-of-center sources like Sidney Hook, Dwight Macdonald, William Barrett, and Bayard Rustin.

American Catholic anticommunism "at the service of peace" would have its own distinct rationale: it would reject communism because of the incompatibility of incarnational humanism with communist understandings of human nature, human history, and human destiny. Such a principled rejection does not foreclose the possibility of a vigorous Christian/Marxist dialogue; it does suggest the impossibility of a final rapprochement between the two world-historical views.[29] Forthright anticommunism is now considered déclassé in many opinion-elite circles that welcomed the Catholic activism that preceded and followed "The Challenge of Peace." American Catholicism could help establish models of the calm, intelligent, cool anticommunism that is badly needed in contemporary American political culture, and for which important precedents exist in the work of John Courtney Murray and the World War II pastoral letters of the American bishops.

Anticommunism, then, is the entry point. But anticommunism does not, and cannot, answer the question "What about the Soviets?" in and of itself. Anticommunism is a necessary, but hardly sufficient, condition for progress toward peace.[30] Since the Second World War, there have been two strategic options for coping with the threat of Soviet power and purpose. "Containment" sought to build a *cordon sanitaire* around the Soviet Union, checking its expansionist tendencies until the day when the internal collapse of the Soviet system rendered it nugatory as an international danger. This approach did much for South Korea, little or nothing for Hungary, and ultimately sank on the rock of Vietnam. Nor does it seem to have had a significant impact on patterns of social control in the U.S.S.R., which, if less crude than in Stalin's day, are hardly less draconian in either intention or effect.

The alternative strategy of détente sought to engage the Soviet Union in a web of economic, political, and cultural relationships with the West, in the hope that these would temper Soviet international activism. The symbolic apogee of this effort was the "Basic Principles of Relations" agreement signed by President Nixon and General Secretary Brezhnev in Moscow in May 1972. Subsequent Soviet actions in the Yom Kippur War, in the Horn of Africa, in support of the overthrow of the shah of Iran, in the invasion of Afghanistan and the proxy repression of Polish Solidarity, coupled with a seemingly endless expansion of Soviet military capability, suggest that, whatever else the Soviet leadership believed détente to mean, it hardly intended anything approaching progress toward peace with freedom.[31]

Moreover, the psychology of détente has been a continuing problem in American political culture, and most specifically in the American peace movement. That the Soviet Union is essentially a "defensive" power, scarred by multiple invasions over the past two centuries; that the roots of U.S./Soviet conflict have to do with problems of misunderstanding and poor communication, rather than with fundamental differences of values and interests; that an accommodating approach to Soviet leadership is most likely to result in arms control or reduction agreements; that human rights abuse in the U.S.S.R. has little to do with agreement on reversing the arms race—these themes, redolent of the psychology of détente, are still received as common wisdom in those parts of American political culture most passionately concerned about the moral problem of war and peace. This is especially true of the religious peace movement.

The psychology of détente appeals to an understandable (and correct) instinct that the world's leading democracy should engage its principal adversary across the full range of issues on the world political and economic agenda. Its desperate and debilitating flaw lies not in the general prescription, but in the analysis of the obstacles to its fulfillment: in the refusal to face squarely the nature of a totalitarian society governed according to Leninist principles by a privileged elite whose primary concern is the maintenance of its own power. This analytic failure typically results in a further failure to comprehend the nature, purpose, and extent of the Soviet global agenda, and the alarmingly impressive array of forces (many of them nonmilitary) that the U.S.S.R. deploys throughout the world for achieving its political ends.[32]

"Containment," on the other hand, although considerably more accurate in its assessment of Soviet intentions (at least as originally formulated by George Kennan in late 1947), also failed to answer satisfactorily the question, "But what about the Soviets?" The failures of containment were the obverse of détente's; a more perceptive analysis was married to a policy of essential *dis*engagement. The West would meet the challenge of Soviet purpose and power only when the U.S.S.R. attempted to expand its *imperium*. Containment was a defensive strategy. It, too, was unlikely to contribute to progress toward the peace of political community. Although it might maintain equilibrium in a dangerous world (no small accomplishment, to be sure), containment could not, by itself, advance the fourfold structure of the politics of *tranquillitas ordinis*—a point well-understood by John Courtney Murray in his critique of U.S. policy toward the U.S.S.R. in the late 1950s.[33]

What Murray hinted at, and what a later generation of American Catholic leaders could help devise, would be an approach to the Soviet issue that combined insights contained in both containment and détente, and constructed a strategy of vigorous engagement with the U.S.S.R. on the basis of classic containment theory's accurately sober analysis of Soviet intentions. Such a strategy would reject the notion that U.S./Soviet conflict is a function of misunderstanding; would actively seek to *change* the present course of Soviet policy; would understand that such changes in Soviet international behavior will inevitably involve change in the internal structure of power in the U.S.S.R.; and would thus take the pluralization of power—the de-Leninization of the Soviet state—as its principal objective.

A number of Sovietologists and political theorists dissatisfied with the options of containment or détente have proposed a strategy of pluralization in recent years.[34] The common threads running through these proposals are the claim that *change* in, rather than accommodation to, the present Soviet international agenda is imperative for progress toward peace, security, and freedom; and the judgment that such change requires alterations in the patterns of power within the Soviet state. The Leninist monopoly of power in the U.S.S.R., on this analysis, is a root cause of the destabilizing activities of the U.S.S.R. in the world. Increase the number of independent centers of power in the U.S.S.R., and the more the Soviet leadership will be required to gain domestic consent for its foreign policy. A leadership even minimally answerable to others than itself would have to engage in the kinds of trade-offs that would restrain foreign adventurism.[35]

According to Aaron Wildavsky, the fundamental lack of accountability between the organs of the Soviet state and the weak Soviet civil society is a principal reason why the Soviet leadership has been able to conduct an aggressive international campaign since World War II. Yet Wildavsky wants to do far more than force the Soviet leadership to turn inward so that it will spend less on defense. The real point is geopolitical, and bears on progress toward peace, security, and freedom:

> The Soviet Union cannot live in peace with nations that are different and independent because it cannot tolerate those self-same qualities within its own regime. Unless there is pluralization in the Soviet Union, it is only a matter of time and opportunity before it seeks to subjugate the only force in the world capable of resisting it.[36]

A policy of pluralization would, Wildavsky argues, complement and extend the meaning of containment. Rather than being defensive, containment would be transformed into an offensive strategy, "aimed at the internal position of the Soviet Union from which the danger emanates."[37]

A strategy of pluralization could also be located within an even broader context: it would seek the pluralization of power within the U.S.S.R. as a necessary condition for the possibility of agreement on the pursuit of the fourfold structure of the politics of *tranquillitas ordinis* in the world. Agreement between the world's two superpowers is essential to progress on that agenda; but agreement is so unlikely as to be virtually impossible given the present structure of power in the U.S.S.R. Thus, the pursuit of peace requires a sustained effort at changing, not merely the external behavior, but the internal dynamics of the Leninist U.S.S.R. A reclaimed and developed Catholic heritage would be as dissatisfied with the idea that we simply "let the Soviets have their rotten system" (one of President Reagan's least felicitous formulations) as with proposals for accommodation to present Soviet international activity. "Their rotten system" is a principal obstacle to the pursuit of peace, security, and freedom in the world. Changing it is essential.

And, of course, dangerous. As Aaron Wildavsky frankly admits, "though pluralization is apparently more pacific than a military line, it is aimed at the jugular—the Communist-Leninist form of political organization—and therefore should be approached with care."[38] Pluralization is, however, quite congruent with a strategy of peace: it uses nonviolent means for achieving needed change; it is "a war of ideas to counter a war of violence."[39] Given the failures of both containment and détente, such an effort seems worth exploring, particularly for a "Church at the service of peace," which seeks needed change in the world through other-than-military instrumentalities. A policy of pluralization does not provide a quick solution to the linked dilemma of war and totalitarianism; but by forthrightly acknowledging, and then seeking nonviolent change in, the totalitarian side of the dilemma, it may help create conditions more conducive to the pursuit of alternatives to war.

What might some of the elements in a strategy of pluralization be? The following possibilities suggest approaches needing careful study and testing.

Inasmuch as one defining characteristic of a Leninist state is its insistence on monopolizing the flow of information within a society, a policy of pluralization would work vigorously to break that monopoly. It would enhance the international broadcasting capabilities of the United States, not to conduct polemics against the Soviet state, but to provide accurate information on conditions within Soviet society (including the *nomenklatura* elite's system of privileges), and on events throughout the world. Aleksandr Solzhenitsyn once described international radio broadcasts into the U.S.S.R. as "the mighty nonmilitary force which resides in the airwaves and whose kindling power in the midst of communist darkness cannot even be grasped by the Western imagination." The Soviet Union now spends more annually on jamming Western broadcasting than the United States spends on both Voice of America and Radio Liberty transmissions into the U.S.S.R. Surely the cause of peace with freedom requires redressing this imbalance. Those who believe that the truth will make us free ought to have a modest confidence that challenging the Leninist monopoly of information in the Soviet Union serves the interests of peace and freedom.

Exchange programs are another potential instrument of pluralization. The U.S.S.R. has aggressively used scholarly, cultural, and, increasingly, religious exchange programs as extensions of its foreign policy. American Catholicism could help lead an exploration of how the yearning to touch another human being across the boundaries of nationality and ideology can be turned toward better ends. It would reject participation in those exchanges and conferences that blatantly promote Soviet foreign policy objectives (the activities of Soviet front organizations like the World Peace Council and the Christian Peace Conference come readily to mind). It would challenge exchange programs that misrepresent the religious situation in the Soviet Union (the 1984 visit to the U.S.S.R. by over 250 American religious leaders organized by the National Council of Churches is a sorry example). Yet it would concurrently help design programs of cultural and scholarly exchange that, by bringing knowledgeable Americans into the U.S.S.R. and Soviet citizens not regime-programed into the United States, would actually foster meaningful dialogue and mutual understanding. And by doing so it would pose a sharp counterpoint to the steady and scurrilous blasts of anti-Western propaganda that are a staple in the Soviet media. The first steps along this path will be small. But small steps aimed at pluralization and change are far preferable to large steps whose primary function (wittingly or unwittingly) is to provide disinformation opportunities to the Soviet state apparatus.[40]

Vigorous defense of religious liberty is another means by which American Catholicism could help foster pluralization and de-Leninization. The Catholic community in the Soviet Union—especially in the Ukraine and Lithuania—is under unremitting pressure from the Soviet state. Yet the American Catholic community has not organized itself in support of Catholics in the U.S.S.R. in anything resembling the degree to which American Judaism has taken up the cause of Soviet Jewry. Given Soviet ratification of the 1975 Helsinki Final Act, the U.N. Charter, and the Universal Declaration of Human Rights, work for religious liberty in the U.S.S.R. is not only an act of obligatory solidarity with suffering fellow believers; it is work in support of international law and of the peace of rightly ordered political community.

The Soviet Union is now undergoing a religious renaissance—a demonstration of the spiritual hollowness of Marxist-Leninist doctrine and an important opening toward pluralization. Yet, somehow, discussion of religious persecution in the U.S.S.R. is often considered gauche in contemporary American Catholicism. Self-education on the current state of religious practice in the Soviet Union; prisoner-of-conscience adoption programs at the parish, diocesan, and national levels; and clear statements by the National Conference of Catholic Bishops and the United States Catholic Conference that religious liberty and peace are indivisible—these, and undoubtedly many other vehicles, are available for meeting an obligation of charity and conducting a strategy of pluralization.

Finally, American Catholicism must deal with the Soviet peace movement, in both its official and dissident forms. The official Soviet "peace" apparatus (both internally, in the Soviet Peace Committee, and externally, in the World Peace Council and the Christian Peace Conference) should be shunned. The shunning can (and should) be polite, and the reasons for it carefully explained: American Catholicism's intellectual and leadership elites must make it clear that they will not make common cause with organizations that reject the peace of *tranquillitas ordinis* and whose sole function is to advance Soviet foreign policy. Neither peace nor freedom, and still less truth, are served by consultations with such agencies or their official representatives (when they are functioning as such; ecumenical religious contacts removed from a "peace" context as understood by the Christian Peace Conference are another matter). Conversely, American Catholics concerned for peace and freedom should be vocal, visible advocates for and defenders of the dissident peace movement in the U.S.S.R., represented by organizations such as the Moscow-based Group to Establish Trust Between the U.S.S.R. and the U.S.A. Such groups are not without their own problems of analysis and prescription; but the very fact of their (threatened and often tenuous) existence is a force for pluralization and change in the U.S.S.R. Work with dissidents will necessarily involve, at the beginning, small steps. But, again, such small steps are more important than large-scale cooperative efforts with the official Soviet "peace" apparatus.

Is a policy of pluralization compatible with progress toward the peace of political community? Does it not involve unwarranted intervention in the internal affairs of the U.S.S.R.? On the latter question, the answer is both yes and no. Yes, a policy of pluralization involves intervention in the affairs of the Soviet state; no, such intervention is not unwarranted, given Soviet obligations under international law. Linkage between the advance of human rights in the Soviet Union and progress toward genuine arms reduction seems virtually unbreakable, both in principle and in practice. It is unbreakable in principle because Christians cannot simply abandon fellow believers and all the other victims of the Soviet system to their fate. It is unbreakable in practice because there is little probability of progress toward a world that is peaceful, secure, and free as long as the U.S.S.R. remains in the iron grip of a Leninist *nomenklatura* elite.

Would the Soviets not retaliate against Western efforts to increase the free flow of information within the U.S.S.R.? Professor Wildavsky's answer is succinct: "Let them." In fact, Wildavsky urges, invitations to reciprocity should be built into a policy of pluralization, for "everything they can say (and probably worse) is already being said here."[41] But the point is surely more than this. Americans have little to fear from an open comparison between our society and the Soviet Union's (a point well understood by the Soviet leadership, as illustrated by its jamming of the Voice of America and its internal media campaign of anti-American propaganda, which often features cartoons of American political leaders festooned with swastikas). An American policy of pluralization through breaking the Leninist monopoly on information within the U.S.S.R. should not be contingent on Soviet acceptance of reciprocal opportunities for stating their case to the American public. But the offer should be made.

Opposition to a strategy of pluralization can also be based on a kind of fatalism, a sense that change in the Soviet Union is so unlikely as to be chimerical as an object of policy.[42] One cogent answer to this fatalism was offered, not by an exponent of détente, but by one of its most stringent critics, Richard Pipes. Pipes readily concedes the difficulties involved in a policy of pluralization. But he also argues that the "system now prevailing in the Soviet Union has outlived its usefulness and . . . *the forces for change are becoming well-nigh irresistible.*"[43] The issue then is not *whether* there will be change in the U.S.S.R., but *what kind* of change. A policy of pluralization aimed at assisting the kind of change that would advance the causes of peace and freedom becomes, on this view, not a luxury, but a necessity.

The deepest taproot of resistance to a policy of pluralization, though, is not the argument over the possibility of change but an even more basic difficulty: "the understandable but debilitating desire not to confront the fundamental problem of American-Soviet relations: the incompatibility of regimes."[44] That debilitating desire is pervasive in the contemporary American peace movement, and particularly in religious peace efforts. Contemporary American Catholicism has been no more invulnerable to this deterioration of imagination and will than any other denomination. Yet the Catholic tradition of moderate realism has always urged that facts be faced, for there is no escape from history and historical responsibility.

The advantage of a policy of pluralization is that it affords the opportunity of acknowledging the incompatibility of the democratic and Leninist systems, and then doing something about it through nonviolent means, rather than being paralyzed by it. Pluralization could also serve an important domestic political purpose: it could offer, not middle ground, but new ground on which those dissatisfied with, but now committed to, the containment or détente models could work together on the linked causes of peace, security, and freedom.

THE QUESTION OF INTERVENTION

Reclaiming and developing the American Catholic heritage on the question of intervention must begin with a principled rejection of the neo-isolationism that came to dominate the Catholic debate on America's international role in the years after the Second Vatican Council. A "soft" neo-isolationism, which emphasized American incapacities in the world and taught the virtual equivalence of "intervention" with "armed force," was a principal theme in the abandonment of the heritage of peace as dynamic political community. Vietnam (understood as *the* paradigm of what happened when America "intervened" in the "internal affairs" of other countries); New Left teachings about American "imperialism" and the impact of the "military-industrial complex" on U.S. foreign policy; and those themes in the theologies of liberation which taught that Third World poverty was a calculated result of American consumerism—all these sources contributed to a form of neo-isolationism that, even if willing to consider the theoretical possibility of a useful American engagement with world politics and economics, was persistently critical of specific "interventions" as they arose. The results of this neo-isolationism—on the large question of U.S. policy in Central America, and on discrete cases like the U.S. action in Grenada—were predictable.

American Catholic neo-isolationism had several curious features, including a counterintuitive insistence on the realities of interdependence. On the one hand, American Catholicism's intellectual and leadership elites taught that American involvement in the world was usually for the worse. On the other hand, these same elites regularly affirmed that this was in fact one world, an interdependent world. "Intervention" would seem a necessary corollary of "interdependence"; but the controlling metaphor of "Vietnam," taken as a generic, rather than specific, noun made such a correlation unpalatable.

The voices of interdependence were correct in their general intuition about the shape of world affairs; the voices of "nonintervention" were not. Reclaiming and developing the Catholic heritage requires that interdependence and intervention be reconceived as two dimensions of the same international reality. This would lead to a new, and considerably improved, spectrum of argument, focused not on the question of *whether* America should intervene in world politics and economics, but *how*: toward what ends, by what means, measured against what standards? Intervention would be a given. In a politically, economically, and, according to Catholic theory, morally interdependent world, intervention is in the nature of things. The important questions to be thought through have to do with the goals of America's inescapable intervention in the world; the obstacles to the achievement of those goals; the means appropriate for the pursuit of our ends, given the obstacles to their achievement; and the criteria by which those ends and means would be judged.

The degree to which the American Catholic debate on intervention deteriorated since the mid-1960s, and failed to address the questions of intervention and interdependence concurrently, was painfully evident in the argument over U.S. policy in Central America. A narrowly focused identification of "intervention" with U.S. military action; distorted empirical perceptions of the actual situation in Central America and its historical antecedents; the "Vietnam" metaphor (epidemic among Catholic activists); and a refusal to face squarely the facts of others' intervention in the region conspired to create a debate centered on the question of *whether*, rather than *how*, America could act for peace, security, freedom, and prosperity within our own hemisphere.

A debate which began with the assumption that intervention in Central America was a geopolitical given and, even more importantly, a moral and strategic responsibility, would have led to a considerably different American Catholic contribution to the public discourse. A developed Catholic heritage would have suggested that questions of security, and economic, social, and political development in Central America, could not be disentangled. Peace, freedom, human rights, and economic well-being would not be advanced by the triumph of the Nicaraguan FSLN and the Salvadoran FMLN. But neither would they be served by maintaining a morally and politically unacceptable status quo in which the people of Central America were disenfranchised serfs. Combining these judgments, rather than falsely choosing between them, would have led to a debate focused on how U.S. intervention could serve the causes of social and economic reform, democratization, and regional security—*together*.

This would not have been an abstract exercise, for there were forces in the region that wished to pursue such a path: those gathered around the Christian Democracy of José Napoleón Duarte in El Salvador, and around the archbishop of Managua, the free trade union movement, *La Prensa*, and the democratic political opposition in Nicaragua. Yet the siren song of "nonintervention" was so seductive in many American Catholic leadership and intellectual circles that these forces for the pursuit of the peace of political community could not even be recognized as such. Nor, given the miasma of anti-anticommunism, could the chief obstacle to progress toward peace, security, freedom, and prosperity in the region be correctly identified.

The reclamation and development of the classic Catholic heritage will therefore involve a sharp about-face on the question of intervention. Acting on the premise of an inescapable American involvement with world affairs, it would challenge New Left-based neo-isolationism, the traditional isolationism that remains just beneath the surface of American political culture, and those forms of libertarian isolationism that detach the value of freedom from the value of order and the pursuit of the common good.[45] The challenge would be both moral and strategic. The moral challenge would emphasize the teaching of *Pacem in Terris* on the "universal common good," which cannot be pursued by an American escape from international activism and responsibility. The strategic meaning of the "pursuit of the international common good" would be spelled out, in terms of goals, by the fourfold structure of the politics of *tranquillitas ordinis* as sketched above; American intervention in the world would be ordered toward the achievement of peace, freedom, and security, not merely for ourselves, but for a dangerously interdependent world that must chart a safe passage from political arena to political community. The obstacle posed to that passage by the intentions and capabilities of the Soviet Union and other totalitarian states would be frankly acknowledged; and the question of intervention in the face of this threat would focus on means to change the present course of Soviet power.

The strategic challenge would thus raise the question of how U.S. intervention in world affairs can help build agreement on the pursuit of peace through political community. Given the realities of international life today, military force in the defense of violated rights must remain one available instrument for intervention. A Church "at the service of peace" would, through just-war theory, vigorously enter the public debate over the ways in which the proportionate and discriminate use of military force might contribute, in certain circumstances, to peace with freedom. The limit case of deterrence, and the more accessible cases of the use of military force in regional conflicts or against the forces of terrorism and chaos, set the boundaries of this discussion.

But the Church would do more. It would also lead an exploration of the many other-than-military means of intervention available to those who would build the peace of political community. These means are relevant not only in U.S./Soviet relations, where the immediate short-term objective is to prevent a prosecution of conflict through direct military confrontation. They are also, as in Central America, immediately relevant to regional crises in which economic and social reform, and the growth of democratic forms of governance, are inextricably linked with security from aggression. The task, in other words, is to challenge the reduction of the notion of "intervention" to "armed force" by complementing necessary military means of security with other-than-military means of pursuing social, economic, and political change congruent with the advance of the universal common good.

The notion of "peace initiatives" is now widespread in American Catholic intellectual and leadership circles. The initiatives concept was originally derived from social psychology. In the work of Charles Osgood, for example, unilateral initiatives were understood as acts taken to signal pacific intentions so that equilibrium was restored between conflicted states and the possibilities of negotiated agreement thereby enhanced. Initiatives were aimed at "tension reduction," which Osgood considered, on his psychological model of international relations, to be the first order of business in reaching agreement.[46]

Although Osgood's formulation of the initiatives concept has some limited applicability, it is also badly flawed. It misconceives the nature of political conflict, and it fails to address the question of how change can be forced through initiative means when change is necessary. Amitai Etzioni argued a more persuasive case for peace initiatives in the wake of the 1963 Partial Test Ban Treaty. Etzioni analyzed the impact of President John Kennedy's independent moratorium on U.S. atmospheric nuclear testing, which was coupled with a requirement that the U.S. moratorium be reciprocated in kind by the U.S.S.R., in breaking the logjam that had previously blocked test ban negotiations.[47]

The initiatives concept holds out promise for developing a new armamentarium of means for America's inescapable intervention in world affairs. The concept must be detached from psychology, though, and inserted into politics. A new discussion of initiatives would acknowledge the occasional importance of symbolic action (as in Anwar Sadat's 1977 trip to Jerusalem), but its principal interest would be in developing means for forcing change through actions taken prior to agreement in conflicted situations. The purpose of such initiatives would not, in the main, be "better understanding," but reciprocal political response. The primary focus would not be on unilateral gestures of good will, but on a coordinated range of independent actions aimed at inducing or compelling reciprocation. What is needed, in other words, is a careful exploration of a *strategy* of initiatives, not the occasional gimmick. Inasmuch as initiatives involve other-than-military means to advance the fourfold structure of the politics of *tranquillitas ordinis*, their development should be of interest to an American Catholicism "at the service of peace." But the prerequisite to developing such an initiatives strategy is its disentanglement from those overpsychologized concepts of political conflict that so often mar American Catholic discussion of war and peace, security and freedom.[48]

The question of nonviolence could also be reopened within a larger American Catholic reconsideration of intervention. The concept of "nonviolence," like "peace initiatives," has been abused since the mid-1960s by the psychologization of conflict. Yet as its principal historian, Gene Sharp, has amply demonstrated, nonviolence, properly understood, is an intensely political concept: it is a means for achieving political purpose, a mechanism for exercising power.[49] Nonviolent sanctions are another alternative means to impel needed change so that progress toward the peace of political community can be achieved. An American Catholicism "at the service of peace" will be rightly interested in the theory and practice of nonviolence. But it will advance the public debate over America's necessary intervention in world affairs if it facilitates a consideration of nonviolent strategies and tactics removed from the neo-isolationist context in which such discussions are often held today, and locates nonviolent action within the more appropriate context of the pursuit of peace as rightly ordered political community.

What nonviolent means are available to resist the gradual Leninization of Nicaragua? Could nonviolent sanctions be brought to bear on the problem of religious persecution in the communist world? What nonviolent means will foster peace and justice through democratization in a country like Chile? These are the kinds of questions in need of immediate address. They are rarely, if ever, encountered in American Catholic leadership and intellectual circles. Raising them would be an important contribution to the reclamation and development of the Catholic heritage.

THE QUESTION OF MILITARY FORCE

The roots of Catholic just-war theory lay in the Church's traditional teaching that politics and morality were not antinomies. No human actions, even in the limit case of war, stood outside the moral universe. Armed force was related to politics; it was not an independent reality, suspended in an amoral ether. Rather, armed force was one band on a spectrum of options for the defense and pursuit of the peace of dynamic political community, all of which were subject to moral scrutiny and judgment. Moral law continued to make demands on the individual conscience and on the rulers of states whenever and wherever military force was invoked, just as it did in all other exercises of power.

The disintegration of the concept of peace as political community has led to a dislocation of just-war reasoning from its central place in the American Catholic debate over peace, security, and freedom. Issues in the renovation of just-war thinking in American Catholicism have been sketched in the previous chapter. Here, a narrower question is at issue: How might the discriminate and proportionate use of military force contribute to the pursuit of political community on an international level?

For some, of course, the question is oxymoronic; "peace" and "military force" are antonyms. Peace cannot be pursued or advanced through military means. Wars will cease when men refuse to fight. This position is incongruent with Catholic social theory. The world will not make its passage from chaotic political arena to rightly ordered political community in the twinkling of an eye, through a great leap of imagination and will. Along that inevitably long and dangerous passage, legitimate interests must be defended, rights secured, and aggression opposed. Nonviolent means may be available to defend interests, secure rights, and oppose aggression. But, then again, they might not; and in the absence of military means of defense, peace and freedom could be irreparably damaged. Moral clarity about the circumstances in which the discriminate and proportionate uses of military force can make their genuine contribution to peace with freedom is thus essential, even set against the horizon of human possibility suggested by *Pacem in Terris*.

Two contemporary dilemmas make such moral clarity even more imperative: international terrorism and guerilla warfare.

No doubt there are some legal and political means available for combatting the curse of terrorism. Airplane hijacking, for example, has been at least partially curtailed by international agreements, even between adversaries like the United States and Cuba. International police and intelligence work, particularly in Western Europe, has been successful in breaking some terrorist organizations. Yet there have been, and undoubtedly will be, occasions when coordinated military action on the part of law-abiding nations is necessary to free innocents from terrorist bondage. The Israeli action at Entebbe airport in Uganda (made possible in part by assistance from the government of Kenya), and the West German action at Mogadishu (undertaken in concert with the government of Somalia) are the two most publicly accessible examples. In both of these cases, military action served the cause of peace, not only in rescuing terrorist-held hostages but in reasserting the rule of law in international life. Such incidents are all too likely to reoccur along the path toward political community and public order in world affairs. Moral clarity on the question of a legitimate military response to them is thus essential.

The most difficult cases are those involving state-sponsored or state-supported terrorism. Yet even here, the Catholic heritage may prove its worth. State-sponsored terrorism is incompatible with the international stability that is one prerequisite for the pursuit of peace, security, and freedom among nations. General principles, under the burden of necessity, may have to be subordinated to choosing lesser evils, but there seems little point in maintaining more than minimal diplomatic intercourse with states that have consciously set themselves outside the framework of orderly international public life. Developing the means to sequester outlaw nations that harbor, sponsor, or give material assistance to international terrorists, and achieving the agreement necessary to make such means effective, would be a step toward, not away from, peace.

If sequestration can be accomplished through existing international instrumentalities such as the United Nations and its relevant functional agencies, fine; if not, then it should be pursued by whatever intermediate associations can be constructed toward the end of vigorously combatting state-sponsored terrorist activity. Here, for example, would be a useful exercise of power by an association of democratic (and perhaps predemocratic) countries. The sequestration of outlaw nations could help lessen the incidence of terrorism in world affairs; and the agreements necessary to achieve it would help establish the "competent authority" that the *ius ad bellum* requires for the resort to military force in resisting terrorist aggression.

In addition to helping facilitate discussion of preventive measures against terrorist activity, a reclaimed and developed Catholic heritage would give serious attention to issues such as the *ad bellum* permissibility of reprisals against terrorists in the aftermath of hijackings, kidnappings, and assassinations, as well as the question of preemptive military action in defense of peace, security, and freedom. The disconcerting themes that dominated the American Catholic debate on Israel's strike against Iraq's Osirak nuclear reactor suggest how badly such an exploration is needed.[50]

Guerilla warfare, a depressingly ubiquitous feature of modern international life, also poses hard questions about the use of military force "at the service of peace." The *ad bellum* issues are complicated by the fact that guerilla wars are often waged in terms of an asserted right to revolution, and *in bello* questions are frequently distorted by guerillas' resort to terrorist activities against civilian and military targets. Considerable work on these questions has been done in the wake of the American experience in Vietnam;[51] less attention has been given to the question of how "just and limited" military response to guerilla insurgencies can be ordered to the dynamic peace of political community.

The American Catholic debate over U.S. policy in Central America exemplifies this deficiency. In El Salvador, the objectives of political, economic, and social reform could not (and cannot) be successfully pursued without a concurrent response to the security threat posed by the guerilla/terrorist forces of the FMLN. Yet the main thrust of American Catholic intellectual, journalistic, and official commentary on El Salvador has been directed toward resistance to the discriminate and proportionate use of military force against Marxist-Leninist guerilla forces.

The Church does not possess the competence to design military actions to combat terrorist or guerilla forces. But neither can it burke the issue of the discrete use of military force against such threats to progress toward peace, security, and freedom. The chief task of an American Catholicism reclaiming and developing its heritage on this contextual question therefore lies in the direction of a more sophisticated reconnection of the *ius ad bellum* and *ius in bello* to that *ius ad pacem* implied by the trajectory of classic just-war reasoning. The pursuit of political community according to the global vision of *Pacem in Terris*, as well as progress toward peace, security, and freedom in today's regional conflicts, demands this exploration. The aim of the discussion, which coheres with the thrust of just-war reasoning over the past fifteen hundred years, is to link the politics of peace to the discriminate and proportionate use of military force once again.

The essential amorality of military force is still asserted by adherents of realpolitik; but that assertion, in tacit form, also undergirds much of contemporary American Catholic activism. Both of these arguments must be challenged, and the most effective challenge will be mounted by a reconstituted Catholic discourse on the relationship between politics, the legitimate use of armed force, and the pursuit of peace with freedom.[52]

THE QUESTION OF THE BOUNDARIES OF POLITICAL OBLIGATION

"Global citizenship" has become a staple in the contemporary American Catholic debate on war and peace, security and freedom. The term is confusing: "citizenship" is a function of membership in a *polis*, and there is, at present, no global *polis* of which citizenship, strictly defined, can be asserted. But the term is more often used as an expression of moral concern than of political self-definition. In that sense, the concept of "global citizenship" reflects those intuitions that led Pope John XXIII to speak of the "universal common good."

The pursuit of the peace of political community in international life requires the development of a sense of moral and political responsibility that transcends national boundaries. This is the clear teaching of the Church's magisterium since World War II. Yet the magisterium has also taught the abiding claims of national political obligation, a theme frequently absent (or drastically minimized) when the claims of "global citizenship" are raised. An American Catholicism reclaiming and developing its heritage has, in this case as on several other contextual questions, the opportunity to create an ecclesiastical and civic debate framed around a "both/and," rather than "either/or," set of propositions.

Taking advantage of this opportunity will require a new defense of national patriotism.

The charge that preconciliar American Catholicism was too fervidly patriotic, too uncritical of U.S. national security policy, was a persistent theme in the abandonment of the classic Catholic heritage over the past twenty years. The charge has some foundation. Even so stout a defender of the American proposition as John Courtney

Murray worried that critical Catholic moral faculties could be displaced by acquiescence to the government policy of the day. Murray thought that this had been the case on the issue of unconditional surrender during World War II; no sustained criticism of the policy was heard from Catholic leaders. "Nor was any substantial effort made to clarify by moral judgment the thickening mood of savage violence that made possible the atrocities of Hiroshima and Nagasaki." Murray thus joined with many other postwar Catholic commentators who believed that "there is place for an indictment of all of us who failed to make the tradition [i.e., of just-war reasoning] relevant" to certain policy questions of that period.[53] But there is, as Murray would be the first to admit, a substantial difference between legitimate criticism set within a context of mature commitment to the American experiment and proposition, and the kind of self-flagellating moral*ism* that led a later generation of Catholic activists to countenance the country's name being spelled "Amerika," and to describe its foreign policy as a venereal disease that greedy merchants of death inflicted on an innocent world.

Moreover, the notion of uncritical patriotism as a dominant leitmotif in the historiography of American Catholicism obscures as well as illuminates. The critical affirmation of the American proposition developed by the bishops of the United States in their pastoral letters on nativism, and the bishops' explicit criticisms of the postwar planning of the Roosevelt and Truman administrations, suggest countercurrents at work. Stephen Decatur's toast to "our country, right or wrong," may have been too frequently invoked. But the often-forgotten preamble to that toast, "our country: in her intercourse with foreign nations, may she always be in the right," was by no means wholly absent from the conscience of American Catholics.

In any event, the choice today is not between mindless jingoism or a psychosis of self-contempt. The issue is the development of a mature patriotism which understands that love and criticism are not mutually exclusive. Maturely critical patriotism can be an expression of incarnational humanism. According to classic Catholic understandings of the means through which God speaks his word to this world, the moral norms that are to shape public life are mediated through the events, persons, and institutions of our lives, in history. All human institutions, and especially political institutions, stand under the judgment of transcendent moral norms. Yet this very assertion contains an important warrant for critically mature patriotism: some polities choose to recognize that they stand under transcendent judgment, and are structured to be permeable to the apprehension of those norms to which they wish to be held accountable.

Here, then, is one important dimension of the moral worthiness of the American experiment: it knows itself to be under the judgment of truths that transcend it, and it has organized itself so that those truths are brought to bear on public life. American democracy can thus foster a moral appreciation for such central Catholic social-ethical values as personalism (i.e., the roots of human rights are in the inherent, God-created dignity of the individual human being, not in the policy of the state) and pluralism. Moreover, democratic structures of governance seem most capable of allowing a society to pursue the common good according to the principle of subsidiarity, through a host of private and public associations. Such claims do not establish the absolute preferability of democratic governance over all other possibly imaginable polities. But the evidence buttressing them suggests that, on the available options, democracy is a morally worthy experiment with legitimate claim on the loyalty of those who benefit from its freedoms and protections.

A mature appreciation of and gratitude for democratic institutions is not only compatible with affirming transcendent moral values; according to Catholic incarnational humanism, democratic patriotism can be an expression of our commitment to those values. There is no contradiction between commitments to the truth claims of Catholicism and commitment to the American democratic experiment. Nor is there any contradiction between commitment to the American proposition and critical moral commentary on American public policy, foreign or domestic. The crucial distinction is between criticism that seeks to hold the experiment accountable to its foundational proposition, and criticism that works to undermine the claims of the experiment to personal moral and political obligation.

Catholic social theory is resolutely antignostic; the truths that shape the Catholic social and political conscience are discerned through encounters with the matter of this world, through particular communities and institutions. A gnostic iconoclasm that urges the abandonment of national obligation for the sake of "global citizenship" fits the Catholic incarnational imagination poorly. The very term "global citizen" might well be abandoned, or, at the very least, its use radically constrained. The issue is not semantic finickiness, but truth: to assert the reality of a global *polis* in the face of its empirical nonexistence is a delusory exercise that can lead to warped and skewed perceptions of reality, and thus to policies that make war more, rather than less, likely. On the other hand, progress toward peace with freedom requires a commitment of political obligation (not merely moral sensitivity) to those who do not share our immediate national community. How might that sense of commitment and obligation be responsibly fostered by an American Catholicism reclaiming and developing its heritage?

Several strategies are worth serious debate. Strengthening a sense of moral and political obligation to democratic forces throughout the world is one avenue to be explored. Building democracy is not simple; but the effort deserves more analytic attention and moral support than has been forthcoming from American Catholic intellectuals and religious leaders since the mid-1960s. The Church could help provide critical support for the efforts of quasi-public agencies like the National Endowment for Democracy, which work through private sector instruments to help build the infrastructure of democracy abroad. But it would go further. The American Church, as part of a major transnational institution, could be a significant force in helping gather the world's existing and nascent democratic countries into forms of association that meet common security, economic, and social needs. The impetus for such action (and the careful thought that must precede it) would reflect the conviction that democratic governance is a morally worthy expression of the core values of Catholic social theory and the historical understanding that the advance of democracy contributes to the cause of peace. Democratic nations do not, on the present historical record, war with each other. There are innumerable conflicts among the world's democracies, some of which involve grave issues of economic interest. But these conflicts have not led to war. Strengthening the existing bonds among the world's democracies, and fostering the development of new democratic experiments, is fully congruent with that "major step forward in civilization and wisdom" that Pope John Paul II has called a human imperative.

The boundaries of American Catholics' political obligation do not stop at the borders of the democratic world. The claims of the "universal common good" transcend the democracies. The boundaries of moral and political obligation extend to those now caught in the vise of tyranny, in either its authoritarian or totalitarian form. In the latter case, the strategy of pluralization (or de-Leninization) expresses a willingness to meet the claims that our common humanity impose on us, and to do so in ways that advance the prospects of peace with freedom. Support for religious dissidents and for the unofficial peace movement in the Soviet Union and its satellites gives flesh to a moral commitment to pursue the "universal common good." The same applies to efforts to facilitate the transition from traditionally authoritarian forms of government to predemocratic or democratic governance in the Third World. Here, great sensitivity is required, particularly where totalitarian opposition forces are in the field. The cause of *tranquillitas ordinis* both within and among nations is not served by a transition from authoritarian to totalitarian rule, as the experience of states such as Cuba and Ethiopia has made painfully clear.

The Church's role is not the micromanagement of U.S. foreign policy. It is to uphold, in and out of season, the idea that open and accountable political processes ordered to the common good offer a this-worldly means to resolve conflict without the organized mass violence of war. And it is to support those forces that seek the resolution of conflict through economic, social, and political reform, rather than through Leninist-inspired revolutionary violence. The commentary of American Catholicism's intellectual and leadership elites on Central America between 1975 and 1985 suggests that there is considerable room for rethinking here.

The issue, then, is not a choice between patriotism and "global citizenship." The issue is how mature patriotism can provide a means for pursuing the dynamic peace of political community in the world. Peace with freedom requires leadership, and that leadership can only come, in this world, from an existing national political community. American Catholics live in a democracy that could, if gathered for the task, make the pursuit of peace and freedom the central purpose of its public life in its third century. Reconstructing a sense of American possibility in the world, shorn of jingoism and moral illusion, but also of masochistic self-deprecation, is an essential part of the reclamation and development of the Catholic heritage that was largely abandoned over the past generation. It is also essential for an American Catholicism "at the service of peace."

THE QUESTION OF AMERICA

Although any number of altered teachings contributed to the abandonment of the heritage of *tranquillitas ordinis* among American Catholic intellectual and religious leaders over the past twenty years, the key theme of abandonment was a deteriorated sense of American possibility in the world, based on a deep skepticism about the American experiment. The skepticism took many forms. At its outer limits America became "Amerika," the source of evil in the world and a prime candidate for revolutionary change. Such extreme formulations did not constitute the mainstream of Catholic commentary; but they were considered within the bounds of reasonable discourse. But even the more moderate critiques were scarcely less intense. Whether the analysis was drawn from dependency theory or the Port Huron Statement, the essential teaching remained—there were fatal flaws in the American experiment, which had led to a morally unworthy organization of power in American society and thence to the evils perpetrated by U.S. foreign policy.

No doubt postconciliar American Catholicism had to reconsider the buoyant optimism about the American experiment that had characterized the preconciliar Church; the world had changed, the country was changing, new understandings were required. And yet one cannot read the literature of the American Catholic debate on war and peace, security and freedom, since the Second Vatican Council without understanding that, by the early 1970s at the very latest, the agenda had been radically revised: rejection, not reform, became a prominent current in the discourse.

Reconsideration of this rejectionism is essential to reclaiming and developing the abandoned Catholic heritage. It is time to take up the task outlined, but never completed, in John Courtney Murray's grand project: the articulation of a moral rationale for and defense of the American democratic experiment. Lacking such a foundation, the American Church will not contribute to the evolution of a national commitment to the peace of political community in the world; it will become, if it has not already, a primary obstacle to that development.

Why did American Catholicism's intellectual and leadership elites fail to take up Murray's grand project? Part of the answer lies in a misapprehension of Murray's intention. If the Murray Project was simply a modern riposte to residual nativism, and if that problem was solved by John F. Kennedy's election to the presidency, then there could seem little point in pursuing Murray's task further. Primary attention could turn (as it did) to the new problem identified by the country's intellectual/cultural elite, the moral failure of American society. But these assumptions are plainly defective, on the evidence of Murray's own writings, as well as on the facts of post-1960 American political culture. Murray was after considerably larger game. Understanding that the American experiment and its foundational proposition were (and understood themselves to be) always under judgment, Murray sought to develop the moral underpinnings of the experiment and the proposition in ways that could provide the civic glue essential to a pluralistic society. Whether Murray's claim that the *philosophia perennis* and its natural law theory were the only candidates for fulfilling that role of moral foundation-building can be argued elsewhere. But it must be admitted that Murray correctly identified a pressing public need, and that addressing it was essential if America's role in world affairs was to serve the causes of peace and freedom.

The moral problem of war and peace cannot be addressed in a historical vacuum; it takes particular shape, and poses distinct problems, given the historical circumstances of the present. And because the question of war and peace is so fundamental a determinant and expression of national self-understanding and purpose, it is impossible to disengage these issues from the larger context of one's understanding of the American experiment, its validity, and its role in a world often hostile to its central values. Reclaiming and developing American Catholic thought on war and peace will necessarily involve a pursuit of Murray's grand project, and vice versa.

Taking up Murray's challenge today requires that American Catholicism's intellectual and leadership elites regain a sense of the originality of the American experiment, of the American experience of being *novus ordo seclorum.* Jacques Maritain's

Reflections on America, originally published in 1958 and now largely forgotten in American Catholic social-ethical circles, is a useful model of the critical affirmation required in and by the present moment.

Maritain believed in the value of first impressions, and was particularly struck by the freshness of the American national experience:

> During my first visit to New York, I was invaded by a kind of thrilling enthusiasm and pleasure in the sudden feeling that here we are freed from history. For a European long immersed in all the rotten stuff of past events, past hatreds, past habits, past glories, and past diseases which compose a sort of overwhelming historical heredity, the first contact with America is thus liable to produce a sort of intoxication, a delight in new-born freedom, as if the old burden of historical necessities were suddenly put aside. It seemed that everything is possible to human freedom.[54]

Maritain knew that there was no escape from the past. But he also knew that however dependent America was on the civilization of Europe, that history was America's prehistory. Almost without knowing it, Americans had "inaugurated a really new phase in modern civilization."[55]

American civilization, to Maritain, was "the least materialist" among modern industrialized peoples.[56] And although American consciousness had, like all modern consciousness, "been infected by the miasmata that emanate from the structures and ritual of our modern civilization,"[57] Maritain refused to acquiesce to the fashionable (then as now) view that gross materialism was the dominant characteristic of American culture. On the contrary, Maritain argued that, although there were greedy individuals in America, there was "no avarice in the American cast of mind." The evidence lay, not only in the great charitable foundations, but in the ordinary rhythm of American civic life, where Maritain quickly noted the importance (and ubiquity) of voluntary organizations. Here, Maritain believed, was the warrant for claiming that "the ancient Greek and Roman idea of the *civis praeclarus*, the dedicated citizen who spends his money in the service of the common good, plays an essential part in American consciousness."[58]

Moreover, Maritain argued, the American experiment had yielded a distinctive and welcome pluralism-amid-community. It had, in the main, defeated those forces of intellectual intolerance that were part of the travail of European history. Human beings were still sinners; the Klan burned crosses. Yet the national character had decisively rejected the use of violence, physical coercion, and lies against dissenters.[59] Pragmatism had played its part in this; the natively pluralistic condition of America posed a stark choice between tolerance and anarchy. Yet American tolerance was much more than pragmatic accommodation. What had been achieved was a *principled* pluralism in community: historical necessity had been turned into an "invaluable gain for civilization," a nation committed, not just required, to a life ordered by tolerance and common respect. "America is the only country in the world where the vital importance of the sense of human fellowship is recognized in such a basic manner by the nation as a whole. . . . There is, in the most existential sense, a strain of Gospel fraternal love deep in the American blood."[60] For all that it was a middle-class nation, America was not a "bourgeois" nation.[61]

Maritain was no romantic emigré, blind to the faults of the land that had given him refuge from Hitler; *Reflections on America* is full of insightful critiques of American society and its polity. Maritain worried that the traditional American disparagement of ideology could lead to sounding too uncertain a trumpet in the world: "You are advancing into the night, bearing torches toward which mankind would be glad to turn; but you leave them enveloped in the fog of a merely experiential approach and mere practical conceptualization, with no universal ideas to communicate. For lack of adequate ideology, your lights cannot be seen."[62] America needed a clearer sense of what it stood *for* in the world; it could no longer lead by the sheer historical fact of what it once arose *against*.

Maritain was also concerned about "American illusions": Enlightenment optimism about human nature and the natural world; a too constricted understanding of success; a nervousness about intellectual life (perhaps rooted in the belief that "if you are a thinker you must be a frowning bore, because thinking is so damn serious"); a tendency to disregard the human importance of leisure.[63] Maritain also understood that racism and "the sex question" remained thorns in the flesh of American society.[64]

But Maritain, no Babbitt, was also reasonably confident about America's capacity to deal with these cultural illusions and their attendant problems of public policy. The root of his confidence was his sense of the distinctive achievement of American democratic pluralism:

> The American body politic is the only one which was fully and explicitly born of freedom, of the free determination of men to live together and work together at a common task. And in this new political creation, men who belonged to various national stocks and spiritual lineages and religious creeds—and whose ancestors had fought the bitterest battles against one another—have freely willed to live together in peace, as free men under God, pursuing the same temporal and terrestrial common good.[65]

This American accomplishment (which, Maritain understood with Murray, was "a continuous process of self-creation," not a finished product[66]) pointed the direction for a distinctive American role in the world. The world looked to America for hope: "What the world expects from America is that she keep alive, in human history, a fraternal recognition of the dignity of man—in other words, the terrestrial hope of men in the Gospel."[67] Maritain closed his book with the prayer that, "please God . . . this crucial fact may never be forgotten here."[68]

Jacques Maritain's prayer went unheeded. The shock that his enthusiasm for the American experiment engenders after one absorbs the past twenty years of Catholic rejection is itself an indication of how far the pendulum has swung, and in which direction. But Maritain was prescient about the future as well as insightful about the present; like Murray, he knew that an America that had lost confidence in itself, that had forgotten its foundational raison d'être, would be incapable of acting for peace and freedom (for "hope") in the world. What even Maritain and Murray could not have anticipated, though, was that American Catholic intellectuals, their successors, would be in the forefront of that radical assault on the moral worth of the American experiment that had as one of its chief aims the withdrawal of American power and influence from world affairs.

The gravamen of that attack can be summed up succinctly: America had failed to build *tranquillitas ordinis*, rightly ordered political community, at home. Why, therefore, expect it to act for peace and freedom in the world? Given the premise, the conclusion follows logically enough. What Maritain and Murray remind us is how questionable the premise is. The issue is not criticism; one could hardly be more critical than Murray, with his claim that America lacked a public philosophy capable of sustaining the experiment over the long haul.

But what Murray and Maritain also understood was that America had in fact accomplished a great good: it had formed and sustained political community among extraordinarily diverse peoples, in a way that met the test of the heritage of *tranquillitas ordinis*. That accomplishment was never, in principle, completed; it was always endangered (and from within the national mind and spirit, as well as from external enemies). But what *had* been accomplished was a good. America demonstrated, by its very existence, the possibility of the dynamic peace of rightly ordered political community in an always conflicted and ineluctably plural world. Yes, a more perfect justice, a more complete charity, and a truer freedom remained to be achieved. But the American experiment had shown that law and governance, rooted in a commitment to the inherent dignity and worth of every individual human being, could provide the means by which justice could be pursued without resort to mass violence. American Catholicism's traditional support of trade unionism expressed this conviction, and illustrated that a critical affirmation of America could come from left of center. In Camus's terms, Americans had learned to be "neither victims nor executioners." That very experience created at least one condition for the possibility of a creative American engagement with the problems of peace and freedom, conflict and pluralism, in the world.

Reclaiming and developing the classic American Catholic heritage on the moral problem of war and peace requires a critically affirming appraisal of the American experiment and the dynamic peace of public order within our own political community. This is not to argue for the replication of the American experience throughout the world, which is impossible. But American Catholicism cannot facilitate a national commitment to pursue peace through political community in the world if its intellectual and leadership elites remain convinced of the fundamental moral dubiousness of the American experiment and experience. Completing Murray's grand project—or, at the very least, recognizing that a debate toward that end must be immediately reopened and vigorously pursued—is essential to an American Catholicism seeking to be "a Church at the service of peace."

Such an approach to the contextual question of America would also help distinguish real from spurious problems in the public debate over war and peace, security and freedom. Charges of American "hegemonism" would be given the short shrift they deserve. The real problems of national complacency and confusion, rooted in traditional isolationism but intensified by the harsh facts of interdependence, could then be addressed in a more thoughtful way. That mirror-imaging of "the superpowers" which was a staple in the nuclear debates of the late 1970s would be similarly rejected for the empirically spurious construct that it is. Those who posed the moral equivalence of the United States and the Soviet Union would no longer set the terms of the discourse within both Church and wider society, and the real questions of American engagement with the problems of Soviet power and purpose could be addressed anew.[69] Reconstructed debates on international human rights and the problems of Third World development would also follow from a more critically affirming appraisal of the American experiment and experience.

In the late 1950s, John Courtney Murray suggested that a "Catholic moment" in the drama of the American proposition had arrived. His call to seize that moment went unheeded. Other voices, often from outside the formal boundaries of the Catholic Church, have repeated Murray's suggestion a generation later.[70] The great question of war and peace is, as Murray himself understood, at the center of any consideration of national life and purpose. Whatever its other manifold flaws of perception and analysis, this point, at least, has been correctly intuited by postconciliar American Catholic activism. That the future of the American experiment in the peace of political community is inextricably bound up with the pursuit of *tranquillitas ordinis* in the world is a point on which many of the combatants in the present debate might be able to agree. Should that agreement form, it will be considerably wiser if its approach to the question of America draws more on the insights of Maritain and Murray than on those themes of rejection that were a powerful force in American Catholic intellectual and leadership circles over the past twenty years.

"Whom you would change, you must first love," Martin Luther King, Jr., used to teach. It is time, in American Catholicism, to love America again, with that critically affirming affection on which true and needed change can be built, so that America's potential for leadership in the pursuit of peace with freedom in the world might be realized.

THE QUESTION OF HUMAN NATURE

Finally, then, it comes down to the question of Man. Is war in the nature of things, given human beings as we know them? The Catholic tradition of moderate realism has persistently taught that war is not inevitable: only likely. The likelihood of war is the easier side of the affirmation to accept, particularly in this bloodiest of human centuries. A Church "at the service of peace" will, however, join John Paul II in urging, not only the necessity, but the possibility of the "major step forward in civilization and wisdom" that would mark the evolution of peace with freedom in world affairs. The conviction of that possibility is strengthened, not weakened, by a clear and sober view of the abiding effects of evil in the world. Catholic incarnational humanism stands firm in its acknowledgment of human propensities for evil. But it stands equally firm in its insistence that, even under the effects of original sin, we remain the image of God in history, creatures capable of hearing and responding to a divine word of invitation and challenge. That conviction is the ultimate basis of the Catholic heritage of *tranquillitas ordinis*. The abandonment of the heritage thus involved a loss, in varying degrees, of the sense of human possibility that has been a hallmark of Catholic theological anthropology.

Reclaiming and developing the Catholic heritage will require a new sense of confidence in the human prospect. Chastened by the harsh realities of war and totalitarianism, American Catholicism must yet resist the temptation to despair. It might well meditate on William Faulkner's 1950 Nobel Prize address in which the novelist, hardly a stranger to the demons of the human spirit, still offered this act of faith:

> I decline to accept the end of man. It is easy enough to say that man is immortal simply because he will endure: that when the last dingdong of doom has clanged and faded from the last worthless rock hanging tideless in the last red and dying evening, that even then there will still be one more sound: that of his puny, inexhaustible voice, still talking. I refuse to accept this. I believe that man will not merely endure: he will prevail. He is immortal, not because he alone among creatures has an inexhaustible voice, but because he has a soul, a spirit capable of compassion and sacrifice and endurance. The poet's, the writer's duty is to write about these things. It is his privilege to help man endure by lifting up his heart, by reminding him of the courage and honor and hope and pride and compassion and pity and sacrifice which have been the glory of his past.[71]

Faulkner's definition of the poet's and writer's duty and privilege is also true of the duty and privilege of a "church at the service of peace." The Church's view of the human prospect, in its grandeur as well as its threatenedness, is the beginning and the end, the alpha and omega, of the heritage of peace as dynamic political community. We have been created for peace and freedom; peace and freedom now stand under mortal threat. Those who pose survival as humanity's highest end suffer from a moral myopia that intensifies the many present dangers that beset us. Those who, facing the full measure of our peril, still work for a future worthy of those who are the images of God in history, make peace and freedom possible, if not guaranteed. That complex proclamation is the enduring task of the Catholic moral imagination in the modern world. American Catholicism bears a special burden of responsibility for the truth of that proclamation. It has not met the test of that burden over the past generation.

It is time to do so.

EPILOGUE

The Moral Necessity of Politics

Kant was right when he said that a state of peace had to be "established." What perhaps even he did not discern was that this is a task which must be tackled afresh every day of our lives.

Michael Howard, *War and the Liberal Conscience*

The peace movement is not a recent phenomenon; it undoubtedly antedates written history, although in its modern form it may well claim Erasmus of Rotterdam as its progenitor. Since Erasmus, the problem of war has been a particularly nagging one for those possessed of what Michael Howard terms the "liberal conscience": those who believe that there is a wide chasm in human affairs between things as they are and things as they ought to be, and that the chasm can be bridged by human effort. Since Erasmus, the liberal conscience has proposed many roads to peace, each based on a distinctive understanding of the reasons for war.

The *philosophes* of the eighteenth century, and the followers of Richard Cobden in the nineteenth, took their cue from the great humanist of Rotterdam and argued that war was the product of an aristocratic ruling class for which militarism was a way of life. Wars would end when the aristocrats were displaced by St. Simon's *les industrieux.*[1] Yet the fall of the aristocracy led, not to peace, but to even more violent war, as the nation-at-arms replaced the small armies of the aristocracy on the fields of battle. *Les industrieux* might be disinterested, on the whole, in foreign affairs. But they could also be "suspicious and xenophobe, prone to paranoia, and passionately vindictive in proportion to the shattering of their peaceful ideals."[2] Nor did the liberal theory of the nineteenth century account for the rise of totalitarianism, and the ways in which skillful demagogues like Lenin, Trotsky, Mussolini, and Hitler could gather whole peoples to battle against what was portrayed as a persistently hostile world.[3]

If the fault did not lie with the aristocrats, then where was it located? The merchants of death, as arms manufacturers came to be called in the early twentieth century, provided a possible scapegoat. No doubt arms competition can worsen international conflict; but it seems, on the whole, more a symptom of conflict than its cause. Moreover, can it be reasonably argued that imperial rivalry and the clash of capitalist interests accounted for the two world wars of this century?[4]

Perhaps, then, the problem lay in the balance of power itself, and the ways in which it was manipulated by diplomats. Inept diplomacy and the politics of the balance of power can cause wars; but power politics is, at bottom, "the politics of not being overpowered." The task of statemanship is to prevent war or, if war must come, to see that one's country enters it under favorable circumstances, and not faced by an overwhelming coalition of enemies. Blaming bad diplomacy for war is like blaming bad driving for traffic accidents; it tells us the obvious, and in doing so tells us too little.[5]

The root of the analytic failure of the liberal conscience, according to Michael Howard, rests in the habit of seeing war as an independent variable, a "distinct and abstract entity about which one can generalize at large."[6] Here, Clausewitz was surely right. War cannot be properly understood outside a political context. If war is not "the continuation of politics by an admixture of other means," then it is simply barbarism. To recognize this is not to indulge in blood lust. On the contrary, it is to locate war within the human universe of reason and will. Doing so is the first, indispensable step toward limiting the sway of war in human affairs. Thus St. Augustine's classic formulation, "We go to war in order to obtain peace," is not a call to the crusade, but an affirmation of the *ius ad pacem* that must orient any just war.

Classic Catholic moral theory understood the symbiotic relationship between war and politics. It not only looked to politics as the context in which the limit case of war could make rational and moral sense; it looked to the politics of *tranquillitas ordinis*—the evolution of dynamic, rightly ordered political community—as the answer to the moral problem of war.

Catholic activists are often accused of having "politicized" the Church. The charge is curiously true and false. The Church in the United States has, in fact, in its intellectual and leadership elites, become a political pilgrim. It has entered the public arena in an unprecedented way. Its bishops now regularly pronounce on fine points of foreign policy. All of this is considered normal, desirable, and in fact a moral obligation by the postconciliar Catholic establishment. The ideological trajectory of the political pilgrimage of the American Church seems clear: from 1965 to 1985 American Catholicism's elites became softly neo-isolationist, anti-anticommunist, and highly skeptical of the moral worthiness of the American experiment.[7]

Yet, for all the "politicization" of the Church at the public policy level, there has also been an odd depoliticization of American Catholic thought on the moral dilemma of war and peace, security and freedom. Conflict is often understood now in primarily psychological, rather than political, categories. So, too, with peace. The spiritual peace of interior serenity before God has been projected, as it were, onto a global stage, and the problem of war reduced to a problem of psychological adjustment, open communication, and better understanding. This post-Freudian (more accurately, Rogerian) phenomenon has been completed by a tendency to temporalize the eschatological peace of *shalom*: to teach that the Kingdom in which conflict is no more can be constructed by human hands. Here, seeds planted by the vulgar Marxism of the Vietnam-era New Left have come to flower.

Both of these constructs are strangely apolitical. They look inward, not to political life, for answers to the moral problem of war. Perhaps more to the point, they are modern forms of Pelagianism. Sin and evil—and war—are epiphenomenal: historical accidents that can be repaired by right intentions and the will to act on them. The transformation of the American Catholic debate on war and peace thus begins to look like the latest episode in the dilemma of the liberal conscience when confronted with the ubiquity of conflict in the world. Aristocrats, merchants of death, and realpolitik diplomats are no longer blamed; the problems lie within us, and can be solved by conversion or therapy.

The reclamation and development of the classic Catholic heritage of thought on war and peace, security and freedom, requires a recommitment to the moral necessity of politics. Politics is a moral necessity because it is the civic expression of incarnational humanism in an always conflicted world. The psychologization of conflict, for all that it condemns the escapism of those with whom it disagrees, is itself escapist. It fails to understand, not only that conflict is a constant of the human condition, but that conflict can be channeled toward human progress when it is mediated through the habits, associations, and institutions that make up *tranquillitas ordinis*, rightly ordered and dynamic political community.

American Catholicism needs, in other words, not to be depoliticized so much as repoliticized. The American Catholic politics of peace and freedom, which must be a politics of virtue, ought to seek the golden mean, which, for Aristotle, was not a point between the extremes, but ahead of them. The issue is not to split the difference between the poles of the existing argument, but to gather the insights that exist at both extremes, and to marry them to what insights exist at the intermediate points along the spectrum. The result, according to Aristotle, was practical wisdom. And practical wisdom is the precondition to the dynamic peace of public order in political community.[8]

The Catholic tradition of moderate realism is like the classic liberal conscience analyzed by Michael Howard, in that it is never satisfied with things as they are. But it also knows, with Chesterton, that "it is futile to discuss reform without reference to form." The failure and the tragedy of American Catholic thought on war and peace over the past generation has not been its passion for reform; it has been the disengagement of reform from form, the abandonment of the classic Catholic heritage as exemplified by Murray.

The great promise of American Catholicism is that it still bears that heritage, and has the means, the opportunity, and the moral responsibility to develop it. The failures of the past generation need not be repeated. More is expected of us. It is time to take up, again, the burden and the privilege of the politics of peace and freedom.

Notes

Prologue

1. Paul Johnson, *Modern Times: The World from the Twenties to the Eighties* (New York: Harper & Row, 1983), p. 1.
2. B. H. Liddell Hart, *History of the First World War* (London: Pan Books, 1972), p. 89.
3. For the sense of foreboding that the Great War aroused, see Barbara Tuchman, *The Guns of August* (New York: Bantam Books, 1976), especially Sir Edward Grey's famous comment: "The lamps are going out all over Europe; we shall not see them lit again in our lifetime" (p. 146).
4. Quoted in Johnson, op. cit., pp. 13–14.
5. Aleksandr Solzhenitsyn, "Men Have Forgotten God," the Templeton Prize Lecture, delivered in London on May 10, 1983; reprinted in *National Review*, July 22, 1983, pp. 872–76.
6. Ibid.
7. New York: Oxford University Press, 1975.
8. Ibid., p. 21.
9. See William Manchester, *The Glory and the Dream: A Narrative History of America 1932–1972* (Boston: Little, Brown, 1974), pp. 280–87. Fussell surveys the decline of chivalric language in op. cit., pp. 22–23.
10. Fussell, op. cit., p. 29; see also pp. 25–29 for a reflection on the decline of that British innocence symbolized by the "sporting spirit" with which the Great War was begun.
11. Ibid., p. 40; see also pp. 131ff.
12. Ibid., p. 58.
13. Ibid., p. 63. The Eliot reference is, of course, to "The Waste Land."
14. Ibid., p. 75.
15. Ibid., p. 79.
16. Ibid., p. 81.
17. Ibid., p. 174.
18. Ibid., p. 176.
19. Ibid., p. 234.
20. Ibid., p. 235.
21. From *The Penguin Book of First World War Poetry*, edited by Jon Silkin, 2nd edition (Harmondsworth: Penguin Books, 1981), pp. 208–9.
22. Ibid., p. 85.
23. Fussell, op. cit., p. 139; see pp. 137–44 for an extended discussion of the use of Bunyan's imagery in the literature of the Great War.
24. Ibid., p. 141.
25. Ibid., p. 320.
26. Ibid., p. 323.
27. Ibid., p. 321.
28. Emile Durkheim, *Suicide* (Glencoe: Free Press, 1951).

29. Hannah Arendt, *Totalitarianism*, part 3 of *The Origins of Totalitarianism* (New York: Harcourt, Brace & World, 1966), p. 176.

30. Ibid., p. 162.

31. Johnson, op. cit., pp. 51–52.

32. See André Malraux, *Man's Fate* (New York: Modern Library, 1936); Ignazio Silone, *Bread and Wine* (New York: Athenaeum, 1962); Arthur Koestler, *Darkness at Noon* (New York: Macmillan, 1941); George Orwell, *1984* (London: Secker & Warburg, 1949).

33. Arendt, op. cit., pp. 162–64.

34. For the Lenin figures, see Johnson, op. cit., pp. 70, 93; for the Stalin period, see Robert Conquest, *The Great Terror* (London: Macmillan, 1968); for the Hitler figures, see William Shirer, *The Rise and Fall of the Third Reich* (New York: Simon and Schuster, 1960); for various estimates of the Mao figures, see Fox Butterfield, *China: Alive in the Bitter Sea* (New York: Times Books, 1982), p. 350. See also Eugene H. Methvin, "20th Century Superkillers," *National Review*, May 31, 1985, pp. 22–29.

35. Johnson, op. cit., pp. 70–71.

36. Arendt, op. cit., pp. 167, 166.

37. Johnson, op. cit., p. 54.

38. Solzhenitsyn, art. cit. p. 874. When he was preparing to leave interrogation for his first camp in the Archipelago, this was the key message that the young internee Solzhenitsyn received: "Trust no one but yourself. . . . The law there is the law of the jungle. There never was and never will be *justice* in Gulag" (Aleksandr Solzhenitsyn, *The Gulag Archipelago 1918–1965: An Experiment in Literary Investigation* [New York: Harper & Row, 1973], volume 1, part 2, pp. 563–64).

39. Arendt, op. cit., p. 175.

40. For Jaegerstaetter's story, see Gordon Zahn, *In Solitary Witness: The Life and Death of Franz Jaegerstaetter* (New York: Holt, Rinehart, and Winston, 1964). See also Armando Valladares, *Against All Hope* (New York, Alfred A. Knopf, 1986).

41. Johnson, op. cit., pp. 50–51.

42. Aleksandr Solzhenitsyn, *One Day in the Life of Ivan Denisovich* (New York: Bantam Books, 1963), pp. 195–203.

43. Peter Berger, *A Rumor of Angels* (New York: Doubleday Image Books, 1969).

44. Johnson, op. cit., p. 58.

45. Albert Camus, *Neither Victims Nor Executioners* (New York: Continuum, 1980). See Robert Pickus's brilliant introduction for a critical application of Camus's standards to the American peace movement during and since the Vietnam War.

46. Ibid., p. 25.

47. Ibid., p. 26.

48. Ibid.

49. Ibid., p. 28.

50. Ibid., pp. 28–30; emphasis added.

51. Ibid., p. 31.

52. Ibid., p. 32.

53. Ibid., p. 33.

54. Ibid., p. 39.

55. Ibid., pp. 44, 46.

56. Ibid., p. 49.

57. Ibid., p. 55.

58. Ibid., p. 61.

59. Albert Camus, "The Unbeliever and Christians," in *Resistance, Rebellion, and Death* (New York: Vintage Books, 1974), pp. 73–74.

60. Ibid., p. 73.

The Catholic Tradition of Moderate Realism

1. New York: Doubleday Image Books, 1964.

2. Ibid., p. 262. A generation later, and almost twenty years after the Second Vatican Council, it might be thought that *that* question, at least, has been settled. It has not. During the discussion after my formal testimony in 1982 before the National Conference of Catholic Bishops' Ad Hoc Committee on War and Peace, which was charged with drafting the bishops' famous pastoral letter "The Challenge of Peace," the chairman of the committee, Archbishop Joseph Bernardin, turned to one of the religious acting as liaison to the Ad Hoc Committee, and asked if she wanted to ask me a question. "Yes," she replied, "I'd like to ask Mr. Weigel what all of this has to do with the Sermon on the Mount." Like Father Murray and his friend, my interlocutor and I "floundered a while in the shallows and miseries of mutual misunderstanding," and then the conversation moved on.

3. For a survey of New Testament themes on the issue of war and peace, written from the author's critically pacifist perspective, see Roland Bainton, *Christian Attitudes Toward War and Peace* (Nashville: Abingdon Press, 1960), pp. 53–65. For a representative view of how New Testament themes influenced the National Conference of Catholic Bishops' pastoral letter, "The Challenge of Peace," see Sandra M. Schneiders, "New Testament Reflections on Peace and Nuclear Arms," in *Catholics and Nuclear War*, Philip Murnion, ed. (New York: Crossroad, 1983); "complexities" are noticeably absent from this essay.

4. Bainton surveys the pacifism of the subapostolic and patristic Church, and the antecedents to Augustine's just-war theory, in op. cit., pp. 66–91. For a view of early Christian pacifism that locates the pacifist ethic of the primitive Church under the rubric of the proscription of idolatry in the First and Second Commandments, rather than under the Fifth Commandment, see Geoffrey Nuttall, *Christian Pacifism in History* (London: Basil Blackwell & Mott, Ltd., 1958), pp. 1–14.

5. Peter Dennis Bathory, *Political Theory as Public Confession: The Social and Political Thought of St. Augustine of Hippo* (New Brunswick: Transaction Books, 1981), pp. 11–12.

6. Ibid., p. 20. For a biographically based overview of Augustine's theological controversies, see Peter Brown, *Augustine of Hippo* (Berkeley: University of California Press, 1969).

7. New York: Harper & Row, 1951, pp. 190–95, 206–17.

8. Bainton, op. cit., p. 98.

9. See Bathory, op. cit., pp. 30–31.

10. Charles McCoy, "St. Augustine," in *History of Political Philosophy*, Leo Strauss and Joseph Cropsey, eds. (Chicago: Rand McNally, 1963), pp. 153–54.

11. Ibid., p. 157.

12. Ibid., p. 154.

13. St. Augustine, *The City of God*, xix, 13.

14. Herbert A. Deane, *The Political and Social Ideas of St. Augustine* (New York: Columbia University Press, 1963), p. 221.

15. John Langan, S.J. "The Elements of St. Augustine's Just War Theory," *Journal of Religious Ethics*, 12:1 (Spring 1984) 29. Langan's article is particularly interesting in its broadening of the typical survey of Augustine's thought beyond *The City of God* to include the bishop of Hippo's correspondence with fellow bishops and Roman officials.

16. Deane, op. cit., p. 221.

17. Ibid.

18. See Bainton, op. cit., pp. 93–95; Deane, op. cit., p. 154.

19. See Deane, op. cit., p. 155.

20. Ibid., pp. 156–57; see also Langan on Augustine's analogical conception of peace, in art. cit., pp. 29–30.

21. Deane, op. cit., p. 156.

22. Ibid., p. 157.

23. Ibid., pp. 162, 164.

24. Bainton, op. cit., p. 97; see also Deane, op. cit., pp. 159, and 310, n. 18.

25. See Langan, art. cit., p. 27.

26. Ibid., p. 25. This overarching interest also explains why Augustine failed to address the question of conscientious objection to war: Old Testament sources on the divine authorization of wars allowed him to maintain his realist view of the prerogatives of political authority. Moreover, given Augustine's concern for the reestablishment of virtue in public life, the necessity of a rightly formed conscience for the warrior must have seemed to him the issue at stake.

27. See Deane, op. cit., pp. 240–41; Langan, art. cit., pp. 33ff.

28. "The punitive conception of war as the restoration of moral order rather than the defense of vital national interest seems too broad and difficult to apply, both because there are so many instances of moral disorder in the world and because the forms of interior evil with which Augustine is pre-occupied cannot be satisfactorily detected or effectively corrected by the ordinary measures of military policy." Augustine's punitive conception of war may also be anachronistic in a world where the chief positive function of military action may be to protect or vindicate "certain rules of international order and the rights they affirm" (Langan, art. cit., pp. 33–34).

29. "The special concern aroused by modern weapons of mass destruction is over the vast increase in the amount of harm they can do to human societies, their members, and their environment. There is also a particular concern over the impersonal character of warfare carried on by advanced technological means, the use of which may be compatible with detached and even indifferent attitudes on the part of the warrior technicians who actually direct and use the weapons. Theirs may well be an evil frame of mind (perhaps of a banal sort) but it is clearly different from the lust for revenge and the craving for power that so troubled Augustine" (Langan, ibid., p. 34).

30. "This is one issue where one could line up Jefferson, Kant, Freud, and Marx along with Bonhoeffer, the Niebuhrs, Vatican II, and Gutiérrez in united opposition to Augustine" (Langan, ibid.). See also Deane on Augustine's "quietism," in op. cit., pp. 151ff.

31. "While no important virtuous disposition can be identified with a single type of action or even with a list of types of actions, virtues and interior dispositions, if really possessed and effective, must make some difference in the conduct of one's life and in one's public behavior. Giving up this connection is both a misunderstanding of our psychology and an invitation to self-deception" (Langan, art. cit., p. 35).

32. Deane, op. cit., p. 226.

33. Ibid., pp. 241–43.

34. In Ronald Knox's sense of the term; see his *Enthusiasm* (New York: Oxford University Press, 1950).

35. That this approach to a positive reinterpretation of Augustine's *tranquillitas ordinis* does not do essential violence to Augustine's theological intentions is a speculation warranted, I believe, by H. Richard Niebuhr's analysis of Augustine in *Christ and Culture.* Niebuhr locates the basic thrust of Augustine's political-cultural project in his (Niebuhr's) fifth model, "Christ the Transformer of Culture," or what Niebuhr calls the "conversionist" perspective (p. 190), rather than within the dialectical perspective of Niebuhr's fourth model, "Christ and Culture in Paradox," where we might well have expected to find the bishop of Hippo:

> Augustine not only describes, but illustrates in his own person, the work of Christ as converter of culture. The Roman rhetorician becomes a Christian preacher, who not only puts into the service of Christ the training in language and literature given him by his society, but, by virtue of the freedom and illumination received from the Gospel, uses that language with a new brilliance and brings a new liberty into that literary tradition. The Neo-Platonist not only adds

to his wisdom about spiritual reality the knowledge of the incarnation which no philosopher had taught him, but this wisdom is humanized, given new depth and direction, made productive of new insights, by the realization that the Word has become flesh, and has borne the sins of the spirit. The Ciceronian moralist does not add to the classical virtues the new virtues of the gospel, nor substitute new law for natural and Roman legislation, but transvalues and redirects in consequence of the experience of grace the morality in which he had been trained and which he taught. In addition to this, Augustine becomes one of the leaders of that great historical movement whereby the society of the Roman empire is converted from a Caesar-centered community into medieval Christendom. Therefore, he is himself an example of what conversion of culture means; in contrast to its rejection by radicals, to its idealization by culturalists, to the synthesis that proceeds largely by means of adding Christ to good civilization, and to the dualism that seeks to live by the gospel in an inconquerably immoral society. . . . Christ is the transformer of culture for Augustine in the sense that he redirects, reinvigorates, and regenerates that life of man, expressed in all human works, which in present actuality is the perverted and corrupted exercise of a fundamentally good nature [pp. 208–9].

Niebuhr makes clear that, notwithstanding this conversionist intention, Augustine did not follow through on a project of transforming the culture of his time (see pp. 215–16); what Niebuhr terms "Augustine's defensiveness" (p. 216) won out over his conversionist intentions. Still, if Niebuhr is correct in his analysis of the basic theological intent of Augustine's lifework, it seems justifiable to interpret *tranquillitas ordinis* as a potentially positive vision of human possibility, rather than a merely punitive construct, as I have suggested here.

36. John Eppstein, *The Catholic Tradition of the Law of Nations* (Washington, D.C.: Catholic Association for International Peace, 1935), p. 66.

37. See H. Richard Niebuhr, op. cit., p. 117.

38. See ibid., p. 120.

39. See ibid., pp. 123ff.

40. Ibid.

41. Ibid., p. 129.

42. Ibid., p. 130.

43. Ibid., p. 131.

44. Ibid.

45. St. Thomas Aquinas, *Summa Theologiae*, II-I, Q. ii., art. viii.

46. Niebuhr, op. cit., p. 135.

47. "'Thou shalt not steal,'" Niebuhr notes, "is a commandment found both by reason and in revelation. 'Sell all that thou hast and give to the poor' is found in the divine law alone. It applies to man as one who has had a virtue implanted in him beyond the virtue of honesty, and who has been directed in hope toward a perfection beyond justice in this mortal existence" (ibid., pp. 136–37).

48. Frederick Copleston, *Medieval Philosophy* (New York: Harper & Row, 1961), p. 168.

49. Ibid., pp. 168–69.

50. Ibid., p. 169.

51. St. Thomas Aquinas, *De Regimine Principum*, cited in Charles McCoy, "St. Thomas Aquinas," in *History of Political Philosophy*, Leo Strauss and Joseph Cropsey, eds. (Chicago: Rand McNally, 1963), p. 205.

52. Frederick Copleston, *History of Philosophy*, volume 2, *Medieval Philosophy: Augustine to Scotus* (Westminister: Newman Press, 1957), p. 415.

53. Ibid., p. 419. For further discussions of St. Thomas's constitutionalism, see Frederick Copleston, *Aquinas* (Harmondsworth: Penguin Books, 1955), pp. 236–42; McCoy, "St. Thomas Aquinas," pp. 208–14. John Courtney Murray, op. cit., pp. 113–25, discusses this in terms of a natural law theory of "consensus" in a democracy. Murray also takes this point up in arguing that some of the earliest conceptual roots of the American democratic experiment lie in medieval political theory, in op. cit., pp. 45ff.

54. Copleston, *History of Philosophy*, volume 2, p. 420.
55. Ibid., p. 418.
56. Copleston, *Medieval Philosophy*, pp. 170–71.
57. St. Thomas Aquinas, *Summa Contra Gentiles*, lib. 3, cap. 130.
58. McCoy, "St. Thomas Aquinas," p. 221.
59. St. Thomas Aquinas, *Summa Theologiae*, II–II, q. 40, art. 1. The following citations from the *Summa* are in the translation by Thomas R. Heath, O.P., in *Summa Theologiae*, volume 35, *Consequences of Charity* (London: Blackfriars, 1972), pp. 80–85.
60. Bainton, op. cit., p. 110.
61. Ibid., pp. 116–18.
62. Ibid., pp. 118–21.
63. Ibid., p. 112.
64. Ibid., p. 114.
65. Ibid., p. 119.
66. Ibid., p. 111.
67. Cited in ibid., pp. 111–12.
68. Ibid., p. 112.
69. Paul Johnson, *A History of Christianity* (Harmondsworth: Penguin Books, 1978), p. 244.
70. R. W. Southern, *Western Society and the Church in the Middle Ages* (Harmondsworth: Penguin Books, 1970), p. 73.
71. Johnson, *A History of Christianity*, p. 249.
72. For relevant documentation, see Eppstein, op. cit., pp. 97–123.
73. Eppstein schematically summarizes the neo-scholastic position as follows:

1. The right of war is odious [as Suárez wrote in his *De caritate: Ius belli est odiosum . . . ergo restrigenda est quoad fieri potest*]. That this right exists in certain circumstances cannot be denied; but it is a regrettable necessity, and the main object of rulers should be to prevent the necessity arising.
2. A declaration of war—other than a war of self-defense against actual attack—is a penal act, justifiable only if an injury has been committed so grave as to require the gravest penalty.
3. The penal justification required for the declaration of war is necessary for the preservation of order in human society and is conferred upon sovereigns in the absence of a higher tribunal.
4. If the possibility of recourse to the arbitration of the Holy See or another competent tribunal exists, this penal jurisdiction may not be exercised by a prince.
5. If any doubt arises concerning the justice of his claims the prince may not resort to arms. In such a case both disputants are under an obligation to choose arbitrators or mediators and to abide by their decision.
6. Even if a certainly just cause of war exists, the prince must refrain from going to war if the losses resulting from war to the belligerent states, the whole community of nations, or the Church are likely to be so great as to outweigh the advantages of repairing the injury done.
7. Soldiers are bound to examine the justice of a proposed war. If uncertain of the justice of the cause their duty is to obey their commanders. If certain of the injustice of the cause they are to refuse to fight.
8. All those who hold office or have any political responsibility in the state are bound in conscience to examine carefully the rights and wrongs of a proposed declaration of war.
9. War may not be waged on the personal authority of the sovereign, but only after the

advice of wise and impartial men has been sought, and after the claims of the opposing side have been carefully weighed [ibid., pp. 122–23].

74. Ibid., pp. 92–93.

75. Bainton, op. cit., p. 134.

76. Ibid. In his famous tract, *The Complaint of Peace*, Erasmus wrote:

Peace enters speaking in her own person and lamenting that she is so little received among men. She marvels at this the more because the heavenly bodies, though inanimate, preserve a happy equilibrium; the very plants cling to each other and the irrational animals do not devour those of the same species. The boar does not bury his tusk in the boar, the lion shows no fierceness to the lion, nor does the serpent expend his venom on the serpent. The wolf is kind to the wolf. And if animals do fight, it is only to assuage their hunger. Why then should not man of all creatures be at peace with man? The more so because he is endowed with reason and gifted with speech, the instrument of social intercourse and reconciliation, and with tears which in a shower dissolve the clouds and suffer the sun to shine again. Man depends for his very existence upon cooperation. . . . Why then should man prey upon man?

Peace, not yet disillusioned, assumes that when she hears the name of man and Christian, there she will find a reception. She approaches hopefully a city begirt by walls and living in accord with laws, only to discover factions. She turns from the common rout to kings and finds them embracing with obsequious flattery while conniving at mutual destruction. . . .

All this is the more amazing when one examines the precepts of the Christian religion. In the Old Testament Isaiah foretold the coming of the Prince of Peace and in the New Testament Christ bequeathed peace as his legacy. The mark by which his disciples should be known is love one for the other. The Lord's Prayer addresses *Our Father*, but how can they call upon a common Father who drive steel into the bowels of their brethren? Christ compared himself to a hen, Christians behave like hawks. Christ was a shepherd of sheep, Christians tear each other like wolves. . . .

And who is responsible for all this? Not the common people, but kings, who on the strength of some musty parchment lay claim to neighboring territory or because of the infringement of one point in a treaty of a hundred articles, embark on war. Not the young, but the greybeards. Not the laity, but the bishops. The very cross is painted on their banners and cannons are christened and engraved with the names of the apostles so that Paul, the preacher of peace, is made to hurl a cannon ball at the heads of Christians.

Consider the wickedness of it all, the breakdown of laws which are ever silent amid the clangor of arms. Debauchery, rape, incest, and the foulest crimes are let loose in war. Men who would go to the gallows in peace are of prime use in war, the burglar to rob, the assassin to disembowel, the incendiary to fire an enemy city, the pirate to sink his vessels.

Consider the cost of it all. . . . When all the damage is taken into account, the most brilliant success is not worth the trouble.

How, then, is peace to be secured? Not by royal marriages, but by cleansing the human heart. Why should one born in the bogs of Ireland seek by some alliance to rule over the East Indies? Let a king recall that to improve his realm is better than to increase his territory. Let him buy peace. The cheapest war would be more expensive. Let him invite the arbitration of learned men, abbots, and bishops. Let the clergy absent themselves from silly parades and refuse Christian burial to those who die in battle. . . .

Above all else let peace be sincerely desired. The populace is now incited to war by insinuations and propaganda, by claims that the Englishman is the natural enemy of the Frenchman and the like. Why should an Englishman as an Englishman bear ill will to a Frenchman and not rather good will as a man to a man and a Christian to a Christian? How can anything so frivolous as a name outweigh the ties of nature and the bonds of Christianity? The Rhine separates the French from the German but it cannot divide the Christian from the Christian. . . .

Let us then repent and be wise, declare an amnesty to all past errors and misfortunes, and bind up discord in adamantine chains which can never be sundered till time shall be no more [Bainton, op. cit., pp. 133–34].

This extraordinary polemic is worth quoting at length, for two reasons. First, it sets in clear juxtaposition a pacifist/evangelical approach to the moral problem of war and peace and the just-war/analytic approach we have seen evolving in Augustine, Aquinas, and the neo-scholastics. Virtually everything is different here: the root nature of the problem ("discord," which must be bound "in adamantine chains" before there can be peace); the solution (repentance and the conversion of hearts); and, of course, the whole tone quality of the exercise. Secondly, given certain adjustments for contemporary national conflicts and modern weapons, Erasmus's *Complaint of Peace* has a remarkably familiar ring to it. As we shall see, Erasmus's reduction of biblical data to the simplest affirmation of *shalom*, his animus against government and the military, and his populism—all wrapped in a rhetoric of "cleansing the human heart"—will become staple features of much of the Catholic pacifist activism of the post-Vietnam period in America.

77. Richard Cox, "Hugo Grotius," in *History of Political Philosophy*, Leo Strauss and Joseph Cropsey, eds., p. 351. For another outline of Grotius's work, see Wolfgang Friedmann, "Hugo Grotius," in *Encyclopedia of Philosophy*, volume 3 (New York: Macmillan, 1967), pp. 393–95. Friedmann cites as the reasons why Grotius took the tack he did his commitment to a natural law theory of politics; his Aristotelianism, and particularly his belief in human rationality; his life experience as a practicing and pragmatic diplomat; and his Erasmian humanism. In the absence of an overarching sovereign, Grotius insisted, according to Friedmann, on the meta-legal principle of *pacta sunt servanda*, "agreements must be kept," as the bedrock foundation of international relations. The theory of *bellum iustum* was thus part of a wider schema of international law, rather than being a separate moral reality: again, a point on which Grotius shows resonances with the medievals and the neo-scholastics (and, in fact, with the whole Augustinian/Thomistic tradition).

78. Cited in Frederick Copleston, *History of Philosophy*, volume three, *Ockham to Suárez* (London: Burns and Oates, 1953), p. 330.

79. Ibid.

80. Cited in ibid., p. 333.

81. Ibid., pp. 333–34.

American Catholicism and the Tradition Received: From John Carroll to the Second Vatican Council

1. For general overviews of American Catholic history, see John Tracy Ellis, *American Catholicism*, 2nd edition, revised (University of Chicago Press, 1969); James Hennesey, S.J., *American Catholics* (New York: Oxford University Press, 1981); Andrew M. Greeley, *The Catholic Experience* (New York: Doubleday, 1967), and Jay P. Dolan, *The American Catholic Experience* (Garden City: Doubleday, 1985). Robert Cross traces the history of an emergent "liberal" caucus in the nineteenth-century Church in *The Emergence of Liberal Catholicism in America* (Chicago: Quadrangle Books, 1968). Gerald Fogarty, S.J., discusses the relationship between the American hierarchy and the Vatican on questions of interest to this study in *The Vatican and the American Hierarchy from 1870 to 1965* (Stuttgart: Anton Hiersemann, 1982, reprinted by Michael Glazier, 1985). William Au's *The Cross, the Flag and the Bomb: American Catholics Debate War and Peace, 1960–1983* (Westport: Greenwood Press, 1985) is an example of the classic historiography applied to more contemporary Catholic activism. The major documentary evidence for the discussion that follows may be found in three volumes: *The National Pastorals of the American Hierarchy: 1792–1919*, edited by Peter Guilday (Washington, D.C.: National Catholic Welfare Council, 1923); *Our Bishops Speak: National Pastorals and Annual Statements of the Hierarchy of the United States, Resolutions of Episcopal Committees, and Communications of the Administrative Board of the National*

Catholic Welfare Conference: 1919–1951, edited by Raphael M. Huber (Milwaukee: Bruce Publishing Company, 1952); and *Documents of American Catholic History*, edited by John Tracy Ellis (Milwaukee: Bruce Publishing Company, 1956).

2. In support of this view, contemporary historians of American Catholicism often point to the private correspondence of two leaders of the liberal wing of the American hierarchy, Archbishop John Ireland of St. Paul, and Msgr. Denis O'Connell, rector of the North American College in Rome, at the time of the Spanish-American War. O'Connell wrote to Ireland on May 24, 1898, that the war was "a question of two civilizations. It is the question of all that is old & vile & mean & rotten & cruel & false in Europe against all this [sic] is free & noble & open & true & humane in America. When Spain is swept of [sic] the seas much of the meanness & narrowness of old Europe goes with it, to be replaced by the freedom & openness of America. This is God's way of developing the world" (cited in Gerald P. Fogarty, S.J., "Public Patriotism and Private Politics: The Tradition of American Catholicism," *U.S. Catholic Historian*, 4:1, p. 23). Ireland (who had, at the Vatican's behest, engaged in some private diplomacy with the McKinley administration, with which he was well connected, with an eye to averting the war) had previously written to O'Connell, on May 11, 1898,

> Well, America is whipping poor Spain. I confess, my sympathies are largely with Spain; but the fact is, she is beaten. Now, Americanism will triumph, and practical application will be given to your pamphlet in Cuba and the Philippines. Cuba, at first independent, will very quickly become American territory. Public opinion is growing in favor of retaining the Philippines. Our national pride is aroused, and we want to be a power *in toto orbe terrarum.* I am not much of an Anglo-Saxon: but, Anglo-Saxonism is to reign, and, if there is wisdom in the Vatican, it will at once seek influence with English-speaking countries, especially America. Unless this is done at once, humanly speaking the Church is doomed. The manifest destiny of the world is Americanism, as you explain the word [cited in John Tracy Ellis, "American Catholics and Peace: An Historical Sketch," pamphlet reprint of an essay that originally appeared in a 1970 collection of essays, *The Family of Nations*, edited by James S. Rausch, and published by Our Sunday Visitor, Inc., in cooperation with the Division of World Justice and Peace, United States Catholic Conference, p. 11].

Fogarty, in "Public Patriotism and Private Politics," adopts the classic historiography in order to draw a rather direct line between the Ireland / O'Connell correspondence and Cardinal Spellman's citation of Stephen Decatur—"My country, may it always be right. Right or wrong, my country"—during Vietnam (p. 47)—a point on which Fogarty, a very able historian, would agree with the dreadful recent biography of Spellman, *The American Pope*, by John Cooney (New York: Times Books, 1984). Fogarty draws out the implications of the classic historiography: "By the 1980s . . . the Church was in danger of becoming too American and less Catholic" (p. 47)—a view that, as Part Two of this study will show, does not bear up under careful examination of the foreign policy opinions of American Catholic elites after Vietnam.

3. Ellis, "American Catholics and Peace," p. 5.

4. Ibid.

5. Ibid., pp. 5–6.

6. The Catholic Worker movement will be discussed in some detail later. For a basic overview, see William D. Miller, *A Harsh and Dreadful Love* (New York: Liveright, 1973), where the tensions within Catholic Worker pacifism are examined. While the public points of argument between the Catholic Worker and the hierarchy may have revolved at times around the pacifism of some key Catholic Worker leaders, the real bone of theoretical contention was over the Worker's anarchism.

7. For an example of an analysis that seems built around such a sense of embarrassment, see Patricia McNeal, *The American Catholic Peace Movement, 1928–1972* (New York: Arno Press, 1978).

8. See George Weigel, "Intellectual Currents in the American Public Effort for Peace," in *The Nuclear Freeze Debate: Arms Control Issues for the 1980s*, Paul M. Cole and William J. Taylor, Jr., eds. (Boulder: Westview Press, 1983).

9. Cited in John Tracy Ellis, *Documents of American Catholic History*, pp. 136–37.

10. Cited in ibid., p. 175.

11. Cited in ibid., p. 176.

12. Cited in Ellis, "American Catholics and Peace, p. 2.

13. For a survey of the argument from Vatican Council I through Vatican Council II, see Fogarty, *The Vatican and the American Hierarchy from 1870 to 1965.*

14. Ellis, "American Catholics and Peace," p. 3.

15. Ibid., p. 4.

16. Ibid.

17. On the Mexican War, see ibid., pp. 7–8; for the Spanish-American War, see ibid., pp. 9–12.

18. See Ellis, *American Catholicism*, pp. 84–123; Fogarty, "Public Patriotism and Private Politics," pp. 13–17.

19. The first six "provincial" councils of Baltimore were, in effect, national councils, inasmuch as the ecclesiastical province of Baltimore comprised the entire United States until the erection of the archdiocese of Oregon City in 1846. Given certain confusions over ecclesiastical boundaries after the Mexican War, as well as the difficulties of travel, the Seventh Provincial Council of Baltimore in 1849 also legislated for the entire country. The First Plenary Council of Baltimore, held in 1852, was thus the first genuinely national council involving a number of provincial jurisdictions. (Peter Guilday sorts out this sometimes confusing history in *The National Pastorals of the American Hierarchy: 1792–1919*, pp. 171–72.) A second Plenary Council of Baltimore was held immediately after the Civil War, in 1866, and the Third Plenary Council of Baltimore was held in 1884. This conciliar tradition in American Catholicism was unique in the Catholic world of the nineteenth century.

20. Cited in Guilday, op. cit., p. 20.

21. Cited in ibid., p. 3.

22. See the bishops' 1833 statement in ibid., p. 78.

23. Cited in ibid., p. 82.

24. The Third Provincial Council also gave the bishops the opportunity to define their place as Catholics within a religiously plural commonwealth:

> We owe no religious allegiance to any State in this Union, nor to its general government. No one of them claims any supremacy over us in our spiritual and ecclesiastical concerns: nor does it claim any such right over any of our fellow citizens, nor would we submit thereto. They and we, by our constitutional principles, are free to give this ecclesiastical supremacy to whom we please, or to refuse it to any one, if we think so proper: but they and we owe civil and political allegiance to the several States in which we reside, and also, to our general government. When, therefore, using our undoubted right, we acknowledge the spiritual and ecclesiastical supremacy of the chief bishop of our universal church, the Pope or bishop of Rome, we do not thereby forfeit our claim to the civil and political protection of the commonwealth: for we do not detract from the allegiance to which the temporal governments are plainly entitled, and which we cheerfully give; nor do we acknowledge any civil or political supremacy, or power over us in any foreign potentate or power, though that potentate might be the chief pastor of our church [ibid., pp. 90–91].

25. Cited in ibid., p. 142. See also the admonitions of the First Plenary Council in ibid., p. 192.

26. Guilday, op. cit., p. 226.

27. The bishops' address to the problems caused by the definition of the doctrine of

infallibility at Vatican I are found in Guilday, op. cit., pp. 232–34. In explaining the definition to Americans in his book, *The Faith of Our Fathers*, Cardinal Gibbons made the analogy between the function of the infallibility doctrine in the Church and the function of the Supreme Court in the United States.

28. Cited in Guilday, op. cit., pp. 234–35; emphasis added. After this ringing affirmation of the American experiment, the bishops immediately sought to reassure the Vatican that their American patriotism did not lessen their devotion to the universal Church, in terms reminiscent of the Augustinian/Thomistic tradition of what constituted a rightly ordered society:

> No less illogical would be the notion, that there is aught in the free spirit of our American institutions, incompatible with perfect docility to the Church of Christ. The spirit of American freedom is not one of anarchy or license. It essentially involves love of order, respect for rightful authority, and obedience to just laws. There is nothing in the character of the most liberty-loving American, which could hinder his reverential submission to the Divine Authority of Our Lord, or to the like authority delegated by Him to His Apostles and His Church. Nor are there in the world more devoted adherents of the Catholic Church, the See of Peter, and the Vicar of Christ, than the Catholics of the United States [p. 235].

The bishops of the Third Plenary Council were thus weaving a path between nativist bigots on one flank and nervous curial Romans on the other. From such tensions, perhaps, do genuine developments of doctrine and understanding occur.

29. See note 2, above.

30. The text of *Longinqua Oceani* may be found in Ellis, *Documents of American Catholic History*, pp. 514–27. The key passage is as follows:

> The main factor, no doubt, in bringing things into this happy state [Leo XIII had just been extolling the growth of the church in the United States] were the ordinances and decrees of your synods, especially those which in more recent times were convened and confirmed by the authority of the Apostolic See. But, moreover (a fact which it gives pleasure to acknowledge), thanks are due to the equity of the laws which obtain in America and to the customs of the well-ordered Republic. For the Church amongst you, unopposed by the Constitution and government of your nation, fettered by no hostile legislation, protected against violence by the common laws and the impartiality of the tribunals, is free to live and act without hindrance. Yet, though all this is true, it would be very erroneous to draw the conclusion that in America is to be sought the type of the most desirable status of the Church, or that it would be universally lawful or expedient for State and Church to be, as in America, dissevered and divorced. The fact that Catholicity with you is in good condition, nay, is even enjoying a prosperous growth, is by all means to be attributed to the fecundity with which God has endowed His Church, in virtue of which unless men or circumstances interfere, she spontaneously expands and propagates herself; but she would bring forth more abundant fruits if, in addition to liberty, she enjoyed the favor of the laws and the patronage of the public authority [pp. 517–18].

For commentary on the impact of this encyclical on the American Church, see Cross, op. cit., pp. 195–96; Fogarty, *The Vatican and the American Hierarchy from 1870 to 1965*, pp. 134, 137, 143, 156, 369–70. Michael Novak's study, *Freedom with Justice: Catholic Social Thought and Liberal Institutions* (San Francisco: Harper & Row, 1984) traces the path by which the Vatican, while maintaining a theoretical hostility to ideological "Liberalism" on the Jacobin-laicist model, has gradually come to see the worthiness of liberal political institutions, a process of conversion for which the American Church can take considerable credit.

31. Cited in Guilday, op. cit., p. 265.

32. For American Catholic activities during the war, see Ellis, *American Catholicism*, pp. 138–39; Hennesey, op. cit., pp. 206–27; John B. Sheerin's biography of the first general secretary of the National Catholic War Council (later the National Catholic Welfare Conference, predecessor to the United States Catholic Conference), John J. Burke, C.S.P., *Never Look Back: The Career and Concerns of John J. Burke* is an informative source for the

tensions that the war posed for the jealously guarded prerogatives of individual bishops who, with the lapse of the nineteenth-century conciliar tradition of American Catholicism, worried about the formation of a national Catholic organization like the NCWC.

There is no little irony in the fact that the pledge of the Third Plenary Council (which hearkens back to John Carroll's letter to George Washington) should have been redeemed before a president not noted for the gentility or ecumenical spirit of his views about Catholicism. Woodrow Wilson in fact exhibited a fair amount of nativist bias during his public life (see Fogarty, *The Vatican and the American Hierarchy from 1870 to 1965*, p. 209).

33. Cited in Guilday, op. cit., p. 267.
34. Cf. ibid., p. 292.
35. Cited in ibid., p. 293.
36. Cited in ibid.
37. Cited in ibid.
38. Cited in ibid., p. 298.
39. Cited in ibid.
40. Cited in ibid., pp. 298–99.
41. Cited in ibid., p. 302.
42. Cited in ibid., p. 304.
43. Cited in ibid.
44. Cited in ibid., p. 305.
45. See ibid., pp. 329–30.
46. See ibid., p. 330. The bishops' argument here was essentially functional, and reflects some of the frustration with the wisdom of "the great" that we saw previously in the literature and poetry of the First World War. The bishops wrote:

> The growth of democracy implies that the people shall have a larger share in determining the form, attributions and policies of the government to which they look for the preservation of order. It should also imply that the calm, deliberate judgment of the people, rather than the aims of the ambitious few, shall decide whether, in case of international disagreement, war be the only solution. Knowing that the burdens of war will fall most heavily on them, the people will be slower to take aggressive measures, and, with an adequate sense of what charity and justice require, they will refuse to be led or driven into conflict by false report or specious argument. Reluctance of this sort is entirely consistent with firmness for right and zeal for national honor. If it were developed in every people, it would prove a more effectual restraint than any craft of diplomacy or economic prudence. . . . Instead of planning destruction, intelligence would then discover new methods of binding the nations together; and the good will which is now doing so much to relieve the distress produced by war, would be so strengthened and directed as to prevent the recurrence of international strife [p. 330].

In 1919, the bishops anticipated many arguments made in the 1980s in favor of the creation of a National Endowment for Democracy (an institution of peace not mentioned in the 1983 pastoral letter of the American bishops, "The Challenge of Peace").

47. Cited in ibid., pp. 330–33.
48. Winston S. Churchill, *The Gathering Storm* (London: Cassell & Co., 1949), p. ix.
49. Hennesey, op. cit., p. 272.
50. Ibid.
51. Ibid., p. 275.
52. For McNicholas's opposition to Roosevelt, see Fogarty, *The Vatican and the American Hierarchy from 1870 to 1965*, pp. 256, 168, 270, 273; Fogarty discusses Curley's isolationism on pp. 271–73. John Tracy Ellis paints a sympathetic portrait of Curley in *Catholic Bishops: A Memoir* (Wilmington: Michael Glazier, 1983), while recounting how Curley's anti-Roosevelt politics led him into a memorable faux pas on the evening of the Japanese attack on Pearl Harbor (pp. 46–52).

53. On Shaughnessy, see Patricia McNeal, op. cit., pp. 111–12. The most vocal Catholic opponent to President Roosevelt was Father Charles Coughlin, the "Radio Priest." For an account of Coughlin's activities, see Alan Brinkley, *Voices of Protest: Huey Long, Father Coughlin, and the Great Depression* (New York: Vintage Books, 1983). The nervousness caused by Coughlin's demagogy extended beyond the White House; for an account of the Vatican's concern, and the means by which Coughlin was eventually silenced by both Archbishop Mooney and the government, see Fogarty, *The Vatican and the American Hierarchy from 1870 to 1965*, pp. 244–45, 251–52, 274–75, 277–78.

54. Hennesey, op. cit., p. 276.

55. For a portrait of the Mundelein/Roosevelt relationship, see Edward Kantowicz, *Corporation Sole* (University of Notre Dame Press, 1983), pp. 217–36. Francis Spellman aspired to a similar intimacy with Roosevelt after Mundelein's death and Spellman's own accession to the archbishopric of New York—a position to which Pius XI had evidently thought to appoint Archbishop McNicholas, which would have had interesting consequences for all concerned. The succession to New York was not only of concern to the individual candidates, but raised diplomatic worries in both the White House and the Vatican. For a summary account of the speculation, now widespread, that Francis Spellman ended up as archbishop of New York because of the death of Pius XI, see Fogarty, *The Vatican and the American Hierarchy from 1870 to 1965*, p. 256.

56. For the still definitive study of Gibbons, see John Tracy Ellis, *The Life of James Cardinal Gibbons: Archbishop of Baltimore: 1834–1921*, 2 vols. (Milwaukee: Bruce, 1952).

57. For the text of these encyclicals, see *The Papal Encyclicals 1909–1939*, Claudia Carlen, I.H.M., ed. (McGrath Publishing Co., 1980), pp. 445–58, 525–35, 537–54. For Pius XI's relations with Mussolini and Hitler, see Carlo Falconi, *The Popes in the Twentieth Century* (London: Weidenfeld and Nicolson, 1967), pp. 151–233.

58. Cited in Huber, *Our Bishops Speak*, p. 226.

59. Cited in ibid.

60. See ibid., pp. 227–28.

61. Cited in ibid., pp. 345–48.

62. Cited in ibid., pp. 230–31.

63. See ibid., pp. 102–9. The distinction between the Soviet regime and the people of the U.S.S.R. had been used earlier to prevent hierarchical opposition to Lend-Lease assistance to the Soviet Union, an issue of great concern not only to the isolationist bishops, but also to the demagogic Fr. Coughlin. In an interesting ecclesiastical strategem, the previously isolationist Archbishop McNicholas (the hierarchy's most respected theologian) was recruited by his more interventionist confreres to write a diocesan pastoral letter making the distinction and thus clearing the way for Catholic support of Lend-Lease, which the episcopal isolationists and Coughlin had argued was proscribed by *Divini Redemptoris*'s prohibition on assisting the communist regime. For the way in which this minor but important *coup* was managed, see Fogarty, *The Vatican and the American Hierarchy from 1870 to 1965*, pp. 271–76.

64. Cited in Huber, op. cit., p. 351.

65. Cited in ibid. (emphasis added).

66. Cited in ibid. President Roosevelt's response was in kind, and would be turned against the administration when the bishops began to criticize the government's modification of Atlantic Charter principles for the sake of keeping Stalin in the United Nations later in the war. Roosevelt wrote to Mooney:

> The letter which you forwarded under date of December 22 as chairman of the Administrative Board, National Catholic Welfare Conference, and in the name of the bishops of the United States, gives me strength and courage because it is a witness to that national unity so necessary in our all-out effort to win the war. Please convey to all of your brethren in the episcopate an assurance of my heartfelt appreciation of the pledge of wholehearted cooperation in the

difficult days that lie ahead. In those days we shall be glad to remember your patriotic action in placing your institutions and their consecrated personnel at the disposal of the government.

We shall win this war and in victory we shall seek not vengeance but the establishment of an international order in which the spirit of Christ shall rule the hearts of men and nations [p. 352].

67. Hennesey, op. cit., p. 280.

68. Ibid., p. 278.

69. For the figures on Catholic conscientious objectors, which are somewhat disputed, see ibid. Gordon Zahn gives a detailed portrait of the World War II Catholic conscientious objector in "Catholic Conscientious Objection in the United States" and "The Social Thought of the Catholic Conscientious Objector," both of which are found in his book *War, Conscience, and Dissent* (New York: Hawthorn Books, 1967). The impact of this small band of conscientious objectors on the subsequent war/peace debate in American Catholicism is discussed below.

70. See Fogarty, *The Vatican and the American Hierarchy from 1870 to 1965*, pp. 279–312.

71. Ibid., pp. 262–312.

72. The full text is in Huber, op. cit., pp. 115–20.

73. Cited in ibid., p. 115.

74. Cited in ibid., p. 116.

75. Ibid.

76. Ibid.

77. Ibid.

78. Cited in ibid., p. 117.

79. See ibid.

80. Cited in ibid.

81. Ibid.

82. Ibid.

83. Ibid.

84. Cited in ibid., pp. 117–18.

85. Cited in ibid., p. 118.

86. Ibid.

87. Cited in ibid., p. 121.

88. Cited in ibid., p. 122.

89. See ibid.

90. Cited in ibid., p. 123 (emphasis added).

91. Ibid. (emphasis added).

92. Cited in ibid., pp. 123–24.

93. Cited in ibid., pp. 124–25.

94. Cited in ibid., p. 125.

95. Ibid.

96. Ibid.

97. See ibid., p. 126.

98. That task would be left to the Jesuit moral theologian John Ford, in his essay "The Morality of Obliteration Bombing," *Theological Studies*, 5 (1944) 261–309.

99. Cited in Huber, op. cit., p. 355.

100. See ibid., p. 356–57.

101. Cited in ibid., p. 357.

102. Cited in ibid., p. 358.

103. Ibid.

104. Ibid.

105. Ibid.

106. Cited in ibid., p. 359.

107. Ibid.

108. Ibid.

109. Cited in ibid., p. 126.

110. Cited in ibid., pp. 126–27.

111. For the Lend-Lease debate, see Fogarty, *The Vatican and the American Hierarchy from 1870 to 1965*, pp. 272–76, 285.

112. Cited in Huber, op. cit., p. 131.

113. Cited in ibid., p. 132.

114. Ibid.

115. Ibid.

116. Ibid.

117. Cited in ibid., pp. 132–33.

118. See the bishops' "Resolution on Compulsory Military Training" of November 17, 1944 (Huber, op. cit., p. 234) and the November 15, 1945, document of the same title in ibid., pp. 237–38. This latter statement, interestingly, locates the problem of conscription firmly within the bishops' internationalist perspective, and asks whether the universal abolition of conscription might be sought, and economic aid used as a lever (or penalty) to that end.

119. See Fogarty, *The Vatican and the American Hierarchy from 1870 to 1965*, pp. 279–312.

120. Fogarty, in "Public Patriotism and Private Politics," claims that "the national pastorals were written by the inner circle of the NCWC, Mooney, McNicholas, and Samuel Stritch, Archbishop of Chicago" (p. 41).

121. For a not unsympathetic portrait of the period, see Garry Wills' essay, "Memories of a Catholic Boyhood," in *Bare Ruined Choirs* (Garden City: Doubleday, 1972), pp. 15–37.

122. See Blanshard's *American Freedom and Catholic Power* (Boston: Beacon Press, 1949).

123. See Fogarty, *The Vatican and the American Hierarchy from 1870 to 1965*, pp. 346–85.

124. See Chapter 4, below.

125. See Fogarty, *The Vatican and the American Hierarchy from 1870 to 1965*, pp. 340–45.

126. Cited in Huber, op. cit., p. 238.

127. Cited in ibid., pp. 239–40.

128. See *Pastoral Letters of the American Hierarchy 1792–1970*, Hugh J. Nolan, ed. (Huntington: Our Sunday Visitor, 1984), pp. 467–68.

129. Cited in ibid., p. 469.

130. Cited in ibid., p. 471.

131. Cited in ibid., pp. 477–78.

132. Cited in ibid., p. 472.

133. See ibid., pp. 481–85.

134. See ibid., p. 492.

135. Cited in ibid., pp. 493–94.

136. Cited in ibid., pp. 515–19.

137. Donald F. Crosby, S.J., *God, Church and Flag: Senator Joseph R. McCarthy and the Catholic Church, 1950–1957* (Chapel Hill: University of North Carolina Press, 1978). Crosby synopsized his documentation and argument in "McCarthyism and Catholicism: A New Look at the Evidence," *Commonweal*, July 21, 1978, pp. 456–68. The citations below are taken from the Crosby article in *Commonweal*.

138. Crosby, art. cit., p. 456. Two of McCarthy's sharpest Catholic critics, James O'Gara and John Cogley of *Commonweal*, shared this judgment.

139. Ibid., p. 458. Crosby also notes that "no documentary evidence of any kind exists to prove that Walsh ever exhorted McCarthy to go on his political crusade" (p. 458).

140. Ibid.

141. Ibid., p. 461. Crosby notes how one prominent Catholic publication, *Our Sunday Visitor*, helped perpetuate the fiction of McCarthy's pull among Catholic voters. After Tydings's defeat, the *Visitor*, which Crosby describes as a "McCarthyite Catholic weekly," reported "enthusiastically that Tydings had been 'almost solidly opposed by Maryland Roman Catholics,' who were the group chiefly responsible for his demise. Noting that McCarthy's enemies had fallen in other states, the *Visitor* concluded modestly that 'the masses of Catholic electors have sensed . . . that they, and perhaps they alone, can save the world.'" (p. 461). A generation later, the *Visitor* would be among those Catholic publications painting an admiring portrait of the Marxist-Leninist Sandinista regime in Nicaragua: a stunning example of the anti-anticommunist syndrome that would develop among American Catholic elites. Although these things are impossible to prove, it does not seem accidental that this syndrome is evident in those Catholic circles that tend to accept the "McCarthyism = Catholicism" thesis, and which thus have a strongly felt stake in distancing themselves from what they perceived to be the rigidities of their past.

142. Ibid., p. 463.

143. Ibid.

144. See Fogarty, *The Vatican and the American Hierarchy from 1870 to 1965*, pp. 340–41. Fogarty, whose study was written after Crosby's, acknowledges that "Catholic opinion was seriously divided on McCarthy's crusade" (p. 340).

145. Crosby, art. cit., p. 468.

146. For the "religious issue" in the 1960 campaign as recollected from well within the Kennedy orbit, see Arthur M. Schlesinger, Jr., *A Thousand Days: John F. Kennedy in the White House* (Boston: Houghton Mifflin, 1965) and Theodore Sorensen, *Kennedy* (New York: Bantam Books, 1966).

147. See John Tracy Ellis, "American Catholics and Peace," p. 13.

148. See Schlesinger, *A Thousand Days*, pp. 106–8. Schlesinger, who also found Kennedy's "ethos . . . more Greek than Catholic," the evidence being JFK's fondness for Aristotle's definition of happiness as the full use of one's powers along lines of excellence, seems wholly ignorant of the Thomistic appropriation of Aristotle and its contemporary expression by Catholic thinkers such as Murray and Jacques Maritain. Schlesinger's chronic use of phrases like "black-and-white moralism," "pietistic rhetoric," and "anti-intellectualism" as descriptions of "parts of the Irish-Catholic community" in America—presumably, the parts that were not Kennedy's—says rather more about genteel Cambridge anti-Catholicism of the secularized sort than it sheds light upon the allegedly benighted background from which Schlesinger so badly wants to distinguish John F. Kennedy. See *A Thousand Days*, p. 107, and *Robert Kennedy and His Times*, pp. 601–2, for the relevant quotes.

149. Peale's role, which was not notable for either its intelligence or its courage under fire, is discussed in Sorensen, op. cit., pp. 212–14.

150. The text of the Houston speech may be found in *The Kennedy Reader* (Indianapolis: Bobbs-Merrill, 1967), pp. 363–67.

151. For a full discussion of this problem, see Alasdair MacIntyre, *After Virtue* (University of Notre Dame Press, 1981), esp. pp. 1–75.

152. See Richard John Neuhaus, *The Naked Public Square: Religion and Democracy in America* (Grand Rapids: Eerdmans, 1984). This judgment on Kennedy is admittedly hard. It is not meant to deny the fact that, at the tactical level of national politics, John F. Kennedy made Catholics "acceptable" in a way that no other figure, lay or clerical, had managed to do before him. But the long-term impact of the terms in which Kennedy expressed the relationship between religiously based values and public policy has been severe, and not merely for

Catholics. Nowhere is this more evident than in the abortion debate after the 1973 Supreme Court decision in *Roe v. Wade*, which has been systematically and, one suspects, deliberately, distorted by continual reference to the matter as a "religious issue"—and, thus, presumably out of bounds in the public policy arena, for precisely the reasons suggested by Kennedy in Houston. The profound constitutional implications of *Roe v. Wade* have thus been obscured, and the issue made into a First Amendment argument. The pressure on Catholic candidates, particularly Democrats, to adopt the Kennedy/Houston line, and thus declare their religiously based moral convictions of no import to their public responsibilities, has been fierce on this question. Governor Mario M. Cuomo of New York, in a speech at Notre Dame University, and Rep. Geraldine Ferraro, Democratic Vice-Presidential candidate, took positions on the issue of religiously based values and public policy that stood in a direct line of descent from Kennedy's Houston speech. Rep. Henry J. Hyde's response to Governor Cuomo, in another Notre Dame speech, offered a different model for understanding the relationship between religiously based values and public policy in a pluralistic democracy. For the documentation on the 1984 debates, see Mario M. Cuomo, "Religious Belief and Public Morality," *New York Review of Books*, 31 (October 25, 1984) 32–37; Henry J. Hyde, "Keeping God in the Closet: Some Thoughts on the Exorcism of Religious Values from Public Life," *Notre Dame Journal of Law, Ethics and Public Policy*, 1:1 (1984) 33–51; Bishop James Malone, "Religion and the 1984 Campaign," *Origins*, 14:11 (August 23, 1984); Cardinal Joseph Bernardin, "Religion and Politics: The Future Agenda," *Origins* 14:21 (November 8, 1984) 321, 323–28. The 1984 debate was particularly interesting, and in this sense different from the nativist battles of 1960, because it ranged mainline Protestants, secularists like Norman Lear, and some Catholics against evangelical Protestants, other Catholics, and some mainline Protestants.

153. Kennedy's commencement address at Yale University in 1962, best remembered today for JFK's quip that he now had "the best of both worlds, a Harvard education and a Yale [honorary] degree," was of a piece with his Houston speech in its claim that the real problems of the age were not philosophical or ideological (and thus deeply enmeshed with questions of meaning and value), but technical and managerial. The text of the address may be found in *Public Papers of the Presidents of the United States: John F. Kennedy, 1962* (Washington, D.C.: United States Government Printing Office, 1963), pp. 470–75. Arthur Schlesinger comments favorably on the speech (predictably), in *A Thousand Days*, pp. 644ff.

American Catholicism and the Tradition in Transition: The Second Vatican Council

1. For a study of Pius IX that explains the events of his reign from the pope's point of view, see E. E. Y. Hales, *Pio Nono: A Study in European Politics and Religion in the Nineteenth Century* (London: Eyre & Spottiswoode, 1954). The summary statement of Pius IX's apologetic Catholicism was the last item in his *Syllabus of Errors*, attached to the encyclical *Quanta Cura* in 1864, which condemned the proposition that "the Roman Pontiff can and should reconcile himself to and agree with progress, liberalism, and modern civilization."

2. For a study of Leo XIII, see Edward Gargan, ed., *Leo XIII and the Modern World* (New York: Sheed and Ward, 1961).

3. For a portrait of Pius X, see Carlo Falconi, *The Popes in the Twentieth Century* (London: Weidenfeld and Nicolson, 1967), pp. 1–88.

4. For portraits of Benedict XV and Pius XI, see ibid., pp. 89–223.

5. For a portrait of Pius XII, see ibid., pp. 234–303.

6. For texts of these encyclicals, see *The Papal Encyclicals 1939–1958*, Claudia Carlen, I.H.M., ed. (McGrath Publishing Co., 1981).

7. For an example of Maritain's neo-Thomism, see his *True Humanism* (London: Geoffrey Bles, 1938); for Gilson, see his *The Spirit of Medieval Philosophy* (London: Sheed and

Ward, 1936) and *The Unity of Philosophical Experience* (New York: Charles Scribner's Sons, 1937). Marechal's sketch of the transcendental Thomistic project may be found in his *Le Point de Départ de la Métaphysique, Cahier V* (Paris: F. Alcan, 1926), parts of which have been translated in *A Marechal Reader*, Joseph Donceel, ed. (New York: Herder and Herder, 1970). Rahner's basic works in transcendental Thomistic philosophy are his *Spirit in the World* (New York: Herder and Herder, 1968) and *Hearers of the Word*, revised by J. B. Metz (New York: Herder and Herder, 1969). An illustrative example of Rahner's application of transcendental Thomism to a central problem in fundamental theology may be found in his essay "Nature and Grace," in *Nature and Grace: Dilemmas in the Modern Church* (New York: Sheed & Ward, 1964). Lonergan's basic works in transcendental Thomism are his monumental study *Insight: A Study of Human Understanding* (New York: Philosophical Library, 1970), and a collection of his essays from the journal *Theological Studies* in the 1940s, gathered in the volume *Verbum: Word and Idea in Aquinas*, David Burrell, ed. (University of Notre Dame Press, 1970).

8. For examples of de Lubac's work, see his *The Mystery of the Supernatural* (New York: Herder and Herder, 1968), and his *Catholicism* (London: Burns, Oates & Washbourne, 1950). De Lubac was widely supposed to be one of the principal targets of the encyclical *Humani Generis*, and one of his great opponents was the head of the Vatican's Holy Office, Cardinal Alfredo Ottaviani. Some thirty years after the *Humani Generis* controversy, Pope John Paul II named de Lubac a cardinal and gave him the late Cardinal Ottaviani's titular church in Rome—an interesting example of papal piquancy in these largely ceremonial matters, but one fraught with meaning for those who would portray John Paul II as a kind of preconciliar mind.

9. For overviews of these theological developments, see Yves Congar, O.P., *A History of Theology* (Garden City: Doubleday, 1968), chapter 6, and T. M. Schoof, O.P., *A Survey of Catholic Theology 1800–1970* (Paramus: Paulist Press, 1970).

10. For a partisan, but interesting, survey of how this organizational energy expressed itself in one prominent American diocese, see Andrew M. Greeley, *The Catholic Experience* (Garden City: Doubleday Image Books, 1969), chapter 8, "The Chicago Experience."

11. William Au surveys some of the intellectual currents moving through the CAIP in *The Cross, the Flag and the Bomb: American Catholics Debate War and Peace, 1960–1983.*

12. Cited in *The Documents of Vatican II*, Walter Abbott and Joseph Gallagher, eds. (New York: America Press, 1966), p. 715.

13. Cited in ibid., pp. 712–13.

14. For histories of Vatican II written during the years of the Council itself, see Henri Fesquet, *The Drama of Vatican II* (New York: Random House, 1967); Xavier Rynne, *Letters from Vatican City* (New York: Farrar, Straus and Giroux, 1963); idem, *The Second Session* (New York: Farrar, Straus and Giroux, 1964); idem, *The Third Session* (New York: Farrar, Straus and Giroux, 1965); and idem, *The Fourth Session* (New York: Farrar, Straus and Giroux, 1966). Guessing the identity of the pseudonymous Rynne, whose works first appeared in *The New Yorker*, was a favorite bit of Council sport; the author apparently bears a striking resemblance to Francis X. Murphy, C.SS.R. The most comprehensive analysis of the Council documents is found in the five-volume series *Commentary on the Documents of Vatican II*, Herbert Vorgrimler, general editor (New York: Herder and Herder, 1969). The commentaries in the Vorgrimler series were written largely by theologians involved in the actual drafting of the Council's formal documents, which is at once a great advantage and a bit of a hermeneutical problem.

15. The text of *Pacem in Terris* that I will follow is the translation issued by the Vatican Polyglot Press, with emendations made by the National Catholic Welfare Conference after publication of the official Latin text in the Vatican journal *Acta Apostolica Sedis*. The question of translation accuracy is particularly critical in dealing with *Pacem in Terris*. In one

key paragraph, #127, the originally-released translation had the pope declaring that, in this atomic age, "it is hardly possible to imagine that . . . war could be used as an instrument of justice." The more accurate translation reads "in an age such as ours which prides itself on its atomic energy, it is contrary to reason to hold that war is now a suitable way to restore rights which have been violated." The originally-released translation seemed to repeal the entire *ius ad bellum*; the more accurate translation holds open the possibility of wars of self-defense, and thus sets the basis for a qualified moral acceptance of deterrence. Paul Ramsey discusses the problems of these translation difficulties in his chapter "*Pacem in terris*" in *The Just War: Force and Political Responsibility* (New York: Charles Scribner's Sons, 1968, reprinted by University Press of America, 1983), p. 78. Latin is not the easiest language in which to discuss contemporary phenomena such as nuclear weapons, deterrence, or, for that matter, traffic laws regulating automobiles, as both *Pacem in Terris* and the key Council document *Gaudium et Spes* ("The Church in the Modern World") illustrated.

16. Thurston N. Davis, S.J., "Pope John's Letter," *America*, May 18, 1963, p. 707.

17. Ibid.

18. *Pacem in Terris* (hereafter "PT"), 40, 41.

19. PT, 44.

20. PT, 45.

21. Ibid.

22. PT, 130.

23. "No era will destroy the unity of the human family," the pope wrote, "since it is made up of human beings sharing with equal right their natural dignity. For this reason, necessity, rooted in man's very nature, will always demand that the common good be sought in sufficient measure because it concerns the whole human family" (PT, 132).

24. PT, 133–34.

25. PT, 135.

26. PT, 137.

27. See PT, 138.

28. PT, 139, 140, 141.

29. "Any human society, if it is to be well-ordered and productive, must lay down as a foundation this principle, namely, that every human being is a person, that is, his nature is endowed with intelligence and free will. Indeed, precisely because he is a person he has rights and obligations flowing directly and simultaneously from his very nature. And as these rights and obligations are universal and inviolable, so they cannot in any way be surrendered" (PT, 9).

30. See PT, 11.

31. PT, 12.

32. PT, 13.

33. PT, 14.

34. PT, 15.

35. See PT, 16.

36. PT, 18–22.

37. PT, 23–24.

38. PT, 25.

39. PT, 26.

40. PT, 27.

41. See PT, 28–34.

42. PT, 34.

43. PT, 158.

44. PT, 159.

45. PT, 160.

46. PT, 128. For a comment, see James E. Dougherty, *The Bishops and Nuclear Weapons* (Hamden: Archon Books, 1984), p. 57.

47. PT, 109.

48. Ibid.

49. "The production of arms is allegedly justified on the grounds that in present-day circumstances peace cannot be preserved without an equal balance of armaments. And so, if one country increases its armaments, others feel the need to do the same; and if one country is equipped with nuclear weapons, other countries must produce their own, equally destructive" (PT, 110).

50. PT, 111.

51. PT, 112–13.

52. PT, 114–16.

53. PT, 127.

54. John Cogley, "How to Read an Encyclical," *America*, May 18, 1963, p. 709.

55. Davis, art. cit., p. 710.

56. Ibid., p. 707.

57. Ibid., pp. 708–9. It is interesting that the classic Cold War phrase "captive nations" did not seem out of place in 1963 to the editor of a journal leaning toward the liberal side of American Catholic opinion.

58. "*Pacem in Terris*," *Commonweal*, April 26, 1963, p. 123.

59. John Cogley, "Peace on Earth," *Commonweal*, May 3, 1963, p. 158.

60. Ibid.

61. See ibid., pp. 158–59.

62. James O'Gara, "Catholics and Isolationism," *Commonweal*, May 17, 1963, p. 219.

63. "Communism and the Pope," *Commonweal*, May 24, 1963, p. 235. For an example of how this perceived "opening to the Left" could lead to distortions of vision on the religious situation in Eastern Europe, see Daniel Callahan, "Seeds Behind the Wall," *Commonweal*, May 31, 1963, pp. 275ff. Callahan admitted that "The Wall [i.e., in Berlin] . . . is a lie. It is no more a 'state frontier' than the walls of Sing Sing up the road from where I live could be called the boundary mark for a suburb of Ossining, New York. *But it is almost as misleading to envision the Christians living in East Berlin as martyrs*. . . . What concerned them . . . was the challenge of being responsible Christians in a society which they see as primarily industrialized and secularized and *only secondarily Communist*" (emphasis added).

Whatever his nods to the reality of The Wall, Callahan managed to make East Berlin sound something like Glasgow or Cleveland: an early example of the kind of parallelism that would bedevil much of the approach of the American Catholic intellectual community to the problem of communism over the next generation. How much of this was a reaction to American Catholicism's alleged responsibility for the crusade of Joseph R. McCarthy is a matter for interesting speculation. To the degree that there is a connection, it seems odd that American Catholic intellectuals would adopt the interpretation of the McCarthy phenomenon favored by anti-Catholic nativists of both the liberal Protestant and secularist varieties.

64. "Pope John," *Commonweal*, June 14, 1963, p. 316. This assumption of a general easing of the condition of believers in the U.S.S.R. and Eastern Europe during the Khrushchev period was a regular feature of the positive appraisals of *Pacem in Terris.* Peter Hebblethwaite continued the trend twenty years later in his biography *Pope John XXIII: Shepherd of the Modern World* (New York: Doubleday, 1985). In fact, however, it is now clear that Khrushchev was in 1963, and had been for some time, embarked on a severe persecution of religious communities in the Soviet Union; see Trevor Beeson, *Discretion and Valour: Religious Conditions in Russia and Eastern Europe*, revised edition (Philadelphia: Fortress Press, 1982), pp. 40, 104, 139. Beeson's book grew out of a study originally commissioned by the British Council of Churches, which, like its American counterpart, is not an institution notable for the

forthrightness of its criticism of Soviet policy toward the churches. But the evidence uncovered by Beeson was simply irrefutable.

65. Michael Novak, "Break with the Past," *Commonweal*, June 28, 1963, pp. 374–75.

66. John Coleman Bennett, "*Pacem in Terris*: Two Views," *Christianity and Crisis*, 23:8 (May 13, 1963) 82.

67. Alan Geyer, "Militancy and Mediation," *Worldview*, June 1963, p. 5.

68. Everett Gendler, "Profession and Practice," *Worldview*, June 1963, p. 8.

69. James W. Douglass, "The Non-Violent Power of *Pacem in Terris*," in *The Non-Violent Cross* (New York: Macmillan, 1968), p. 83. Throughout his commentary, Douglass assumes (and cites) the inaccurate translation of PT #127 noted above, n. 15.

70. Ibid., p. 84.

71. Ibid., p. 85.

72. Ibid., p. 87.

73. See ibid., pp. 87–88.

74. See ibid., pp. 94, 98.

75. John Courtney Murray, S.J., "Things Old and New in *Pacem in Terris*," *America*, April 27, 1963, p. 612.

76. Ibid. (emphasis added).

77. See ibid.

78. Ibid., pp. 612–13. On this analysis, Murray argued that the pope's distinction between "false philosophical teachings" and the "historical movements" that bore these teachings referred to nineteenth-century continental liberalism rather than communism.

79. See ibid., p. 613.

80. Ibid., p. 614.

81. Ibid.

82. Ibid.

83. Ibid.

84. William V. O'Brien, "Balancing the Risks," *Worldview*, June 1963, p. 10.

85. Ibid.

86. See ibid.

87. Ibid.

88. Ibid., p. 11.

89. Ibid.

90. Ibid.

91. Ibid., p. 11.

92. Ibid.

93. Ibid.

94. Cited in *Worldview*, June 1963, p. 4. Herberg's comments on the encyclical were published in the May 7, 1963, issue of *National Review*.

95. Reinhold Niebuhr, "*Pacem in Terris*: Two Views," *Christianity and Crisis*, 23:8 (May 13, 1963) 81.

96. Ibid., p. 83.

97. Ibid.

98. Ibid. Would Niebuhr have modified this assertion had he lived to see the influence of Pope John Paul II on the Polish trade union *Solidarnosc*, and its impact on the course of world events?

99. Paul Ramsey, "*Pacem in terris*," in Ramsey, op. cit., p. 70. This is a revised version of an article that Ramsey originally published in *Religion in Life*, 33 (Winter 1963–1964) 116–35.

100. Ibid., p. 71. The words, as Ramsey notes, are those of the midnight visitor to Ivan Karamazov in Dostoevsky's *The Brothers Karamazov*.

101. Ibid., pp. 71–72.

102. Ibid., p. 75.
103. Ibid. (emphasis added).
104. Ibid., p. 77.
105. Ibid., p. 84.
106. Ibid., p. 85.
107. Ibid., p. 86.
108. Ibid., p. 85.
109. See Peter Berger, "Are Human Rights Universal?," *Commentary*, March 1977. Robert Goldwin discusses the problem of adding entitlements to a list of basic human rights in "The Cause of Human Rights: Human Rights and American Foreign Policy," *Current*, (February 1985):

> The question is, do we strengthen or weaken the protection of the right to life and liberty by adding to legal documents a list of the necessary supports of life? There are two dangers immediately apparent in the codification of such rights: first, this can give a false sense that asserting the right addresses the problem of obtaining food, water, air, and work; second, it can have the effect of increasing the power of government agencies in new and unwelcome ways.
>
> Declaring the right to an adequate diet does not increase the supply of food or improve its distribution. If we think of the plight of desperate peoples, with hundreds of thousands dying of malnutrition, can it help to insist that these people have a right to an adequate diet? The problem is not whether they have the right, but how to get food to them, how to correct the problems that led to the lack of food, and how to put their agricultural production on a sounder footing. Surely no one seriously asserts that a significant part of the problem of hunger in the world stems from the doubt that people have the right to an adequate diet.
>
> It distresses me, thinking of the plight of the hungry millions in the world, most of them victims of misconceived government policies more than of natural deficiencies, that their well-fed diplomatic representatives in international organizations devote so much time and money to creating the cruel illusion that asserting the right will somehow provide more food for the hungry. It is either a conscious fraud, or a naive faith in the magic of words, to assert that recognizing the human right to enough food will resolve the problem. Whether fraud or folly, it is deplorable and shameful.

Goldwin then makes a larger point, of which the "right to food" is but an important example:

> There is a real danger in enumerating rights, especially if the list is long and affirmative and guaranteed. If one guarantees education, housing, and jobs, for instance, the state will end up being the only schoolmaster, landlord, and boss; extend the guaranteed rights to health care, resorts, retirement pensions, and such, as is done in many constitutions, and it is easy to see that government, if it takes these guarantees seriously, must control almost every aspect of life—another way of saying totalitarianism. It is not an accident that rights are most secure where the constitutional list is short, negative, and free of guarantees [pp. 32–33].

Defenders of Pope John's catalogue of human rights in *Pacem in Terris* might argue that there is a difference between a religious leader's adumbration of those things necessary for a life to be truly human, and the drawing up of constitutional lists of "rights." But such a rejoinder would fail to take account of the fact that the pope explicitly endorsed the Universal Declaration of Human Rights in *Pacem in Terris* (143–44), a model for the "catalogue" approach to the definitions of human rights.

This entire issue will be discussed more fully in Part Three. For a more positive appraisal of Pope John's approach to human rights, see David Hollenbach, S.J., *Claims in Conflict: Retrieving and Renewing the Catholic Human Rights Tradition* (New York: Paulist Press, 1979), esp. chapter 2.

110. *Gaudium et Spes* (hereafter "GS"), 1. The translation I will follow here is that of Joseph Gallagher in *The Documents of Vatican II*, Walter Abbott and Joseph Gallagher, eds.

(New York: America Press, 1966). A later, and in some ways more accurate, translation may be found in *Vatican Council II: The Conciliar and Post-Conciliar Documents*, Austin Flannery, general editor (Northport: Costello Publishing Co., 1975). Readers may wish to compare the two translations. The official Latin text may be found in *Sacrosanctum Oecumenicum Concilium Vaticanum II: Constitutiones, Decreta, Declarationes*, prepared by the office of the secretary general of the Council, and issued by the Vatican Polyglot Press in 1966. I follow the Gallagher translation here because its phraseology helped shape the American Catholic debate on the morality of war and peace for a full decade after the Council.

111. Cited in Vincent A. Yzermans, *American Participation in the Second Vatican Council* (New York: Sheed and Ward, 1967), p. 186.

112. Ibid.

113. For details on the drafting of GS and the history of the debates on it, see Charles Moeller, "History of the Constitution" in *Commentary of the Documents of Vatican II, Volume Five, Pastoral Constitution on the Church in the Modern World*, Herbert Vorgrimler, general editor (New York: Herder and Herder, 1969), pp. 1–77. The Vorgrimler commentary also includes a history of the text of chapter 5 ("The Fostering of Peace and the Promotion of a Community of Nations") by Willem J. Schuijt, and commentary on chapter 5 by René Coste. Further discussion of the evolution of *Gaudium et Spes* may be found in Xavier Rynne, *The Third Session* (New York: Farrar, Straus, and Giroux, 1965) and idem, *The Fourth Session* (New York: Farrar, Straus, and Giroux, 1966). Vincent Yzermans sketches the American role in the drafting of GS in op. cit., pp. 185ff.

114. GS, 6.

115. GS, 80 (emphasis added). Flannery, op. cit., translates this key sentence as follows: "All these factors force us to undertake a completely fresh appraisal of war" (p. 989). The Latin text is *Quae omnia nos cogunt ut de bello examen mente omnino nova instituamus.* Gallagher's translation, "with an entirely new attitude," though arguably somewhat less accurate in conveying the meaning of *mente omnino nova*, was the phrase that seized the imaginations of many American Cartholics in the subsequent debate over the morality of war and peace, particularly those who looked at *Gaudium et Spes* as a decisive break with the Church's traditional approach to the just-war theory. The difference between a "completely fresh appraisal" of war and an evaluation of war "with an entirely new attitude" is not insignificant: the Flannery translation suggests a new intellectual exercise, whereas the Gallagher translation evokes the evangelical/personal conversion themes of *Gaudium et Spes.* But, in any case, *mente omnino nova*, on whatever translation, suggests a tradition in transition, which is my main point here.

116. GS, 77 (emphasis added).

117. Ibid.

118. GS, 78.

119. Ibid.

120. Ibid.

121. Ibid.

122. GS, 79.

123. Ibid.

124. Ibid.

125. Ibid. This commendation of military service was a result of the interventions of several American bishops, notably Philip M. Hannan, former auxiliary bishop of Washington, D.C., and then archbishop of New Orleans. Hannan, who was most noted at the time as the man whom the Kennedy family had asked to be the eulogist at the funeral of President Kennedy, was a former chaplain of the 101st Airborne Division. His role in the drafting and amending of *Gaudium et Spes* is analyzed in the historical commentaries listed in note 113

above, as well as in James Douglass, *The Non-Violent Cross*, chapter 5. It should be remembered that all of these commentators on Hannan's role were on what can accurately be described, in this case, as "the other side of the argument."

126. GS, 80.
127. Ibid.
128. GS, 81.
129. Ibid.
130. Ibid.
131. GS, 82.
132. Ibid. Archbishop Hannan was a key figure in the formulation of this text in *Gaudium et Spes*, as well as in the document's modest affirmation of deterrence. The latter was a source of considerable controversy at the council, described in the historical commentaries noted above. See particularly Rynne's description of the argument in *The Fourth Session*, p. 225–30. The Moeller, Schuijt, and Coste commentaries in Vorgrimler, op. cit., make it clear that several senior Council fathers wished the Council to flatly condemn nuclear weapons and, thus, deterrence. This issue would emerge again, and the terms of the debate would be strikingly similar, in the process leading up to the publication of the National Conference of Catholic Bishops' 1983 pastoral letter, "The Challenge of Peace."
133. GS, 82.
134. Ibid.
135. GS, 83.
136. Ibid.
137. Ibid.
138. GS, 84.
139. GS, 85.
140. Ibid.
141. GS, 86.
142. GS, 88.
143. GS, 90.
144. George G. Higgins, "Commentary on 'The Pastoral Constitution on the Church in the Modern World,'" in Yzermans, op. cit., p. 264.
145. Ibid., p. 265.
146. Ibid.
147. Ibid., p. 266.
148. Higgins argued that, had the Council fathers "decided to concentrate exclusively on the internal reform of the Church and to say nothing at all about the Church's relation to the modern world, the Council would have been, not a failure, but, at best, only a partial success. It would also have been a grave disappointment to the world, which is probably more eager today than ever before in modern times to begin a fruitful dialogue with the Church. . . . Thanks be to God, the Council not only elected to begin this belated dialogue with the modern world, but did do, as Pope Paul VI remarked on the closing day of the Council, 'with the accommodating friendly voice of pastoral charity.'" It was this spirit, Higgins suggested, that had led the majority of Council fathers to reject an amendment to GS, posed by some 450 bishops, that would have explicitly named communism as a grave problem of the modern world in the GS discussion of atheism. See ibid., p. 269.
149. Douglass, op. cit., p. 102.
150. See ibid., pp. 105–6.
151. See ibid., pp. 109–10.
152. See ibid., pp. 113–14.
153. Cited in ibid., p. 118. "How," Ritter had asked, "are we able to condemn every intention of destroying cities and at the same time at least in part approve the balance of

terror?" The only solution, Ritter believed, was "a clear and distinct declaration that the moral law requires that all urgently and without delay collaborate in the elimination of the possession of such armaments, no matter how grave the difficulties are which are feared and which must be overcome" (ibid.).

154. See ibid., pp. 124–25.

155. Ibid., p. 126.

156. Ibid.

157. Ibid., p. 128. Douglass does not mention those Catholics who, in Lithuania, Cuba, the Peoples Republic of China, and other lands under communist regimes, had already made, lived out, and often died because of this "discovery."

158. See ibid., p. 129.

159. Paul Ramsey, "The Vatican Council and Modern War," *Theological Studies*, 27:2 (June 1966) 179.

160. Ibid., p. 180.

161. Ibid., pp. 180–81.

162. See ibid., p. 183.

163. Ibid., p. 186.

164. See ibid., p. 187.

165. Ibid., p. 188. Ramsey also wrote:

> It needs to be pointed out that the Council says nothing that removes the morality from deterrence, or removes responsibility for this from among possible Christian vocations. It is true that the peace that deterrence assures is not the peace that passes understanding. It is not even a very good worldly peace; it is only a peace of sorts. The Council does not shirk its responsibility for calling attention to the fact that, in terms of worldly peace, deterrence is "not a safe way to preserve a steady peace. Nor is the so-called balance from this race a sure and authentic peace." Still, nothing in all this says that responsible decision and action in regard to deterrence falls below the floor of the morally permissible. . . . The expression "peace of a sort," which deterrence insures, was well chosen . . . to bend even this lower good toward the better that has yet to be done [ibid., pp. 188–89].

166. See ibid., p. 192.

167. Ibid., p. 193.

168. Ibid.

169. Ibid., p. 194.

170. Ibid., p. 199.

171. Ibid., pp. 199–200.

172. Ibid., pp. 201–2.

173. Ibid., p. 200. Ramsey's claim that *Gaudium et Spes* had vindicated just-war theory in its most comprehensive form was not only attacked by those like James Douglass who wanted the Council to reject just-war theory on pacifist grounds. Robert W. Tucker was also unimpressed, but for a different reason: he saw no way to think responsibly about the role of a great power in the nuclear age, given the limitations imposed by just-war categories of analysis. In a biting critique of "The Church in the Modern World," Tucker concluded that the Council had tried "to reconcile the irreconcilable, the requirements of *bellum justum* and the necessities of a nuclear Power. And thus does the Council demonstrate once again that there is no way by which the circle can be squared, that there is no way by which the injunction against doing, or threatening, evil that good may come can be reconciled with the constituent principle of statecraft," by which Tucker meant the classic Athenian argument in Thucydides' "Melian Dialogue," *raison d'état*. Ramsey's response was noteworthy, in an essay entitled "Tucker's *Bellum Contra Bellum Justum*":

> A statement of Jacques Maritain has recently become quite a consolation to me. "Moralists," he wrote in *Man and the State*, "are unhappy people. When they insist on the immutability of

> moral principles, they are reproached for imposing unliveable requirements on us. When they explain the way in which those immutable principles are to be put into force, they are reproached for making morality relative. In both cases, however, they are only upholding the claims of reason to direct life."
>
> For which aspect of the professional work of a moralist Professor Tucker has the greater distaste, it is difficult to tell. Any attempt to show how moral principles work in political practice and in the conduct of war is like trying to "square the circle." The constituent elements of statecraft, Tucker believes, must revolve in a closed circle. The quest to define *legitimate* military necessity or *legitimate* reasons of state is bound, therefore, to eventuate either in the renunciation of statecraft or in the renunciation of morality.

Ramsey's worry here was remarkably prescient. If there was a central claim in the Catholic tradition of moderate realism, it was that politics and morality were not antinomies. Right reason could discover, and the virtue of prudence could apply, moral principles even amid the inevitable tangles, complexities, and ambiguities of political life. This was true even in the moral limit case of war. This assertion would continue to be attacked by men like Tucker, operating on classic premises, as well as by contemporary systems analysts and others in the key points where American academic life meets the policy world. But it would also be attacked by men like James Douglass who, operating on entirely different presuppositions from Tucker's, would make common cause with the inheritors of Thucydides in a pacifist attack on the very possibility of just-war reasoning. The evangelical criteria used by *Gaudium et Spes* played a significant role in this attack, as they did in Douglass's commentary on "The Church in the Modern World," which amounts to a confirmation of Ramsey's claim that the abandonment of just-war reasoning would lead some to an abandonment of statecraft.

The Tucker/Ramsey exchange may be found in *Just War and Vatican Council II: A Critique*, a booklet published in 1966 by the Council on Religion and International Affairs.

174. The Council's discussion of development looks particularly weak in light of the evidence of recent history. There is not, it now seems reasonably clear, any direct relationship between amounts of aid money available and the degree of economic development achieved in a Third World country. Financial assistance, technical assistance, and all the rest of the armamentarium of development economics are important, but the most crucial factors seem to be political and cultural. Does the government get out of the way of economic progress, or does it seek to use development assistance for its own ends? Are the virtues of deferred gratification, common enterprise, thrift, work, honesty, and responsibility inculcated by the culture? The Church's role in this process of culture formation should be apparent; but it is not discussed at any length in GS, which leans toward the now discredited notion that the poor are poor because the rich are rich. There have been too many Tanzanias, too many Ethiopias, and too many Chads, on the one hand, and too many Singapores, Taiwans, and South Koreas on the other, not to see the weakness of the Council's analysis.

175. See Martin Work, "Introduction" to the "Declaration on the Apostolate of the Laity," in Abbott and Gallagher, *The Documents of Vatican II*, pp. 486–88.

176. See "Dogmatic Constitution on the Church," in Abbott and Gallagher, op. cit., chapter 2 (pp. 24ff.).

177. Ibid., 33. "Through their baptism and confirmation, all are commissioned to that apostolate by the Lord Himself. Moreover, through the sacraments, especially the Holy Eucharist, there is communicated and nourished that charity toward God and man which is the soul of the entire apostolate" (ibid.).

178. Ibid.

179. "Decree on the Apostolate of the Laity," in Abbott and Gallagher, op. cit., 7.

180. Ibid., 14 (emphasis added). The translation in Flannery, *Vatican Council II: The Conciliar and Post-Conciliar Documents*, makes this point even more strongly: "On the national and international planes the field of the apostolate is vast; and it is there that the laity more than others are the channels of Christian wisdom."

181. See H. Richard Niebuhr, *Christ and Culture*, chapter 6.

182. This is not to deny a role for the clergy, religious, and the hierarchy in these matters. But the fact remains that, in the mind of the Council, the primary responsibility for what we might call the "apostolate of public policy" rests with the laity. Abrogation of this lay prerogative is a new form of clericalism.

183. For a portrait of the controversies in which Murray was involved, see Donald Pelotte, *John Courtney Murray: Theologian in Conflict* (New York: Paulist, 1975). Gerald Fogarty traces some of the same history, from the standpoint of the relations between the Vatican and the American bishops, in *The Vatican and the American Hierarchy from 1870 to 1965*, pp. 346–403.

184. "Declaration on Religious Freedom," in Abbott and Gallagher, op. cit., 1.

The John Courtney Murray Project

1. John Deedy, "John Courtney Murray: The Jesuit from Olympus," in *Seven American Catholics* (Chicago: Thomas More Press, 1978), p. 125.

2. Walter J. Burghardt, "He Lived With Wisdom," *America*, September 9, 1967, pp. 248–49. For other portraits of Murray, the man and the mind, see John Cogley, "John Courtney Murray," *America*, September 2, 1967, pp. 220–21; Emmet John Hughes, "A Man for Our Season," *The Priest*, 25:7 (July–August 1969) 389–402; and Albert Broderick, "From a Friend Who Never Met Him," *America*, September 9, 1967, pp. 246–49. My own favorite Murray story is from the Cogley reminiscence: "One time we were having dinner at a roadside place near Baltimore. We lingered over drinks and took our time over the after-dinner coffee. The manager kept eyeing us nervously, but Fr. Murray was not about to leave. Finally, when he retired to the rest-room, the manager came over and asked if he might talk to me privately. 'Maybe you didn't know we have entertainment here,' he said. 'But the stripper refuses to go on as long as the priest is here, and we are an hour behind schedule already.' When Fr. Murray returned, I explained. 'The girl might get fired,' he said. 'Let's get out of here'" (p. 221).

3. See Gerald Fogarty, *The Vatican and the American Hierarchy from 1870–1965*, pp. 368ff.

4. See Deedy, art. cit., pp. 128–31. See also the *Time* magazine cover story on Murray, "City of God & Man," December 12, 1960, pp. 64ff.

5. See John Murray Cuddihy, *No Offense: Civil Religion and Protestant Taste* (New York: Seabury, 1978).

6. See ibid., pp. 89ff. The best refutation of the charge that Murray somehow "accommodated" Catholicism to a Protestant civil religion of "good taste" may be found in Murray's biting rejoinder to the church/state worries of Union Theological Seminary President W. Russell Bowie, "The Catholic Position—A Reply," *American Mercury*, 69:309 (September 1949) 274ff. The *American Mercury* of November 1949 contains Bowie's attempt at a response, and Murray's rejoinder (pp. 636–39). As this exchange graphically illustrates, Murray was a formidable controversialist. For Murray's response to Blanshard's new nativism, see Murray's essay "Paul Blanshard and the New Nativism," *The Month* (new series), 5:4 (April 1951) 214–25. A more polemical rejoinder to Blanshard may be found in Murray's review of *American Freedom and Catholic Power* in *The Catholic World*, 169:1011 (June 1949) 233–34.

7. David Hollenbach, "Public Theology in America: Some Questions for Catholicism after John Courtney Murray," *Theological Studies*, 37:2 (June 1976) 292.

8. Ibid., p. 293.

9. Ibid., pp. 298, 302. It is hard to understand precisely what Hollenbach means here. Robert Bellah, in his classic essay "Civil Religion in America," argued precisely the opposite: that American public history was saturated with biblical imagery. See Bellah's essay in his collection *Beyond Belief* (New York: Harper & Row, 1970). As for the present, the rise of the

religious new right, whatever else it may or may not portend, hardly suggests the absence of a sense of the sacred in American public life.

10. Hollenbach, art. cit., p. 302.

11. Ibid., p. 291.

12. David Hollenbach, ed., "Theology and Philosophy in Public: A Symposium on John Courtney Murray's Unfinished Agenda," *Theological Studies*, 40:4 (December 1979) 701. Hollenbach's conclusion that the Marxist theory of economic and social "rights" is one of those traditions that "make major contributions to the development of an adequate theory of human rights" may be at play here. (See his *Claims in Conflict: Retrieving and Renewing the Catholic Human Rights Tradition* [New York: Paulist Press, 1979], p. 33.)

13. John Coleman, "A Possible Role for Biblical Religion in Public Life," in Hollenbach, ed., "Theology and Philosophy in Public," p. 702.

14. Ibid., pp. 705–6. In the same symposium, J. Bryan Hehir defended "The Perennial Need for Philosophical Discourse" (pp. 710–13). Hollenbach concluded his report on the symposium by noting that "the most recent efforts" to develop a "fundamental political theology" in the United States "have not addressed the critical relationship between Christian tradition and prevailing forms of *American* political and social discourse in a serious way. Though Murray's suppositions about the compatibility of these two traditions may be too simple, he took the American secular political tradition much more seriously than have most contemporary American theologians. Creative development of American Catholic social thought will occur when Murray's lead is followed in this regard" (p. 715). Indeed.

15. John Courtney Murray, *We Hold These Truths: Catholic Reflections on the American Proposition* (Garden City: Doubleday Image Books, 1964), p. 7. John Deedy describes the genesis of *We Hold These Truths* (hereafter "WHTT") in art. cit., p. 143. WHTT was a gathering (and modest reediting) of papers that Murray had written over several years. The book was received with lavish praise from many quarters when published in 1960.

16. WHTT, p. 7.

17. Ibid.

18. George F. Will, for example, portrays James Madison as a kind of political mechanic, in *Statecraft as Soulcraft* (New York: Simon & Schuster, 1983).

19. WHTT, p. 8.

20. Ibid., p. 9.

21. Ibid., pp. 17–18.

22. See Charles Frankel, *Morality and U.S. Foreign Policy* (New York: Foreign Policy Association, 1975), p. 52.

23. WHTT, p. 21.

24. Ibid.

25. Ibid. (emphasis added).

26. See ibid., p. 19. Murray wrote that the modern barbarian "may wear a Brooks Brothers suit and carry a ballpoint pen. . . . In fact, even beneath the academic gown there may lurk a child of the wilderness, untutored in the high tradition of civility, who goes busily and happily about his work, a domesticated and law-abiding man, engaged in the construction of a philosophy to put an end to all philosophy. . . . This is perennially the work of the barbarian, to undermine rational structures of judgment, to corrupt the inherited intuitive wisdom by which people have always lived, and to do this not by spreading new beliefs but by creating a climate of doubt and bewilderment in which clarity about the larger aims of life is dimmed and the self-confidence of the people is destroyed." This passage from WHTT opened *Time*'s Murray cover story, art. cit., p. 64. Alasdair MacIntyre reached a similar conclusion at the end of his study *After Virtue* (University of Notre Dame Press, 1981), a generation after Murray. Wrote MacIntyre, reflecting on the possibility of a new Dark Age, "This time,

however, the barbarians are not waiting beyond the frontiers; they have already been governing us for quite some time" (p. 245).

27. WHTT, pp. 22–23.
28. Ibid., p. 23.
29. See ibid., p. 10.
30. Ibid., pp. 30, 33.
31. Ibid., p. 33.
32. Ibid., pp. 33–35.
33. WHTT, p. 39.
34. Ibid., p. 40.
35. Ibid., p. 43.
36. Ibid.
37. Ibid., p. 44.
38. Ibid.
39. Ibid., p. 45.
40. Ibid.
41. Ibid.
42. Ibid., pp. 45–46.
43. Ibid., p. 47.
44. Ibid., p. 48.
45. Ibid. Murray knew that economics, too, involved virtue: beyond material gain, economic achievement was a matter of the accomplishment of a human good.
46. Ibid., p. 48.
47. Ibid., p. 50.
48. Cited in ibid.
49. Ibid., p. 51.
50. Ibid., p. 52.
51. Ibid. A generation later, Richard John Neuhaus echoed Murray's sense that there was a "Catholic moment" at hand in the task of American culture-formation; see his *The Naked Public Square*, p. 262. Neuhaus sees more hope for an ecumenical endeavor in the task of reclaiming the American proposition than Murray was willing to admit in WHTT. Neuhaus is particularly intrigued by the potential culture-forming role of nonfundamentalist evangelical Protestantism, which, whatever its visceral distaste for the words "natural law," operates on a parallel set of assumptions about the relationship between virtue (personal and civic) and the life of the commonwealth.
52. WHTT, p. 53.
53. Ibid., p. 19.
54. Ibid., p. 93.
55. Ibid., p. 113.
56. Ibid., pp. 113–14.
57. Ibid., p. 114.
58. Ibid., pp. 114–15.
59. Ibid., pp. 116–17.
60. Ibid., p. 281.
61. Ibid., p. 282.
62. Ibid., p. 302.
63. See ibid., p. 303.
64. Ibid., p. 185.
65. Ibid., p. 187.
66. Ibid.

67. See ibid., p. 188.
68. Ibid., p. 204.
69. Ibid., p. 262.
70. Ibid.
71. Ibid., pp. 263–64.
72. Charles Frankel sums up the realist case in op. cit., from which this brief (and simplified) portrait is drawn.
73. Ibid., p. 46.
74. Ibid., pp. 39–40.
75. WHTT, p. 271.
76. Ibid., p. 272. Murray's comments on Jimmy Carter's campaign promise to give the country a government as good as its people, on Carter's alleged dependence for social-ethical insight on Reinhold Niebuhr, and on Mario Cuomo's invocation of "family" as the paradigm for public policy would have been interesting.
77. See ibid., p. 273.
78. Ibid. (emphasis added). "I am, of course, much troubled by the question of the national interest," Murray concluded, "but chiefly lest it be falsely identified in the concrete, thus giving rise to politically stupid policies. But since I do not subscribe to a Kantian 'morality of intention,' I am not at all troubled by the centrality of self-interest as the motive of national action. From the point of view of political morality, as determined by the purposes inherent in the state, this motive is both legitimate and necessary" (ibid., pp. 272–73).
79. Ibid., p. 267.
80. Ibid., p. 274.
81. Ibid.
82. Ibid., p. 270.
83. Ibid., p. 275.
84. See WHTT, pp. 238ff. This essay appeared in many fora. It was published as a pamphlet by the Council on Religion and International Affairs under the title "Morality and Modern War" in 1959; the essay also appeared in *Theological Studies*, 20:3 (March 1959) under the title "Remarks on the Moral Problem of War." Donald Pelotte gives the various bibliographic references for this essay, originally a paper delivered before the Catholic Association for International Peace on October 24, 1958, in *John Courtney Murray: Theologian in Conflict* (New York: Paulist Press, 1975), p. 195.
85. See WHTT, pp. 238–39. Murray's counsel on avoiding a single-question entry point for the discussion was not followed in the American bishops' 1983 pastoral letter, "The Challenge of Peace."
86. WHTT, p. 240.
87. Ibid.
88. Ibid.
89. Ibid., p. 241.
90. Ibid., p. 242.
91. Ibid.
92. Cited in ibid. Pius's words were from his 1950 Christmas Message.
93. WHTT, p. 242.
94. See ibid., pp. 243–45.
95. Ibid., p. 246.
96. Ibid.
97. Ibid.
98. Ibid.
99. Ibid., p. 247. Murray was citing a statement of Pius XII's in 1953.
100. Ibid., p. 249.

101. Ibid., pp. 249–50 (emphasis added).
102. Ibid., p. 251.
103. Ibid., pp. 251–52.
104. Ibid., p. 253.
105. Ibid.
106. Ibid., p. 254.
107. Ibid., p. 258.
108. The technology of limited war has advanced far beyond the state of the art in 1959. Precision guidance systems have made possible nuclear weaponry that would result in less blast damage than conventional high explosives. The advent of feasible strategic defense systems could dramatically alter the terms of the moral/strategic debate. Murray's call for an ethics of "limited war" was rejected, in some respects, by the American bishops in 1983; but the questions he raised, and the reasons for which he raised them, remain. See Albert Wohlstetter, "Bishops, Statesmen, and Other Strategists on the Bombing of Innocents," *Commentary* (June, 1983), pp. 15–35, and the exchange of letters on Wohlstetter's article in *Commentary* (December, 1983), pp. 4–22. See also my essay "The New Nuclear Debates," in *Catholicism in Crisis*, June 1984.
109. WHTT, p. 260.
110. Ibid., pp. 260–61.
111. John Courtney Murray, "World Order and Moral Law," *Thought*, 19:75 (December 1944) 581–86.
112. Ibid., p. 581.
113. Ibid.
114. Ibid., pp. 581–82.
115. Ibid., p. 582.
116. Ibid.
117. Ibid.
118. Ibid., p. 583 (emphasis added).
119. Ibid.
120. Ibid.
121. Ibid. The "trust" of which Murray wrote here was quite different from the "trust" that was enjoined by Pope John XXIII in *Pacem in Terris.* The issue for Murray was one of reasonable confidence in essentially just political institutions, which was not, as we have seen, what the pope intended nineteen years later.
122. Ibid., pp. 585–86 (emphasis added).
123. See WHTT, pp. 214–15.
124. Ibid., p. 215.
125. Ibid., pp. 216–17.
126. Ibid., p. 218.
127. Ibid.
128. Ibid., p. 227.
129. Ibid., p. 228.
130. See ibid., p. 229.
131. Ibid., p. 233.
132. Ibid., p. 234.
133. John Courtney Murray, "Selective Conscientious Objection," pamphlet published in Huntington, Indiana, by Our Sunday Visitor Press, p. 5.
134. Ibid., p. 6.
135. Ibid.
136. Ibid., pp. 6–7.
137. Ibid., p. 7.

138. Ibid., p. 8.

139. Ibid., p. 11.

140. Ibid., p. 14.

141. See John Courtney Murray, "The Issue of Church and State at Vatican Council II," *Theological Studies*, 27:4 (December 1966) 599–600.

142. See, for example, MacIntyre, op. cit., and Neuhaus, op. cit.

143. See Max Weber, "Politics as a Vocation," in *From Max Weber: Essays in Sociology*, H. H. Gerth and C. Wright Mills, eds. (New York: Oxford University Press, 1946), pp. 77–128.

144. Cited in WHTT, p. 196.

145. Unspoken nervousness about Murray's Gelasian sympathies may be found, for example, in the symposium on the "unfinished agenda" of Murray, noted above (n. 12).

146. James Hennesey, *American Catholics*, p. 303.

147. Again, see Neuhaus, op. cit., and MacIntyre, op. cit.

Five Portraits

1. David J. O'Brien, "The pilgrimage of Dorothy Day," *Commonweal*, December 19, 1980, p. 711.

2. See William D. Miller, *A Harsh and Dreadful Love: Dorothy Day and the Catholic Worker Movement* (New York: Liveright, 1973); William D. Miller, *Dorothy Day: A Biography* (San Francisco: Harper & Row, 1982); Mel Piehl, *Breaking Bread: The Catholic Worker and the Origin of Catholic Radicalism in America* (Philadelphia: Temple University Press, 1982); Jim Forest, *Love is the Measure* (Mahwah: Paulist Press, 1986); and Dorothy Day's three volumes of autobiography, *The Long Loneliness*, *Loaves and Fishes*, and *On Pilgrimage: The Sixties*, all published in paperback by Curtis Books.

3. John C. Cort, "Dorothy Day at 75," *Commonweal*, February 23, 1973, p. 475.

4. Ibid.

5. Miller, *A Harsh and Dreadful Love* (hereafter "HDL"), p. 3.

6. Evelyn Waugh, *Scott-King's Modern Europe* (Boston: Little, Brown, 1948), p. 89.

7. See Miller, HDL, pp. 5–6.

8. Cited in Miller, HDL, p. 278.

9. William Miller, "Dorothy Day, 1897–1980: 'All Was Grace,'" *America*, December 13, 1980, p. 386.

10. Cited in ibid.

11. Miller describes these tensions in HDL, pp. 154–84. See also James O'Gara, "The rock of contention," *Commonweal*, May 6, 1983, pp. 275ff.

12. Anne Fremantle, "Tributes and Recollections," *America*, November 11, 1972, p. 389.

13. Cited in Miller, HDL, p. 9.

14. Cited in Miller, HDL, p. 162.

15. Miller, HDL, p. 190.

16. Miller writes:

> "Fidel Castro says he is not persecuting Christ, but Churchmen who have betrayed him. . . . After all, Castro is a Catholic" [Dorothy Day wrote]. American Catholics had not persecuted churchmen, but were they not guilty of ignoring Christ? Castro's revolution had been *for* the poor, and if one *had* to choose between the violence done to the poor by the acquisitive bourgeois spirit of many Americans (among whom were churchmen) and the violence of Castro, which was aimed at helping the poor, then she would take the latter. "We do believe," she wrote, "that it is better to revolt, to fight, as Castro did with his handful of men . . . than to do nothing." Did this mean that Dorothy Day had at last been forced to admit that there was some ultimate point where circumstances became so desperate that only violence could resolve the problem? It sounded like it, but she had not. "We affirm our belief in the ultimate victory of

good over evil, of love over hatred, and we believe that the trials which beset us in the world today are for the perfecting of our faith" [HDL, pp. 305–6].

Dorothy Day was not, of course, the only American Catholic who fundamentally misread the moral and political meaning of Castro's revolution. But it is particularly disheartening to look back on her gullibility in light of Armando Valladares's memoir of twenty years in Castro's political prisons, *Against All Hope* (New York: Alfred A. Knopf, 1986). One cannot read Valladares's story without being reminded of the more gruesome scenes in the *Inferno*.

Shortly after his release, Valladares (whose prison poetry, often written in his own blood, led French president Mitterand to intercede with Castro for the Cuban's freedom) described the most heartbreaking experience of his imprisonment:

> "During those years, with the purpose of forcing us to abandon our religious beliefs and to demoralize us, the Cuban communist indoctrinators repeatedly used the statements of support for Castro's revolution made by some representatives of American Christian churches. Every time that a pamphlet was published in the United States, every time that a clergyman would write an article in support of Fidel Castro's dictatorship, a translation would reach us, and that was worse for the Christian political prisoners than the beatings or the hunger. While we waited for the solidarity embrace from our brothers in Christ, incomprehensibly to us, those who were embraced were our tormentors. Castro's political police have used these statements of support for Castro with such skill and for such a long time to confuse the prisoners and the population in general, that today the Christians in Cuba's prisons suffer not only the pain of torture and isolation, but also the conviction that they have been deserted by their brothers in faith." (Quoted in *Religion & Democracy*, August/September 1983, pp. 3–4.)

Valladares's testimony on this point has to weigh in any evaluation of the work of those who gave—and those who are still giving—active aid and comfort to the persecutors of their fellow believers, no matter what else they may have thought they were doing.

17. Miller, HDL, p. 303.

18. Dorothy Day, "A Reminiscence at 75," *Commonweal*, August 10, 1973, p. 424.

19. Fremantle, art. cit., p. 389.

20. Alex Avitabile, "'America' and the Early Years," *America*, November 11, 1972, p. 399.

21. For what follows, see Gordon Zahn, "Catholic Conscientious Objection in the United States," and "The Social Thought of the Catholic Conscientious Objector," in Zahn, *War, Conscience, and Dissent* (New York: Hawthorne Books, 1967).

22. Zahn, "The Social Thought of the Catholic Conscientious Objector," p. 163.

23. Ibid.

24. Ibid., p. 166.

25. Ibid., p. 174.

26. James Finn, "Gordon Zahn," in *Protest: Pacifism and Politics* (New York: Vintage Books, 1968), p. 62.

27. Gordon C. Zahn, "A Religious Pacifist Looks at Abortion," *Commonweal*, May 28, 1971, p. 282.

28. Gordon C. Zahn, "The Future of the Catholic Peace Movement," *Commonweal*, December 28, 1973, p. 338.

29. Ibid., p. 339.

30. Ibid.

31. Ibid., p. 340.

32. Ibid.

33. Ibid., p. 341.

34. See Gordon C. Zahn, *German Catholics and Hitler's Wars* (New York: Sheed & Ward, 1962); idem, *In Solitary Witness: The Life and Death of Franz Jaegerstaetter* (New York: Holt, Rinehart & Winston, 1964). The 1968 Beacon Press paperback reprint of the

Jaegerstaetter biography, through a new preface, explicitly ties Jaegerstaetter to the Vietnam-era draft resistance movement.

35. Gordon C. Zahn, "Pacifism and the Just War," in *Catholics and Nuclear War*, Philip Murnion, ed. (New York: Crossroad, 1983), p. 127.

36. See Robert Pickus, Preface, in Albert Camus, *Neither Victims Nor Executioners* (Chicago: World Without War Publications, 1972), p. 8.

37. Edward Rice, *The Man in the Sycamore Tree: The Good Times and Hard Life of Thomas Merton* (Garden City: Doubleday Image Books, 1972), p. 187.

38. Merton books have become a veritable cottage industry since his death in 1968. Among more recent studies may be cited, John Howard Griffin, *Follow the Ecstasy* (Fort Worth: Latitude Press, 1983); Michael Mott, *The Seven Mountains of Thomas Merton* (Boston: Houghton Mifflin, 1984); Monica Furlong, *Merton: A Biography* (New York: Harper & Row, 1980); and Elena Malits, *The Solitary Explorer: Thomas Merton's Transforming Journey* (San Francisco: Harper & Row, 1980). Merton's letters on issues of war and peace are now available in *The Hidden Ground of Love: The Letters of Thomas Merton on Religious Experience and Social Concerns*, William H. Shannon, ed. (New York: Farrar, Straus, Giroux, 1985); this volume includes several of Merton's famous "Cold War Letters," privately circulated to friends when his Cistercian superiors became nervous about Merton's public notoriety on peace issues.

39. Cited in Rice, op. cit., p. 95.

40. Thomas Merton, "Peace: Christian Duties and Perspectives," in *Thomas Merton on Peace*, edited with an introduction by Gordon C. Zahn (New York: McCall, 1971), p. 13 (hereafter "TMP").

41. Thomas Merton, "The Christian in World Crisis: Reflections on the Moral Climate of the 1960s," TMP, pp. 36–37.

42. Cited in Gordon C. Zahn, "Original Child Monk: An Appreciation," TMP, p. xiii.

43. Thomas Merton, "Note on Civil Disobedience and Nonviolent Revolution," TMP, p. 230.

44. See Zahn, "Original Child Monk," pp. xvi–xvii.

45. Ibid., p. xvii.

46. See ibid.

47. Merton, "Peace: Christian Duties and Responsibilities," TMP, pp. 14–15.

48. See Zahn, "Original Child Monk," p. xx.

49. Ibid., p. xxi.

50. Thomas Merton, "Christian Ethics and Nuclear War," TMP, p. 83.

51. Ibid. Who was proposing such a first strike, and who accepted such a proposition with moral ease, Merton did not specify. Merton also characterized Western political thought during the Cold War as marked by "thoughtless passivity . . . crude opportunism, and astonishing lack of discernment" (ibid.). Some leaders might even entertain, among the "apocalyptic temptations" of the twentieth century, the notion of trying "to wipe out Bolshevism with H-bombs" ("Peace: A Religious Responsibility," TMP, p. 114). Again, who was entertaining such an idea was not disclosed.

52. Merton, "Christian Ethics and Nuclear War," TMP, p. 87. On the relationship between morality and politics, Merton made an interesting comparison in a commentary on *Pacem in Terris*, where he wrote that "the great difference between Pope John and Machiavelli is not that the Pope believed in God and Machiavelli did not (as far as I know Machiavelli was, in his own way, a 'practicing Catholic') but rather that Pope John believed in *man* and Machiavelli did not." Merton seems here on the brink of an appropriation of the tradition of *tranquillitas ordinis*. But in the very next sentences, he takes up what were, on Paul Ramsey's and John Courtney Murray's analyses, the softest and least persuasive themes of the encyclical: "Because he had confidence in man, Pope John believed in love and peace. Because he

lacked this confidence, Machiavelli believed in force and deceit" ("The Christian in World Crisis," TMP, p. 60). For Murray and Ramsey, the appropriate move was from *man* to *politics*; for Merton, it often seemed to be from *man* to *psychology*. The implicit debasement of the political, and the tendency to psychologize political conflict (especially the superpower conflict), were Mertonesque themes that would have considerable influence on the subsequent Catholic debate.

53. Ibid.

54. Thomas Merton, "Christianity and Defense in the Nuclear Age," TMP, p. 90.

55. Ibid.

56. Ibid., p. 93.

57. Zahn, "Original Child Monk," p. xxix.

58. Ibid., p. xxviii.

59. Cited in ibid., p. xxvii.

60. Ibid., p. xxiv. In Merton's words, "The tragic thing about Vietnam is that, after all, the 'realism' of our program there is so unrealistic, so rooted in myth, so completely out of touch with the needs of the people whom we know only as statistics and to whom we never manage to listen, except where they fit in with our psychopathic delusions. Our external violence in Vietnam is rooted in an inner violence which simply ignores the human reality of those we claim to be helping. The result of this at home has been an ever-mounting desperation on the part of those who see the uselessness and inhumanity of the war, together with an increasing stubbornness and truculence of the part of those who insist they want to win, regardless of what victory may mean" ("Faith and Violence," TMP, p. 195).

61. Ibid., p. 194.

62. Thomas Merton, "Note for *Ave Maria*," TMP, pp. 232–33.

63. See Zahn, "Original Child Monk," p. xxvii.

64. From the press statement of the Catonsville Nine, released at their burning of draft records on May 17, 1968; cited in Francine du Plessix Gray, *Divine Disobedience: Profiles in Catholic Radicalism* (New York: Vintage Books, 1971), p. 47. Gray's chapter on the Berrigans originally appeared in *The New Yorker*.

65. Cited in ibid., p. 69.

66. Cited in ibid., pp. 88–89.

67. From Daniel Berrigan, *The Trial of the Catonsville Nine* (Boston: Beacon Press, 1970), pp. 117–18.

68. Ibid., p. 114.

69. Cited in ibid., p. 125.

70. Philip Berrigan, "An Open Letter to a Bishop," *Commonweal*, May 26, 1972, pp. 283–84.

71. The editors of *Commonweal*, "'It's not enough to be sympathetic:' An Interview with Daniel Berrigan," *Commonweal*, July 14, 1972, pp. 377, 382.

72. Reprinted in John Deedy, "News and Views," *Commonweal*, January 16, 1976, p. 34.

73. Philip Berrigan and others, "Letter to the Editor of *Commonweal*, July 21, 1978, p. 479.

74. Gray, op. cit., pp. 56–57.

75. Cited in ibid., p. 142.

76. Cited in ibid., p. 146. Gray also wrote: "All his ideas came straight from the Gospel. Philip Berrigan said he hadn't read theology in years. He only read Scripture, and everything he could find about politics, history, economics. He smiled, a sly look in his kind impatient eyes, and added: 'It's an explosive mixture'" (ibid., p. 147). It was indeed an explosive mixture. But its volatility came, not so much from a repristinated view of the Gospel imperative, but from the "politics, history, and economics" through which Berrigan read scripture. This will be a recurrent theme in American Catholicism from the mid-1960s on: the notion that political

imperatives were derived from a "pure" reading of "the Gospel," unmediated by theology. Of course, as a generation of Catholic biblical scholarship has taught, there is no unmediated appropriation of the biblical witness; one has to be clear, and critical, about the hermeneutical tools (explicit or tacit) that one brings to exegesis. In this respect, it may be of interest that the books brought to the imprisoned Catonsville Nine before their trial by Jesuit visitors from Loyola College in Baltimore included "the latest volumes of Che Guevara, Herbert Marcuse, Regis Debray: all books on revolution" (ibid., p. 50). Even were the books in question written by James Burnham, Whittaker Chambers, and William F. Buckley, Jr., the essential point would remain: the Berrigans' claim that their politics devolved from a pure reading of the Gospel cannot be taken as an adequate statement of the roots of their political judgments.

77. Cited in Gray, op. cit., p. 87.

78. Cited in ibid., p. 73.

79. Cited in Gray, op. cit., p. 120.

80. Cited in ibid., p. 55.

81. Daniel Berrigan, "'It's not enough to be sympathetic,'" p. 378.

82. Ibid.

83. Cited in their letter to *Commonweal*, October 10, 1980, p. 546.

84. Daniel Berrigan, "How to Make a Difference," p. 386.

85. For three studies of the Berrigans' impact on American Catholic activism, all written from points of view sympathetic to the Berrigan ideology, see Gray, op. cit.; Patricia McNeal, *The American Catholic Peace Movement 1928–1972* (New York: Arno Press, 1978); and Charles Meconis, *With Clumsy Grace: The American Catholic Left 1961–1975* (New York: Seabury, 1979).

86. See *The New Left: A Documentary History*, Massimo Teodori, ed. (Indianapolis: Bobbs-Merrill, 1969).

87. Revisionist historiography of the Cold War had a major influence on New Left ideology (and vice versa). The basic text of the revisionist school was William Appleman Williams, *The Tragedy of American Foreign Policy* (Cleveland: World Publishing Co., 1959). Williams followed this volume with several others, among which may be noted *The Contours of American History* (Cleveland: World Publishing Co., 1961), *The Roots of the Modern American Empire* (New York: Random House, 1969), and *Empire as a Way of Life* (New York: Oxford University Press, 1980). Williams's impact derived, not only from his own works, but from those of a host of young disciples, including Gabriel Kolko, Barton Bernstein, and Diane Shaver Clemens. The origins of the Cold War were of primary interest to this younger generation of revisionist scholars, whose research was used to support the New Left contention that Vietnam was an inevitable expression of the imperialist bias in U.S. foreign policy, a bias that Williams traced back to the Open Door Policy of the late 1890s. For critiques of revisionism, see Robert James Maddox, *The New Left and the Origins of the Cold War* (Princeton University Press, 1973) and Robert Tucker, *The Radical Left and American Foreign Policy* (Baltimore: Johns Hopkins University Press, 1971). For an analysis of the impact of revisionist historiography on the Vietnam-era and post-Vietnam American peace movement, see my "Intellectual Currents in the American Public Effort for Peace," in *The Nuclear Freeze Debate*, Paul M. Cole and William J. Taylor, Jr., eds. (Boulder: Westview Press, 1983), especially pp. 116–19. Part of the power of revisionist historiography and the neo-isolationism that derived from it was its ability to appeal to older American isolationist sentiments, even as it inverted the moral analysis of traditional isolationism by focusing on American, rather than European, malevolence.

88. I say "moral" and political tasks because the power of the New Left is understandable only if its ideology is seen as proceeding from a moral impulse. That this was the case from the beginning is demonstrated by the seminal "Port Huron Statement," which may be found in *The New Left: A Documentary History*, pp. 163–82. For an analysis and critique of New Left

themes by one who first broke with the cultural liberalism of the 1950s and early 1960s, and then broke with the New Left critique of that liberalism, see Norman Podhoretz, *Breaking Ranks* (New York: Harper & Row, 1979). In a later essay making similar arguments, Podhoretz noted that the old liberalism was attacked and largely destroyed, not from the Old or New Right, but from the New Left:

> Liberal self-assurance grew throughout the Eisenhower years, and finally reached heights of dizzying exaltation after the election of Kennedy. Nor did conservative criticism have much to do with the collapse of this self-confidence a few years later. It was an assault from the left rather than a challenge from the right that undermined the liberals. To be sure, the New Left would probably have failed in this campaign if not for the fact that the three wars launched by liberalism in the sixties—the one against Communism in Vietnam, and the two at home against racial discrimination and poverty—were all felt to have been lost, thus discrediting in the eyes of most liberals themselves their own ideas about foreign policy, about social policy, and about economic policy. Into this vacuum of demoralization the ideas not of the right but of the New Left came pouring in. The result was the conversion of liberalism from containment to neo-isolationism, from anti-discrimination to preferential treatment, and from economic growth to redistributionism" [Norman Podhoretz, in a symposium on the future of American liberalism, in *The American Spectator*, April 1985, p. 17].

These three "conversions" would have remarkably similar parallels in American Catholic official, intellectual, and activist circles, some years after they had become fixed in the political culture of the American liberal and left-liberal world.

89. From James Finn's interview with Daniel Berrigan in Finn, op. cit., p. 150.

90. See Gray, op. cit., p. 129.

91. Daniel Berrigan, "How to Make a Difference," p. 385.

92. See Gray, op. cit., p. 169.

93. Ibid., p. 56. These views were not exclusive to the Berrigans, of course. Gray wrote of a rally before the trial of the Catonsville Nine:

> Religious passion returned to the podium in a talk by Baptist minister Harvey Cox, the militant Harvard theologian who had started the fashion, the previous year, of accepting draft cards on collection plates at his war-protest services. Cox . . . compared the act of the Nine to Jeremiah destroying the clay pots on the steps of the Temple; to William Lloyd Garrison's public burnings of the Constitution in protest against slavery; to Martin Luther's burning of Canon Law ("the bad side of the good news") in front of the University of Wittenberg. "Catholic priests," Cox exclaimed to more thunderous applause, "have a special task of carrying out sacrificial acts which lead to redemption!" [ibid., p. 163].

Over the space of almost twenty years, this rhetoric sounds almost inconceivable, a caricature of revolutionary chic. So I may be permitted a brief personal reminiscence. I remember attending the rally at which Cox and Daniel Berrigan spoke, as a curious high-school senior. Two things stick in my mind as I think back on that fall evening in Baltimore. First, I remember the sense of righteous anger in St. Ignatius Hall, almost palpable in the tension it created in the room, and which at one point drove me out onto Calvert Street for a breath of fresh air. Secondly, and perhaps more important, was the equally palpable sense of community and fellow feeling in the room. That this sense of community seems, in retrospect, to have been built around negative themes—American evil in particular—is, in some respects, less important than the sense itself. One should not underestimate the impact that the Berrigans' ability to create what many profoundly religious people believed to be a community of caring had on the reception of their ideas. Given the inadequacies in their thought and analysis of events (Daniel Berrigan, for example, was still insisting in 1972 that American prisoners were well-treated in Hanoi, a judgment based on his own contacts with the North Vietnamese government; see "'It's not enough to be sympathetic,'" p. 379), the power of the "resistance community" precisely *as community* accounts for at least some of the Berrigans' impact.

94. Gray, op. cit., p. 117.

95. See Daniel Berrigan, *The Trial of the Catonsville Nine* (Boston: Beacon Press, 1970), pp. 36–37. See also the letter of Catonsville Nine member Brother David Darst, F.S.C., in *America*, November 16, 1968, pp. 456–58: "If lesser warnings have fallen upon deaf ears, one must not cease calling out; one must find ways of crying louder, in the hope that maybe, this time, those implementing the death policy will stop what they are doing. Thus if the *only* way to make a desperate cry is to destroy draft files, then obviously they will just have to be destroyed" (p. 458).

96. Daniel Berrigan, "Notes from the Underground," p. 265.

97. Ibid., p. 264.

98. Cited in Robert Pickus, "Support the Berrigans?" (unpublished manuscript of a radio commentary on KPFA radio, April 15, 1971). Father Blase Bonpane, a Maryknoll priest, made an even stronger statement at the St. Ignatius Hall rally before the trial of the Catonsville Nine. In Gray's telling, "'Nonviolence is an imperialist solution,' [Bonpane] clamored with a pointed look at Dorothy Day. 'Only guerilla warfare will alleviate the misery of the masses in the underdeveloped countries. . . . The peasants do not start violence . . . It is inflicted upon them and they have the Christian right to retaliate.' Even the SDS militants were taken aback at that. 'Is he really a priest?' one asked. Father Bonpane finished clamoring and stepped down, a syrupy smile on his face, looking, off the speaker's stand, like the servile sextons who sell holy water, phosphorescent rosaries, candied skulls in the basements of Mexican churches" (Gray, op. cit., pp. 164–65). Over a decade later, Bonpane, by this time (1981) out of Maryknoll, was still arguing that rage was a Christian prerogative. Haranguing a meeting of nuns in Los Angeles who "had just finished reciting 'Mary's Song of Liberation' ('It sounds a little bland to say "Magnificat," explained one of the nuns in the group')," Bonpane would say that "[Archbishop] Romero [of El Salvador] never used the concept of non-violence as hypocrisy. . . . You speak to power about non-violence. You speak to the president of the United States about non-violence. You speak to the Pentagon about non-violence. You do not speak to peasants about nonviolence. . . . Ask yourself, how many times in the New Testament is Jesus in a state of rage—many times—read it" (Gregory Parnell and Cecilia Hidalgo, "Rage as a Christian Motif," *National Catholic Register*, May 17, 1981, p. 1).

99. See Daniel Berrigan, "'It's not enough to be sympathetic,'" p. 378. Several years earlier, Berrigan told James Finn, "But I think realistically, especially after this tour of Latin America, that we are *not* going to have an end to certain kinds of limited warfare, at least in our lifetime. I don't see any realistic possibility of it. Nor do I clearly see the alternatives to it, and this is what makes me very tentative and qualified in my own pacifism. Because while I see clearly a Gospel ideal, I think we must also deal with our history and our times. And I talked down there to so many *good men* and revolutionary men, in the *best* sense, who themselves did not see a way out by means of nonviolence. This, of course, throws new light upon my own thinking, at the same time a new perplexity, you know" (Finn, op. cit., pp. 145–46).

100. Cited in Gray, op. cit., p. 157.

101. Percy wrote in full:

> I think the Berrigans are wrong.
>
> They have violated federal law, destroyed public property and terrorized government employees.
>
> These actions they justify as the moral expression of their convictions about U.S. foreign policy.
>
> It would follow, by the same logic, that a Catholic opposed to the use of public funds to promote population control could with equal propriety destroy the files of the Internal Revenue Service.
>
> No society could long endure if many people resorted to the same violent, not to say illegal, means of translating belief into action.

> But perhaps that is what the Berrigans want.
>
> You [the editors of *Commonweal*] and the Berrigans consider United States policy in Southeast Asia to be criminal. It is hardly necessary to point out that a great many people, perhaps as decent, as courageous, as equally distressed by the Vietnam war, do not agree with you and the Berrigans. Shall the issue be determined then by the more successful stratagem of violence?
>
> In these parts, the Ku Klux Klan burns churches and tries to scare people in various ways. Their reasons are, to them, the best: they do it for God and country and to save us from the Communists. I would be hard pressed to explain to a Klansman why he should be put in jail and the Berrigans set free.
>
> As it happens, I stand a good deal closer to the Berrigans than to the Klan. The point is however: God save us all from the moral zealot who places himself above the law and who is willing to burn my house down, and yours, providing he feels he is sufficiently right and I sufficiently wrong [*Commonweal*, September 4, 1970, p. 431].

102. Robert Pickus, art. cit., pp. 1–4.

103. James Douglass, *The Non-Violent Cross* (New York: Macmillan, 1968).

104. Ibid., pp. vii–viii.

105. James Douglass, *Resistance and Contemplation: The Way of Liberation* (New York: Delta Publishing Co., 1972), pp. 17–18.

106. Douglass, *The Non-Violent Cross*, p. xvii.

107. Douglass, *Resistance and Contemplation*, p. 33.

108. Archbishop Raymond G. Hunthausen, "Faith and Disarmament," address to the Pacific Northwest Synod of the Lutheran Church in America, Pacific Lutheran University, Tacoma, Washington, June 12, 1981; pamphlet reprint available from the archdiocese of Seattle.

109. Archbishop Raymond G. Hunthausen, Pastoral Letter to the Archdiocese of Seattle, January 28, 1982; pamphlet reprint available from the archdiocese of Seattle.

110. Archbishop Raymond G. Hunthausen, "Finding Our Way Back," address at the University of Notre Dame, January 29, 1982; photocopy reprint available from the archdiocese of Seattle. James Douglass's influence on the ideas expressed by Archbishop Hunthausen in these three statements was confirmed to me by the Very Rev. Michael G. Ryan, vicar general and chancellor of the archdiocese of Seattle.

An autobiographical word is in order here. For two years, I taught in the graduate school of St. Thomas Seminary in the archdiocese of Seattle, and for seven years was a regular columnist in the Archdiocesan newspaper, the *Progress*. In the latter, I did not hesitate to express, forcefully, my disagreement with Archbishop Hunthausen's formulation of the moral problem of war and peace. It is significant, I believe, that not once during those seven years did the archbishop ask me to modify my views as they were expressed in the newspaper of which he was the publisher. This was not a popular position in certain archdiocesan circles, but the archbishop stuck to it, and allowed me an unfettered, critical voice. For this, I have been, and remain, profoundly grateful to him. I can imagine few other dioceses in the United States where this situation would have obtained.

Themes for the Abandonment of the Heritage

1. Kenneth Waltz analyzes the historical evolution of contexts in *Man, the State, and War* (New York: Columbia University Press, 1959). Michael Howard also uses a "history of contexts" approach in his splendid book *War and the Liberal Conscience* (New Brunswick: Rutgers University Press, 1978).

2. See my essay "Intellectual Currents in the American Public Effort for Peace" in *The Nuclear Freeze Debate*, Paul M. Cole and William J. Taylor, Jr., eds. (Boulder: Westview

Press, 1983), pp. 107–48, for an example of how this matrix of questions sheds light on the intellectual history of the American peace movement over the past fifty years.

3. See Gabriel Almond, *The American People and Foreign Policy* (New York: Praeger, 1960).

4. "End of a Year, End of a Decade," *America*, December 27, 1969, p. 628.

5. The neo-Thomistic philosophical work of Jacques Maritain, and the transcendental Thomistic philosophical and theological studies of Karl Rahner, were influential expressions of this trend. For Maritain, see his *True Humanism* (London: Geoffrey Bles, 1938). Rahner's basic anthropological position may be found in his two large studies, *Hearers of the Word* (New York: Herder and Herder, 1969) and *Spirit in the World* (New York: Herder and Herder, 1968). Among the key essays in Rayner's evolving theological anthropology may be cited "Christology Within the Setting of Modern Man's Understanding of Himself and of His World," *Theological Investigations*, 11 (New York: Seabury, 1974), pp. 215–29; "Hominisation: The Evolutionary Origin of Man as a Theological Problem," *Quaestiones Disputatae*, 13 (New York: Herder and Herder, 1965); "Theology and Anthropology," *Theological Investigations*, 9 (London: Darton, Longman and Todd, 1972), pp. 28–45. A brief overview may be found in Rahner's entry, "Man," in the theological encyclopedia *Sacramentum Mundi*, 3 (Montreal: Palm, 1968), pp. 365–70.

6. Cited in Francine du Plessix Gray, *Divine Disobedience* (New York: Vintage Books, 1971), p. 117.

7. The *fons et origo* of liberation theology is Gustavo Gutiérrez's extraordinarily influential book, *A Theology of Liberation* (Maryknoll: Orbis Books, 1973).

8. These themes will be discussed in detail in chapter ten, below.

9. For Freire's view, see his *Pedagogy of the Oppressed* (New York: Herder & Herder, 1970). For Peter Berger's salient critique, see his *Pyramids of Sacrifice* (New York: Basic Books, 1975).

10. See Paul Johnson, *Pope John Paul II and the Catholic Restoration* (New York: St. Martin's Press, 1981), p. 65.

11. William V. Kennedy, "The Problem of Peacemaking," *America*, June 18, 1966, p. 855.

12. See ibid.

13. See ibid., p. 856.

14. Ibid.

15. Ibid.

16. "The Search for Peace with Justice," *America*, October 1, 1966, p. 376.

17. "The Churches and Peace," *America*, November 19, 1966, p. 641.

18. Dan and Rose Lucey, "A Nation Directed to Peace?," *America*, August 3, 1969, pp. 120–21.

19. Ibid., p. 121.

20. The citations above are from Most Rev. Carroll T. Dozier, "Peace: Gift and Task—Pastoral Letter to the People of the Diocese of Memphis," December 1971. Pamphlet reprint available from the diocese of Memphis.

21. For another view of early Christian pacifism, see Geoffrey Nuttall, *Christian Pacifism in History* (London: Basil Blackwell and Mott, 1958), pp. 1–14. Roland Bainton, himself sympathetic to the pacifist perspective, agrees that the question of idolatry bore heavily on primitive Christian pacifism. See *Christian Attitudes Toward War and Peace* (Nashville: Abingdon, 1960), pp. 73–74.

22. "Peace Depends on You, Too," *America*, December 22, 1973, p. 474.

23. Eugene C. Bianchi, "A National Peace Academy," *America*, December 2, 1978, p. 405.

24. William V. Shannon, "The Marines Have Landed," *Commonweal*, May 21, 1965,

p. 279. William V. O'Brien, in *U.S. Military Intervention: Law and Morality* (Beverly Hills: SAGE Publications, 1979), argues on classic just-war grounds that military intervention to interdict the imposition of a communist regime is in principle sustainable under the criterion of "just cause," but that such a threat was not sufficiently demonstrable in the Dominican Republic (pp. 39–61).

25. Ronald Steel, "Cold War Myths: The Futility of Attempts to Live in the Past," *Commonweal*, February 18, 1966.

26. Ibid., p. 577.

27. See ibid.

28. Ibid., pp. 576–78.

29. Andrew Christiansen, "Issues of War and Peace," *America*, March 21, 1970, p. 302.

30. Ibid.

31. Ibid.

32. Ibid.

33. Cited in ibid.

34. William Pfaff, "A Grand Strategy . . . for 1960," *Commonweal*, March 27, 1970, p. 55.

35. "The Archetypical American," *America*, February 3, 1973, p. 78.

36. Michael T. Klare, "Intervention Game-Plans for the '70s," *Commonweal*, March 14, 1975, p. 447.

37. "Regional Arms Races," *Commonweal*, April 11, 1975, p. 35.

38. "The Mayaguez Incident," *Commonweal*, June 6, 1975, pp. 163, 164.

39. Murray Polner, "Opening Pandora's Box," *America*, October 12, 1979, p. 553.

40. "The Carter Doctrine," *America*, February 9, 1980, p. 93.

41. "Carter, Kennedy, and Afghanistan," *Commonweal*, February 15, 1980, pp. 67–69. On this analysis, Senator Henry Jackson was a greater threat to world peace than those who had invaded Afghanistan. The "exhumed Munich analogy" also worried J. Patrick Dobel, who argued that the United States should let the U.S.S.R. get stuck to the "tar baby" of Third World entanglements. See J. Patrick Dobel, "Let the Russians Do It," *Commonweal*, June 20, 1980, pp. 368, 369.

42. See, for example, "The Myth of Neo-Isolationism," *Commonweal*, March 26, 1971.

43. As in the Dobel essay, n. 41, above.

44. William V. Shannon, "A Word to My Critics," *Commonweal*, May 7, 1965, p. 207.

45. Ibid.

46. Ibid.

47. Ibid.

48. Ibid.

49. "Israel's Surprise Attack," *America*, June 20, 1981, p. 496.

50. Ibid.

51. Ibid.

52. "The Grenada Invasion," *America*, November 12, 1983, p. 281. That *neither* the Maurice Bishop "faction" nor the Hudson Austin "faction" was an expression of the will of the Grenadian people was not noted by the editors of *America*. Their charge about the "implausibility" of restoring democracy to Grenada has, of course, been refuted by events.

53. Ibid. They have been made public; they are persuasive. But no subsequent *America* editorial has taken account of these very interesting records of the strategy of gaining totalitarian control over a population—and this despite the fact that the Church in Grenada was a primary target of the Bishop regime.

54. Ibid.

55. "Grenada's Gain, Our Loss," *Commonweal*, November 18, 1983, p. 611.

56. Ibid.

57. Ibid., p. 612.

58. J. Bryan Hehir, "The Case of Grenada," *Commonweal*, December 16, 1983, p. 681.

59. Ibid. Was General Jaruzelski a "key actor" in precisely the same sense as Prime Minister Eugenia Charles of Dominica?

60. Ibid., pp. 681, 698.

61. John Courtney Murray, *We Hold These Truths*, p. 19.

62. William V. O'Brien, "Balancing the Risks," *Worldview*, June 1963, p. 11.

63. See Howard G. and Harriet B. Kurtz, "War Safety Control," *America*, October 8, 1960, p. 43.

64. Ibid., p. 45.

65. L. C. McHugh, "The Troika Doctrine," *America*, July 7, 1961, p. 503. Thus Father Robert Graham praised U.S. Ambassador Adlai Stevenson's performance at the U.N., particularly his ability to meet the ideological thrust of the U.S.S.R. Stevenson, according to Graham, had shown how to "defend U.S. interests on the plane of international collective security." See Robert A. Graham, "America's Apostle of Peace," *America*, June 2, 1962, p. 344.

66. "Decade of Development and Disarmament," *America*, October 4, 1969, p. 255.

67. For Solzhenitsyn, see *Solzhenitsyn at Harvard*, Ronald Berman, ed. (Washington, D.C.: Ethics and Public Policy Center, 1980, pp. 3ff. *America*'s editorial of November 1, 1980, "The National Security Debate" (pp. 260–61), argued the priority of development issues over "any supposed lag in nuclear weapons."

68. Patricia A. Mische, "Weaving a Future in Guatemala," *America*, September 12, 1981, p. 116. Mische's work exemplifies the ways in which the psychologization of conflict can distort what appears, on the surface, to be an analysis within the classic Catholic framework. See also *Toward a Human World Order* (New York: Paulist Press, 1972).

69. See Suzanne C. Toton, "Peacemaking Put in Context," *America*, August 18, 1983, pp. 67–69.

70. See Clifford P. Hackett, "Endowing Democracy: An Idea That Came and Went," *Commonweal*, October 7, 1983, pp. 521–23. Hackett focused his criticism on the research process then underway to determine the modus operandi of a National Endowment for Democracy. The telling point, though, was that he did not address the relationship between democracy and peace, an acknowledgment that might have given his critique of the research a more useful frame of reference.

71. "International Terrorism," *America*, May 6, 1978, pp. 356–57.

72. Francis X. Winters, "Four Worlds—One Globe," *America*, March 13, 1976, p. 203.

73. "The Battle for Berlin: First Phase," *America* editorial, March 12, 1960, p. 699.

74. See Douglas Mattern, "The War Syndicate," *Commonweal*, December 20, 1974, pp. 263–66. Mattern affirmed the necessity of international organization for peace, but saw cutting the Pentagon budget and "curtailing the flow of U.S. arms to other countries" as the priorities for activists. Mattern's essay did not mention Soviet, Czech, or East German participation in the international arms trade, or the role of Soviet-armed proxies in Third World regional conflicts. *America* acknowledged the Soviet role in arms transfers in a December 16, 1978, editorial, "The Arms Plague" (pp. 448–49).

75. Brendon Caulfield-James, "Disarmament at the United Nations," *America*, August 28, 1982, p. 92.

76. See Robert A. Graham, "A Legacy of Conscience," *America*, October 24, 1970, pp. 312–15, and John A. Lucal, "The United Nations and the Holy See," *America*, October 24, 1970, pp. 315–17.

77. "World Law, World Court," *Commonweal*, March 27, 1970, pp. 51–52.

78. "World Organizations: 25 Years," *America*, June 13, 1970, p. 624.

79. "The UN at 25," *Commonweal*, June 26, 1970, p. 306. See also "The UN Celebrates its 25th Anniversary," *America*, September 26, 1970, p. 194. In an October 2, 1970, editorial,

Commonweal would write, "Today more than ever it is obvious that the only sane direction in international affairs is that toward a rule of law which is worldwide in scope and to which individual nations will have to submit for crucial decisions. If this goal is to be attained, the world organization that now exists must be made stronger, and the only way this can be done is by a willingness on the part of every nation to yield a certain amount of national sovereignty. If the United States is really serious about wanting to honor the United Nations on its 25th birthday, serious steps in this direction would be the best way" ("UN with Teeth," p. 4). The editorial did not discuss the ideological politics of the U.N., nor the need for democratization in Third World countries, in its call for a strengthened U.N. system. The failure of the U.N. was one of will, not politics. Similar weaknesses were evident a year later in the *America* editorial "That Elusive Internationalism" (October 2, 1971, p. 222).

80. See Homer A. Jack, "Terrorism: Another UN Failure," *America*, October 20, 1973, pp. 282–85, and "The UN and Israel," *Commonweal* editorial, February 20, 1974, pp. 251–52.

81. "The View from the UN," *America*, December 28, 1974, p. 419. The editorial did not discuss the assembly's selectivity of outrage in focusing on Rhodesia and the Republic of South Africa while keeping a discreet silence about the depredations of Idi Amin in Uganda.

82. See ibid.

83. "A Call for a New American Diplomacy," *America*, March 22, 1975, p. 203.

84. Thomas Powers, "The Zionism Resolution," *Commonweal*, December 19, 1975, p. 628.

85. "A New World Information Order," *America*, November 11, 1978, p. 325. The editorial would note "a large measure of hypocrisy in support of the draft [NWIO] declaration by the Soviet Union and some of the small dictatorships of the Third World. Racism is not the only injustice in the world today; dictatorships come in all colors and sizes. The UNESCO statement would do well to recognize this."

86. See John J. Logue, "The Law of the Sea at low tide," *Commonweal*, August 28, 1981, pp. 460–63.

87. Douglas Roche, "The World: Can Religion Make for its Peace?," *America*, October 12, 1974, p. 192.

88. Ibid., p. 193.

89. Ibid.

90. George G. Higgins, "Human Rights in the Catholic Tradition," *America*, July 28, 1979, p. 37.

91. Ibid.

92. Ibid.

93. Ibid., pp. 37–38.

94. Thomas P. Melady, "Selective Outrage," *America*, February 8, 1975, pp. 88–91. Melady, who had served as U.S. ambassador to Burundi and to Uganda, was especially angered at the U.N.'s failures to cite gross human rights violations in these two black African states.

95. "Détente and Human Rights," *America*, February 12, 1977, p. 119.

96. George G. Higgins, "Human Rights at Belgrade," *America*, March 11, 1978, p. 186.

97. James Brockman, "Human Rights in the Americas," *America*, June 7, 1980, p. 485.

98. See, for example, Benjamin L. Masse, "Pope John's *Mater et Magistra*," *America*, July 29, 1961, pp. 565–68; idem, "Pope John's Essay on Inequality," *America*, March 16, 1963, pp. 368–71; Victor C. Ferkiss, "The Struggle over Foreign Aid," *America*, April 13, 1963, pp. 488–90; and Benjamin L. Masse, "Why Foreign Aid?," *America*, May 4, 1963, pp. 637–39.

99. Denis Goulet, "A Missing Revolution," *America*, April 2, 1966, p. 438.

100. Ibid., p. 439.

101. Ibid.

102. See David French, "Does the US Exploit the Developing Nations?," *Commonweal*, May 19, 1967, pp. 257–59.

103. See "Eclipse of an Idea," *Commonweal*, May 14, 1971, pp. 227–28.

104. James R. Jennings, "*Quadragesimo Anno*—Forty Years After," *America*, September 25, 1971, p. 209.

105. George H. Dunne, "Development—A Christian Concern?", *America*, December 2, 1972, p. 466.

106. George H. Dunne, "Development—A Critique," *America*, December 9, 1972, pp. 490–93.

107. See Dan and Rose Lucey, "To Tanzania With Love," *America*, July 13, 1974, p. 14.

108. See Edward J. Cripps, "A New World Order?," *America*, October 11, 1975, pp. 200–204; John Madeley, "A Planetary Bargain?," *Commonweal*, September 26, 1980, pp. 516–17.

109. "The Churches' Concern for Peace," *America*, April 2, 1966, p. 434.

110. Ibid.

111. "Catholics and Peace," *Commonweal*, January 7, 1972, p. 316.

112. See "Investment in War," *Commonweal*, January 28, 1972, p. 388, and "The Church and Socially Responsible Investment," *America*, January 29, 1972, p. 82.

113. Gordon C. Zahn, "In Praise of Individual Witness," *America*, September 8, 1973, p. 145.

114. James F. Donnelly, "America: Metaphor and Method for Social Commitment," *America*, October 12, 1974, p. 189.

115. Ibid., p. 188.

116. Ibid., p. 189.

117. Network played an important role in the contextual debate through workshops, symposia, training sessions, and other fora for education and "consciousness-raising." Yet its fundamental criterion for success was that of the lobbyist: legislation passed, weapons-systems defeated, etc. On this issue of tactics, the Center of Concern opted for a different role in shaping the contextual debate; those in charge of the Center of Concern understood the importance of cadre-building in the politics of witness.

118. Cited in Paul J. Weber, "Bishops in Politics: The Big Plunge," *America*, March 20, 1976, p. 220.

119. Ibid., p. 221.

120. See ibid.

121. Ibid., p. 223.

122. John Coleman, "American Bicentennial, Catholic Crisis," *America*, June 26, 1976, p. 552.

123. Ibid., p. 553.

124. See ibid., p. 554.

125. Ibid., p. 556.

126. Ibid., p. 551.

127. John Coleman, "Toward a World More Human: Five Years of Concern," *America*, May 14, 1977, p. 444.

128. See, for example, Cronin's critique of the Birchers in "Anti-Communism and Freedom," *America*, April 22, 1961, pp. 172–74. Cronin concluded with this warning: "Communism is an external danger that demands from us the utmost in vigilance and sacrifice. But let us not be blind to the danger involved in policies of unbounded suspicion and the use of ruthless methods in so-called anti-Communist activities. If we become a nation of hate and distrust, then spiritually we are like the Communists. In fighting for the faith, we have lost charity. In defending our freedom, we have ceased to be free men" (p. 174).

129. Wilson Carey McWilliams, "Ending the Cold War," *Commonweal*, January 6, 1967, p. 363.

130. Georges Morel, "The Meaning of Karl Marx," *America*, October 28, 1967, p. 468.

131. William Pfaff, "The Revolution in Retrospect," *Commonweal*, October 20, 1967, p. 72.

132. William Pfaff, "Czechoslovakia Invaded," *Commonweal*, September 6, 1968, p. 581.

133. See Thomas Gannon, "Chile Turns to Marxism," *America*, October 24, 1970, p. 321.

134. "A.D. 1972: Balancing the Books," *America*, December 30, 1972, p. 558.

135. Frank Getlein, "Downstream from Watergate," *Commonweal*, November 29, 1974, pp. 212–13.

136. Tom Dorris, "Hard Choices for Soviet Christians," *America*, May 31, 1975, p. 423.

137. Leo Friedmann and Steven Philip Kramer, "The Myth of Soviet Aggression," *Commonweal*, June 4, 1976, pp. 359–60, 362.

138. Frank Getlein, "Holy Russia Rises in Vermont," *Commonweal*, July 7, 1978, p. 422.

139. John C. Cort, "Another Round," *Commonweal*, July 21, 1978, pp. 454–55.

140. "Talking to the Russians," *America*, June 17, 1978, p. 476.

141. See Thomas Powers, "A Question of Intentions," *Commonweal*, December 19, 1980, pp. 709–10.

142. See "Lessons from the Brezhnev Era," *America*, November 27, 1982, p. 323.

143. J. Bryan Hehir, "Kennan and the U.S.-Soviet Debate," *Commonweal*, January 28, 1983, p. 63. Kennan, at this point in his career as a public commentator, was arguing that the United States had little cause to worry about Soviet nuclear expansion while we tolerated pornography here at home. See "Selections from Interviews," in George F. Kennan, *The Nuclear Delusion: Soviet-American Relations in the Atomic Age* (New York: Pantheon Books, 1982), p. 74.

144. "Student Exchange for Peace," *America*, August 13, 1983, p. 62.

145. John MacDougall, "An American Among the Soviets," *America*, December 3, 1983, p. 352. MacDougall's essay is as pure an expression as one can find of the view that American antipathy toward the Soviet state is the result of misunderstanding and ignorance.

146. "The Lessons of Flight 007," *Commonweal*, September 23, 1983, p. 484.

147. John Garvey, "The Uses of Tragedy," *Commonweal*, October 17, 1983, p. 524.

148. Richard J. Krickus, "Persecution in Lithuania," *Commonweal*, July 16, 1976, p. 459.

149. Josiah G. Chatham, "Toast to America," *America*, May 23, 1964, p. 719.

150. Ronald Steel, "The American Empire," *Commonweal*, June 9, 1967, pp. 335–39. Steel's formulation here was rather mild. But he would also write that the American empire represented an "intoxication with power," and that it involved a "massive postwar interventionism." Steel's conclusion was that the age of empires, ours or others, was past: "We are in an age of nationalism, in which both communism and capitalism are ceasing to be ideologically significant, and in which the preoccupations of our diplomacy are often irrelevant. The last of the ideologues, we are clinging to political assumptions that have been buried by time and circumstances. . . . [The] political condition that has dominated our lives—the cold war—may now be over" (ibid.).

151. "A New Revolution," *Commonweal*, December 29, 1967, p. 397.

152. David Burrell, "The Violent American," *Commonweal*, September 6, 1968, pp. 586–87.

153. Jane Strouse, "A Taste of Fascism," *Commonweal*, September 20, 1968, p. 618.

154. Thomas M. Conrad, "Do-It-Yourself A-Bombs," *Commonweal*, July 25, 1969, p. 457.

155. Edward R. Cain, "The Stars and Stripes Forever," *Commonweal*, March 27, 1970, pp. 62, 61.

156. Richard H. Miller, "A Verbal Mugging," *Commonweal*, August 24, 1973, pp. 457–58.

157. Richard H. Miller, "Roots of the Cold War," *Commonweal*, January 21, 1972, p. 380.

158. See Philip C. Rule, "The Decline of Flower Power," *America*, February 13, 1971, p. 145.

159. Wes Barthelmes, "Cry, Our Beloved Country," *Commonweal*, April 30, 1971, p. 186.

160. See Raymond A. Schroth, "Back to Mylai," *Commonweal*, March 17, 1972, p. 34.

161. James Finn, "A Vision of America," *Commonweal*, December 1, 1972, p. 201.

162. Richard Du Boff and Edward Herman, "Corporate Dollars and Foreign Policy," *Commonweal*, April 21, 1972, p. 163.

163. Cited in John Deedy, "News and Views," *Commonweal*, January 12, 1973, p. 314.

164. Peter Steinfels, "Reflections on Soviet Dissent," *Commonweal*, October 12, 1973, p. 30.

165. William J. Byron, "'Seeking a Just Society': An Agenda for Americans," *America*, March 29, 1975, pp. 229, 233.

166. "Faith and Justice in the American Third Century," *America*, January 10, 1976, p. 3.

167. "Human Rights" and "Peace in 1976," *America*, January 10, 1976, p. 4.

168. "New Myths for the Third Century," *America*, January 22, 1977.

169. Frank Getlein, "The American Disease," *Commonweal*, September 24, 1976, p. 613.

170. Michael Novak, "Shall We Soon Praise Podhoretz?," *Commonweal*, May 3, 1974, p. 222.

Vietnam: Vehicle for the Abandonment of the Heritage

1. Sandy Vogelgesang, *The Long Dark Night of the Soul: The American Intellectual Left and the Vietnam War* (New York: Harper & Row, 1974).

2. See, for example, Benjamin L. Masse, "The Revolt in Vietnam," *America*, November 26, 1960, pp. 300–304; Frank N. Trager, "Dilemma in Laos," *America*, July 8, 1961, pp. 506–10; "Diem the Mandarin," editorial in *America*, August 3, 1963, pp. 111–12; "Tiger by the Tail," editorial in *America*, August 31, 1963, pp. 207–8; and Vincent S. Kearney and Francis J. Buckley, "Vietnam Dilemma," *America*, September 7, 1963, pp. 237–40.

3. William V. Kennedy, "The War in Vietnam," *America*, November 2, 1963, p. 504.

4. "The Saigon Coup," *America*, November 16, 1963, pp. 624–25.

5. William V. Kennedy, "The Lesson in Vietnam," *America*, March 28, 1964, p. 401.

6. "Dissent and Realism," *America*, May 8, 1965, p. 657.

7. "President Johnson's Diplomatic Blitz," *America*, January 15, 1966, p. 65.

8. "Just as Dangerous as the Left," *America*, January 22, 1966, p. 118.

9. Howard Zinn, "Negroes and Vietnam," *Commonweal*, February 18, 1966, p. 579.

10. John C. Bennett, "Christian Realism in Vietnam," *America*, April 30, 1966, pp. 616–17.

11. See Paul Ramsey, "Farewell to Christian Realism," *America*, April 30, 1966, pp. 618ff.

12. "The Heart of the Matter in Vietnam," *America*, February 26, 1966, p. 283.

13. Christopher Emmet, "Vietnam: Agonizing Reappraisal," *America*, March 12, 1966, p. 352.

14. "If the American People Lose Heart," *America*, July 9, 1966, p. 25.

15. "Getting Out," *Commonweal*, December 23, 1966, pp. 335–36.

16. See ". . . the U.S. Dilemma," *America*, December 24–31, 1966, p. 818.

17. Robert McAfee Brown, "An Open Letter to the U.S. Bishops," *Commonweal*, February 17, 1967, pp. 547–48.

18. James P. Shannon, "Catholic Bishops and Vietnam," *Commonweal*, March 17, 1967, pp. 671–72. *America*'s editors shared some of Bishop Shannon's resentment at Dr. Brown's accusation of indifference:

> Bishop Shannon has brought out into the open an issue that is in need of airing—the tendency of the dissenters to label as deficient in Christian moral conscience any who disagree with their views on the Vietnam war. There are Christians in this country who do not regard U.S. involvement in Vietnam as immoral and who, while they deplore war as much as the next Christian, are not quite prepared to condemn Administration policy out of hand. Meanwhile, an added question for Dr. Brown: Just what has rallying, or failing to rally, to the banner of "Clergy and Laymen Concerned" to do with ecumenism? ["The Bishops Taken to Task," *America*, March 11, 1967, p. 334].

19. Robert McAfee Brown, "Reply," *Commonweal*, March 17, 1967, p. 673. Brown's assumption that his disagreement with Shannon and other Catholic bishops had to do with "style" was itself a telling point.

20. Sidney Lens, "Double Standard in Vietnam," *Commonweal*, March 17, 1967, p. 669.

21. James O'Gara, "Treadmill to Disaster," *Commonweal*, September 22, 1967, pp. 569, 572.

22. See John Moriarty, "Policies of Delusion," *Commonweal*, September 22, 1967, p. 578.

23. Peter Steinfels, "The Case for Withdrawal," *Commonweal*, September 22, 1967, pp. 585, 587.

24. Philip Berrigan, "The United States and Revolution," *Worldview*, November 1967, pp. 10–11.

25. Vincent Kearney, "The Human Tragedy in Vietnam," *America*, October 7, 1967, p. 380.

26. William V. Shannon, "The Case for the War," *Commonweal*, December 8, 1967, pp. 326–27.

27. Robert McAfee Brown, "The Church and Vietnam," *Commonweal*, October 13, 1967, pp. 53–55.

28. "The War in Vietnam is Immoral," *US Catholic*, December 1967.

29. "Let's Both Call a Halt—Now," *America*, January 20, 1968, p. 68.

30. See "Fanaticism on the Rampage," *America*, February 17, 1968, p. 212. For a systematic refutation of the American media portrait of Tet as a massive defeat, see Peter Braestrup, *Big Story* (Boulder: Westview Press, 1977). Braestrup argues that the American media were simply incompetent to handle a story as complicated and bizarre as the war in Vietnam. He doubts whether ideological filters were at work in the reporting on Tet. Yet the reception of the American press presentation of Tet in the "resistance movement" and in other American Catholic circles critical of the war was surely colored by contextual pre-judgments about the nature and meaning of the war.

31. See "Reflections on the Right to Dissent," *America*, March 9, 1968, p. 311.

32. "The High Price of the Paris Talks," *America*, June 1, 1968, p. 725.

33. N. Khat Huyen, "Vietnam: A Proposed Solution," *America*, May 11, 1968, pp. 633–35.

34. "Chances for Peace," *Commonweal*, January 12, 1968, pp. 427–28.

35. See *America*'s editors' interview with Daniel Berrigan in *America*, March 9, 1968, and Berrigan's own articles, "Mission to Hanoi" and "Mission to Hanoi, Part II," *Worldview*, April and May 1968.

36. "Vietnam: Gordian Knot," *Commonweal*, March 21, 1969, pp. 3–4.

37. Stanley Hoffmann, "A great power cannot treat every tremor as a major threat," in "After Vietnam, What?," *Commonweal*, March 21, 1969, pp. 13–14. Hoffmann's themes would shape the American bishops' commentary on Central America a decade later. See chapter 10, below.

38. "Cold-Blooded Aggression," *Commonweal*, May 15, 1970, pp. 211–13.

39. "The United States Moves Into Cambodia," *America*, May 16, 1970, p. 517.

40. "Kent and Cambodia," *Commonweal*, May 22, 1970, pp. 235–36.

41. "John Wayne in Vietnam," *Commonweal*, December 11, 1970, p. 268.

42. See Gordon Zahn, "The Church as Accomplice: Reflections on My Lai," *Worldview*, March 1971, p. 5.

43. Gordon Zahn, "The Scandal of Silence," *Commonweal*, October 23, 1971, p. 84.

44. Cited in John Deedy, "News and Views," *Commonweal*, March 12, 1971.

45. "Easter and the American Conscience," *America*, April 10, 1971, p. 364.

46. Todd Gitlin, "Ellsberg and the New Heroism," *Commonweal*, September 3, 1971, pp. 447–49, 451.

47. Marvin Bordelon, "The Bishops and Just War," *America*, January 8, 1972, pp. 18–19.

48. See "Harrisburg: Politics, Crime, and Morality," *America*, April 22, 1972, p. 416.

49. "A Religious Test?," *Commonweal*, September 29, 1972, p. 517.

50. "'Peace! peace! and there is no peace'," *America*, January 13, 1973, p. 9.

51. "The Challenge to Congress," *Commonweal*, January 26, 1973, p. 363.

52. "After the Cease-Fire," *Commonweal*, February 9, 1973, p. 411. By "outside influences," *Commonweal* understood the United States, preceded by the French. The editorial did not mention the Soviet Union as an outside influence, nor North Vietnam as an outside influence in the south.

53. "Amnesty Now," *Commonweal*, March 16, 1973, pp. 27–28.

54. Cited in John Deedy, "News and Views," *Commonweal*, April 27, 1973, p. 172.

55. Bernard Sklar, "America and the POWs," *Commonweal*, June 1, 1973, pp. 303–6.

56. "Honor in Vietnam," *Commonweal*, March 15, 1974, pp. 27–28.

57. Gabriel Kolko, "The Other South Vietnam," *Commonweal*, March 15, 1974, pp. 38, 35.

58. Edward S. Herman, "On 'Helping' South Vietnam," *Commonweal*, February 15, 1975, p. 394.

59. "Involvement's Last Hour," *Commonweal*, April 25, 1975, p. 68.

60. Susan Abrams, "The Vietnam Babylift," *Commonweal*, September 24, 1976, p. 617.

61. Tissa Balasuriya, "Theological Reflections on Vietnam," *Commonweal*, September 26, 1975, p. 428.

62. See Sesto Quercetti, "The Vietnamese Church: Conflict and Courage," *America*, September 18, 1976, pp. 138–43; idem, "Vietnam": Four Years Later," *America*, May 26, 1979, pp. 435–37.

63. Robert Drinan, "Asia's Refugees, the World's Conscience," *America*, September 15, 1979, p. 110.

64. See Bernard Gwertzman, "Antiwar Activists Appeal to Hanoi," *New York Times*, December 21, 1976. The text of the appeal may be found in James Finn, "Fighting Among the Doves," *Worldview*, April 1977, p. 5.

65. See James Finn, "Fighting Among the Doves," pp. 4–9. See also James Forest, "The parable of the unresponsive witness," *Commonweal*, December 22, 1978, pp. 814–16. Forest deserves great credit for initiating the "Appeal to the Government of Vietnam." Yet in his later reflection on the death of a Vietnamese Buddhist monk, who had met his fate without any serious protest from American activists, Forest could not help finding a parallel between Thich Thien Minh and Kitty Genovese. The urge to drive home American evil thus remained irresistible, even in honorable quarters.

66. Joseph O'Hare, "Of Many Things," *America*, March 14, 1981, p. 189.

67. See David Hoekema, "A Wall for Remembering," *Commonweal*, July 15, 1983, p. 399.

68. Michael Novak, "A Switch to Reagan: For a Strong America," *Commonweal*, October 24, 1980, p. 590.

69. See, for example, William V. O'Brien, "Wars of National Liberation IV," *Worldview*, May 1966, pp. 4–8; Quentin L. Quade, "Religion, Moral Authority, and Intervention," *Worldview*, December 1966, pp. 10–13; James V. Schall, "Wars Will Cease When . . . ," *Worldview*, May 1967, pp. 9–11.

70. See, for example, Gordon Zahn, "Wars Will Occur As Long As . . . ," *Worldview*, July–August 1967, pp. 8–12; idem, "The Church as Accomplice: Reflections on My Lai," *Worldview*, March 1971, pp. 5–8.

71. See Mary Temple, letter to the editors of *Commonweal*, June 27, 1969, pp. 413–14.

72. See ibid., p. 415.

73. See Thomas F. Ritt, "The Bishops and Negotiation Now," in *American Catholics and Vietnam*, Thomas E. Quigley, ed. (Grand Rapids: Eerdmans, 1968), p. 112. Bishop Wright joined Negotiation Now when he was assured that the organization "recognized and condemned Viet Cong terrorism and systematic political assassination of village chiefs who did not support them" (Mary Temple, art. cit., p. 414.)

74. See Pickus's critique of Vietnam-era activism in his "Violence and the Peace Movement: The Relevance of Camus," *Worldview*, December 1968, pp. 11–13.

75. Thomas Ritt, art. cit., p. 118.

76. Thomas Ritt, letter to the editors of *Commonweal*, June 27, 1969, p. 415.

77. See Mary Temple, art. cit., p. 414.

78. Ibid.

79. Stanley Karnow, *Vietnam: A History* (New York: Viking Press, 1983), p. 331. Norman Podhoretz discusses this in *Why We Were In Vietnam* (New York: Simon & Schuster, 1982):

> Those who said that Hanoi was not a Chinese proxy also tended to think that the Vietcong was not a North Vietnamese proxy, and about this *they* were wrong. To be sure, the National Liberation Front, true to its character as a front, included elements that were not Communist, as well as Communists and sympathizers who thought they were fighting for "democracy, freedom, and peace," and who believed that the fall of Saigon would lead to "a domestic policy of national reconciliation, without risk or reprisal, and a foreign policy of non-alignment." But Doan Van Toai—whose words I have just quoted, who was arrested and jailed many times for leading student demonstrations against the Thieu regime and against American involvement, and who never joined the Vietcong only because the NLF felt he could be more useful as an intelligence agent—was deceived about this, as were what he calls "the most prestigious intellectuals in the West." Shortly after the North Vietnamese army conquered the South, the Provisional Revolutionary Government (PRG, into which the NLF had been transmuted) was disbanded; and far from being appointed to positions of power in the new Vietnam, many of its members were arrested. "Today," wrote Doan Van Toai in 1981, "among 17 members of the Politburo and 134 members of the Vietnamese Communist party, not a single one is from the NLF" (though there were a few "who had been North Vietnamese Communist Party representatives with the NLF"). Although not all members of the NLF knew it then, they had indeed been acting as proxies for Hanoi, and those who persisted in thinking otherwise were eliminated in the end as unreliable [pp. 174–75].

80. Cited in Podhoretz, op. cit., pp. 198–99. See also Truong Nhu Tang, *A Vietcong Memoir* (New York: Harcourt, Brace, Jovanovich, 1985).

81. For commentary on Tet, see Braestrup, op. cit.; Karnow, op. cit.; and Guenter Lewy, *America in Vietnam* (New York: Oxford University Press, 1978).

82. On the Christmas bombing of Hanoi, see Podhoretz, op. cit.; Martin F. Herz, *The*

Prestige Press and the Christmas Bombing, 1972 (Washington, D.C.: Ethics and Public Policy Center, 1980); and Lewy, op. cit. The death toll in Hanoi, according to the *Economist* of London, was "smaller than the number of civilians killed by the North Vietnamese in their artillery bombardment of An Loc in April or the toll of refugees ambushed when trying to escape from Quang Tri in the beginning of May. That is what makes the denunciation of Mr. Nixon as another Hitler so unreal" (cited in Podheretz, op. cit., p. 122).

83. See "Human Life in Our Day," pastoral letter of the American hierarchy, November 15, 1968, in Hugh J. Nolan, ed., *Pastoral Letters of the American Hierarchy, 1792–1970* (Huntington: Our Sunday Visitor, 1971), esp. ##135ff., and "Resolution on Southeast Asia," issued by the National Conference of Catholic Bishops, November 1971, in *Pastoral Letters of the United States Catholic Bishops, Volume III: 1962–1974*, Hugh J. Nolan, ed. (Washington, D.C.: United States Catholic Conference, 1983), pp. 289–91.

84. Norman Podhoretz. op. cit., p. 189.

85. See "Resolution on Imperatives of Peace," issued by the National Conference of Catholic Bishops, November 16, 1972, in *Pastoral Letters of the United States Catholic Bishops, Volume III: 1962–1974*, Hugh J. Nolan, ed. (Washington, D.C.: United States Catholic Conference, 1983), #4.

The Moral Horizon of "An Entirely New Attitude"

1. See H. Richard Niebuhr, *Christ and Culture*, chapter 6.

2. Among Yoder's most influential books are *The Politics of Jesus* (Grand Rapids: Eerdmans, 1972), *Nevertheless* (Scottdale: Herald Press, 1972), and *The Original Revolution* (Scottdale: Herald Press, 1972). Yoder argues in these and other works that Troeltsch's classic church/sect distinction is an inadequate model for ecclesiology. I would argue that, although the Troeltschian distinction should not be absolutized—it is, after all, a model for analysis, not a snapshot of reality—it still sheds considerable light on trajectories of thought in ecclesiology.

3. Richard John Neuhaus, Foreword to *The Pope and Revolution*, Quentin L. Quade, ed. (Washington, D.C.: Ethics and Public Policy Center, 1982), pp. ix–x.

4. The basic text of religious secularization theorists in America was, of course, Harvey Cox's *The Secular City: Urbanization and Secularization in Theological Perspective* (New York: Macmillan, 1965). For a refutation of Cox that blends theoretical analysis with empirical evidence, see Andrew M. Greeley, *Unsecular Man* (New York: Schocken, 1972). Theodore Caplow's study, *All Faithful People* (University of Minnesota, 1983), shows that, contrary to Robert and Helen Lynd's expectations in the original Middletown survey, the residents of Muncie had become *more*, rather than less, religious between the 1920s and the 1980s. Peter Berger said, with admirable and typical succinctness, that the Caplow study "has put the final nail in the coffin of the theory that modernity means secularity" (quoted in Richard John Neuhaus, "What Do the Fundamentalists Want?," *Commentary*, May 1985, p. 43).

5. Neuhaus, Foreword, p. x.

6. For several representative examples of this new "sectarian sensibility" in the American Catholic debate, see Thomas Merton, "Is the World a Problem?," *Commonweal*, June 3, 1966, pp. 305–9; Paul Velde, "Guerilla Christianity," *Commonweal*, December 13, 1968, pp. 371–73; Joseph Bishop, "Liturgy as a Subversive Activity," *Commonweal*, December 25, 1970, pp. 324–27; Dorothee Sölle, "The Gospel and Liberation," *Commonweal*, December 22, 1972, pp. 270–75; Alfred T. Hennelly, "'Church and World' and Theological Developments," *America*, February 28, 1976, pp. 153–56; and John Garvey, "Christianity and Power: Was Jesus a Realist?," *Commonweal*, September 9, 1983, pp. 455–56.

Counterarguments, defending American Catholicism's classic transformationalist approach to the church/world dialectic, may be found in Edward A. Marciniak, "Catholic Social Action: Where Do We Go From Here?," *America*, December 12, 1970, pp. 511–16; Andrew M. Greeley, "Catholic social activism—real or rad/chic?," *National Catholic Reporter*, February 7, 1975, pp. 7–11; Avery Dulles, "Finding God and the Hartford Appeal," *America*, May 3, 1975, pp. 334–37; "Assessment and Anticipation," *Commonweal* editorial, December 23, 1977; "A Chicago Declaration of Christian Concern," *Commonweal*, February 17, 1978, pp. 108–10; and Patrick Glynn, "Pulpit Politics," *The New Republic*, March 14, 1983, pp. 11–14.

7. John Courtney Murray, *We Hold These Truths*, p. 263.

8. For useful summaries of contemporary Catholic exegetical methodology, see Raymond E. Brown, "Hermeneutics" in *The Jerome Biblical Commentary*, Raymond E. Brown et al., eds. (Englewood Cliffs: Prentice-Hall, 1968), pp. 605–23; John S. Kselman, "Modern New Testament Criticism," *The Jerome Biblical Commentary*, pp. 7–20; and Alexa Suelzer, "Modern Old Testament Criticism," *The Jerome Biblical Commentary*, pp. 590–604.

9. Bishop Carroll T. Dozier, "Peace: Gift and Task"; Archbishop Raymond G. Hunthausen, "Faith and Disarmament."

10. Joseph Gallagher, "The American Bishops on Modern War," *America*, November 5, 1966, p. 549.

11. "Karl Rahner in New York," an interview by Eugene C. Bianchi, *America*, June 12, 1965, pp. 860–61.

12. Cited in Peter Hebblethwaite, "Relaxing with Karl Rahner," *Commonweal*, November 1, 1974, p. 112.

13. Wilson D. Miscamble, "American Catholics and Foreign Policy," *America*, December 8, 1979, p. 370.

14. Robert McAfee Brown, "The Church in Vietnam," *Commonweal*, October 13, 1967, p. 55.

15. Hunthausen, "Faith and Disarmament."

16. See Eileen Egan, "The Beatitudes, the Works of Mercy, and Pacifism," in *War or Peace? The Search for New Answers*, Thomas A. Shannon, ed. (Maryknoll: Orbis, 1980), pp. 169–87.

17. See, for example, King's splendid and influential "Letter from Birmingham Jail," in Martin Luther King, Jr., *Why We Can't Wait* (New York: Harper & Row, 1963), pp. 77–100.

18. See Gene Sharp, *Gandhi as a Political Strategist* (Boston: Porter Sargent, 1979). For the most rigorous scholarly historical discussion of the theory and practice of nonviolence, see Sharp's three-volume study, *The Politics of Nonviolent Action* (Boston: Porter Sargent, 1973), and his collection of essays, *Social Power and Political Freedom* (Boston: Porter Sargent, 1980).

19. See Robert Pickus, "Support the Berrigans?"

20. Cf., for example, Archbishop Raymond G. Hunthausen, "Faith and Disarmament," and "Finding Our Way Back."

21. Cited in Charles Frankel, "Morality and US Foreign Policy," p. 44.

22. A synopsis of *Speak Truth to Power* was published in a special October 1955 issue of *The Progressive*, which also included commentary on the AFSC proposal by Dwight Macdonald, Norman Thomas, Reinhold Niebuhr, Karl Menninger, and George Kennan, and a "Reply to the Critics" by the two men most responsible for the original study, Robert Pickus and Stephen Cary.

23. I have searched in vain, through both the periodical literature and more extended essays of postconciliar American Catholic pacifists, for a single reference to *Speak Truth to Power*.

24. See James F. Childress, "Just War Theories: The Bases, Interrelations, Priorities, and Functions of Their Criteria," *Theological Studies*, 39 (September 1978) 427–45; Ralph B. Potter, *War and Moral Discourse* (Richmond: John Knox Press, 1969).

25. See James Turner Johnson, *Ideology, Reason, and the Limitation of War: Religious*

and Secular Concepts, 1200–1740 (Princeton University Press, 1975); idem, *Just War Tradition and the Restraint of War* (Princeton University Press, 1981); idem, *Can Modern War Be Just?* (New Haven: Yale University Press, 1984); LeRoy B. Walters, Jr., "Five Classic Just-War Theories: A Study in the Thought of Thomas Aquinas, Vitoria, Suárez, Gentili, and Grotius," Ph.D. dissertation, Yale University, 1971; and F. H. Russell, *The Just War in the Middle Ages* (London: Cambridge University Press, 1977).

26. Bishop Walter Sullivan, quoted in the *National Catholic Reporter*, December 11, 1981.

27. See William V. O'Brien, *The Conduct of Just and Limited Wars* (New York: Praeger, 1981).

28. See John Courtney Murray, "Selective Conscientious Objection," commencement address at Western Maryland College, June 4, 1967, reprinted in pamphlet form by Our Sunday Visitor, Huntington, Indiana.

29. See United States Catholic Conference, "Declaration on Conscientious Objection and Selective Conscientious Objection," October 21, 1971, in *Pastoral Letters of the United States Catholic Bishops, Volume III: 1962–1974*, Hugh J. Nolan, ed. (Washington, D.C.: United States Catholic Conference, 1983).

30. See James Finn, "The Amnesty Issue," *Commonweal*, November 3, 1972, p. 107.

31. Ibid., p. 108.

32. Ibid., p. 107.

33. Ibid., pp. 107–8. Points similar to Finn's were also raised by Walter Jeffko in "An Ethics of Amnesty," *America*, May 10, 1975, pp. 359–61: "When I disobey a law on moral grounds, I have a simultaneous obligation to act in such a way that I strengthen—or at least maintain—the law in general, not weaken it. For, since law as such is a necessary means to justice in the indirect relation of persons, we must always act in such a way that shows regard for the law, even when, indeed, especially when, we violate from law" (p. 360).

34. *Gaudium et Spes*, #80.

35. See Hehir's discussion in "The Just War Ethic and Catholic Theology: Dynamics of Change and Continuity," in Shannon, op. cit., pp. 26–29. The journal *Continuum* was an important vehicle for the development of nuclear pacifist thought. Its editor, Justus George Lawler, gathered much of his *Continuum* material on the subject in his *Nuclear War: The Ethic, the Rhetoric, the Reality—A Catholic Assessment* (1965). See also J. Bryan Hehir's view in Hehir and Robert A. Gessert, *The New Nuclear Debate* (New York: Council on Religion and International Affairs, 1976).

36. Paul Ramsey, "*Pacem in terris*," in *The Just War: Force and Political Responsibility* (New York: Charles Scribner's Sons, 1968), p. 85.

37. Pope Paul VI, "1972 Annual Message for World Day of Peace," in *Origins*, 1:29 (January 6, 1972) 491.

38. National Conference of Catholic Bishops, "To Teach as Jesus Did," pastoral letter of November 1972, in *Pastoral Letters of the United States Catholic Bishops, Volume III, 1962–1974*, Hugh J. Nolan, ed. (Washington, D.C.: United States Catholic Conference, 1983), #26.

39. Cited in Michael True, "Persisters for Peace," *Commonweal*, April 26, 1974, p. 180.

40. "Justice in the World," statement of the 1971 International Synod of Bishops (Boston: St. Paul Editions), p. 4.

41. See Charles M. Murphy, "Action for Justice as Constitutive of the Preaching of the Gospel: What Did the 1971 Synod Mean?," *Theological Studies*, 44:2 (June 1983) 298–311.

42. Pope Paul VI, "1972 Annual Message for World Day of Peace," in *Origins*, 1:29 (January 6, 1972) 490.

43. Ibid.

44. Pope Paul VI, "Reconciliation: The Way to Peace," 1975 Annual Message for World Day of Peace, in *Origins*, 4:27 (December 26, 1974) 430–32.

45. Pope Paul VI, "Contrary and Positive Weapons of Peace," 1976 Annual Message on World Day of Peace, in *Origins*, 5:28 (January 1, 1976) 450–52; Pope Paul VI, "If You Want Peace, Defend Life," 1977 Annual Message for World Day of Peace, in *The Pope Speaks*, 22:1 (Spring 1977) 38–45.

46. See, for example, the argument of Edward Luttwak in *Strategic Power: Military Capabilities and Political Utility* (Washington, D.C.: Center for Strategic and International Studies, 1976).

"The Challenge of Peace": American Catholicism and the New Nuclear Debate

1. This interpretation of "The Challenge of Peace" (hereafter "TCOP") pervades Jim Castelli's history of the document, *The Bishops and the Bomb: Waging Peace in a Nuclear Age* (Garden City: Doubleday Image Books, 1983) and, with some exceptions, the semiofficial anthology of commentary on TCOP, *Catholics and Nuclear War*, Philip Murnion, ed. (New York: Crossroad, 1983).

2. Three examples of this interpretation, written in quite differing degrees of dispassion, may be found in William V. O'Brien, "A Just-War Deterrence/Defense Strategy," *Center Journal*, 3:1 (Winter, 1983) 9–29; John Tagg, "Breakfast with the Bishops: An Anniversary," *National Review*, June 15, 1984, pp. 26–35; and Tom Bethell, "The Bishops' Brain," *American Spectator*, July 1983. Analyses of the pastoral that fit neither hermeneutic category comfortably include Stanley Hauerwas, "On Surviving Justly: An Ethical Analysis of Nuclear Disarmament," *Center Journal*, 3:1 (Winter 1983) 123–52; my own "Beyond 'The Challenge of Peace:' *Quaestiones Disputatae*," *Center Journal*, 3:1 (Winter 1983) 101–22; and "In Defense of Deterrence," *The New Republic*, December 20, 1982. An interesting ecumenical approach to the pastoral may be found in Dean C. Curry, ed., *Evangelicals and the Bishops' Pastoral* (Grand Rapids: Eerdmans, 1984); see in this last the Foreword by Archbishop John J. O'Connor.

3. See "Human Life in Our Day," pastoral letter of the National Conference of Catholic Bishops, in *Pastoral Letters of the American Hierarchy 1792–1970*, Hugh J. Nolan, ed. (Huntington: Our Sunday Visitor, 1971), #103.

4. Ibid., #106.

5. See ibid., #107.

6. Ibid.; the bishops were citing *Gaudium et Spes*, #82.

7. See ibid., #110.

8. Ibid., #111.

9. See ibid., #113.

10. "To Live in Christ Jesus: A Pastoral Reflection on the Moral Life" (Washington, D.C.: United States Catholic Conference, 1976), p. 34. Hehir's comment, and the relevant passage from the bishops' text, may be found in his essay "The Just War Ethic and Catholic Theology: Dynamics of Change and Continuity," in Shannon, *War or Peace? The Search for New Answers*, pp. 28–29.

11. "Human Life in Our Day," #93; the bishops were citing *Gaudium et Spes* #77.

12. Cardinal John J. Krol, "SALT II: A Statement of Support," testimony before the Senate Foreign Relations Committee on September 6, 1979, reprinted in *Origins*, 9:14 (September 20, 1979), p. 195.

13. Ibid., p. 196.

14. Ibid.

15. Ibid.

16. Ibid.

17. Ibid.

18. Ibid., p. 197.

19. Ibid.

20. Ibid.

21. "The cost of using nuclear weapons on the part of those who employ them has deprived them of much of their strategic utility and has made their political usefulness equally problematic" (ibid.).

22. Ibid., p. 198.

23. Cardinal John J. Krol, "The Churches and Nuclear War," *Origins*, 9:15 (September 27, 1979), p. 236.

24. See Daniel Patrick Moynihan, "Cold Dawn, High Noon," in *Counting Our Blessings* (Boston: Little, Brown, 1980), pp. 277–336.

25. The Gumbleton/Hehir exchange before the USCC administrative board is replayed in Gumbleton's article, "Chaplains blessing the bombers," *Commonweal*, March 2, 1979, pp. 105–7, and Hehir's rejoinder, "Limited but substantial achievements," *Commonweal*, March 2, 1979, pp. 108–10. *America* criticized the Pax Christi/Gumbleton approach to SALT II in an editorial of February 24, 1979.

26. Joseph J. Fahey, "SALT II: The Arms Race to End All Arms Races," *America*, February 24, 1979, p. 127.

27. Ibid., p. 129.

28. Ibid., p. 130.

29. Ibid.

30. Ibid.

31. David Dillon, "Vision of Shalom," *Commonweal*, June 22, 1979, p. 358.

32. Ibid., pp. 358–59.

33. Ibid., p. 359.

34. Ibid.

35. Ibid.

36. Ibid.

37. Thomas Powers's essays in *Commonweal* were emblematic of the attentive public Catholic debate on nuclear weapons; these essays were subsequently gathered in a collection entitled *Thinking About the Next War* (New York: Alfred A. Knopf, 1982). See also "A Presidential Dove?," *Commonweal*, November 5, 1976, pp. 707–8 (for the theme of the "madness of the arms race"); Douglas Mattern, "With Humanity as Pawns," *Commonweal*, May 27, 1977, pp. 329–31 (an essay stressing the role of the "military-industrial complex," and the need for a populist revolt against the course of current policy); Frank Getlein, "Down in Flames," *Commonweal*, August 5, 1977 (an essay celebrating the cancellation of the B-1 bomber, but failing to note the effect of cruise missiles on arms control); John Deedy, "News and Views," *Commonweal*, August 5, 1977 (for a typical description of the neutron bomb as an "Orwellian" weapon); Frank Getlein, "Kill Him, Save It," *Commonweal*, August 19, 1977 (the neutron bomb, again); David Riesman, "Human Rights: Conflicts Among Our Ideals," *Commonweal*, November 11, 1977 (a crucial essay on how the need for nuclear accommodation with the U.S.S.R. should take precedence over human rights concerns); "Disarmament and Survival," *Commonweal*, January 20, 1978 (an editorial labelling the MX as a "first strike counterforce weapon which . . . could jolt the Soviets into first-strike postures of their own," with no mention of the Soviet SS-18); "A World Waging Peace," *America*, May 27, 1978 (an editorial on the First U.N. Special Session on Disarmament, stressing the need for a "return to sanity—a restoration of human values as the guide to our priorities"); "Give SALT a Chance," *Commonweal*, November 24, 1978 (an editorial critical of "linkage" between Soviet human rights violations and arms control); "Curbing the Traffic in Death," *Commonweal*, January 19, 1979 (an editorial supporting the "commendably dovish Mr. [Paul] Warnke"); "SALT II: Rules for Debate," *America*, May 26, 1979 (an editorial lamenting "those popular

chauvinistic sentiments that are still preoccupied with the myth of the United States as the most powerful nation in the world," and decrying the notion of "linkage," which would "distract dangerously from the overriding importance of the danger to the world posed by nuclear weapons"); "MX Missiles and SALT," *America*, June 23, 1979 (an editorial claiming that the "highly accurate and fast MX missile means that the United States wants a first-strike capability"); "SALT II: A Step Toward Sanity," *America*, June 30, 1979 (an editorial describing the evolution of the deterrence system as "this exercise of lunacy"); Daniel Berrigan, "The Nightmare of God: A Diary of Sorts," *America*, June 30, 1979 (an essay noting that "at the Pentagon, and radiating outward across the land, is a vast pulsing network of special interest, megacupidity, cost overruns, padding, buccaneering, dangerous and futile labor"); Francis X. Meehan and William Mattia, "The Arms Race and the American Parish," *America*, September 22, 1979 (an essay arguing, *inter alia*, that only by resisting "the arms race" can the church "resist being absorbed into the secular culture of American militarism. . . . Resisting armaments and all that they signify today can be a way of purifying our very Eucharists so that they may truly become a worship in spirit and in truth"); Frank Getlein, "Name of the Game," *Commonweal*, October 26, 1979 (an essay describing the "racetrack" basing mode for the MX missile as "boyhood's dream plated with adult armor against the real world"); and Patricia Mische, "Young People and SALT II," *America*, December 15, 1979 (an essay foreshadowing the fear-mongering among children that would become one of the most distasteful elements in the nuclear freeze movement).

Counterthemes were occasionally struck. See "A Collapse in Moscow," *America*, April 16, 1977 (an editorial supporting the Carter administration's March 1977 "deep cuts" proposal to the U.S.S.R.); "Balancing the Equations of Terror," *America*, July 23, 1977 (another editorial in support of early Carter administration arms control efforts); John Langan, S.J., "Below the SALT . . . some ethical dilemmas," *Commonweal*, December 7, 1979 (an essay noting that the pre–World War I arms race analogy had to be balanced by the analogy from the 1930s); and James F. Reid, "SALT: the case against," *Commonweal*, December 7, 1979.

38. Quinn's homily is reprinted in full in my small book, *The Peace Bishops and the Arms Race* (Chicago: World Without War Publications, 1982), pp. 10–16.

39. Cited in ibid., p. 18.

40. Cited in Castelli, op. cit., pp. 28–29.

41. Cited in ibid., p. 29.

42. Cited in ibid., p. 37.

43. Cited in ibid., p. 36.

44. Cited in ibid., p. 37.

45. Cited in ibid.

46. Cited in ibid., pp. 37–38.

47. Theodore M. Hesburgh, Foreword, in Castelli, op. cit., pp. 11–12.

48. Castelli, op. cit., p. 15.

49. Cited in ibid., p. 43.

50. Jonathan Schell, *The Fate of the Earth* (New York: Alfred A. Knopf, 1982).

51. For an analysis of freeze movement ideology and politics, see Adam M. Garfinckle, *The Politics of the Nuclear Freeze* (Philadelphia: Foreign Policy Research Institute, 1984).

52. See Judith Dwyer, "The Role of American Churches in the Nuclear Weapons Debate," in *The Nuclear Freeze Debate*, Paul M. Cole and William J. Taylor, Jr., eds. (Boulder: Westview Press, 1983), pp. 77–92.

53. See Lawrence Freedman, *The Evolution of Nuclear Strategy* (London: St. Martin's Press, 1983) for a historical overview.

54. One of the important influences on the "consensus" approach to "The Challenge of

Peace" was the spirit of collegiality within the American bishops. See David M. Byers, "The American Bishops at Collegeville," *America*, August 28, 1982, pp. 87–90, for a portrait of an important retreat attended by all the bishops during the drafting of TCOP.

55. Castelli, op. cit., is a partisan tract, useful only for elementary chronology and some telling quotes.

56. See ibid., pp. 13–18.

57. Ibid., pp. 78–80.

58. The author of this study was one of the formal witnesses before the NCCB committee. The full list of witnesses may be found appended to the second draft of the pastoral letter in *Origins*, 12:20 (October 28, 1982) 326.

59. Castelli, op. cit., p. 79.

60. Ibid., p. 135.

61. Cited in ibid., p. 84.

62. TCOP, 2.

63. TCOP, 3.

64. TCOP, 4.

65. See TCOP, 9–10.

66. TCOP, 13.

67. See TCOP, 15.

68. See TCOP, 16–17.

69. See TCOP, 27.

70. See ibid.

71. See TCOP, 38, 54.

72. TCOP, 55.

73. TCOP, 56.

74. See TCOP, 61, 66.

75. TCOP, 68, 70.

76. TCOP, 72, 73.

77. TCOP, 75.

78. TCOP, 77–78.

79. TCOP, 78.

80. TCOP, 80–120, 121.

81. TCOP, 122.

82. TCOP, 123.

83. TCOP, 140.

84. TCOP, 145.

85. TCOP, 147, 148.

86. TCOP, 150, 153.

87. TCOP, 156.

88. TCOP, 161.

89. TCOP, 159. For sources of Hehir's comments on the "centimeter of ambiguity," see William V. O'Brien, "A Just War Deterrence/Defense Strategy," p. 26, n. 2.

90. See TCOP, 163.

91. Ibid.

92. TCOP, 173.

93. See TCOP, 177 and 178–185.

94. TCOP, 186.

95. See TCOP, 188.

96. TCOP, 190 and footnote 84.

97. See TCOP, 190.

98. TCOP, 191. The pastoral could not resist a gratuitous slap at Reagan administration

intentions here, writing that "U.S. proposals like those for START (Strategic Arms Reduction Talks) and INF (Intermediate-range Nuclear Forces) negotiations in Geneva are *said* to be designed to achieve deep cuts; our hope is that they will be pursued in a manner that will realize these goals" [emphasis added].

99. TCOP, 196, 197, 198.

100. TCOP, 204, 205.

101. TCOP, 205. I urged the consideration of independent initiatives in my formal testimony before the NCCB ad hoc committee on war and peace, but as a means for forcing needed change in Soviet policy. The bishops located the notion of "independent initiatives" in a primarily psychological context (i.e., as a "confidence-building measure"), and thereby stripped the notion of much of its potential power. In my testimony, and in subsequent correspondence and conversation with members of the ad hoc committee and other bishops, I insisted that independent initiatives would not be effective were they seen as a means of shoring up "détente," for such an approach would serve only to confirm, rather than change, the Soviet agenda.

102. See TCOP, 207.

103. TCOP, 208–18.

104. See TCOP, 220.

105. TCOP, 222–27.

106. See TCOP, 229.

107. See TCOP, 231–33.

108. TCOP, 244.

109. Ibid.

110. TCOP, 249–58.

111. TCOP, 264.

112. TCOP, 270.

113. See TCOP, 268.

114. TCOP, 270.

115. TCOP, 285.

116. TCOP, 286, 289.

117. See TCOP, 290–98.

118. "We realize that different judgments of conscience will face different people, and we recognize the possibility of diverse concrete judgments being made in this complex area" (TCOP, 318). Yet, two and a half years later, Archbishop Roger Mahony of Los Angeles would suggest that Bishop Mathiessen's call to leave the defense industry was merely premature, because of the unfortunate state of Catholic peace education. See "Archbishop Mahony: A Frank First Interview," *Los Angeles Times Magazine*, November 17, 1985, p. 17.

119. TCOP, 326.

120. See TCOP, 331.

121. TCOP, 332.

122. TCOP, 334, 335, 336.

123. See TCOP, 337.

124. TCOP, 339.

125. Among the influential essays giving shape to the debate, and thus to the bishops' reflections, were Francis X. Winters, "The Bow or the Cloud? American Bishops Challenge the Arms Race," *America*, July 25, 1981, pp. 26–30; John J. O'Connor, *In Defense of Life* (Boston: St. Paul Editions, 1981); Bishop Roger Mahony, "The Catholic conscience and nuclear war," pastoral letter to the diocese of Stockton, reprinted in *Commonweal*, March 12, 1982, pp. 137–43; David J. O'Brien, "An open letter to the bishops," *Commonweal*, May 21, 1982, pp. 295–301; Michael Novak, "Nuclear Morality," *America*, July 3, 1982, pp. 5–8; John R. Connery, "The Morality of Nuclear Warpower," *America*, July 17, 1982, pp. 25–28;

Francis X. Winters, S.J., "Nuclear Deterrence Morality: Atlantic Community Bishops in Tension," *Theological Studies*, 43:3 (September 1982) 428–46; John Langan, S.J., "The American Hierarchy and Nuclear Weapons," *Theological Studies*, 43:3 (September 1982) 447–67; Francis X. Winters, S.J., "Catholic Debate and Division on Deterrence," *America*, September 18, 1982, pp. 127–31; "The bishops and the bomb—nine responses" (i.e., to the first draft of TCOP; the respondents were John Langan, S.J., William V. Shannon, Joan Chittester, Philip Odeen, Thomas J. Downey, James Finn, Gordon Zahn, William J. Nagle, and Charles A. Curran), *Commonweal*, August 13, 1982, pp. 424–40; David Hollenbach, S.J., "Nuclear Weapons and Nuclear War: The Shape of the Catholic Debate," *Theological Studies*, 43:4 (December 1982; Hollenbach's essay was later expanded into a small book, *Nuclear Ethics: A Christian Moral Argument* [New York: Paulist Press, 1983]); "The Bishops on Nuclear Weapons," *America* editorial, December 4, 1982; "A Milestone for the Bishops," *Commonweal* editorial, December 3, 1982; James Finn, "Nuclear Terror: Moral Paradox," *America*, February 19, 1983, pp. 126–29; Michael Novak, "Moral Clarity in the Nuclear Age," *National Review*, April 1, 1983 (this essay was subsequently gathered with others in a small book of the same title, published in 1983 by Thomas Nelson, Inc.). The Novak essay was particularly important in shaping the themes of the Vatican consultation discussed below. William V. O'Brien's essay, "Just-War Doctrine in a Nuclear Context" (*Theological Studies*, 44:2 [June 1983] 191–220), appeared after the adoption of TCOP, but contains central themes in O'Brien's argument to and with the NCCB ad hoc committee. Among my own contributions to this debate may be cited, "The Catholic 'Peace' Bishops," *Freedom at Issue*, 67 (July–August 1982), pp. 11–14; the small book *The Peace Bishops and the Arms Race* (Chicago: World Without War Publications, 1982); and "An Open Letter to Archbishop Bernardin," *Catholicism in Crisis*, 1:2 (January 1983) 14–19.

The new journal *Catholicism in Crisis* was itself a child of the TCOP debate, although addressing many other issues as well.

126. The full list of participants in the Vatican consultation may be found in "A Vatican Synthesis," *Origins*, 12:43 (April 7, 1983) 691–92.

127. Ibid., p. 691.

128. Ibid.

129. Ibid., p. 693.

130. Ibid.

131. Ibid.

132. Ibid.

133. Ibid., p. 694.

134. Ibid.

135. Ibid.

136. Ibid.

137. Ibid.

138. TCOP, draft #2, in *Origins*, 12:20 (October 28, 1982) 317.

139. "A Vatican Synthesis," p. 694.

140. Ibid., pp. 694–95.

141. Ibid., p. 695 (emphasis added).

142. Ibid.

143. Ibid.

144. "A Positive and Helpful Exchange of Views," *Origins*, 12:43 (April 7, 1983) 696.

145. That the NCCB leadership was concerned to bring TCOP into line with the position of the Holy See was evident at the May 1983 special NCCB meeting to consider the final draft of the document, where a key vote was reversed when Cardinal Bernardin invoked "Rome." See Castelli, op. cit., pp. 165–70. According to Castelli's reconstruction of the May 1983

meeting, one can infer that the Vatican was interested in rather more than a "centimeter of ambiguity" on the matter of deterrence.

146. TCOP, draft #2, p. 312.

147. See "Out of Justice, Peace: Joint Pastoral Letter of the West German Bishops," 130, in *Bishops' Pastoral Letters*, James V. Schall, S.J., ed. (San Francisco: Ignatius Press, 1984). This volume also contains the joint pastoral letter of the French bishops, "Winning the Peace," and "Towards a Nuclear Morality," a statement by Cardinal Basil Hume, O.S.B., of Westminister, England. All three statements make interesting counterpoints to TCOP. The West German statement is far and away the most developed of the three.

148. See, as a representative example, the *New York Times* coverage of the May 1983 NCCB meeting, April 30 and May 1–5, 1983. The *Times*, and virtually every other American media source, saw the debate over whether the bishops would call for a "curb" to the arms race or a "halt" to the arms race as the central issue, "halt" (the term eventually adopted) being widely and accurately interpreted as support for the nuclear freeze. On this point, see Castelli, op. cit., pp. 155–62. Bishop James Malone of Youngstown, who would later be elected president of the NCCB, urged that the bishops change from "curb" to "halt" because this would "set the tone for the rest of the document." The shift to "halt" passed overwhelmingly, which may be taken as a signal of what the bishops' primary intentions by May 1983 had become.

The bishops' position on deterrence was heavily criticized by some strategists for proposing what the critics called a "bluff deterrent." See, in particular Albert Wohlstetter, "Bishops, Statesmen, and Other Strategists on the Bombing of Innocents," *Commentary*, 75:6 (June 1983) 15–35, and the exchange of correspondence on Wohlstetter's essay in *Commentary*, 76:6 (December 1983) 4–22. On the "bluff deterrent," see also William V. O'Brien, "A Just War Deterrence/Defense Strategy."

149. On TCOP's muddled approach to the relationship between pacifism and just-war theory, see James Finn, "Pacifism and Just War: Either or Neither," in *Catholics and Nuclear War*, pp. 132–45. Gordon Zahn argued the opposite case in the same volume, in an essay entitled "Pacifism and the Just War," pp. 119–31.

150. This was one of the disquieting curiosities of the May 1983 NCCB meeting. One can only assume that bishops like Gumbleton and Hunthausen, who fundamentally disagreed with key themes of TCOP (e.g., its teaching on the immorality of a pacifist stance by governments, its refusal to espouse unilateral disarmament) voted for the document because of what they assumed its eventual political impact would be—that is, that, over time, it would help build a constituency in favor of their prescriptions, even if they had been rejected by the NCCB at the present stage of the debate. Archbishop Hannan of New Orleans and Auxiliary Bishop Austin Vaughan of New York, persistent and vocal critics of TCOP before and during the May 1983 meeting, on the grounds that it took entirely too negative a view of deterrence and entirely too hopeful a view of U.S.-Soviet relations, held to their convictions, and voted against the document as a whole. To note this is not to argue in favor of Hannan's or Vaughan's approach. But the unilateralists' inconsistency says rather a lot about the politics of TCOP.

151. TCOP, 78.

152. *Pacem in Terris*, 113.

153. Cardinal Joseph Bernardin, "A Consistent Ethic of Life: An American Catholic Dialogue," the Gannon Lecture, Fordham University, December 6, 1983, reprinted in *Thought*, 59 (1984) 101.

154. See my essay, "The Bishops' Pastoral Letter and American Political Culture: Who Was Influencing Whom?," in *Peace in a Nuclear Age*, Charles Reid, ed. (Washington: Catholic University of America Press, 1986).

155. The bishops' approach to the MX debate may be gleaned from the front page of the

New York Times for March 16, 1985. Bishop Malone's letter to all members of Congress urging a no vote on MX funding may be found in the *Congressional Record* (Senate), March 18, 1985, p. S3003. Cardinal Bernardin and Archbishop O'Connor also issued statements supporting the Malone letter; see *The Progress* (newspaper of the archdiocese of Seattle), March 21, 1985, pp. 10–11. On the Strategic Defense Initiative, see Cardinal Bernardin's "Keynote Address" to the Conference on American Religion and International Relations, University of Missouri, March 7, 1985, and the cardinal's panel remarks to the same conference, "The Role of the Churches in International Affairs." Both texts are available from the archdiocese of Chicago.

156. See Matthew Murphy, *Betraying the Bishops* (unpublished manuscript); also my article "Nuke Discussion Guides Get Disappointing Grades," *National Catholic Register*, May 6, 1984, pp. 1, 8. See also Chester E. Finn, Jr., "Catholic Schools Veer Toward Pacifism," *Wall Street Journal*, December 27, 1983, for a view of how the pastoral letter and some themes in the argument preceding it had begun to influence certain Catholic educational circles. Finn was particularly struck by Catholic educators making common political cause on the nuclear weapons issue with groups that had been in the forefront of opposition to independent education and tuition tax credits.

American Catholicism and the Dilemma of Peace and Freedom in Central America

1. Phillip E. Berryman, "Latin American Liberation Theology," *Theological Studies* 34:3 (September 1973) 358.

2. See ibid., p. 359.

3. Ibid., p. 360.

4. Cited in ibid., p. 361.

5. Cited in ibid., pp. 366–67.

6. Berryman, art. cit., p. 361.

7. Cited in ibid., p. 361.

8. See ibid.

9. Cited in Richard John Neuhaus, "Liberation Theology and the Captivities of Jesus," *Worldview*, June 1973, p. 46. For other discussions of dependency theory, see Joseph Ramos, "Dependency and Development: An Attempt to Clarify the Issues," in *Liberation North, Liberation South*, Michael Novak, ed. (Washington, D.C.: American Enterprise Institute, 1981), pp. 61–68; and idem, *The Spirit of Democratic Capitalism* (New York: Simon & Schuster, 1982), pp. 298–314.

10. Cited in Berryman, art. cit., p. 363.

11. Ibid.

12. Cited in Berryman, art. cit., p. 366. Assman had reported that "his most theological work [was] his reporting on the guerilla action in Teoponte, Bolivia" (ibid., n. 36).

13. Ibid., p. 367.

14. Ibid.

15. Ibid.

16. James Finn, "Pacifism and Justifiable War," in Shannon, op. cit., p. 12.

17. Cited in Michael Novak, "Liberation Theology and the Pope," *Commentary*, June 1979, p. 62. On this question, Phillip Berryman wrote:

> It has been suggested that liberation theology can be seen as an overcoming of the Marxist critique of religion by way of a new theological praxis. Juan Luis Segundo indicates how Medellín has taken over elements of Marxist analysis which then "by their own right enter to form part of theology." The theologians did not set out to become Marxists; for some it was mediated by contact with Paulo Freire and "conscientization," for others by reading of economists and sociologists or by contact with political activists. In any case, it is the reality

itself which impels Christians to go back to Marx. Many Christians have found that Marxism is not only a system of thought but a "synthesis of reasons for living, a mobilizing doctrine." More significantly, Marxism today is not simply the position of the "other," heard out with sympathy, but is becoming the body of categories with which one lives his political commitment. Not so much an "external encounter," it is for many "a way of relating with oneself, a new way of thinking and living one's faith" [art. cit., pp. 374–75].

18. See Berryman, art. cit., pp. 373–74. Some of the confusions in this discussion were related by Michael Novak in his study *Freedom with Justice: Catholic Social Thought and Liberal Institutions* (San Francisco: Harper & Row, 1984): "Cardinal Arns [of Brazil, a prominent defender of liberation theologies] was once heard to say that it was not for the Church to say that capitalism or socialism is better; only, in his eyes, capitalism seems morally inferior, since each farmer must be allowed *to own his own land and keep his profits for himself.* This is what the Cardinal calls socialism. But is it not a basic principle of capitalism?" (p. 192). In "Liberation Theology and the Pope," Novak noted that "what Latin Americans persist in calling 'capitalism' is, in Latin America, largely a form of syndicalism or corporatism, which descends from the rights given by the Spanish and Portuguese crown to certain large landholders or adventurers and constitutes virtual monopoly or state mercantilism" (p. 62).

19. Berryman, art. cit., p. 371, citing Medellín.

20. See ibid., p. 379.

21. Ibid., p. 386.

22. Ibid., p. 387.

23. Cited in ibid.

24. Ibid.

25. Ibid., p. 389.

26. Ibid., citing Gutiérrez's *Theology of Liberation.*

27. See ibid.

28. Neuhaus, art. cit., p. 48.

29. Cited in ibid.

30. See ibid.

31. See Berryman, art. cit., p. 387.

32. See Neuhaus, art. cit., p. 42.

33. See Novak, *The Spirit of Democratic Capitalism*, pp. 276–82.

34. Neuhaus, art. cit., p. 46.

35. Ibid., p. 47.

36. Cited in Novak, "Liberation Theology and the Pope," p. 60.

37. Cited in ibid., p. 61.

38. Cited in ibid.

39. Cited in ibid., p. 62.

40. Ibid., pp. 63–64.

41. See ibid., p. 64.

42. See Leszek Kolakowski, *Main Currents of Marxism*, volume 3, *The Breakdown* (Oxford: Oxford University Press, 1981), pp. 523–30.

43. Novak, "Liberation Theology and the Pope," p. 64. See also Michael Novak, "Liberation Theology in Practice," *Thought*, 59:233 (June 1984) 136–48.

44. See "Instruction on Certain Aspects of the 'Theology of Liberation'," *Origins*, 14:13 (September 13, 1984) 193–96; 200.

45. See ibid., pp. 198–99.

46. See ibid., p. 196.

47. See ibid., p. 197.

48. Ibid., p. 202.
49. See ibid., pp. 199, 200.
50. See ibid., p. 199.
51. See ibid.
52. See ibid., p. 201.
53. Ibid.
54. See ibid.
55. Ibid., p. 202.
56. Ibid., p. 203.
57. Ibid., p. 204. The Instruction continued:

The defenders of orthodoxy are sometimes accused of passivity, indulgence, or culpable complicity regarding the intolerable situations of injustice and the political regimes which prolong them. Spiritual conversion, the intensity of the love of God and neighbor, zeal for justice and peace, the gospel meaning of the poor and of poverty, are required of everyone and especially pastors and those in positions of responsibility. The concern for the purity of the faith demands giving the answer of effective witness in the service of one's neighbor, the poor and the oppressed in particular, in an integral theological fashion. By the witness of their dynamic and constructive power to love, Christians will thus lay the foundations of this "civilization of love" of which the conference of Puebla spoke, following Paul VI [p. 204].

58. See *New York Times*, September 4, 1984, pp. 1, 10; *New York Times*, September 6, 1984, pp. 1, 14; *Newsweek*, September 17, 1984, pp. 83–85. Avery Dulles, S.J., struck a different note in "Liberation Theology: Contrasting Types," *America*, September 22, 1984, pp. 138–39.

59. Charles H. Savage, Jr., "After Castro," *America*, November 24, 1962, pp. 1130.

60. Gerard J. Mangone, "Latin America Progress Report," *America*, July 4, 1964, pp. 11–12.

61. Ibid., p. 12.

62. See ibid., p. 14.

63. For a portrait of Illich, see Francine du Plessix Gray, *Divine Disobedience*, pp. 231–32.

64. Illich, art. cit., p. 88.

65. Ibid., p. 89.

66. Ibid., p. 90.

67. Ibid.

68. Ibid.

69. Ivan Illich, "Violence: A Mirror for Americans," *America*, April 27, 1968, p. 568.

70. Thomas G. Sanders, "Brazil's Catholic Left," *America*, November 18, 1967, p. 598.

71. See ibid., p. 601.

72. Henrique C. de Lima Vaz, "The Church and Conscientização," *America*, April 27, 1968, p. 579.

73. Ibid., pp. 580, 579.

74. Juan Luis Segundo, "Social Justice and Revolution," *America*, April 27, 1968, p. 575.

75. Juan Luis Segundo, "Has Latin America a Choice?," *America*, February 22, 1969, p. 215.

76. Gary MacEoin, "President Carter's Choices," *Commonweal*, September 30, 1977, p. 616.

77. Ibid., p. 618.

78. Jeffrey L. Klaiber, "Freedom Equals Liberation," *America*, June 6, 1970, p. 607.

79. Ibid.

80. Gary MacEoin, "Latin America: Who Is To Blame?," *Commonweal*, June 25, 1971, p. 334.

81. Ibid.

82. Ibid., p. 336.

83. Phillip Berryman, "Camilo Torres: Revolutionary-Theologian," *Commonweal*, April 21, 1972, p. 166.

84. Ibid.

85. See James Brockman, S.J., "The Land Divided," *America*, August 30, 1975, pp. 88–89.

86. See "New Treaties and New Symbols," *America*, February 18, 1978, p. 112.

87. Paul A. Fitzgerald, S.J., "The Panama Canal: Use and Ownership," *America*, December 31, 1977, p. 476.

88. Vincent Kearney, S.J., at least raised the question of internationalization in "Panama and Suez," *America*, December 31, 1977, pp. 477–78.

89. See "Nicaragua on the Brink," *America*, December 9, 1978, p. 421.

90. Ibid.

91. See James Brockman, S.J., "Nicaragua in January," *America*, February 24, 1979, p. 138.

92. See Chris Gjording, S.J., "Nicaragua's Unfinished Revolution," *America*, October 6, 1979, p. 170.

93. See ibid., p. 168.

94. Ibid., p. 170.

95. Ibid., p. 171.

96. Ibid.

97. Tennent C. Wright, S.J., "Ernesto Cardenal and the humane revolution in Nicaragua," *America*, December 15, 1979, p. 388. That Cardenal had something other than the Church in Acts in mind when he described himself as a "Christian communist" was graphically illustrated in an interview he gave to Kenneth Woodward, in which Cardenal claimed that became a Marxist by reading the Gospel and visiting Cuba in 1970, where he "found out that the government there was actually putting the Gospel into practice: giving the people food, clothing help, education—and on an equal basis" (cited in the *National Catholic Register*, March 11, 1984, p. 3).

98. Cited in ibid.

99. Robert F. Drinan, S.J., "Nicaragua after Somoza," *America*, February 9, 1980, p. 101.

100. Ibid.

101. See ibid.

102. See ibid.

103. Ibid., p. 102.

104. See Arthur McGovern, S.J., "Nicaragua's Revolution: A Progress Report," *America*, December 21, 1981, p. 378.

105. See ibid.

106. Ibid., p. 379.

107. Ibid.

108. Ibid., p. 380.

109. Ibid.

110. See Philip Land, S.J., "Military Aid to El Salvador," *America*, March 22, 1980, p. 245.

111. Ibid., pp. 245–46.

112. Ibid., p. 246.

113. Ibid.

114. "Blood of Martyrs. . . ," *Commonweal*, April 25, 1980, p. 229.

115. See "Change in El Salvador?," *America*, December 27, 1980, p. 421.

116. See "The Salvadoran Tragedy," *America*, January 31, 1981, p. 74.

117. "Into El Quagmire," *Commonweal*, March 13, 1981, p. 133.

118. Ibid.

119. "Oligarchs, Old and New," *Commonweal*, October 9, 1981, pp. 550, 549.

120. "America's Poland," *Commonweal*, February 26, 1982, pp. 102–3.

121. John Garvey, "A Politics of Silence," *Commonweal*, May 7, 1982, p. 265. Garvey also wrote, "We have been raised on the belief that our ability to choose between the limited options offered us at election time is a proof we are free. . . . The 'will of the people' becomes sacred, as if there were such a simple thing, or, even if there were, as if it would be a good thing." Garvey conceded that "the solution does not lie in the repudiation of democracy," but the whole thrust of his article was to demean, if not simply dismiss, the significance of those hundreds of thousands of Salvadorans who had, in many cases, walked miles and braved long lines in the blazing sun in order to vote.

122. "Salvadoran Ballots—And Bullets," *Commonweal*, June 4, 1982, pp. 323–24.

123. Jim Chapin and Jack Clark, "The El Salvador labyrinth," *Commonweal*, June 3, 1983, pp. 329–35.

124. "Faith in Fire Power," *Commonweal*, January 14, 1983, p. 4.

125. "Getting the Message," *America*, August 27, 1983, p. 81.

126. See "The View from Below," *America*, February 18, 1984, p. 101. See also Philip Land's commentary on the Kissinger Commission Report in *Center Focus* (monthly newsletter of the Center of Concern), March 1984.

127. See Gerald F. Seib, "Catholics and Other U.S. Church Groups Oppose Reagan's Hard-Line Policy on Central America," *Wall Street Journal*, December 8, 1983, p. 56.

128. See E. F. Sweeney, "Missionary Society of St. James the Apostle," *New Catholic Encyclopedia* (New York: McGraw-Hill, 1967), volume 9, p. 924; J. J. Considine, "Papal Volunteers for Latin America (PAVLA)," *New Catholic Encyclopedia*, volume 10, p. 978. For an essay capturing the flavor of these enterprises, see Gerald Mische, "AID Goes to Latin America," *America*, May 26, 1962, pp. 298–300.

129. Joseph A. O'Hare, S.J., "The Summer of '72," *America*, September 30, 1972, pp. 228–32.

130. Cited in Gerald M. Costello, "Latin America: The Mission Transformed," *America*, January 20, 1979, pp. 27–28 (emphasis added).

131. "Statement of the United States Catholic Conference on Central America" (Washington, D.C.: United States Catholic Conference, 1982).

132. Ibid., p. 2. The USCC statement did not mention the tensions between Medellín and Puebla implicit in John Paul II's sharp critique of themes in the theologies of liberation at the Puebla meeting.

133. Ibid., p. 3. The USCC statement did not cite death in El Salvador of another American Catholic, Michael Hammer, of the American Institute for Free Labor Development, who had been a key consultant to the Salvadoran land reform program.

134. Ibid.

135. Ibid., p. 4.

136. Ibid.

137. "Steps Toward New US-Central America Relationship," *Origins*, 13:21 (November 3, 1983) 381.

138. Ibid.

139. Ibid.

140. Ibid., p. 382.

141. Ibid.

142. Ibid., p. 383 (emphasis added).

143. See ibid.

144. Ibid.

145. Ibid.

146. "Redirecting U.S. Policy in Central America," *Origins*, 13:43 (April 5, 1984) 716.

147. Ibid., p. 717.

148. Ibid., p. 718.

149. "U.S. Military Aid: El Salvador," *Origins* (April 3, 1980), p. 671.

150. "Romero: Voice for the Poor," *Origins* (April 3, 1980), p. 671.

151. See "Action Asked on U.S. Policy in El Salvador," *Origins*, 10:27 (December 18, 1980) 419.

152. "USCC Testimony on Central America," *Origins*, 12:41 (March 24, 1983) 653.

153. "U.S. Military Aid to El Salvador Resumed," *Origins*, 10:30 (January 29, 1981) 526.

154. "Focus of U.S. Church Concern for El Salvador," *Origins*, 10:40 (March 19, 1981) 627–29.

155. Ibid., p. 626.

156. "USCC Opposes Military Aid to El Salvador," *Origins*, 11:39 (March 11, 1982) 617.

157. "Focus of U.S. Church Concern for El Salvador," p. 631.

158. "USCC Opposes Military Aid to El Salvador," p. 618.

159. Ibid.

160. Ibid., pp. 618–19.

161. "Human Rights and U.S. Foreign Policy," *Origins*, 12:27 (December 16, 1982) 438.

162. "Steps Toward New U.S.-Central America Relationship," p. 382.

163. "Redirecting U.S. Policy in Central America," p. 717.

164. "U.S. Policy and Central America," *Origins*, 14:45 (April 25, 1985) 736.

165. "Statement of the United States Catholic Conference on Central America," pp. 4–5.

166. "Nicaragua: Attacks Against Clerics Protested," *Origins*, 12:15 (September 23, 1982) 227–28.

167. "Church and State in Nicaragua," *Origins*, 12:27 (December 16, 1982) 432.

168. "Archbishop Responds to Nicaraguan Official," *Origins*, 12:27 (December 16, 1982) 433.

169. "Human Rights and U.S. Foreign Policy," p. 437 (emphasis added).

170. "USCC Testimony on Central America," p. 654.

171. Humberto Belli, former editorial page editor of the independent Managua paper *La Prensa*, described the FSLN handling of the pope's visit in *Nicaragua: Christians Under Fire* (San José, Costa Rica: Instituto Puebla, 1984):

> The Sandinista government's handling of the visit of Pope John Paul II to Nicaragua offered additional evidence of the kind of approach it intends to take toward religion. Leaving aside the points on which a consensus does not exist, the following list of events offers a description on which there is ample agreement by a relatively impartial observer. The list is drawn from reports published in major media outlets, in particular *The Washington Post*, the *New York Times*, and the major U.S. television networks.
>
> 1. The Sandinistas did not allow Nicaraguans the freedom to assemble to greet the Pope. Traffic was halted throughout most of the country; only the Sandinista Defense Committees were entitled to transportation to meeting places. Thousands of Catholics were made to walk great distances from the surrounding cities in order to see the Pope. Many could not make it. John Paul II, aware of the circumstances, greeted "the thousands of Nicaraguans who have not found it possible to come to the meeting places as they might have wanted."
>
> 2. The Sandinistas prevented people from gathering ahead of time at the sites where the

Pope was scheduled to appear. In Managua police fired automatic weapons over the heads of worshippers who attempted to get early places. Sandinista partisans were thus able to pack the front rows in the plaza.

3. An ABC-TV crew from the United States was detained and roughed up and their video tapes confiscated by police.

4. John Paul II was interrupted during his sermon in Managua, and then for the remainder of the Mass, by heckling and the chanting of slogans. During communion, to cite a single incident, a Sandinista agitator with a powerful magnifier cried, "Holy Father, if you are truly the representative of Christ on earth, we demand you to side with us." The police who were assigned to control the crowds frequently led the chants. Members of the Papal entourage later stated that they had never before seen such behavior on a papal trip.

5. Government technicians connected microphones distributed among pro-government troops to the main loudspeaker system, amplifying the cry of the agitators and the chants of "people's power."

6. During the celebration of the Mass all nine members of the Sandinista National Directorate—including Tomás Borge and Daniel Ortega—joined the crowd in waving their left fists and shouting "people's power."

7. The Catholic Church and the government had agreed that Papal appearances in both Managua and León were to be wholly religious and apolitical. The church itself warned parishioners against politicking and exhorted them not to carry partisan symbols or placards. Sandinista political supporters, however, carried political banners and posters and chanted political slogans through megaphones. The Sandinista leadership ended the Papal mass by singing the party anthem.

What is important to notice regarding these incidents is that they were not spontaneous outbursts of popular indignation but well prepared actions carried out by FSLN partisans. Also, these incidents were not the excesses of some overly fervent Sandinistas but acts in which the full directorate of the FSLN participated. . . .

It is significant to note also that on this occasion the Sandinistas acted in full view of the global media. People who are prepared to engage in such ugly treatment of a revered religious leader are hardly to be expected to act with more restraint when dealing with less well-known believers outside the glare of international publicity. The treatment of the Pope thus effectively symbolizes the direction which the Sandinista government intends to continue to pursue in its struggle against religion [pp. 52–53].

See also Belli's *Breaking Faith: The Sandinista Revolution and Its Impact on Freedom and Christian Faith in Nicaragua* (Westchester: Crossway Books, 1985).

172. "Bishops' Fact-Finding Mission to Central America," *Origins*, 12:38 (March 3, 1983) 610.

173. "Redirecting U.S. Policy in Central America," p. 718.

174. By 1984 the following sources were available in print: "Comandante Bayardo Arce's Secret Speech before the Nicaraguan Socialist Party (PSN)," (U.S. Department of State); "A Secret Sandinista Speech," *The Economist: Foreign Report*, August 23 and September 6, 1984; Juan Tamayo, "Nicaraguan Decries Need to Vote," *Washington Post*, August 8, 1984; "Interview with Archbishop Miguel Obando y Bravo," *National Catholic Register*, July 29, 1984; Robert Leiken, "Nicaragua's Untold Stories," *The New Republic*, October 8, 1984, pp. 16–23; Arturo J. Cruz, "Nicaragua's Imperiled Revolution," *Foreign Affairs*, 61:5 (Summer, 1983) 1031–47; Paul Hollander, "Sojourners in Nicaragua: A Political Pilgrimage," *National Catholic Register*, May 29, 1983; and Geraldine O'Leary de Macias, "Foreign Christian Are a Problem," *National Catholic Register*, October 16, 1983. Given the USCC's persistent claims that it drew its information from local Nicaraguan sources, it is inconceivable that the kind of information available on FSLN ideology in these articles are not available to the USCC long before the fall of 1983.

175. On the ideological configuration of the Salvadoran FMLN, see Gabriel Zaid, "Enemy colleagues: a reading of the Salvadorian tragedy," *Dissent*, 29 (Winter 1982) 13–40.

176. See Arturo Cruz and Arturo Cruz, Jr., "A Peace Plan for Nicaragua," *The New Republic*, March 18, 1985, pp. 17–18; "U.S. Policy and Central America," *Origins*, 14:45 (April 25, 1985) 734–36.

177. "U.S. Policy and Central America," p. 734.

178. Gary MacEoin continued to defame Obando y Bravo in the pages of *America*; see his "Nicaragua: A Church Divided," *America*, November 10, 1984. Humberto Belli responded in "Nicaragua's Bishops: A Response to Gary MacEoin," *America*, February 23, 1985, pp. 145–48. Bishop Rene Gracida of Corpus Christi was a distinctive voice among the American bishops on the Nicaraguan issue; see his "Brotherhood and Fratricide in Central America," *Origins*, 14:45 (April 25, 1985) 737–39. Gracida broke ranks with the USCC both in his analysis of the Sandinista regime and in his sense of the possibilities of a useful American role for peace and freedom in Central America.

179. See, for example, William C. Doherty, Jr., *Sandinista Repression of Nicaraguan Trade Unions* (Washington, D.C.: American Institute for Free Labor Development, 1985). The letter to the NCCB leadership from the Permanent Human Rights Commission is available in the commission's English-language report for November 1985.

180. The tawdriest episode in this sad process was *America*'s publication, without qualification or comment, of an interview with Foreign Minister Miguel D'Escoto in which D'Escoto accused Cardinal Obando of "treasonous" activity and of being the principal Central American asset of the CIA. See *America*, November 16, 1985. See also the responses to D'Escoto from Michael Novak, Mario Paredes, James Finn, and George Weigel, in *America*, January 18, 1986.

181. For counterpoints to the dominant themes in the American Catholic debate over Central America, see Bishop Marcos McGrath, "Development for Peace," *America*, April 27, 1968, pp. 562–67 (McGrath was particularly critical of the corruption of the term "violence" in the debate); James Finn, "Catholic revolutionaries: part of the solution or part of the problem?," *Commonweal*, September 3, 1971, pp. 456–58 (a review of books by Thomas and Marjorie Melville, and Camilo Torres). On the Melvilles, former Maryknoll missionaries, Finn wrote:

> What becomes gradually clear is that the Melvilles have changed the pattern of their lives but not their mindset. They have adopted a new political catechism next to which the old Baltimore catechism is a model of subtlety, discrimination, and intellectual refinement. But they are still missionaries, eager to convert. Unable, apparently, to distinguish between the spiritual and the social—"is there a difference?" Tom asks—they now bring to social matters the same moral arrogance and simple-mindedness they initially brought to matters religious.

See also Peter Steinfels, "Name Your Socialism," *Commonweal*, September 12, 1980, pp. 484–86, and the subsequent exchange of correspondence on Steinfels's essay in *Commonweal*, November 21, 1980, pp. 654–61 (Steinfels, a democratic socialist, was worried about the too easy accommodations with Leninist politics that he perceived underway in North American liberation theology circles); "The Risks in Central America," *Commonweal*, April 22, 1983 (an editorial criticizing American religious apologists for the FSLN for not "subjecting that regime to the 'social analysis' on which they pride themselves" [p. 228]); "The First Casualty," *Commonweal*, August 12, 1983 (an editorial critical of the whitewashing of Sandinista repression by the Intercommunity Center for Justice and Peace with funding from the New England Holy Cross Fathers, the Sisters of Loretto, and Maryknoll).

Other voices, then, did exist. But their impact was minimal in comparison, for example, with the framework for analysis and prescription created by what was perhaps the single most influential book in the American Catholic debate on Central America, Penny Lernoux's *Cry of the People: The Struggle for Human Rights in Latin America—The Catholic Church in Conflict with U.S. Policy* (New York: Penguin Books, 1982).

182. These themes still shaped Catholic commentary on Nicaragua as Sandinista repression intensified in 1986, and Catholic publicists continued to support the FSLN regime while deprecating the democratic credentials of the political leadership of the Nicaraguan resistance. For representative essays, see Frank M. Oppenheim, S.J., "A Report on Nicaragua," *America*, March 8, 1986, pp. 183–85; and Robert J. Henle, S.J., "The Great Deception: What We Are Told About Central America," *America*, May 24, 1986. In an editorial on April 26, 1986, *America*'s editors instructed the bishops of Nicaragua in their responsibilities:

> "Nicaragua's Catholic hierarchy would be well-advised to heed their own words from an eight-page statement, issued on April 7, that called the Sandinista-contra conflict 'a fratricidal war that is not only killing our young people on the battlefield, but is killing and destroying our best moral and human values.' The bishops need to exercise pastoral leadership within their own church and rein in rhetorical excesses to serve as a true 'light to the nations,' and provide a palpable example of Christian reconciliation in order to end the fratricidal suffering of their people" (p. 334).

A rather different perspective on the crisis of the Nicaraguan Church was offered by Cardinal Obando y Bravo in an essay in the May 12, 1986, *Washington Post* (p. A15), in which he protested that the Sandinistas had "gagged and bound" the Church by seizing the Church's printing press, stealing files (including the cardinal's personal seal), censoring the bishops' Holy Week pastoral letter from *La Prensa*, and proscribing broadcasts by Radio Catolico. Less than two months later, the Sandinistas shut down *La Prensa*, forbade the re-entry into Nicaragua of the cardinal's communications director, Msgr. Bismarck Carballo, and expelled Bishop Pablo Antonio Vega from the country.

The USCC continued to oppose military assistance to the Nicaraguan resistance. In a letter to all members of Congress, USCC general secretary Msgr. Daniel Hoye claimed that the USCC opposed "military aid to any party in Nicaragua," but once again failed to answer the question of how Soviet and Cuban arms shipments into Nicaragua could be stopped. Msgr. Hoye also reiterated USCC support for the Contadora negotiations, which he described as "a very useful Latin American initiative." (Cited in *Origins* 15:42, April 3, 1986, pp. 694–95.)

Seven weeks after Msgr. Hoye's letter, Nicaraguan president Daniel Ortega told a group of factory workers in Managua that "not one rifle will leave Nicaragua in any negotiation." (*New York Times*, May 12, 1986, p. A3.)

Portrait—J. Bryan Hehir

1. On the ideological currents at work in the Institute for Policy Studies, see Joshua Muravchik, "The Think Tank of the Left," *New York Times Magazine*, April 26, 1981, pp. 36ff.

2. Cited in Charlotte Hays, "The Voice in the Bishops' Ear," *Washington Post Magazine*, April 3, 1983, p. 7.

3. For Hehir's debts to Congar, Rahner, and Murray, see Hays, art. cit., p. 11. Hoffmann's key books include *Gulliver's Troubles: Or the Setting of U.S. Foreign Policy* (New York: McGraw-Hill, 1968); *Primacy or World Order: American Foreign Policy Since the Cold War* (New York: McGraw, 1978); *Duties Beyond Borders: On the Limits and Possibilities of an Ethical International Politics* (Syracuse University Press, 1981); and *Dead Ends: American Foreign Policy in the New Cold War* (Cambridge: Ballinger, 1983). Hoffmann is also a frequent commentator on current issues for *The New York Review of Books.*

4. See J. Bryan Hehir, "A Challenge to Theology: American Wealth and Power in the Global Community," *Proceedings of the Thirtieth Annual Convention of the Catholic Theological Society of America*, Luke Salm, F.S.C., ed. (Catholic Theological Society of America, 1975), pp. 143–45.

5. See ibid., pp. 146–47.

6. Ibid., p. 149.
7. Ibid., p. 150.
8. Ibid., p. 152.
9. See ibid., p. 155.
10. Ibid.
11. See, for example, J. Bryan Hehir, "Religious Organizations and International Affairs," *Network Quarterly*, 4:3 (Summer 1976) 2.
12. J. Bryan Hehir, "On the Role of the Pastoral Letters," interview in *World Policy Journal*, Fall 1984, p. 192.
13. On the Soviet role in the Middle East, see J. Bryan Hehir, "Foreign Policy in the 1980s: Seeing the Issues in Context," *Christianity and Crisis*, September 29, 1980, p. 269; on illustrative human rights concerns, see his "Religious Organizations and International Affairs," p. 3; on linkage, see his "Human Rights and the National Interest," *Worldview*, 25:5 (May 1982) 20; on détente, see his "Foreign Policy in the 1980s," p. 268.
14. Hehir, "Foreign Policy in the 1980s," p. 267.
15. Ibid. (emphasis added).
16. See, for example, Hehir, "Human Rights and the National Interest," p. 21.
17. Hehir, "Religious Organizations and International Affairs," p. 6 (emphasis added).
18. See J. Bryan Hehir, "Human Rights Factors in U.S. Foreign Assistance," *Origins*, 7:46 (May 4, 1978) 733.
19. See J. Bryan Hehir, "The Catholic Church and the Arms Race," *Worldview*, July–August 1978, pp. 13ff; see also Hehir, "On the Role of the Pastoral Letters," p. 183.
20. See Hehir, "On the Role of the Pastoral Letters," p. 183.
21. See J. Bryan Hehir, "SALT II: Limited but substantial accomplishments," *Commonweal*, March 2, 1979, pp. 108–10.
22. See Hehir, "The Catholic Church and the Arms Race," pp. 15–16.
23. See ibid., p. 16.
24. See ibid., p. 17.
25. See Hehir, "On the Role of the Pastoral Letters," p. 187.
26. See ibid., p. 186.
27. See J. Bryan Hehir, "Can nuclear protest change policy?," *Commonweal*, June 18, 1982, p. 360.
28. See ibid., pp. 360, 383.
29. By "traditional arms control community" I mean that complex of theorists, academics, *quondam* government officials, and publicists who have abandoned work for general and complete disarmament (and, in most cases, nuclear disarmament), and have opted for arms control as a means for managing the inevitable weapons competition between the United States and the Soviet Union so that it becomes as safe as possible under prevailing technological and political circumstances. Lawrence Freedman traces the history of this school of thought in *The Evolution of Nuclear Strategy* (London: St. Martin's Press, 1982). The arms control community is not monolithic, but it does reflect several key common themes, among them the postulate that weapons which attack weapons are destabilizing, whereas weapons that threaten people are, paradoxically, stabilizing. Bryan Hehir does not share the traditional arms control community's acceptance of Mutual Assured Destruction as a guiding strategic concept; yet his prescriptions on nuclear weapons policy tend to fall rather neatly within the ambit of one wing of this strategic school. The SALT I antiballistic missile treaty may be taken as the quintessential expression of the policy perspective of the traditional arms controllers.
30. On Hehir's concept of a "bluff deterrent," see Rodger Van Allen, "Pax Christi in Dayton," *Commonweal*, December 19, 1975, p. 613; for critiques, see William V. O'Brien, "A Just War Deterrence/Defense Strategy," *Center Journal*, 3:1 (Winter 1983) 9–29; Albert

Wohlstetter, "Bishops, Statesmen, and Other Strategists on the Killing of Innocents," *Commentary*, 75:6 (June 1983) 15–35; and the exchange of correspondence on Wohlstetter's essay in *Commentary*, 76:6 (December 1983) 4–22.

31. See Hehir, "On the Role of the Pastoral Letters," pp. 185–86.

32. See Hehir, "Can nuclear protest change policy?" p. 360.

33. J. Bryan Hehir, "Simple slogans or moral reasoning?," *Commonweal*, April 23, 1982, p. 234. Fr. Hehir did not discuss how calling just-war reasoning "mental gymnastics, casuistry of the worst sort" (Bishop Sullivan), or describing the Trident submarine as "the Auschwitz of Puget Sound" (Archbishop Hunthausen, the target of Lehman's anger), contributed to the level of reasonable discourse.

34. Hehir, "Can nuclear protest change policy?" p. 360.

35. Ibid.

36. J. Bryan Hehir, "Mobilizing opinion, curbing technology," *Commonweal*, May 20, 1983, p. 298.

37. Hehir, "On the Role of the Pastoral Letters," pp. 184–85.

38. Ibid., p. 192.

39. Hehir, "Foreign Policy in the 1980s," p. 268.

40. Ibid., pp. 269–70.

41. Ibid., p. 270.

42. J. Bryan Hehir, "Wanted: a policy for Central America," *Commonweal*, February 25, 1983, p. 106.

43. J. Bryan Hehir, "The case of Grenada," *Commonweal*, December 16, 1983, pp. 681, 698.

44. Ibid., p. 681.

45. Hehir, "Wanted: a policy for Central America," p. 106.

46. See, for example, Hehir's 1984 commencement address at the Catholic University of America, "A Public Church," in *Origins*, 14:3 (May 31, 1984) 40–43, especially p. 43.

47. See J. Bryan Hehir, "The Perennial Need for Philosophical Discourse," in "Theology and Philosophy in Public: A Symposium on John Courtney Murray's Unfinished Agenda," David Hollenbach, S.J., ed., *Theological Studies*, 40:4 (December 1979) 710–13.

48. See J. Bryan Hehir, "Non-Violence, Peace, and the Just War," *Worldview*, June 1969, pp. 17–20.

Beyond Abandonment

1. Cited in William D. Miller, *A Harsh and Dreadful Love*, p. 221.

2. Pope John Paul II, "Address to Scientists and Scholars," *Origins*, 10 (March 24, 1981) 621, #4.

3. Pope John Paul II, "War Is Death," *Origins*, 10 (March 24, 1981) 619–20.

4. Ibid., p. 420.

5. Ibid. (emphasis added).

6. Ibid.

7. Robert Nisbet, *Prejudices: A Philosophical Dictionary* (Cambridge: Harvard University Press, 1983), p. 243.

8. See Pope Paul VI, "Address to the General Assembly of the United Nations," October 4, 1965, #1, in *The Pope Speaks* 11 (1966), p. 49.

9. See Andrew M. Greeley, *The American Catholic: A Social Portrait* (New York: Basic Books, 1977), pp. 126–51, for relevant survey research data on American Catholic attitudes toward Vatican II–mandated liturgical and penitential changes.

10. Andrew M. Greeley, "American Catholicism: 1909–1984," *America*, June 30, 1984, p. 490.

11. See Greeley, *The American Catholic: A Social Portrait*, pp. 130–31.

12. Norman Podhoretz, "The Future of Liberalism," *The American Spectator*, April 1985, p. 17.

13. "The Challenge of Peace," #23.

To Be a *Church* at the Service of Peace

1. Cited in Jeffrey Hart, "The Mission of a University," *National Review*, October 3, 1980.

2. G. K. Chesterton, *Orthodoxy* (Garden City: Image Books, 1959), pp. 47–48.

3. Pope John Paul II, "Address to Scientists and Scholars," #6.

4. "The Challenge of Peace," 277.

5. George Huntston Williams, *The Mind of John Paul II: Origins of his Thought and Action* (New York: Seabury Press, 1981), p. 265.

6. Pope John Paul II, *Redemptor Hominis*, 21–22, in Claudia Carlen, I.H.M., ed., *The Papal Encyclicals 1958–1981* (McGrath Publishing Co., 1981), pp. 250–51 (the words in italics in the pope's text cite *Gaudium et Spes*, 22).

7. See Williams, op. cit., p. 307.

8. Pope John Paul II, *Redemptor Hominis*, 2.

9. Cited in Williams, op. cit., p. 390, n. 73.

10. Karol Wojtyla in 1964, cited in ibid., p. 307.

11. Ibid.

12. Paul Johnson, *Pope John Paul II and the Catholic Restoration*, p. 189.

13. Pope John Paul II, *Redemptor Hominis*, 2.

14. See Karl Rahner, *Hearers of the Word* (New York: Herder and Herder, 1969).

15. See Karl Rahner, "Christology Within an Evolutionary View of the World," *Theological Investigations*, 5 (London: Darton, Longman, Todd, 1966), pp. 135–53; "Christology Within the Setting of Modern Man's Understanding of Himself and of his World," *Theological Investigations*, 11 (New York: Seabury, 1974), pp. 215–29.

16. See Karl Rahner, "Theology and Anthropology," *Theological Investigations*, 9 (London: Darton, Longman, Todd, 1972), pp. 28–45.

Conciliar theological anthropology also formed the foundation for the March 1986 "Instruction on Christian Freedom and Liberation" (Vatican Polyglot Press, 1986. Parenthetical numbers are references to numbered paragraphs in the Instruction.) issued by the Vatican Congregation for the Doctrine of the Faith as a complement to the Congregation's earlier "Instruction on Certain Aspects of the 'Theology of Liberation.'"

The 1986 Instruction emphasized that redemption is "liberation in the strongest sense of the word, since it has freed us from the most radical evil, namely, sin and the power of death." (3) Human beings learned the true nature and depth of their freedom in the Gospel's call to communion with God. (5) That freedom lay, not in the "liberty to do anything whatsoever," but in the freedom "to do good," in which alone would happiness be found. (26) Totalitarianism was evil precisely because of its violation of the radical freedom of the human person before God, who "wishes to be adored by people who are free." (14,27,44)

Sin, which is "alienation from the truth of . . . being . . . a creature loved by God," was the fundamental alienation of modern times (indeed, of any times), and the basic obstacle to true human freedom. (38) Rejecting God made men and women into slaves who destroyed "the momentum of [their] aspiration to the infinite and of [their] vocation to share in the divine life." (40) Because sin, which was at the root of unjust social situations, was a "voluntary act," one could only speak accurately of "social sin" in a "derived and secondary sense." (75) The great biblical symbol of the Exodus thus referred, not to political emancipation alone, but to a political freedom set in the context of the greater liberation that is life in communion with the creator and sustainer of life. (44)

The Resurrection revealed our "definitive liberation" and freed us for a life of service to others. (51) "There is no gap between love of neighbour and desire for justice,' the Instruction argued. "To contrast the two is to distort both love and justice. Indeed, the meaning of mercy completes the meaning of justice by preventing justice from shutting itself up within the circle of revenge." (57)

The Instruction rejected the concept of a "partisan church," claiming that "the political and economic running of society is not a direct part of [the Church's] mission." (61) It was precisely by being the Church that the Church made its most basic contribution to the life of the earthly City—for by being the Church, and not simply one more partisan political actor, the Church could witness to human dignity, resist attempts to expel God from human life, and judge the claims of political and ideological movements, which despite their professed intentions, drive human beings farther away from freedom in all its dimensions. (65) The Church indeed had a "love of preference" for the impoverished and the unfree, but this love "excludes no one," since it is fundamentally a witness to the God-given dignity of all men and women. Therefore, the Church's love for the dispossessed cannot be reduced to "sociological and ideological categories which would make this preference a partisan choice and a source of conflict." (68)

The Instruction also argued that the Church's social teaching was "constantly open" to new questions as they emerge in history; rejected the recourse to violence as a first option for redressing social injustice; criticized those who discredited the notion of "reform" for the sake of "revolution"; and rejected class struggle as "the structural dynamism of social life. (72, 76–77) The Instruction endorsed nonviolent action for meeting the demands of justice, and highlighted the role of the laity as agents of freedom and liberation in the world. (79–80). The Instruction also balanced the classic Catholic social-ethical principle of subsidiarity with a "principle of solidarity": the freedom of the individual before God required action on behalf of liberation and freedom for all. (89) "Authentic development" in poor countries, the Instruction suggested, can only come about in open political systems, where there is a "real separation between the powers of the State," which helps guard against human rights abuses by governments. (95)

In the months after the Instruction was issued, many prominent theologians of liberation (notably, Fr. Leonardo Boff, O.F.M., of Brazil) claimed that the Instruction had vindicated their interpretation of Christian liberation as contained in the documents of Medellin and the subsequent expansion of the theologies of liberation. It seems a difficult case to sustain. The Instruction clearly rejects core Marxist concepts (class struggle, structural evil, "redemptive" violence) that have been important influences in the theologies of liberation. The Instruction endorses Christian "base communities," as long as they remain in communion with the local bishop and the universal Church. But the Instruction also flatly rejects the concept of the "partisan Church" and the exclusive "preferential option for the poor"—two other concepts at the heart of the theologies of liberation. And the Instruction gives more weight to the relationship between democracy (or, at the least, predemocratic institutions and societal arrangements) and development than one tends to find in the various theologies of liberation.

Thus the Instruction looks considerably more like a development of the classic heritage of *tranquillitas ordinis* than it does a vindication of the theologies of liberation, at least as they have been developed to date by Fr. Boff and his colleagues. It may well be that some theologians of liberation are discovering the radical limitations of dependency theory and the "Marxist analysis" that underlies it. Perhaps some theologians of liberation are discovering, particularly in light of the once-lauded Sandinista regime's grim record of repression in Nicaragua, the greater prospects of democratic theory and *praxis*. Perhaps. But the burden of proof remains on Fr. Boff and his colleagues to prove, by word and deed, that their work is congruent with the understanding of Christian freedom and liberation contained in the remarkable Instruction of March 1986.

17. Chesterton, *Orthodoxy*, p. 20.

18. Pope John Paul II, cited in E. J. Dionne, Jr., "Determined to Lead," *New York Times Magazine*, May 12, 1985, p. 30.

19. See "The Challenge of Peace," 24–25.

20. See H. Richard Niebuhr, *Christ and Culture*, chapter 6, "Christ the Transformer of Culture."

21. On "emotivism," see Alasdair MacIntyre, *After Virtue*, pp. 1–34.

22. Jim Castelli, *The Bishops and the Bomb*, p. 143.

23. J. Bryan Hehir, in *An American Catholic Catechism*, George J. Dyer, ed. (New York: Seabury Press, 1975), p. 276.

24. James Rosenau, "Fragmegrative Challenges to National Strategy," paper presented to the National Security Affairs Institute, National Defense University, October 9, 1982.

25. See the memoirs of two key Carter administration figures, who themselves represent something of the problem of fragmegration: Zbigniew Brzezinski, *Power and Principle* (New York: Farrar, Straus, Giroux, 1983), and Cyrus Vance, *Hard Choices* (New York: Simon & Schuster, 1983). The Brzezinski and Vance books demonstrate how contextual disagreement precedes disagreement over policy choice.

26. John Courtney Murray, *We Hold These Truths*, p. 18.

27. Ibid., p. 19.

28. On the presently available evidence, there seems little inclination to conceive the Church's roles in these terms at the United States Catholic Conference. Changes in the mandate of the USCC, and in its staff's understanding of the agency's priority role, would thus be necessary. For an examination and critique of the USCC's present role, see J. Brian Benestad, *The Pursuit of a Just Social Order* (Washington: Ethics and Public Policy Center, 1982). Benestad's book is, unfortunately and unfairly, too often dismissed in Catholic activist circles as an argument against the Church's involvement in questions of public policy. It is nothing of the sort.

29. See my article, "A New Ecumenism," *Eternity*, July/August 1985, pp. 33–37.

30. For a discussion of these various images of the Church, see Avery Dulles, *Models of the Church* (Garden City: Doubleday, 1974). The image of the Church as the "People of God" was prominent in the Second Vatican Council "Dogmatic Constitution on the Church" (*Lumen Gentium*); see especially 9–17.

31. Henry de Lubac, *Catholicism* (London: Burns, Oates, and Washburne, 1950), p. 29.

32. *Lumen Gentium*, 1.

33. See Avery Dulles, S.J., "The Emerging World Church: A Theological Reflection," *Proceedings of the Thirty-Ninth Annual Convention of the Catholic Theological Society of America*, George Kilcourse, ed., pp. 9–11.

34. John Macquarrie, *The Concept of Peace* (New York: Harper & Row, 1973), p. 81.

To Be a Church at the Service of *Peace*

1. For one expression of Catholic realist resistance to even a tempered pursuit of the vision of *Pacem in Terris*, see my exchange with William V. O'Brien in *Just War Theory in the Nuclear Age*, John D. Jones and Marc F. Griesbach, eds. (New York: University Press of America, 1985), pp. 153–86, 189–90.

2. A classic statement of arms control theory may be found in Paul Warnke, "Apes on a Treadmill," *Foreign Policy*, 18 (Spring 1975) 12–29. For an overview of the entire debate, see

Lawrence Freedman, *The Evolution of Nuclear Strategy.* For an application of key arms control themes to the possibility of strategic defense, see McGeorge Bundy et al., "The President's Choice: Star Wars or Arms Control," *Foreign Policy*, 63, pp. 264–78.

3. See "Questions of Politics, Strategy, and Ethics," testimony before the House Foreign Affairs Committee on June 26, 1984, by Cardinal Joseph Bernardin and Archbishop John J. O'Connor, reprinted in *Origins*, 14:10 (August 9, 1984) 154–58; and the letter to members of Congress by Bishop James Malone, president of the NCCB/USCC, urging a vote against funding of the MX missile, reprinted in *Origins*, 14:41 (March 28, 1985) 668–70.

4. See Helen Caldicott, *Missile Envy: The Arms Race and Nuclear War* (New York: William Morrow, 1984).

5. See my discussion of this point in *Just War Theory in the Nuclear Age*, pp. 154–64.

6. See ibid., pp. 158, 178, 188.

7. See, for example, Zbigniew Brzezinski, Robert Jastrow, and Max M. Kampelman, "Defense in Space is Not 'Star Wars,'" *New York Times Magazine*, January 27, 1985, pp. 28–51, in which the authors discuss the possibility of an "alternative strategy of mutual security" (pp. 46, 48). A similar point was made to a bipartisan seminar in the House of Representatives in early 1984 by Michael Howard, the Regius Professor of Modern History at Oxford. Howard urged the possibility of a mutual security or common security approach to what he regarded as the inevitable development of defensive weaponry. See a summary of Howard's comments in *Exploring Soviet Realities: Problems in the Pursuit of Peace—Report of a Bipartisan Seminar in the U.S. House of Representatives 1984–1985* (available from the offices of the World Without War Council, Inc., 1730 Martin Luther King, Jr. Way, Berkeley, California 94709).

8. The common security approach to strategic defense has been mentioned by President Reagan, National Security Adviser Robert McFarlane, and Secretary of Defense Caspar Weinberger. The January 1985 White House publication, "The President's Strategic Defense Initiative," puts the matter this way:

> The United States does not view defensive measures as a means of establishing military superiority. Because we have no ambitions in this regard, deployments of defensive systems would most usefully be done in the context of a cooperative, equitable, and verifiable arms control environment that regulates the offensive and defensive developments and deployments of the United States and the Soviet Union. Such an environment could be particularly useful in the period of transition from a deterrent based on the threat of nuclear annihilation, through deterrence based on a balance of offensive and defensive forces, to the period when adjustments to the basis of deterrence are complete and advanced defensive systems are fully deployed. During the transition, arms control agreements could help manage and establish guidelines for the deployment of defensive systems" [p. 5].

This alteration in the strategic regime was located, by the administration, within a common security framework: "In pursuing the Strategic Defense Initiative, the United States is striving to fashion a future environment that serves the security interests of the United States and our allies, *as well as the Soviet Union.* Consequently, should it prove possible to develop a highly capable defense against ballistic missiles, we would envision parallel United States and Soviet deployments, with the outcome being enhanced mutual security and international stability" (p. 4; emphasis added). For another suggestive administration comment, see Robert C. McFarlane, "Strategic Defense Initiative," U.S. Department of State *Current Policy*, #670 (March 7, 1985), especially p. 3.

9. The classic argument for the prospects of a legally based international security system may be found in Grenville Clark and Louis Sohn, *World Peace through World Law* (Cambridge: Harvard University Press, 1966). The counter case may be found in sharpest form in *A World Without a U.N.*, Burton Yale Pines, ed. (Washington, D.C.: Heritage Foundation, 1984). Daniel Patrick Moynihan's essay, "The United States in Opposition" (*Commentary*,

59:3 [March 1975] 31–44), provides a more measured critique of the evolution of the U.N. system, a critique carried out at greater length in Moynihan's book *A Dangerous Place* (Boston: Atlantic–Little, Brown, 1978).

10. See Moynihan, *A Dangerous Place*; Jeane J. Kirkpatrick, *The Reagan Phenomenon* (Washington, D.C.: American Enterprise Institute, 1984). The forthrightness of American U.N. representatives like Moynihan and Kirkpatrick is not only compatible with, but quite probably an essential component of, work for the reform of international legal and political institutions according to the vision of *Pacem in Terris.* The speeches of U.S. representatives Michael Novak and Richard Schifter during the 37th and 38th sessions of the United Nations Commission on Human Rights in Geneva provide further examples of how commitments to the rule of law can be combined with forthright critiques of present U.N. practice; see *Rethinking Human Rights* and *Rethinking Human Rights II: One Standard, Many Methods* (Washington, D.C.: The Foundation for Democratic Education, 1981, 1982). Ambassador Max M. Kampelman's speeches to the Madrid Review Conference on the Helsinki Accords demonstrate how human rights and security concerns may be linked; see *Three Years at the East/West Divide*, Leonard Sussman, ed. (New York: Freedom House, 1983).

11. The University of Chicago's Gidon Gottlieb, for example, argues for the necessity of international law, but urges that it be pursued outside the present U.N. system, in "How to Rescue International Law," *Commentary*, 78:4 (October 1984) 46–50.

12. Eugene V. Rostow, "Peace as a Problem of Law," in *Peace in the Balance: The Future of U.S. Foreign Policy* (New York: Simon & Schuster, 1972), p. 303. Rostow's comments are reminiscent of Jacques Maritain's inaugural address to the second international conference on UNESCO in 1947:

> The first questions which present themselves to one who meditates seriously on the conditions for a just and enduring peace are obviously those called forth by the idea of a supranational organization of the peoples of the world. Everyone is aware of the obstacles to carrying such an idea into effect; they are even greater today than immediately after victory. At the present time, a truly supranational world organization is beyond the realm of possibility. A philosopher, however, would fail in his duty if he did not add that this very thing which is today impossible, is nevertheless necessary, and that without it the creation of a just and enduring peace cannot be conceived. Hence, it follows that the first obligation incumbent upon the men of today is that they work with all their forces to make possible what is thus necessary" ["The Possibilities for Cooperation in a Divided World," in *The Social and Political Philosophy of Jacques Maritain*, Joseph W. Evans and Leo R. Ward, eds. (New York: Charles Scribner's Sons, 1955), p. 125].

Maritain believed that those who were ideologically divided could still cooperate at the level of practical reason, and saw such cooperation as an essential building block of the political community that would have to precede international political organization (see art. cit., pp. 130–36).

13. "Address of His Holiness Pope John Paul II to the XXXIV General Assembly of the United Nations Organization, 2 October 1979" (Vatican Polyglot Press).

14. See *Pacem in Terris*, 9–30. For a more recent defense of a comprehensive definition of "human rights," see David Hollenbach, S.J., *Claims in Conflict.*

15. "Address of His Holiness Pope John Paul II to the XXXIV General Assembly of the United Nations Organization," pp. 13–14.

16. See my essay, "A Preliminary Examination of Religious Freedom," in *Freedom in the World 1982*, Raymond D. Gastil, ed. (Westport: Greenwood Press, 1982), pp. 121–41.

17. On the priority of religious liberty as a human rights issue, as well as the relationship between religious liberty and peace, see my essays "Religious Freedom: The First Human Right," *Catholicism in Crisis*, 1:11 (October 1983) 14–18; and "Religion as a Human Right," *Freedom at Issue*, March/April 1984, pp. 3–6.

18. Peter L. Berger, "Are Human Rights Universal?," *Commentary*, September 1977, p. 62.

19. For representative critiques of dependency theory, see P. T. Bauer, "Foreign Aid for What?," *Commentary*, 66 (December 1978); Joseph Ramos, "Dependency and Development: An Attempt to Clarify the Issues," in *Liberation South, Liberation North*, Michael Novak, ed. (Washington, D.C.: American Enterprise Institute, 1981), pp. 68–72; Michael Novak, *The Spirit of Democratic Capitalism* (New York: Simon and Schuster, 1982), pp. 273–76; 299–307; and P. T. Bauer, *Dissent on Development* (Cambridge: Harvard University Press, 1972), especially chapter 1, "The Vicious Circle of Poverty and the Widening Gap," and chapter 9, "The Spurious Consensus and its Background."

20. On the importance of "human capital" in development, see Thomas Sowell, *The Economics and Politics of Race: An International Perspective* (New York: William Morrow, 1983). Sowell makes similarly intriguing points in his *Ethnic America* (New York: Basic Books, 1981).

21. Peter L. Berger, "Underdevelopment Revisited," *Commentary*, 78:1 (July 1984) 42.

22. On a related point, Berger writes,

> In focusing on this particular criterion for defining successful development I am invoking, of course, the ideal of equity; but I am *not* invoking "equality," a utopian category that can only obfuscate the moral issues. It is inequitable and immoral that, next door to each other, some human beings are starving while others gorge themselves. To make this situation more equitable and thus morally tolerable, the starvation must stop and the poor must become richer. This goal can be attained without the rich becoming poorer. In other words, I do not assume the need for a leveling of income distribution. Western societies (including the United States) have demonstrated that dramatic improvements are possible in the condition of the poor without great changes in income distribution; the poor can get richer even while the rich get richer too. And there are good economic grounds for thinking that income-leveling policies in the Third World inhibit growth, with the poor paying the biggest price for this inhibition. "Equality" is an abstract and empirically murky ideal; it should be avoided in assessing the success or failure of development strategies [ibid., pp. 42–43].

23. Ibid., p. 43.

24. Ibid.

25. See Nick Eberstadt, "Health Crisis in the U.S.S.R.," *New York Review of Books*, 28 (February 19, 1981) 23–31.

26. I am grateful to Nick Eberstadt for suggesting this important definition of development. For a seminal discussion of these issues, see William Douglass, *Developing Democracy* (Washington, D.C.: Heldref Publications, 1972). See also Michael Novak and Peter Berger, *Speaking to the Third World: Essays on Democracy and Development* (Washington, D.C.: American Enterprise Institute, 1985).

27. National Conference of Catholic Bishops, "Pastoral Letter on Marxist Communism" (Washington, D.C.: United States Catholic Conference, 1980).

28. See, for example, the arguments depicted in William L. O'Neill, *A Better World: The Great Schism—Stalinism and the American Intellectuals* (New York: Simon & Schuster, 1982), and William Barrett, *The Truants: Adventures Among the Intellectuals* (Garden City: Anchor Press/Doubleday, 1982).

29. For another view, see Arthur McGovern, S. J., *Marxism: An American Christian Perspective* (Maryknoll: Orbis Books, 1980). McGovern argues that there is "no essential, necessary connection between atheism/materialism and Marxist socialism" (p. 268).

30. On this crucial point, see Robert Pickus, "Liberal Anti-Communism Revisited," *Commentary*, September 1967, pp. 58–62. Alfonso Robelo, a former member of the Sandinista junta in Nicaragua, made the point cogently in a June 15, 1985, speech to Social Democrats U.S.A.: "I do not want to quarrel here with those who make elegant geo-political

and balance-of-power arguments in favor of containing communism, nor with those who have asserted that traditional authoritarianism is a lesser evil than Communist totalitarianism. My point is simply this: mere anti-Communism lacks the moral energy and popular appeal that is required of any movement which hopes to withstand the force of Communist expansion" (cited in the *Wall Street Journal*, July 5, 1985, p. 6).

31. As Leonid Brezhnev said in 1973 to a Warsaw Pact summit meeting: "We are achieving with détente what our predecessors have been unable to achieve using the mailed fist. . . . Trust us, comrades, for by 1985, as a consequence of what we are now achieving with détente, we will have achieved most of our objectives . . . and a decisive shift in the correlation of forces will be such that, come 1985, we will be able to extend our will wherever we need to" (cited in Elias M. Schwartzbart, "Negotiating with the Russians," *Freedom at Issue*, 84 [May/June 1985], p. 5).

32. That there is no Soviet global agenda is a frequent theme taught by those caught up in the psychology of détente. For a persuasive rebuttal, see Herbert J. Ellison, "Soviet Global Strategy," paper presented to a conference, "Spotlight on the Soviet Union," Oslo, Norway, April 25–27, 1985. See also Ellison's remarks in *Exploring Soviet Realities: Problems in the Pursuit of Peace—Report of a Bipartisan Seminar in the U.S. House of Representatives.*

33. *We Hold These Truths*, chapter 10.

34. Representative examples include Wildavsky's essay, "Containment plus Pluralization," in *Beyond Containment*, Aaron Wildavsky, ed. (San Francisco: ICS Press, 1983), pp. 125–45; and Richard Pipes's book, *Survival Is Not Enough* (New York: Simon & Schuster, 1984). For a critique of such approaches, see Seweryn Bialer and Joan Afferica, "Reagan and Russia," *Foreign Affairs*, 61:2 (Winter 1982/1983); Wildavsky's rebuttal is in art. cit., pp. 143–45.

35. See Wildavsky, art. cit., p. 128.

36. Ibid.

37. Ibid., pp. 130–31.

38. Ibid., p. 130.

39. Ibid., p. 138.

40. See American Nobel laureate Paul Flory, "Science in a Divided World" (paper presented to the Arizona State University Centennial Symposium, February 5, 1985).

41. Wildavsky, art. cit., p. 141. "Reciprocity," as Wildavsky understands the term, should not be confused with the way in which American television and print journalists have tended to treat Soviet spokesmen and propagandists like the ubiquitous Vladimir Posner. To present Posner as a "journalist" is an exercise in disinformation, since Mr. Posner is a functionary of the Soviet state. Ideally, American print and electronic reporters would seek out and present independent (that is, non-regime-programmed) opinion in the Soviet Union. But under present circumstances, it seems more likely that men like Posner will be the only voices regularly available. Still, there is no reason why Soviet government *apparatchiki* shouldn't be brought onto American television to make their case: *if* they are correctly identified and *if* they are subjected to the kind of grilling to which our own political leaders must submit. Observing these two provisos would help ensure that the long-term ends of pluralization, rather than the short-term goals of Soviet disinformation, were being served. As "spacebridge" technology and direct satellite television broadcasting make communication between the United States and the Soviet Union easier, public and governmental commitment to the goal of pluralizing the Soviet Union will become increasingly important.

42. See ibid.

43. Pipes, op. cit., p. 280 (emphasis added).

44. Wildavsky, art. cit., p. 142.

45. On this latter point, see the platforms of the Libertarian Party in the 1980 and 1984 presidential elections.

46. Osgood called his plan GRIT, for "graduated reciprocation in tension-reduction"; see his *An Alternative to War or Surrender* (Urbana: University of Illinois Press, 1962).

47. See Amitai Etzioni, "The Kennedy Experiment," *Western Political Quarterly*, 20:2 (June 1967) 361–80.

48. Initiatives were proposed in "The Challenge of Peace," ##205–6, but the bishops failed to distinguish their concept of "initiative" from the inadequate understandings then dominant among many religious peace activists. "The Challenge of Peace" does not, for example, make clear that the Kennedy moratorium on nuclear testing was conditional on Soviet reciprocation in kind, but in fact describes the Kennedy action (inaccurately) as "unilateral." For a more politically sharpened discussion of an initiatives strategy, see Robert Woito, *To End War* (New York: Pilgrim Press, 1982), pp. 493–502.

49. See Gene Sharp, *The Politics of Nonviolent Action*, 3 volumes (Boston: Porter Sargent, 1973).

50. Probes toward such an exploration may be found in William V. O'Brien, *The Conduct of Just and Limited War* (New York: Praeger, 1981), pp. 17–19, 68, 161, 183, 194–95. See also Michael Walzer, *Just and Unjust Wars* (New York: Basic Books, 1977), pp. 197–222.

51. See the O'Brien and Walzer books cited in n. 50, above; Guenter Lewy assesses the charges of war crimes against American troops during the guerilla war in Southeast Asia in *America in Vietnam*, pp. 223–373.

52. Although the American bishops, in "The Challenge of Peace," seemed to believe that too little attention had been paid to the moral problem of the use of nuclear weapons, I would argue a somewhat different case: that too exclusive a focus on the problem of nuclear weapons, their possession and use, within those American Catholic leadership, activist, and intellectual circles most concerned with the problem of war and peace has led to a situation in which we lack sufficient moral clarity on those issues of the use of military force that are before us virtually every day—e.g., the use of military force against terrorism, and in the context of guerilla war. This imbalance needs to be redressed.

Attention to problems of conventional war, terrorism, and guerilla war does not preclude, of course, a vigorous debate on the issue of the possible proportionate and discriminate use of low-yield, high-accuracy nuclear weapons. That debate, well underway in the aftermath of "The Challenge of Peace," must be linked to the debate over nuclear force modernizations (e.g., de-MIRVing) and the possible development of strategic defense systems, so that all these issues are considered in light of an evolution in the strategic regime that promotes the prospects of, not merely arms control, but genuine and verifiable arms reduction. That debate, in turn, must be linked to the question of U.S. strategy vis-à-vis the Soviet Union. Linkage at this level is unavoidable, according to the fourfold structure of the politics of *tranquillitas ordinis.*

53. *We Hold These Truths*, p. 253.

54. Jacques Maritain, *Reflections on America* (New York: Charles Scribner's Sons, 1958), p. 19.

55. See ibid., pp. 26, 23.

56. Ibid., p. 29.

57. Ibid., p. 31.

58. Ibid., pp. 33–35. Michael Novak makes a similar (and more detailed) argument in *The Spirit of Democratic Capitalism*, pp. 143–70, in which he suggests that the image of the "communitarian individual" is a considerably more apt description of the reality of American society than "bourgeois individualism."

59. See Maritain, op. cit., p. 71.

60. Ibid., pp. 71–72, 86.

61. Ibid., p. 87.

62. Ibid., p. 118. Maritain was not arguing here for ideological propaganda. The problem he identified "does not mean, of course, that it would be advisable to manufacture an ideology for the sake of propaganda, God forbid! It means that the development of a greater general interest in ideas and universal verities is a presupposed condition without which no genuine possibilities of intellectual communication can emerge" (p. 118).

63. Ibid., pp. 131, 132, 134–35, 156.

64. See ibid., pp. 49–64.

65. Ibid., p. 168.

66. Ibid., p. 169.

67. Ibid., p. 199.

68. Ibid.

69. That the mirror-image problem extends beyond the activist community and, under the weight of the nuclear peril, distorts the analysis of eminent scholars is made painfully clear by the example of George F. Kennan, who once argued as follows:

> Isn't it grotesque to spend so much of our energy on opposing . . . Russia in order to save a West which is honeycombed with bewilderment and a profound sense of moral decay? Show me first an America which has successfully coped with the problems of crime, drugs, deteriorating educational standards, urban decay, pornography, and decadence of one sort or another; show me an America that has pulled itself together and is what it ought to be, then I will tell you how we are going to defend ourselves from the Russians. But as things are, I can see very little merit in organizing to defend from the Russians the porno shops in central Washington. In fact, the Russians are much better in holding pornography at bay than we are [George F. Kennan, "Selections from Interviews," in *The Nuclear Delusion: Soviet-American Relations in the Atomic Age* (New York: Pantheon Books, 1982), p. 74].

Fr. J. Bryan Hehir advised the readers of *Commonweal* that the book that includes this extraordinary statement "provides a standard of excellence against which both the [Reagan] administration and its critics can be tested on their views of U.S.-Soviet relations" ("Kennan and the U.S.-Soviet debate," *Commonweal*, January 28, 1983, p. 39).

70. See, for example, Richard John Neuhaus, *The Naked Public Square*, p. 262.

71. William Faulkner, "Acceptance Speech," in the *Nobel Prize Library: William Faulkner, Eugene O'Neill, John Steinbeck* (New York: Alexis Gregory, 1971), p. 8.

Epilogue

1. See Michael Howard, *War and the Liberal Conscience* (New Brunswick: Rutgers University Press, 1978), p. 130.

2. Ibid., p. 131.

3. Ibid.

4. Ibid., pp. 131–32.

5. Ibid., p. 132.

6. Ibid.

7. For some antecedents to the ideological pilgrimage of the American Catholic elite, see Paul Hollander, *Political Pilgrims: Travels of Western Intellectuals to the Soviet Union, China, and Cuba* (New York: Oxford University Press, 1981). See also idem., "Political Tourism in Cuba and Nicaragua," *Society*, May/June 1986, pp. 28–37.

8. For a discussion of this point, see John W. Cooper, *The Theology of Freedom: The Legacy of Jacques Maritain and Reinhold Niebuhr* (Macon: Mercer University Press, 1984), pp. 40–41.

Index

DATE DUE	BORROWER'S NAME